World Sustainability Series

Due to its scope and nature, sustainable development is a matter which is very interdisciplinary, and draws from knowledge and inputs from the social sciences and environmental sciences on the one hand, but also from physical sciences and arts on the other. As such, there is a perceived need to foster integrative approaches, whereby the combination of inputs from various fields may contribute to a better understanding of what sustainability is, and means to people. But despite the need for and the relevance of integrative approaches towards sustainable development, there is a paucity of literature which address matters related to sustainability in an integrated way.

Notes on the quality assurance and peer review of this publication

Prior to publication, the works published in this book are initially assessed and reviewed by an in-house editor. If suitable for publication, manuscripts are sent for further review, which includes a combined effort by the editorial board and appointed subject experts, who provide independent peer-review. The feedback obtained in this way was communicated to authors, and with manuscripts checked upon return before finally accepted. The peer-reviewed nature of the books in the "World Sustainability Series" means that contributions to them have, over many years, been officially accepted for tenure and promotion purposes.

Walter Leal Filho · Golda Edwin ·
Nandhivarman Muthu
Editors

Curriculum Development for Sustainable Development

Theory and Practice

Springer

Editors
Walter Leal Filho
Sustainable Development and Climate Change Management Research and Transfer Centre (FTZ NK)
Hamburg University of Applied Sciences
Hamburg, Germany

Golda Edwin
Research and Outreach
Association for Promoting Sustainability in Campuses and Communities (APSCC)
Puducherry, India

Nandhivarman Muthu
Pondicherry University
Puducherry, India

ISSN 2199-7373 ISSN 2199-7381 (electronic)
World Sustainability Series
ISBN 978-3-032-22145-2 ISBN 978-3-032-22146-9 (eBook)
https://doi.org/10.1007/978-3-032-22146-9

This Springer imprint is published by the registered company Springer Nature Switzerland AG
The registered company address is: Gewerbestrasse 11, 6330 Cham, Switzerland

Preface

The rethinking and reformulation of education towards a sustainable future is a major challenge faced by curriculum makers today. This challenge demands moving beyond simply adding environmental themes. It requires a fundamental shift towards systems thinking, fostering an understanding of the deep interconnections between ecological, social, and economic well-being. The process of evolution of curricula needs to cultivate critical thinking, empathy, and actionable skills for resilience and regeneration. This involves reimagining pedagogies to be learner-centered, experiential, and focused on real-world problem-solving. The goal is to empower a generation not merely to navigate the complexities of the twenty-first century, but to become active, collaborative architects of a just and flourishing world for all.

The task of reimagining education for a sustainable future is another pressing challenge facing curriculum developers today. Simply refining existing models is not enough. There is a need to build new ones, capable of nurturing systems thinkers, ethical leaders, and compassionate problem-solvers in our interconnected world.

Embedding sustainability into curricula requires systemic alignment—from policy frameworks and institutional strategies to classroom practices. Curriculum development for sustainable development must therefore engage educators, policymakers, and learners in co-creating transformative educational experiences that foster responsibility, critical awareness, and action.

Against this background, the aim of this volume is to attempt to narrow the gap between high-level theories and interventions that can be implemented in classrooms. It offers a strong theoretical base for teaching and learning in ESD, as well as examples of the application in practice for education professionals. The chapters contained herein detail a complex weaving of pieces of the sustainability conversation into the fabric of their curriculum, re-creating it from a loose collection of topical courses to an interdisciplinary and lively learning framework.

The volume brings together diverse perspectives from educators, researchers, and practitioners across disciplines and geographies, illustrating how collaboration and interdisciplinarity enrich curriculum design for sustainable development.

Grounded in the principles of the UN Sustainable Development Goals—particularly SDG 4 (Quality Education) and SDG 15 (Life on Land)—this book underscores education's central role in advancing ecological literacy, social justice, and planetary well-being.

We thank all authors, whose varied expertise and groundbreaking ideas make up the intellectual spine of this book. Their dedication to advancing the field of ESD, often in the face of institutional inertia, is a constant source of inspiration. We are grateful to the international reviewers for their valuable support.

We hope this book serves not only as a reflection of current progress but also as a catalyst—inspiring continued innovation, dialogue, and action toward rewriting learning for a sustainable and equitable world.

Hamburg, Germany — Walter Leal Filho
Puducherry, India — Golda Edwin
Puducherry, India — Nandhivarman Muthu

Contents

Envisioning Eco-Centric Frameworks for Curriculum Development and Implementation in the Anthropocene

Bruno Borsari

Abstract Whether the Anthropocene is classified as a new, geologic era, that dominated by human activities has been determining planetary changes on climate and the environment, or simply an event, the need for adjusting curricula to these startling conditions and geographies is compelling and urgently needed. Reforming and transforming modern education have become essential to prepare society to dealing with increasing anthropogenic challenges. Therefore, this chapter aims to enthuse educators to employ eco-centric strategies in their curricula, capable of placing living systems at the center of instruction. Secondly, this chapter wants to emphasize the need for shifting approaches in education and learning by re-examining curricula and modes of implementation methods to fortify pedagogical tactics in education for sustainability. Although many frameworks may be adoptable, or adaptable for the design and implementation of study programs in sustainability, the underpinnings of an education for sustainable development could benefit from eco-centrism as the foundation of every curriculum, for the present and foreseeable future. Hence, this evaluation study employed a mixed-method approach to assess the viability of eco-centric frameworks for curriculum design, which included a critical review of the current literature and case studies (n = 6). The former method consisted in a two-pronged approach where the first focused on the Anthropocene and its implications for sustainability education. The second one instead was directed at shedding light on curricula for sustainable development and education reform. The six case studies showcase the author's work in higher education during his three and a half decades career. This was accomplished in Sierra Leone (West Africa), in North America (Louisiana, Pennsylvania and Minnesota, in the U.S.) and in Panama. A combination of qualitative data from the literature review and quantitative data from the case studies enabled a triangulation of the results. This assessment substantiated the validity of framing curricula in sustainable development within eco-centric theories and concepts. In addition, an original model concludes this work. This chapter will empower educators by adjusting curricula to the whims generated by the Anthropocene a step further towards sustainability.

B. Borsari (✉)
Biology Department, Winona State University, Winona, MN, US
e-mail: bborsari@winona.edu

W. Leal Filho et al., *Curriculum Development for Sustainable Development*, World Sustainability Series, https://doi.org/10.1007/978-3-032-22146-9_1

Keyword Anthropocene · Curriculum · Ecocentrism · Education reform · Sustainability

1 Introduction

The Anthropocene—a geological event demarcated by human impacts upon Earth's ecosystems—has radically altered our planet and urges for educational transformation (Kunnas & Borsari, 2024). **Advances in sustainability education (SE) expand quickly, unraveling new research questions and needs for this knowledge to become convertible into reproducible actions for sustainable living (Barth et al.,** 2023)**. Hence, learning cannot remain entangled into reductionist paradigms of instruction at the expense of planetary stability. As climate change, biodiversity loss, and ecological degradation intensify, schools and higher education institutions (HEIs) should not only strive for adapting but rather focus their efforts by embedding sustainability into new study programs to prepare students coping with a changing world (Sterling,** 2021**).** Designing and implementing curricula, in response to current, global conditions and trends is crucial (Blake et al., 2013; Sterling, 2011). While numerous frameworks for SE have emerged over recent decades, a critical gap may persist because many of these underpinnings operate within anthropocentric schemes that continue struggling to recognize the intrinsic value beyond the more-than-human dimension (Chigbu & Makapela, 2025; Borsari & Kunnas, 2023). Despite a growing consensus about the relevance of SE, the frameworks employed across various institutions often perpetuate linear philosophies that are only partially capable of tackling current sustainability crises. For example, many SE approaches are grounded in human development indices, economic growth models, or excessive reliance on technology and its applications to human endeavors. In addition, an overall optimism, often sidelining the ethical and relational dimensions of human-nature interactions hinders the recognition of sustainability deficits (Onwueme & Borsari, 2007). **This trend may highlight clear contradictions when education seeking to promote sustainability remains structured within mechanistic paradigms that antagonize eco-humanistic approaches (Javed et al.,** 2025)**. Hence, the purpose of this study focuses on:**

- **Inspiring educators to employ eco-centric strategies in their curricula, capable of placing living systems at the center of every learning process, because fostering ecological knowledge from microorganisms to the ecosphere is essential to thrive in the Anthropocene.**
- **Shedding light on the need for a paradigm shift in education and learning approaches that by re-examining their study programs and modes of implementation methods, could strengthen pedagogical strategies in SE towards holism.**

An eco-centric worldview could benefit from ethical reorientations that shift from domination over nature towards amplifying connectedness and reciprocity with the Earth and all living beings (Capra & Luisi, 2014). After all, eco-centrism celebrates the inherent value of all life forms and the interdependence between humanity and ecosystems. It views nature as a community of which humans are a part, moving away from anthropocentric notions that for too long have been considering nature just as a resource for human use and disposal (Naess, 1989). **Eco-centrism shifts the focus from human-centered goals to ecosystems' health and resilience (Saha,** 2025**).** Thus, integrating this worldview into curriculum development is philosophically compelling and educationally critical.

There is a need for this assessment study because there is a growing disconnect between educational goals and the ecological realities they are meant to address. Climate reports and planetary boundaries research underscore the urgency of transforming how societies learn and act (Rockström et al., 2009a & 2009b; IPCC, 2023). Secondly, current pedagogical models often lack to engage students in systems thinking, ecological ethics, or in place-based education (Orr, 1992; Borsari, 2012). Also, more barriers like institutional resistance or socio-political constraints may deflect curriculum trajectories from the exact paths from which they were designed. **Hence, measuring the effects of eco-centric curricula remains a challenge, although from the health sciences it is known that just by spending time outdoors especially in nature-rich environments, improves people's well-being (White et al.,** 2024**).** Yet ecological literacy, ethical transformation, and systems thinking are complex and long-term outcomes that are difficult to quantify, or attribute directly to curricular design. As such, the study recognizes the need for assessing their effectiveness in various learning environments. Also, enacting SE framed in eco-centrism may require more than curricular models while reassessing approaches about how knowledge is gained, taught, and lived. As educators and curriculum designers, grapple with the implications of the Anthropocene, the need to cultivate a new educational *ethos* rooted in respect for life, ecological justice, and planetary stewardship, has never been more urgent. By placing eco-centrism at the center of curriculum development and implementation, education can become a spurring force not only for sustainability but also for resilience. Hence, this paper unravels the philosophical foundations of eco-centrism, by reviewing existing SE frameworks, while proposing strategies for integrating ecological principles across curricula for every educational level. In doing so, it seeks to contribute to a transformative vision of education—one that engages learners to co-create a more just and sustainable world for them and all living beings.

The critical assessment of curricular frameworks that this chapter aims at accomplishing was not free from limitations however, and these consisted of:

- **Focusing on higher learning curricula in industrialized countries, this study may be of limited applicability to Indigenous education thus, excluding contexts that already embody eco-centric values in ways that western paradigms could learn from.**

- **Considering a reduced number of case studies, emphasizing mostly on the agriculture and conservation aspects of SE, that derived exclusively from the author's personal teaching experience.**
- **The evaluative nature of this endeavor like any assessment, is descriptive and impedes making broad generalizations.**

Despite these constraints however, the author strived to provide readers with salient ideas that could empower educators for designing and implementing innovative curricula in sustainability.

2 Methodology

A multi-methods strategy consisting of literature review and case-study analyses was employed in this work to evaluate frameworks of eco-centrism for transforming existing curricula and current instruction approaches into an education for sustainable development. The literature review served the purpose of producing qualitative data to test the hypothesis for justifying, or refuting the need of eco-centric curricula in boosting education in sustainability. In addition, it aimed at identifying possible research voids in this distinctive, research topic.

2.1 Bibliographic Sources Search Strategy

The literature review was conducted through Google Scholar, known for its all-inclusive indexing of peer-reviewed and gray literature. The search was conducted on June 6, 2025, and considered the publication of peer-reviewed articles, books and book chapters, from 1989 to 2025. All sources were published in English. Boolean operators were used in this search to reflect crucial research dimensions as follows:

Dimension a) The Anthropocene and its implications for sustainability education.

"Anthropocene", "ecocentrism", "sustainable development", "education", "assessment".

Dimension b) Curricula in sustainability, education reform and evaluation.

"Curricula in sustainability", "education reform", "Case-Study Research", "Learning", "Teaching".

The Preferred Reporting Items for Systematic Reviews and Meta-Analyses (PRISMA) was employed as suggested by Page et al. (2021), for locating and selecting the sources. Thus, 1340 bibliographic sources were identified and after reading their title, 402 were selected for additional consideration. The reading of the abstract, introduction and literature review narrowed further this collection of references to 68 items (Fig. 1).

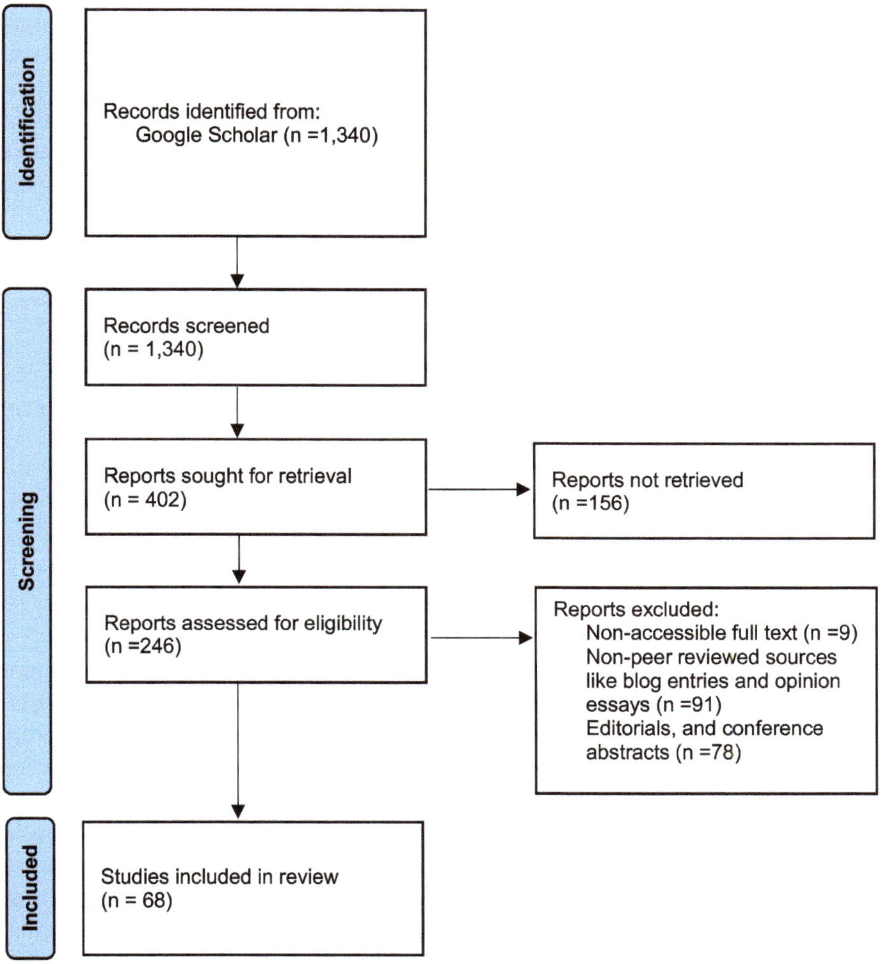

Fig. 1 Identification and selection scheme of studies via google scholar accomplished through PRISMA. Adapted from: page et al. (2021)

Although non-experimental, a review of the literature as methodology yields reliable data to describe a precise research context, which could foster inspiration for additional studies along the same trajectory, with appropriate adaptations, depending on the characteristics of each new, study context (Bryman, 2016).

Selected case studies (n = 6) yielded quantitative and additional qualitative data, that allowed for a triangulation of the study results, as suggested by Patton (2002) for any education assessment report. These were chosen from geographic areas around the world because this is where the author lived and worked during his professional career. For instance, he resided in Sierra Leone prior to relocating to the U.S., and initially, he lived in southwestern Louisiana and western Pennsylvania prior to settling in southeastern Minnesota. A few years ago, he was granted a Fulbright award that brought him to Panama to conduct curriculum research. The author sought correlations between the qualitative data derived from the literature review and those gathered from the case studies for this assessment. The cases added quantitative data, enriching the results analysis, while enhancing accuracy when drawing conclusions. Also, Zahle (2019) pointed out that case-study research is an effective, investigative method, which made it become a well-established research strategy mainly employed by social scientists (Widner et al., 2022).

3 Results and Discussion

The data derived from the literature review and case studies analysis are presented here, under separate subheadings.

4 The Anthropocene and its Implications for Sustainability Education

For a quarter century, the "Anthropocene" concept emerged as a powerful descriptor coined by Crutzen and Stoermer (2000), to explain that the totality of human activities upon Earth has become so huge, that we have departed from the climate stability of the Holocene and entered a new geologic era they named the Anthropocene (Lewis & Maslin, 2015). Hence, the Anthropocene represents humanity's gigantic and accelerating impact to the planet's geology and biogeochemical cycles. As such, it presents a scientific framework of current environmental crises that also raises unrelenting ethical and pedagogical questions (Kunnas & Borsari, 2024). Essential among these is the role of education in fostering a sustainable future, arguing that curricula must undergo transformative changes to align with the complexities of the Anthropocene via ecological literacy, systems thinking, critical reflection, and socio-political engagement (Borsari & Kunnas, 2023).

In the meantime, the Anthropocene construct has gained widespread acceptance however, its proposed onset remains a point of contention within the scientific community, with some experts suggesting it started somewhere between the 17th and mid-20th centuries. Instead, others have even posited earlier origins, extending back over two million years to the discovery of fire and stone tools-making by early hominins (Williams et al., 2015), or to 12,000 years ago, when people began to farm (Ellis et al., 2021). Instead, according to Zalasiewicz and team (2014) and to Steffen and his colleagues (2015), the 1950s characterized by an intensification of industrial activities should be considered the onset of the Anthropocene.

Kunnas and Borsari (2024) contended that espousing a larger temporal framework offers a more balanced depiction of human agency, one that encompasses both detrimental and potentially beneficial impacts as early interpretations of anthropogenic climate change might not have been negative after all. For example, since the early twentieth century, the Swedish chemist Arrhenius expounded that increasing concentrations of carbon gases in the atmosphere could result in global warming, which he envisioned benign for boosting the yields of agricultural crops. About a century later, Ruddiman (2003) reported that without early humans' interventions, that triggered deforestation on a large-scale to expand their farming scale, Earth might have cooled excessively over the past several millennia, determining a minor glaciation. In line with this hypothesis CO_2 levels began to rise about 8000 years ago, due to deforestation and associated burning habits in Asia, while the rise of methane (CH_4) emissions, which began 5000 years ago is related to expansions of rice farming and the husbandry of livestock species (Ellis, 2021).

However, industrialization, urbanization, deforestation, species extinction, and climate change mark a clear departure from the stable Holocene epoch that began at the end of the last glaciation (Rockström & Gaffney, 2021). Yet, the Anthropocene is often dated from the late eighteenth century, coinciding with the Industrial Revolution and an intense use of coal and more fossil fuels. Nonetheless, Ellis et al. (2021) support the hypothesis of an earlier onset, like the start of agriculture, whereas others favor the mid-1950s 'Great Acceleration' as landmark, when population size began to grow exponentially, and resource extraction expanded globally, on an industrial scale (Steffen et al., 2015). However, for Chakrabarty (2009) the Anthropocene is not only the scale of human impacts but also an entanglement of natural and social systems because this event hazes the boundary between human history and geologic time, forcing a rethinking of human agency, responsibility, and possibilities for shaping the future.

The 'ecosystem' and 'ecosphere' concepts conceived as networks provided a distinctive character to nature's structure because their points of intersection are complex systems of additional networks. These ideas were inspired by Odum's ecosphere construct (1971), which replaced the biosphere concept because the former includes abiotic factors that become crucial to support life on Earth. Taking this approach, ecology made a big jump forward for molding the attributes of a more holistic worldview. Also, recognizing endosymbiosis as the theory explaining the mutualistic relationship of mitochondria and chloroplasts, that two billion years ago established themselves within the cytoplasm of early eukaryotes, denotes a most significant, evolutionary landmark in the history of the biological sciences (Archibald, 2011).

From an educational angle, a multitude of implications intertwine with education in the Anthropocene. Collectively, these demand veering pedagogies toward critical consciousness as proposed by Freire's concept of critical pedagogy (1970), which inspires learners to question dominant ideologies, that preserve environmental and social inequalities. Through this approach, education becomes a robust empowering vehicle for students to engage with moral, political, and existential discourses. Within this frame education is not just a mean for transferring information! Rather, the enquiries about desirable futures and/or who benefits, and who suffers from current systems become focal to this education framework and most relevant for curricula in sustainability.

Similarly, Jickling and Wals (2008) argued for an education that goes beyond instrumental goals and creates space for dialogue, disagreement and democratic participation. In the end, this kind of schooling challenges technocratic, totalitarian narratives to acknowledge that teaching and learning in the Anthropocene must face values, worldviews, and power structures.

Therefore, Wals and Corcoran (2012) advocated for a learning mode that engages multiple perspectives and ways of knowing, including traditional ecological knowledge and the arts. Addressing the complex challenges of the Anthropocene requires moving beyond the rigidity of specific, learning compartments, toward interdisciplinary and transdisciplinary instructional approaches. Interdisciplinarity integrates knowledge across traditional academic fields, while transdisciplinary strategies require collaboration between academic and non-academic practitioners, including local communities, indigenous peoples, and policymakers (Stein et al., 2023). This exposure enriches understanding among students and fosters empathy, creativity, and innovation, which constitute keystone competencies for navigating the uncertainties brought by the Anthropocene (Borsari & Kunnas, 2023).

In the Anthropocene the future is not conceived as a continuation of the past but rather a new, unknown context of disruption, which demands for urgent transformation and education reform (White et al., 2024). As such, the latter should cultivate futures thinking, or the ability to envision and shape multiple potential futures. This involves scenario planning, speculative design, and foresight exercises that help learners engage with questions of long-term sustainability and resiliency (Milojevic, 2005). This framework encourages proactive, engagement rather than passive fatalism. It aligns with the transformative aims of SE, which seek not only to inform but to inspire all learners and citizens to act and participate in sustainability endeavors (Borsari, 2016). Resilience education, as part of this shift emphasizes the capacity to adapt, learn, and thrive amid change, preparing learners to be both critical thinkers and compassionate actors, especially during tempestuous times (Reid et al., 2012).

A final implication of the Anthropocene is about decolonizing education because the dominant paradigms that contributed to the planetary crisis—colonialism, capitalism, and extractive economies—are embedded in most education systems. Their intersectionality with learning and similar models of domination and control are unfit for education to empower students, rather these indoctrinate them in fulfillment of hegemonic governments' needs and expectations (Stein et al. 2023). Therefore, decolonizing education becomes sacrosanct yet, enacting this arduous process demands recognizing the historical injustices and epistemic traumas perpetrated by Eurocentric curricula and pedagogies. To this end, Andreotti (2016) called for an education in the Anthropocene that confronts privilege, grants space to marginalized voices, and supports epistemological diversity. This means valuing indigenous knowledge systems, unconventional ontologies, and alternative terminologies of well-being that challenge dominant development paradigms. In this way, decolonizing education expands the ethical and intellectual objectives of SE, making it more inclusive, impartial, and 'connected to place'.

Ultimately, the Anthropocene is not a solitary, scientific framework urging humanity to envision sustainable futures. Rather, it calls to rethink how we live, relate, and learn in a rapidly changing world. Its implications for education are far-reaching, demanding urgent changes that span from content-based instruction to transformative, holistic, and unprejudiced-oriented learning. Essentially, it invites humanity to imagine and shape potential futures while engaging in speculative thinking, that requires moving beyond modern frameworks—beyond flawed concepts of humanity, nature, and environmental education—because the solutions we seek can't be found within the same mental frames that created the problems (Le Grange, 2019). An education reform supportive of sustainable development in the Anthropocene fosters ecological literacy, critical reflection, interdisciplinarity, futures thinking, and decolonial awareness. Only through such comprehensive and inclusive approaches can education fulfill its role as a catalyst for creating more resilient, equitable and sustainable societies (Leal Filho et al., 2018).

5 Curricular Emphases for Sustainability, Education Reform and Evaluation

In a timeframe defined by geopolitical crises, inequalities and ecological instability, SE has become a crucial component of curriculum development across all educational levels. From early childhood education diverse frameworks can inform how sustainability is embedded into teaching and learning practices (Weldemariam et al. 2017). These frameworks range from experiential and place-based learning to systems thinking and critical pedagogy, each contributing uniquely to fostering ecological literacy (Sterling 2001; Borsari 2012), and responsible citizenship (Sterling 2011).

At many levels of instruction SE often emphasizes experiential and cooperative learning approaches, that are framed within constructivism theory (Lord, 1998). These strategies prioritize direct interaction with natural habitats (Borsari & Vidrine, 2023), while fostering early ecological awareness, spurring in children strong environmental values, thus overcoming ecophobia through repeated, meaningful encounters in nature (Sobel, 2004). Curricula at this level often employ outdoor learning, gardening and/or local ecological projects, reinforcing concepts of stewardship and interdependence with all living beings and the environment. The same could occur at college-level-learning institutions where systems thinking becomes most relevant because this framework boosts students learning about complex interactions, at the juncture between ecological and socio—economic systems (Borsari, 2016).

For over twenty years systems thinking has been employed as a holistic worldview, which is indispensable for tackling sustainability challenges (Sterling, 2001). The International Baccalaureate (IB) curriculum exemplifies this approach, blending global issues and sustainability across curriculum subjects, encouraging all learners to consider the long-term consequences of their actions within interrelated, complex systems (IBO, 2013). In addition, higher education institutions (HEIs) adopt these and even more diverse and interdisciplinary frameworks, combining critical pedagogy, transformative learning, and futures thinking. The latter has been pointed out by Wals and Corcoran (2012), who reiterated that postgraduate programs often utilize planning scenarios to enable students' visions of feasible, sustainable alternatives to current, unsustainable lifestyles. At the same time, critical pedagogy stimulates students to question dominant narratives and power structures that contribute to environmental and socio-economic degradation (Freire, 1970). This approach fosters critical thinking and agency a step further, positioning students not only as 'consumers' of knowledge, but also as change-makers. In line with this framework is Merizow's theory of transformative learning (1997), intersects with SE for emphasizing personal growth, while being acceptive of various viewpoints, and ethical reflections.

Moreover, curricula in sustainability increasingly draw upon the UN's Education for Sustainable Development framework, which promotes competencies such as critical thinking, collaboration, and adaptability (UNESCO, 2017). This framework also provides a common language and sets of goals that align curriculum efforts with global sustainability objectives, from the Sustainable Development Goals (SDGs) to regional policy implementations (Leal Filho et al., 2021), including students' participation in implementing the SDGs (Filho et al., 2024). Despite the diversity of approaches, challenges remain in aligning these frameworks across educational institutions, as departmental turf protection, scarcity of teachers' learning opportunities, and administrative constraints can hinder coherent curriculum development (Tilbury, 2011).

Undoubtedly, the Anthropocene poses profound challenges and opportunities for educational theory and practices that challenge inadequate models of instruction. These are often rooted in disciplinary compartments, anthropocentrism, and reductionism, that remain affected by inadequacy for addressing the systemic nature of socio-ecological crises (Borsari & Mora, 2020). Meeting the demands of the Anthropocene requires fostering integrative, reflexive, and transformative learning approaches in education. Thus, ecological literacy becomes keystone requirement for education in the Anthropocene, because this entails the capability for understanding the principles of interconnectedness among ecosystems, in addition to the consequences of human actions within them.

Orr (2004), emphasized that ecological literacy should be foundational for all education, not confined to science disciplines, and mainly experienced outdoors. It involves grasping concepts such as biogeochemical cycles, or feedback loop mechanisms, carrying capacity, energy flows, ecological resilience and associated themes (Vidrine et al., 2024). Systems thinking complements ecological literacy by enabling learners to recognize the complexity, interdependence, and dynamic nature of environmental challenges. This is a century-old holistic approach, which commenced with biologists who began addressing complex problems by focusing on the interconnectedness and relationships within a system, rather than its isolated parts, recognizing living organisms as cohesive units (Capra & Luisi, 2014). Ecology had already established as a respected science at that time, combining the study of living organisms within their communities and ecosystems. It became most relevant to support a holistic worldview when the concepts of complexity and networks were introduced through ecology (Odum, 1971).

As Sterling (2011) argued, systems thinking allows students to perceive patterns, relationships, and long-term consequences, encouraging holistic rather than reductionist approaches to problem-solving. For this reason, educators should design curricula that transcend subject boundaries and emphasize relational thinking, framed within eco-centrism (Lozjanin et al., 2025).

6 Case-Studies Analysis

The analysis of selected case studies with unique curricular features each curriculum possesses, in various education contexts, complemented this evaluation. A summary of descriptors and goals for these curricula from the six cases is reported in Table 1.

7 Sierra Leone

In Sierra Leone, SE is a curricular trait that despite the challenges of climate change continues to experience difficulties in getting traction due to its educational infrastructure, which endures a lack of schools and qualified teachers. This problem was exacerbated by the 1991–2001 civil war and further affected by more recent public health crises such as the Ebola epidemic of 2015 (Mai, 2019), and the Covid-19 pandemic of 2020. Due to these tragic circumstances teaching activities were interrupted at SJVI and with these also students' graduations, which in the agriculture curriculum had been capable of preparing about 20 students per year, since 1985. Yet, food production remains a primary activity in this country to fulfill the needs of local communities, and most Sierra Leonians work in agriculture, whose profits account for 58% of the national income (Wren Media, 2007). Therefore, SE focuses on farming approaches that support staple crops like rice, which is now cultivated through sustainable rice intensification (SRI) practices, an agroecological method that has been taught to many farmers in Sierra Leone (Borsari, 2025a).

8 Panama

Also in this context, food security remains a priority for most people residing in rural areas and among indigenous communities. Hence, feasible strategies in SE focused on producing food through experiential workshops modeling the 'peasant-to-peasant' approach. These were designed by the author in collaboration with his agroecology students, attracting about 200 subsistence agriculturists at the university farm, in 2015 (Borsari et al., 2016).

Table 1 Distinctive curricular attributes from the case studies considered in this study

Country/State	Higher Education Institution (HEI)	Curriculum goals	Distinctive curriculum traits	References
Sierra Leone	Saint Joseph vocational institute (SJVI)	–Prepare students of agriculture understand and practice regenerative farming	–Apply every day agroecological knowledge at the SJVI farm	Borsari (2025a)
Panama	Universidad Santa Maria La Antigua (USMA)	–Engage agroecology students in developing workshops for subsistence farmers	–Assisting with the deployment of hands-on workshops for local farmers	Borsari et al. (2016)
Louisiana	Louisiana State University at Eunice	–Expose students in science courses to SE with emphasis on habitat restoration	–Apply SE learning through prairie restoration projects on campus and beyond	Borsari and Vidrine (2023)
Pennsylvania	Slippery rock university (SRU)	–Guide graduates to better resolve pressing environmental and social challenges-employ principles of sustainability as framework for action	–First SE graduate curriculum in the U.S. (MSc.)— Core set of mandatory courses and three emphasis tracks: Agroecology, built environment, sustainable resource management	Borsari (2024)
Minnesota	Winona State University (WSU)	–Instill SE principles in an introductory course in biology for non-science majors	–SE application via: Case-study pedagogy, students' mini-talks, lab exercises, art exhibits	Borsari (2025b)
Minnesota	Minnesota state college southeast (MNCS)	–Introduce SE principle with emphasis on food production and security	–Apply knowledge gained from classes at the campus garden	Borsari and Borsari (2025)

9 United States of America

At Louisiana State University at Eunice the focus of SE is about conservation and restoration of coastal prairie habitats for its implications for veering agriculture curricula towards sustainability (Borsari, 2003). Hundreds of volunteers were involved in restoring prairies in and around Eunice through the years, including several LSU-E students (Borsari & Vidrine, 2023).

Slippery Rock University of Pennsylvania began enrolling students in 1991 for the Master of Science in Sustainable Systems, a curriculum with three emphasis tracks (Agroecology, Built Environment, Sustainable Resource Management), exemplifying SE in a multidisciplinary fashion (Borsari, 2024). Although SRU terminated this curriculum, in 2004 a survey of MS3 alumni counted 44 graduates who are sustainability leaders in the U.S, northeast region.

At Winona State University freshmen students (circa 7000) were exposed to SE in an introductory biology course (BIO 118), through a variety of pedagogies as described by Borsari (2025b). A distinctive feature of this curriculum is the 'connections seminar' course taught by the author and a colleague from the English Department, titled: Whose Planet is this Anyway? (RESC 201). This course enrolled 59 freshmen who learned approaches for remediating to current environmental degradation, climate change and unsustainable living issues (Borsari & Vidrine, 2023). Instead, at Minnesota State College Southeast (MSCS), SE emerged with the design of a 60-credit curriculum "Sustainable Foods and Farming", which began enrolling students in 2019. This degree program (Associate of Applied Science), was conceived to prepare students succeed in the rapidly evolving, agriculture sector, especially for students interested in agroecology (Borsari & Borsari, 2025).

10 Conclusion and Future Prospects

This chapter evaluated SE and its contributions to sustainability curricula, in the Anthropocene. The latter requires locating safe pathways within planetary boundaries for people and the planet (Rockström et al., 2021), making education crucial to enhance society's resilience. A recent, international assessment confirms that universities are making progress in employing SE at institutions of higher learning (Leal Filho et al., 2025), while engaging students alongside (Leal Filho et al., 2024). The transformative role of SE has been substantiated for decades as the necessary shift to occur in modern education in support of sustainability (Sterling, 2011; Leal Filho et al., 2018). Nonetheless, barriers continue affecting a conversion of curricula towards eco-centrism (Sterling, 2021; Ávila et al., 2017). These can be multiple and diverse, ranging from waning administrative support, students' interests, access to facilities like living labs, resources, instructors' preparedness and more (Leal Filho et al., 2023).

These issues hamper an establishment of ecological values yet, they will be enhanced by SE and its supportive frameworks to replace anthropocentric education, which remains antagonistic to sustainability. A model illustrating the contrasting forces affecting the development of curricula in sustainability is reported (Fig. 2).

Opposing frameworks determine the inclination of the central axis upon which curricula for sustainable development stand. Ethics is a keystone feature of the proposed model because values in alignment with the frameworks of eco-centrism

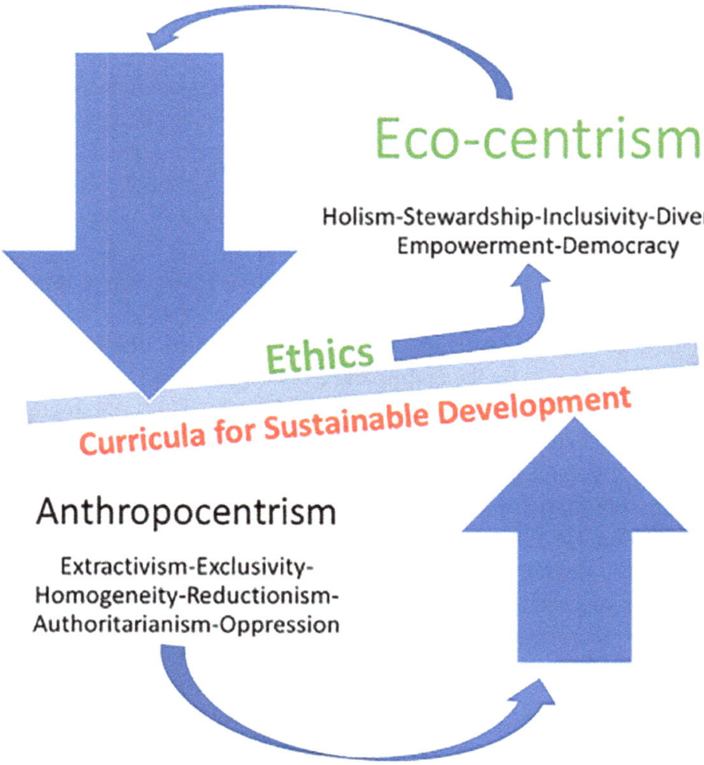

Fig. 2 A new ethics inspired by eco-centric frameworks supersede anthropocentrism and its mechanistic underpinnings, spurring a new equilibrium that enhances eco-centric curricula

will exert a strong force against anthropocentrism. Consequently, the slope of the central axis will increase in favor of eco-centrism, and this will improve further sustainability curricula and their outcomes. Therefore, the role played by ethics is pivotal in this cyclical model because stronger ethics develops, weakening the anthropocentric force at every cycle, thus becoming unable to counteract the force exerted by expanding, eco-centric curricula.

In sum, SE is much more than an assemblage of teaching modules but rather a holistic and pragmatic approach needed to address declining resources and increasing uncertainties the Anthropocene bring with it. Thus, boosting ethical principles in SE is essential to envision sustainable living systems and their future. This approach will make resounding the argument that sustainable development is not just a new trend, but rather the basis for long-term futures achievable through cautionary planning and ecologically informed decision-making, with democracy as a vital component of sustainability education and development.

References

Andreotti, V. d. O., (2016). Education, knowledge and the righting of wrongs. In M. A. Geo-JaJa & S. Majhanovich (Eds.), *Effects of globalization on education systems and development* (pp. 131–141). The World Council of Comparative Education Societies. Sense Publishers. https://doi.org/10.1007/978-94-6300-729-0-8

Archibald, J. M. (2011). Origin of eukaryotic cells: 40 years on. *Symbiosis, 54*, 69–86. https://doi.org/10.1007/s13199-011-0129-z

Ávila, L. V., Leal Filho, W., Brandli, L., Macgregor, C. J., Molthan-Hill, P., Özuyar, P. G., &.Moreira, R. M. (2017). Barriers to innovation and sustainability at universities around the world. *Journal of Cleaner Production*, 164. https://doi.org/10.1016/j.jclepro.2017.07.025

Barth, M., Jiménez-Aceituno, A., Lam, D. P. M., Bürgener, L., & Lang, D. J. (2023). Transdisciplinary learning as a key leverage for sustainability transformations. *Current Opinion in Environmental Sustainability, 64*, 101361. https://doi.org/10.1016/j.cosust.2023.101361

Blake, J., Sterling, S., & Goodson, I. (2013). Transformative learning for a sustainable future: An exploration of pedagogies for change at an alternative college. *Sustainability, 5*(12), 5347–5372. https://doi.org/10.3390/su5125347

Borsari, B. (2003). The coastal prairie habitat as a model to promote sustainability in the agricultural curriculum. In S. Fore (Ed.), *proceedings of the 18th north American prairie conference: Promoting prairie* (pp. 17–23). Truman state university press.

Borsari, B. (2025a). Climate change education through agroecology curricula in Sierra Leone. In M. F. Mbah, P. Molthan-Hill, & E. L. Molua (Eds.), *practices, perceptions and prospects for climate change education in Africa* (pp. 249–267). Spring. https://doi.org/10.1007/978-3-031-84081-4-13

Borsari, B. (2025b). Student-centered teaching for sustainability education in an introductory biology course at Winona State University: A case-study. In W. Leal Filho, L. V. Trevisan, P. J. Pace, & M. Mifsud (Eds.), *Education for sustainable development: The contribution of universities* (pp. 39–58). Springer. https://doi.org/10.1007/978-3-031-86985-3_3

Borsari, B., & Borsari, G. (2025). Contributions of food systems curricula in preparing graduates to fathom climate change uncertainties and challenges. In W. Leal Filho, M. Sima, A. Lange Salvia, M. Kovaleva, & E. Manolas (Eds.), *University initiatives on climate change education and research* (pp. 1–27). Springer. https://doi.org/10.1007/978-3-031-25960-9121_1

Borsari, B. (2024). Ecological design as framework of the Sustainable Systems (MS3) curriculum: A historical review of achievements, challenges and community outreach in architecture for sustainability. In W. Leal Filho, T. Dibbern, S. R. de Maya, M. C. Alarcón-del-Amo, & L. M. Rives (Eds.), *The contribution of universities towards education for sustainable development* (pp. 125–144). Springer. https://doi.org/10.1007/978-3-031-49853-4_9

Borsari, B., & Vidrine, M. F. (2023). Envisioning prairies as curricular focus for climate change education. In W. Leal Filho, M. Sima, A. Lange Salvia, M. Kovaleva, & E. Manolas (Eds.), *University initiatives on climate change education and research* (pp. 1–25). Springer. https://doi.org/10.1007/978-3-031-25960-9_56-1

Borsari, B., & Kunnas, J. (2023). Historical memory and eco-centric education: Looking at the past to move forward with the 2030 Agenda for Sustainable Development. In W. Leal Filho, A. M. Azul, F. Doni, & A. L. Salvia (Eds.), *Handbook of sustainability science in the future* (pp. 985–999). Springer. https://doi.org/10.1007/978-3-031-04560-8_40

Borsari, B., & Mora, C. (2020). Education reform through a systems approach for sustainable development. In W. Leal Filho, A. Azul, L. Brandli, P. Özuyar, & T. Wall (Eds.), *Quality education (Encyclopedia of the UN Sustainable Development Goals)* (pp. 1–10). Springer. https://doi.org/10.1007/978-3-319-69902-8_4-1

Borsari, B. (2016). Sustainability education as a curriculum emphasis. *Academic Exchange Quarterly, 20*(4), 31–37.

Borsari, B., Peréz de Gracia, N., & Castro Peralta, J. (2016). Students' engagement in an extension program in agroecology for subsistence farmers at the Universidad Católica Santa Maria La

Antigua (USMA), Panamá. In W. Leal Filho & P. Pace (Eds.), *Teaching education for sustainable development at university level* (pp. 147–161). Springer. https://doi.org/10.1007/978-3-319-32928-4_10
Borsari, B. (2012). Curriculum framework for sustainability education. *Academic Exchange Quarterly, 16*(1), 74–78.
Bryman, A. (2016). *Social research methods* (5th ed.). Oxford University Press.
Capra, F., & Luisi, P. L. (2014). *The Systems View of Life: A unifying vision.* Cambridge University Press.
Chakrabarty, D. (2009). The climate of history: Four theses. *Critical Inquiry, 35*(2), 197–222.
Chigbu, B. I., & Makapela, S. L. (2025). Data-driven leadership in higher education: Advancing sustainable development goals and inclusive transformation. *Sustainability, 17*(7), 3116. https://doi.org/10.3390/su17073116
Crutzen, P. J., & Stoermer, E. F. (2000). The 'Anthropocene'. In L. Robin, S. Sörlin, & P. Warde (Eds.), *The future of nature: Documents of global change* (pp. 479–490). Yale University Press. https://doi.org/10.12987/9780300188479-041
Ellis, E. C., Gauthier, N., Goldewijk, K. K., Bliege Bird, R., Boivin, N., et al. (2021). People have shaped most of terrestrial nature for at least 12,000 years. *Proceedings of the National Academy of Sciences, 118*(17), e2023483118. https://doi.org/10.1073/pnas.2023483118
Ellis, E. C. (2021). Land use and ecological change: A 12,000-year history. *Annual Review of Environment and Resources, 46*, 1–33. https://doi.org/10.1146/annurev-environ-012220-010822
Filho, W. L., Trevisan, L. V., Dinis, M. A. P., et al. (2024). Fostering students' participation in the implementation of the Sustainable Development Goals at higher education institutions. *Discover Sustainability, 5*, 22. https://doi.org/10.1007/s43621-024-00204-7
Freire, P. (1970). *Pedagogy of the oppressed*. Seabury Press.
International Baccalaureate Organization (IBO). (2013). IB and the United Nations Sustainable Development Goals. https://www.ibo.org/
Intergovernmental Panel on Climate Change (IPCC). (2023). AR6 synthesis report: Climate change 2023. https://www.ipcc.ch/report/sixth-assessment-report-cycle/
Javed, M. H., Ahmad, A., Nizami, A. S., Gastaldi, M., & D'Adamo, I. (2025). Sustainable development at the crossroads: Navigating eco-humanism and eco-modernism. *Current Opinion in Green and Sustainable Chemistry, 53*(101), 018. https://doi.org/10.1016/j.cogsc.2024.101018
Kunnas, J., & Borsari, B. (2024). Planetary health and climate change in the Anthropocene. In W. L. Filho, D. G. Vidal, & M. A. P. Dinis (Eds.), *Planetary health and climate change: Understanding the impacts of climate change to the well-being of our planet* (pp. 311–326). Springer. https://doi.org/10.1007/978-3-031-72740-5_14
Le Grange, L. (2019). The Anthropocene: Becoming-imperceptible of environmental education.On Education. *Journal for Research and Debate, 2*(4). https://doi.org/10.17899/on_ed.2019.4.4
Leal Filho, W., Trevisan, L. V., Sivapalan, S., et al. (2025). Assessing the impacts of sustainability teaching at higher education institutions. *Discover Sustainability, 6*, 227. https://doi.org/10.1007/s43621-025-01024-z
Leal Filho, W., Özuyar, P. G., Dinis, M. A. P., et al. (2023). Living labs in the context of the UN Sustainable Development Goals: State of the art. *Sustainability Science, 18*, 1163–1179. https://doi.org/10.1007/s11625-022-01240-w
Leal Filho, W., Raath, S., Lazzarini, B., Vargas, V. R., de Souza, L., Anhlon, R., Quelhas, O. L. G., Haddad, R., Klavins, M., & Orlovic, L. V. (2018). The role of transformation in learning and education for sustainability. *Journal of Cleaner Production*. https://doi.org/10.1016/j.jclepro.2018.07.017
Leal Filho, W., Frankenberger, F., Lange Salvia, A., Azeiteiro, U., Alves, F., Castro, P., Will, M., Platje, J., Orlovic, L. V., Brandli, L., Price, E., Doni, F., Mifsud, M., & Veiga Ávila, L. (2021). A framework for the implementation of the Sustainable Development Goals in university programmes. *Journal of Cleaner Production, 299*(126), 915. https://doi.org/10.1016/j.jclepro.2021.126915

Lewis, S. L., & Maslin, M. A. (2015). Defining the Anthropocene. *Nature, 519*, 171–180. https://doi.org/10.1038/nature14258
Lord, T. (1998). Cooperative learning that really works in biology teaching: Using constructivist-based activities to challenge student teams. *The American Biology Teacher, 60*(8), 580–588.
Lozjanin, A., Chhabra, G., & Mehdian, N. (2025). Exploring green pedagogy for eco-centric praxis-based learning in higher education. *Journal of Applied Learning and Teaching, 8*(1), 66–79. https://doi.org/10.37074/jalt.2025.8.S1.12
Jickling, B., & Wals, A. E. J. (2008). Globalization and environmental education: Looking beyond sustainable development. *Journal of Curriculum Studies, 40*(1), 1–21. https://doi.org/10.1080/00220270701684667
Mai, T. T. (2019). *Project Information Document–Sierra Leone Free Education Project –P 167897*. The World Bank. http://documents.worldbank.org/curated/en/711051560267527870/Project-Information-Document-Sierra-Leone-Free-Education-Project-P167897
Mezirow, J. (1997). Transformative learning: Theory to practice. *New Directions for Adult and Continuing Education, 74*, 5–12. https://doi.org/10.1002/ace.7401
Milojević, I. (2005). *Educational futures: Dominant and contesting visions*. Routledge.
Naess, A. (1989). *Ecology, community and lifestyle: Outline of an ecosophy*. Cambridge University Press.
Odum, E. P. (1971). *Fundamentals of ecology* (3rd ed.). Saunders.
Onwueme, I., & Borsari, B. (2007). The sustainability asymptogram: A new philosophical framework for policy, outreach and education in sustainability. *International Journal of Sustainability in Higher Education, 8*(1), 44–52. https://doi.org/10.1108/14676370710717580
Orr, D. W. (2004). *Earth in mind: On education, environment, and the human prospect*. Island Press.
Orr, D. W. (1992). *Ecological literacy: Education and the transition to a postmodern world*. SUNY Press.
Page, M. J., McKenzie, J. E., Bossuyt, P. M., Boutron, I., Hoffmann, T. C., Mulrow, C. D., Shamseer, L., Tetzlaff, J. M., Akl, E. A., Brennan, S. E., Chou, R., Glanville, J., Grimshaw, J. M., Hróbjartsson, A., Lalu, M. M., Li, T., Loder, E. W., Mayo-Wilson, E., McDonald, S., & Moher, D. (2021). The PRISMA 2020 statement: An updated guideline for reporting systematic reviews. *BMJ, 372*, n71. https://doi.org/10.1136/bmj.n71
Patton, M. Q. (2002). Qualitative research and evaluation methods (3rd edn.). Sage Reid, W., Walker, B., & Salt, D. (2012). *Resilience thinking: Sustaining ecosystems and people in a changing world*. Island Press.
Rockström, J., & Gaffney, O. (2021). *Breaking boundaries: The science of our planet*. Dorling Kindersley.
Rockström, J., Gupta, J., Lenton, T. M., Qin, D., Lade, S. J., Abrams, J. F., et al. (2021). Identifying a safe and just corridor for people and the planet. *Earth's. Future, 9*(3), e2020EF001866. https://doi.org/10.1029/2020EF001866
Rockström, J., Steffen, W., Noone, K., et al. (2009a). A safe operating space for humanity. *Nature, 461*(7263), 472–475. https://doi.org/10.1038/461472a
Rockström, J., Steffen, W., Noone, K., Persson, Å., Chapin, F. S., III, Lambin, E., et al. (2009b). Planetary boundaries: Exploring the safe operating space for humanity. *Ecology and Society, 14*(2), 32. http://www.ecologyandsociety.org/vol14/iss2/art32/
Ruddiman, W. F. (2003). The anthropogenic greenhouse era began thousands of years ago. *Climatic Change, 61*, 261–293.
Saha, S. (2025). A call for transitioning to an ecocentric approach to further the sustainability paradigm focusing on nature–human relationship. *Jindal Global Law Review, 16*, 335–356. https://doi.org/10.1007/s41020-025-00270-6
Sobel, D. (2004). *Place-based education: Connecting classrooms and communities*. The Orion Society.
Stein, S., Andreotti, V., Ahenakew, C., Suša, R., Valley, W., Huni Kui, N., & McIntyre, A. (2023). Beyond colonial futurities in climate education. *Teaching in Higher Education, 28*(5), 987–1004. https://doi.org/10.1080/13562517.2023.2193667

Sterling, S. (2021). Concern, conception, and consequence: Re-thinking the paradigm of higher education in dangerous times. *Frontiers in Sustainability, 2*(743), 806. https://doi.org/10.3389/frsus.2021.743806

Sterling, S. (2001). *Sustainable education: Re-visioning learning and change*. Green Books.

Sterling, S. (2011). Transformative learning and sustainability: Sketching the conceptual ground. *Learning and Teaching in Higher Education, 5*, 17–33.

Steffen, W., Broadgate, W., Deutsch, L., et al. (2015). The trajectory of the anthropocene: the great acceleration. *The Anthropocene Review, 2*(1), 81–98.

Tilbury, D. (2011). *Education for sustainable development: An expert review of processes and learning*. UNESCO. https://unesdoc.unesco.org/ark:/48223/pf0000191442

UNESCO. (2017). *Education for sustainable development goals: Learning objectives.*. https://unesdoc.unesco.org/ark:/48223/pf0000247444

Vidrine, M. F., Borsari, B., Pastorek, M., & Allen, C. M. (2024). Rewilding the Cajun Prairie:Efforts in land rehabilitation ensure a sustainable future in southwestern Louisiana and beyond. In W. Leal Filho, A. L. Salvia, & C. R. Portela de Vasconcelos (Eds.), *An agenda for sustainable development research* (pp. 91–110). Springer. https://doi.org/10.1007/978-3-031-65909-6_6

Wals, A. E. J., & Corcoran, P. B. (2012). *Learning for sustainability in times of accelerating change*. Wageningen Academic Publishers. https://doi.org/10.3920/978-90-8686-757-8

Weldemariam, K., Boyd, D., Hirst, N., et al. (2017). A critical analysis of concepts associated with sustainability in early childhood curriculum frameworks across five national contexts. *International Journal of Early Childhood, 49*, 333–351. https://doi.org/10.1007/s13158-017-0202-8

White, P. J., Ardoin, N. M., Eames, C., & Monroe, M. C. (2024). Agency in the anthropocene: Education for planetary health. *The Lancet Planetary Health, 8*(2), e117–e123. https://doi.org/10.1016/S2542-5196(23)00271-1

Widner, J., Woolcock, M., & Ortega Nieto, D. (Eds.). (2022). *Strategies for social inquiry: The case for case study (Methods and Applications in International Development)*. Cambridge University Press. https://doi.org/10.1017/9781108688253

Williams, M., Zalasiewicz, J., Haff, P. K., et al. (2015). The Anthropocene biosphere. *The Anthropocene Review, 2*(3), 196–219.

Media, W. (2007). Settling for a future in Sierra Leone. *New Agriculturist.*. https://web.archive.org/web/20200804200645/www.new-ag.info/en/focus/focusItem.php?a=291

Zahle, J. (2019). Data, epistemic values, and multiple methods in case study research. *Studies in History and Philosophy of Science, 78*, 32–39. https://doi.org/10.1016/j.shpsa.2018.11.005

Zalasiewicz, J., Williams, M., & Waters, C. (2014). Can an Anthropocene Series be defined and recognized? *Geological Society, London, Special Publications, 395*, 39–53.

Bruno Borsari is Professor Emeritus in the Biology Department at Winona State University in Winona, Minnesota. An agroecologist by trade, he spent 34 years teaching, researching, and consulting in sustainable development, restoration agroecology, soil/water conservation, and food security. He earned his Laurea in agricultural sciences from the University of Bologna, Italy and his PhD from the University of New Orleans, LA, USA. He taught biology and agroecology in west and central Africa, in Panama, Italy, and USA. His research agenda is multidisciplinary and focuses on the intersectionality between food systems, ecosystems and sustainability education.

A Baseline for Creating Sustainability Learning Outcomes: A Framework Informed by International Meta-Competencies' Prevalence

Olivia DiPrinzio and Peter Buck

Abstract Critical risks from the anthropogenic climate crisis, nitrogen cycle disruption, biodiversity and ecosystem loss, and their subsequent effects on human well-being and security have prompted calls for significant innovations in higher education. In light of these risks, scholars and practitioners have proposed competencies-based curriculum frameworks to establish graduate profiles equipped to spearhead sustainable development in years to come. Reviewing the progress of such proposals, this study sought to discern if there was a relationship between accepted sustainability meta-competencies and stated sustainability learning outcomes of HEIs within the United States and around the world, as no such review at this level currently exists. This research examines the frequency and expression of six prominent meta-competencies in over 50 institutions' sustainability learning objectives worldwide. The team utilized a hermeneutic approach to develop a flat, deductive coding framework for 400 reviewed objectives from international institutions. Of the six meta-competencies, systems thinking was the most prominently expressed for American institutions while strategic competence surpassed systems thinking in non-U.S. institutions. Creativity was found to not be present as a primary meta-competency but was observed as secondary or tertiary. As this research cross-references international and domestic findings to establish an internationally relevant framework, a practicable keyword and theme bank for each meta-competency was produced for institutions' future curricular development. The products of this research are novel to sustainability curriculum development in two key ways. First, this work develops an understanding of the current relative importance of each meta-competency, commonly referred to as a key competency, through an international lens, as expressed through learning outcomes. Second, this work creates an internationally relevant base framework of keywords and themes for HEIs to utilize, irrespective of location, when curating desirable graduate profiles of future sustainable development leaders. In the

---

O. DiPrinzio
Department of Geoscience, The Pennsylvania State University, State College, University Park, PA, USA

P. Buck (✉)
Penn State Sustainability, State College, University Park, PA, USA
e-mail: PeterBuck@psu.edu

W. Leal Filho et al., *Curriculum Development for Sustainable Development*, World Sustainability Series, https://doi.org/10.1007/978-3-032-22146-9_2

end, the goal of developing the framework was such that it could be utilized by HEIs looking to either bolster pre-existing curriculum or create new curriculum, which supports the proposals highlighted above.

Keywords Curriculum · Learning outcomes · Meta-competencies · Sustainability · Comparative education · Educational frameworks

1 Introduction

Critical risks from the anthropogenic climate crisis and their subsequent effects have prompted calls for significant innovations in higher education. In light of these risks, scholars and practitioners have proposed competencies-based curriculum frameworks to guide education for sustainable development. At the same time, higher education institution (HEI) networks and the Organization for Economic Cooperation and Development have encouraged the higher education field to create consistent associated learning outcomes for sustainability (Tremblay et al., 2012). Today, there is a gap in sustainability education research as it relates to how accepted meta-competencies are being translated into institutional learning outcomes on an international scale. In light of the dual calls for sustainability meta-competencies and learning outcomes, this study sought to discern if there was a relationship between established sustainability meta-competencies and subsequent learning outcomes of HEIs within the United States and around the world. Furthermore, this study aimed to determine trends underlying how these meta-competencies have been translated into institutional learning outcomes.

This publication details a hermeneutic analysis of learning outcomes, presents resultant keywords and themes by meta-competency, and discusses the developed internationally relevant framework. The products of this research are novel to sustainability education in two key ways. First, this work develops an understanding of the current relative importance of each meta-competency, commonly referred to as a key competency, through an international lens, as expressed through learning outcomes. Second, this work creates an internationally relevant base framework of keywords and themes for HEIs to utilize when developing the graduate profiles synonymous with positioning oneself as a leader in sustainable development. In the end, the goal of developing the framework introduced here was that it could be utilized by HEIs looking to either create new or bolster pre-existing curriculum.

2 Literature Review

Decades of observation show that humanity is dangerously altering planetary systems. Beginning in the 1970s, the Stockholm Declaration recognized that humanity was both "a creature and moulder" of its environment and that through the acceleration of technological innovation, economic development, and population growth, that we had reached a point in history where we could "do massive and irreversible harm to the earthly environment on our life and well-being depend" (United Nations, 1972). To thrive, humanity would have to "defend and improve the human environment for present and future generations" (United Nations, 1972)—a finding reinforced by the World Commission on Environment and Development's Brundtland Commission (1987) and again at the Rio Earth Summit in 1992 (United Nations Commission on Sustainable Development, 1992). In response, the Earth Charter Commission called for "creative leadership" to "[i]ntegrate into formal education and life-long learning the knowledge, values, and skills needed for a sustainable way of life" (Earth Charter Commission, 2000).

In tandem with these efforts, educators began developing meta-competencies: necessary higher-order abilities related to being able to learn, anticipate, create and adapt to many areas of work, including sustainability (Brown and McCartney, 1995). The challenges of sustainability described above feature specific and distinct characteristics that have prompted figures such as Wiek et al. to propose "a particular set of interlinked and interdependent key competencies" for sustainability that separate it from basic, other disciplinary, or professional competency models (Wiek et al., 2011). A robust literature has developed around these competencies (see: Bartlett et al., 2020; Brundiers et al., 2021; Engle et al., 2017; de Haan, 2006; Mochizuki and Fadeeva, 2010; Redman and Wiek, 2021; Rieckmann, 2012; Schrand et al., 2013; Wiek et al., 2011). This literature has reached agreement on a suite of sustainability meta-competencies: interpersonal literacy, anticipatory (temporal) thinking, normative (ethical) literacy, systems thinking, and strategic competence (Brundiers et al., 2021; Engle et al., 2017; Wiek et al., 2011). In conjunction with these five, other meta-competencies such as creativity, intrapersonal, implementation, and integration competence, among others, have been presented as potential additions (Brundiers et al., 2021; Engle et al., 2017; Redman and Wiek, 2021).

Paralleling this evolution, learning outcomes were identified as the primary mechanism for the incorporation of these meta-competencies into general education. This is supported by numerous organizations including the Organization for Economic and Cooperative Development (OECD). Such organizations argue that learning objectives have become essential to a meaningful education within institutions. Most importantly at this point, organizations are highlighting the need for comparative learning objectives to be developed "in response to a range of higher education trends, challenges and paradigm shifts (Tremblay *et al.*, 2012)."

Supporting this stipulation, the creation of the Tbilisi Declaration, recognized as the first document to develop sustainability learning outcomes, continued to accelerate the development of outcomes. Subsequent research has propelled this work forward (Svanström et al., 2008). For example, Sandri et al. leads the current research on the concept of measuring sustainability learning outcomes' success and the measures to do so while Svanström et al. compared common competencies and skills as observed in a select number of case study HEI outcomes (Sandri, 2013; Svanström et al., 2008). This work expands upon the spirit of Svanström et al.'s work by providing an updated review on competencies, pursuing an international comparison, and posing uniform keywords and themes (Svanström et al., 2008). No other large body of literature was found to conduct such research, thus leading to an inability to compare obtained results.

This leads to today's state of affairs in sustainability education. As many sustainability educators know, a sufficient body of literature supports the benefits of such education, often referred to as Education for Sustainable Development (ESD) (Geli de Ciurana and Filho, 2006; Mokski et al., 2023). Yet, discussion brews over the current shortcomings of ESD, with suggested reasonings including institutional structural inadequacies and subject matter misinterpretations (Filho, 2011; Holst, 2023; Jones et al., 2008). As a collective call exists for the further study, development, and institutional adoption of sustainability curriculum, sustainability learning outcomes have the potential to provide such structure (Builes-Vélez et al., 2024; Geli de Ciurana and Filho, 2006; Mokski et al., 2023; Svanström et al., 2008). As a result, this paper's research meets the growing call for more developed focus on and analysis of learning outcomes for sustainability (Svanström et al., 2008).

3 Establishing Research Methodology

In summary, the established research methodology can be broken down into three key facets. First, the authors prepared for the coding process by generating initial expectations and criteria for each meta-competency, as detailed in sect. 3.1. Next, described in sect. 3.2, the sampling of learning outcomes from both American and international HEIs was completed. Finally, the framework was established using a hermeneutic approach for the code, which was flat and deductive, as detailed in sect. 3.3.

3.1 Preparation for Coding

Following the review and selection of six prominent meta-competencies from the literature, initial criteria were established for each meta-competency's code, as detailed in Table 1. Of note, five of the six meta-competencies selected are currently widely accepted by those working in the sustainability sphere today, with creativity as a relative outlier (Engle et al., 2017). Initial criteria utilized for qualitative coding included the original definition along with precursory related keywords and themes.

3.2 Sampling and Data Collection

With the meta-competencies selected, researchers identified instituted sustainability learning outcomes at American HEIs. This was then expanded to include outcomes

Table 1 Established initial expectations and criteria for each meta-competency

Meta-competency	Keywords	Themes
Interpersonal literacy	Interdisciplinary, teamwork, collaborative, relationships	Ability to work on a team, acceptance of others' backgrounds, personal awareness, skills promoting collaboration
Temporal thinking	Time, space, generations, time identifiers (future, past), historical, anticipatory	Understanding of the problem across time and/or space, consideration of generations' wellbeing, application of historical precedent
Ethical literacy	Ethical, justice, equality, consequences, civics, responsibility, cross-cultural, rights, community engagement, diversity, values, moral, vulnerable	Cultural awareness, diversity equity and inclusion (DEI), development of just sustainabilities, use of morality
Systems thinking	Systems, local, regional, international, integrate, interconnected	Analyzing contextual dimensions (location, politics, economics, ecosystem), connection of disciplines' functionality
Creativity	Innovation, imagination, adaptive	Use of ingenuity, unique problem solving
Strategic competence	Implementation, develop, execute, administer	Proposing solutions, development process, implementation plans

from international HEIs that were signatories of the Talloires Declaration (Association of University Leaders for a Sustainable Future, 1990). The process consisted of researching individual courses, majors, and comprehensive HEI graduate profiles for corresponding learning outcomes. The sample population included a diverse array of institutional characteristics including enrollment figures, public or private status, and research, liberal arts, or community college designation. No limitations were put in place regarding any criteria listed.

Following compilation, thirty-four domestic HEIs were found to have clearly defined and publicly accessible sustainability learning outcomes.[1] "Domestic" HEIs are classified as institutions in the United States, as it relates to the authors' location. The selected list of HEIs included Arizona State University, Elmhurst University, Dartmouth College, Allegheny College, University of Pittsburgh, and UC Irvine, among others. For international HEIs, twenty-one signatories of the Talloires Declaration were found to have such outcomes and included Hong Kong University, Unicaf University, the University of Zagreb, and the University of Balamand, among others.

[1] A comprehensive list of the reviewed institutions, domestic and international, and their institutional characteristics is available upon request.

3.3 Code Framework Development

To develop the framework, a hermeneutic approach was utilized for the framework's development as it allowed for "interpretative approaches" when examining institutional responses to sustainability in education (George, 2020). In using a hermeneutic approach, researchers could meet the subject matter's demand for interpretation regarding the relationship between the concepts defining meta-competency literature and the articulation of learning outcomes. In developing each meta-competency's code, it was decided to adopt a deductive coding style to include the existing meta-competency definitions from accepted literature. The choice of a deductive coding style additionally allowed for the inclusion of the preliminary expectations set, as defined in Table 1. The coding frame was chosen as flat due to all six meta-competencies being accepted by the authors as necessary for sustainability learning, thus no meta-competency was treated as more important than the others nor acting as the root of all others (Buchanan, 2022).

The framework assumed that basic proficiencies, such as reading, writing, and arithmetic, were already possessed by all students attending each HEI. The authors accepted that a word used in one HEI has the same textbook definition in another place, such that others using the same or a similar method would come to similar conclusions. However, there was an acknowledgement of the limitation that such decision does not account for cultural variations influencing the application of words or concepts relating to sustainability. Finally, prior to beginning the coding process, it was agreed that the data set was not large enough such that technical manipulation using coding software, such as N-vivo, was required.

With two groups—domestic and international—in place, development began on domestic learning outcomes with a review of the first random sample group to produce a baseline for the themes and keywords of each meta-competency. The second random sample group was reviewed using the precursory criteria in Table 1 in addition to the new criteria from the first sample's results. New keywords and themes produced were then used to restructure, if deemed necessary, the code for each meta-competency. This process was repeated for each sequential randomized group reviewed until each code was complete for the domestic pool. All sample groups were re-evaluated with the finalized meta-competencies' codes to ensure defensibility, which included the finalization of keywords, themes, and outliers. Finally, for each meta-competency, additional keywords were added to each bank as necessary, due to being synonymous. This process was then repeated for the international outcomes, starting each of the meta-competencies' codes from scratch for the first random sample group and utilizing the precursory criteria.

4 Results

4.1 Coding Results

In developing a qualitative framework for sustainability meta-competencies, certain trends emerged. Of the 400 learning outcomes, 109 were unable to be sorted into the six meta-competencies and their respective codes. The removed learning outcomes consisted of strict definitions, contextually specific outcomes, and those that embodied basic competencies, thus leaving 291 learning outcomes for coding. As the authors' objective was to review and utilize implemented trends, all themes and keywords are inspired by identified themes from HEIs' outcomes, thus credit goes to reviewed institutions.

After categorizing each learning outcome by applicable meta-competency, creativity was identified as the primary meta-competency for only three domestic and zero international outcomes, which established a basis for its exclusion as a primary meta-competency. This meta-competency also yielded only five keywords and one theme of assessing and developing creative solutions with resilience. Due to the low frequency of outcomes clearly meeting these criteria, the creativity meta-competency was removed from the code and the learning outcomes were redistributed to the remaining five meta-competencies.

Below, each meta-competency is explored in both a domestic and international context before investigating the next meta-competency. Following the exploration of each meta-competency's code, there is a final comparison that produced cross-referenced banks for keywords and themes based on domestic and international results. When compiling the results, the frequency of each term and theme was recorded. In the following tables, any term or theme that occurs three or more times is denoted with an asterisk, and those that occur more than seven times is denoted with two asterisks.

4.2 Interpersonal Literacy

As guided by definitions originating from the identified literature, interpersonal literacy outcomes evaluated or emphasized communication skills, teamwork, leadership, and interpersonal knowledge, such as empathy, inclusive behaviors, and respect. Interpersonal knowledge was included in the code, as sustainability and environmental work has an established history of exclusionary practices that harm individuals and communities (Thomas, 2022). Expanding on basic interpersonal skills, the meta-competency included various engagement modes including cross-disciplinary, community, and civic, as they exercise a student's relational knowledge and communication skills. Finally, in this meta-competency's code, there was an engrained practice of cultural acknowledgement through the inclusion of topics such as cultural awareness, diversity of collaborators and listeners, and global perspectives.

For this meta-competency's expression, twenty-two of the 158 domestic outcomes and eighteen of the 133 international outcomes were classified as interpersonal literacy focused (Tables 2, 3, 4, 5, 6, 7, 8, 9, 10, 11, 12, 13, 14, 15, 16, 17, 18, 19, 20 and 21).

Table 2 Domestic keywords for interpersonal literacy in outcomes

Attitudes	Diverse stakeholders	Personal actions
Audiences	Diversity (audiences)	Perspective
Civic engagement*	Empathy	Power relations
Civic responsibility	Engage in public dialogue	Relations*
Collaboration**	Engaged	Relationships
Commitment	Global awareness	Respect
Communicate*	Interpersonal	Societal
Community engagement	Leadership	Teamwork
Conflict	Multidisciplinary teams	
Cross-disciplinary team	Multiple perspectives	

Table 3 Domestic themes for interpersonal literacy in outcomes

Communicate with diverse audiences*	Personal and community-wide engagement with sustainability	Civic engagement's impact on sustainability
Experience with multidisciplinary teams*	Power relations knowledge	Teamwork and collaboration skills
Institutional relationships	Practice and development of empathetic leadership	Understanding social forms of interconnectedness
Partaking in culturally appropriate community engagement	Respect for others and their perspectives	

Table 4 International keywords for interpersonal literacy in outcomes

Actors	Community partnership	Multi-disciplinary team
Advocate	Cross-disciplinary team	Professional
Civic engagement	Explain	Teams*
Collaboration	Inter-disciplinary team	Teamwork*
Communication*	Interpersonal	

Table 5 International themes for interpersonal literacy in outcomes

Ability to join and work on diverse teams**	Effective communication skills to address diverse stakeholders*	Teams of varying scope: multidisciplinary, cross-disciplinary, interdisciplinary
Advocate for and partake in community engagement*	Integration of inclusive practices	
Develop a personal sustainability commitment	Professional development	

Table 6 Domestic keywords for temporal thinking in outcomes

Anticipatory	Futures	Past
Contemporary	Historical	Present
Current	History	Temporal
Future*	Over time	Trace

Table 7 Domestic themes for temporal thinking in outcomes

Conceptualize, form, and evaluate future possibilities based on trends and influences*	Past methods of and their influences on data collection
Examine sustainability within social systems across time periods	Understanding historical and present forces' influence*

Table 8 International keywords for temporal thinking in outcomes

Actual*	Historical*	History
Current		

Table 9 International themes for temporal thinking in outcomes

Define and evaluate problems from the past and present*	Evaluate various policies from the past and present*	Consider environmental history and humanity's effects on it

Table 10 Domestic keywords for ethical literacy in outcomes

Cross-cultural	Impact	Professional values
Dignity	Just	Protect future generations
Disproportionately impacted	Moral dilemmas	Responsibility*
Diversity	Non-violence	Rights
Equity	Normative	Service
Ethical**	Norms	Social justice
Ethos	Peace	Stewardship
Humane	Personal values	Values
Human-environment relations*	Professional responsibility	Vulnerable communities

Table 11 Domestic themes for ethical literacy in outcomes

Ability to debate and resolve moral issues in the social sustainability sphere	Development of ethos through experience	Stewardship for and commitment towards just development
Assessment of norms and values guiding actions	Effects on living systems*	Sustainability's foundation in peace
Connections to social justice and rights, including environmental justice*	Examination of the ethics defining the sustainability development goals	Understand ethical links between humans and theirs ecosystem
Consideration of alternative views or mindsets	Incorporation of health in equity considerations	
Develop and practice social responsibility at both the personal and professional level*	Relationship between humanity and the environment**	

Table 12 International keywords for ethical literacy in outcomes

Code of ethics	Honesty	Responsible
Cultural values	Human rights	Social justice
Engage critically	Human-environment relation	Social responsibility
Equity	Personal agency	Social values
Ethical	Public service	Theoretical
Ethics	Responsibility	

Table 13 International themes for ethical literacy in outcomes

Critical engagement with and use of sustainable frameworks*	Forms of responsibility (professional, social, and democratic)	Social justice and equity
Emphasis of professionalism	Human-environment relationships and interactions	Values including ethics, responsibility, and democracy*

4.3 Temporal Thinking

The research evaluated outcomes' inclusion of purposeful consideration for the future, ability to critically evaluate historical trends, and acknowledging past actions' influence on sustainability's current state. These facets were assessed in the context of disciplinary-specific principles and applications, such as scientific, economic, and political. Specific to future-related concepts, learning outcomes were evaluated on current considerations for future generations. Furthermore, each learning outcome's focus on the ability to conceptualize an implemented solution's impact was evaluated. Finally, there was a minor theme of students being capable of tracing and recording

Table 14 Domestic keywords for systems thinking in outcomes

Complex*	Interconnecting*	Roles
Complexity*	Interdependence*	Scales
Dimensions	Interdisciplinary**	Spheres
Domains	Interrelationship*	Systemic
Embedded	Intersections	Systems**
Globalization	Location scale**	Systems thinking*
Identities	Multidisciplinary	Three dimensions of sustainability*
Impact	Multiple dimensions	Tradeoffs
Integrate*	Multiple perspectives	Transdisciplinary
Interactions*	Nexus approach	

Table 15 Domestic themes for systems thinking in outcomes

Ability to deal with complex systems and frameworks*	Identification of interdisciplinary applications and decisions	Practicing collaboration for interdisciplinary concepts
Articulate relationships and scales within disciplines in sustainability**	Impact of globalization on social and ecological systems	Recognition of the role and interdependence of the three dimensions of sustainability**
Evaluating challenges across scales	Institutional evaluation	Understanding other disciplines' role in sustainability
Examining issues on local, regional, and global scales*	Integration of sustainability in one's professional or personal roles and its impact*	
Explain complex issues from various perspectives	Multidisciplinary approaches*	

Table 16 International keywords for systems thinking in outcomes

Affect	Integrate	Sectors
Analyze*	Interaction	Society
Complexity*	Interdisciplinary	Socio-cultural
Connection	Interrelated	Socio-environmental
Context	Local government, urban, and environmental policies*	Systems thinking
Disciplinary	Location scale (local, regional, global)**	Systems*
Dynamic structure	Mechanisms	Three dimensions of sustainability**
Ecosystems	Organization of space	
Factors	Scale	

Table 17 International themes for systems thinking in outcomes

Interconnection of systems and/or sectors*	Triple bottom line with a heavy focus on application in policy**	Understanding organizations and interrelationships
Knowledge about and analysis of technology and energy systems	Understanding multidisciplinary issues and systems*	Understanding of socioenvironmental dealings*
System analysis and complexity*	Understanding of sociopolitical dealings	
To think and operate in an interdisciplinary manner	Understanding of systems based on geographic and geospatial scale**	

Table 18 Domestic keywords for strategic competency in outcomes

Action	Design	Plan
Action plan	Develop	Possible pathways
Address	Formulate	Practice
Agent of change	Implement	Processes
Apply	Institutional change	Propose
Articulate	Integrative approach	Solution
Build	Key stakeholders	Strategic change agent
Conduct*	Paradigms	Strategic competency

Table 19 Domestic themes for strategic competency in outcomes

Ability to implement solutions with key stakeholders in mind	Creation of institutions focusing on sustainability	Developing strong tracks for sustainability prioritization
Acting as an advocate for institutional change	Design and presentation of research solutions	Execution of fieldwork research
Active local sustainability application	Designing just transitions	Identification of possible solutions
Continuous adaptability	Developing and implementing sustainable solutions*	Propose integrative approaches to solution development

impacts over time. In sum, these abilities are easily married with and elevated by other skill sets in achieving comprehensive solutions and navigating existing sociopolitical frameworks.

As the second explored meta-competency, ten domestic and seven international outcomes were categorized as having temporal thinking as the primary meta-competency.

Table 20 International keywords for strategic competency in outcomes

Agent of change	Development*	Manage
Apply*	Devise	Organize
Assessment	Environmental management	Plan*
Challenges	Evaluate	Problem solving
Conduct	Generate	Project
Contribute*	Identify*	Propose
Create	Implement**	Research and development
Demonstrate	Innovation	Research**
Design	Lead influence	Review
Determine problems	Learn	Solution*
Develop*	Maintain	Strategy

Table 21 International themes for strategic competency in outcomes

Acting as an agent of change*	Identification of opportunities	Taking charge on strategies or projects*
Application of skills, methods, and technology**	Mitigation and prevention of impacts*	
Create and conduct research or fieldwork**	Introduce or produce processes, strategies, solutions, or actions**	

4.4 Ethical Literacy

Next, learning outcomes were evaluated on promoting sustainable responsibility, the ability to address moral issues, and focusing on equitable behaviors and practices. A prominent theme in coding for ethical literacy was the development of one's personal values, which stems back to cultivating responsibility. An overall trend exists regarding the nurturing of personal sustainability commitments to ensure both authentic engagement and dissemination of sustainable practices by example. Supporting this, moral reasoning skills serve as a pre-requisite for a well-developed sense of ethical literacy, including the ability to apply ethics within the context of any discipline. The application of such skills additionally includes the maintenance of human-environment relationships, which led to the concept's inclusion in this meta-competency's code.

Another significant theme was a focus on equity, justice, and protection of rights. Since distinct historical trends and sociopolitical forces have created a system of injustices for communities, a detailed understanding of ethical frameworks assists in analyzing former influences and identifying inequities in modern contexts. As a result, students should be educated on not only the implementation of equity and justice, but also the current state of injustices experienced by various cultural, spatial, and demographic groups. This is particularly relevant as environmental justice continues to solidify its place as a fundamental facet of addressing the climate emergency (EPA, 2021; Intergovernmental Panel on Climate Change (IPCC), 2023; Thomas et al., 2019). Furthermore, equity considerations and environmental justice have permeated into the HEI sphere through equity commitments, which can aptly be supported via increased inclusion throughout curriculum.

For ethical literacy, thirty-two domestic and fifteen international learning outcomes were classified as embodying this meta-competency as the primary.

4.5 Systems Thinking

This research coded systems thinking with a focus on scales, including temporal, geospatial, and intensity. This naturally flows from the definition of the "triple bottom line," which is the framing of sustainability as an issue defined by "economic prosperity, environmental quality, and social justice (Elkington, 2008)." Expanding on this, each learning outcome was assessed on the incorporation of scaled thinking, intersection of disciplines, and understanding of globalization. An understanding of scales assists in the proper evaluation of location-specific issues, prevents the repetition of historical mistakes, and allows anticipation of likely future scenarios.

Looking more intimately at the pertinent scales, globalization impacts emerged as a key component, as several HEIs cited an awareness of its existence and acknowledged its limitations as fundamental to sustainability education. Moving to the individual scale, systems thinking was coded to include the roles that individuals hold in relation to sustainability. The nurturing and development of sustainability-conscientious students leads to responsible professionals and community members. These individuals would not only be aware of their impact but also have analysis skills essential to leaders and change makers in these roles.

Finally, in highlighting the intersection of disciplinary knowledge within sustainability, this meta-competency was found to heavily emphasize the requirement of multi-, inter-, and transdisciplinary approaches to solutions. This stems from systems thinking being grounded in the ability to understand complex interactions between spheres of knowledge and their subsequent interdependence. One type of interdependence exists between disciplines housed under the umbrella of sustainability, whether longstanding or newly introduced. The second type of interdependence exists between "unrelated" disciplines in the cooperation on and development of sustainable solutions. This approach to sustainable development leads teams, businesses, and global leaders to incorporate their expertise in creating resilient solutions.

Looking domestically, there were seventy-two domestic and forty-two international identified learning outcomes, thus making systems thinking the largest meta-competency outcome set across the sampled domestic HEIs.

4.6 Strategic Competence

Building upon the base literature and Potter et al., 2023, strategic competence was defined by both the development and implementation of actions. Thus, learning outcomes were evaluated based on solution development and an assessment of their applicability. In evaluating a student's ability to conceptualize and design solutions, the meta-competency explores the identification of viable possibilities, assessment of key stakeholders, scoping of impacts, and, in some cases, fieldwork and research to support a project.

A secondary focus for this meta-competency was the implementation of and communication about solutions. As one may deduce, the meta-competency thus integrates interpersonal literacy skills due to multidisciplinary team collaborations and a capacity to address diverse audiences. Within this focus, there was an additional sub-theme on the increased use of technology to solve sustainability-related problems and their subsequent implementation, particularly in international outcomes. Finally, there was a minor theme of advocating for institutional change in promoting just adaptation.

Looking at this meta-competency's expression, twenty-two domestic and fifty-one international learning outcomes were identified, thus making it the most frequent meta-competency in international pedagogy. Of note, an emphasis on verbs distinguished the international keywords from the domestic sample.

4.7 Comparison of Domestic and International Trends

Following the codes' completion, a cross-referencing of domestic and international keyword banks continued to develop the full picture necessary for an internationally relevant framework. As shown in Table 22, no more than ten keywords were found to be in both keyword banks for any of the meta-competencies. Furthermore, in all cases excluding strategic competence, the number of identified vocabulary terms for each meta-competency's code was higher in the domestic analysis compared to the international. As a result, the ratio of international keywords that overlapped with domestic keywords for each meta-competency was determined.

Temporal thinking and interpersonal literacy had 50% or more of international keywords matching those in the corresponding domestic keyword bank. Systems thinking had a keyword similarity rate of 38%, but ethical literacy and strategic competence had a rate equal to approximately 30% of words. This review suggests that expectations surrounding ethical literacy are variable in relation to the environment. Additionally, the keyword findings once again supported the supposition that international HEIs hold action-based learning and growth in a higher regard when discussing strategic competence. A visual comparison of each term differentiated by international versus domestic frequency can be seen in Fig. 1.

Table 22 Matching keywords between domestic and international learning outcomes

Interpersonal	Temporal	Ethical	Systems thinking	Strategic competence
Civic engagement	Current	Equity	Complexity	Agent of change
Collaboration	Historical	Ethical	Integrate	Apply
Communication	History	Human-environment relations	Interaction	Conduct
Cross-disciplinary team		Responsibility	Interdisciplinary	Design
Multidisciplinary team		Social justice	Location scale (local, regional, global)	Develop
Interpersonal			Multidisciplinary	Implement
Teamwork			Scale	Plan
			Systems	Propose
			Systems thinking	Solution
			Three dimensions of sustainability	

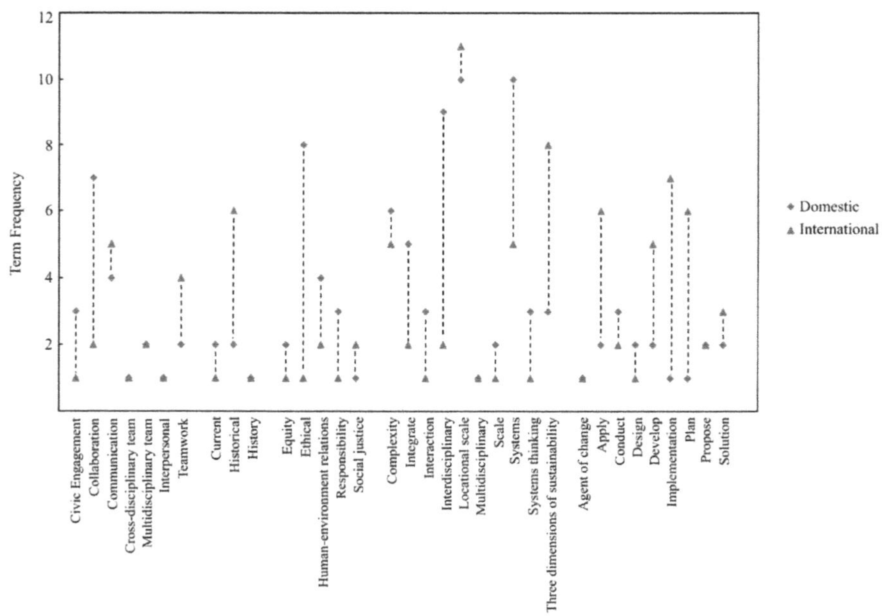

Fig. 1 Term frequency differences between domestic and international pools for the cross-referenced terminology defining each sustainability meta-competency's code, property of the authors

Table 23 Matching themes between domestic and international learning outcomes

Interpersonal	Temporal	Ethical	Systems thinking	Strategic competence
Capability to adapt to and work on diverse teams	Comprehension of historical and present impacts on the environment	Engagement with various types of responsibility	Acknowledgement and integration of the triple bottom line	Becoming an agent of change in one's chosen setting
Participation in and promotion of community engagement		Interactions and relationships between humanity and the environment.	Comprehension of multidisciplinary concepts	Deal with strategies, solutions, and action-based items
Strong communication skills, especially with diverse stakeholders		Practicing social justice and equity	Understanding of geographically and spatially diverse systems	Development and completion of research or fieldwork
Teams of varying scope			Understanding the intertwined nature of sectors and systems	Siezing of opportunities
			Grasp inherent complex nature of systems	

Finally, domestic and international resultant sets of themes were cross-referenced for each meta-competency, as listed in Table 23. Similar to the keywords, the number of themes identified per meta-competency for the domestic results outnumbered, in this case, all those from the international results. Differing from the keywords' analysis, the themes for strategic competence demonstrated the greatest level of overlap followed by ethical literacy at 50%, systems thinking at 45%, and temporal thinking at 25%. Notably, the difference in ethical literacy's frequency of matched themes compared to matched keywords may support any claim regarding the impact of using translation software. Yet, with only three of the six themes from the international outcome set overlapping with the domestic set, it is still supported that ethical systems shaped by culture could have a strong influence on ethical literacy's manifestation in the curriculum.

5 Discussion

Sustainability learning outcomes have proliferated across American and international institutions with significant correlations in terminology, themes, and certain meta-competencies' expression rate. When considering these similarities, the first notable case of uniformity originates from both samples having omitted creativity as a primary meta-competency in outcomes. Lacking a clear link to creativity and its synonyms, this research excluded it in the framework's creation but still acknowledges creativity's role, as current literature stipulates it to be an integral aspect of sustainability and the corresponding subject's pedagogy (Cheng, 2019; Engle et al., 2017; Sandri, 2013). Furthermore, meaningful objections arise regarding creativity as being an inherent aspect of other meta-competencies such as strategic competence or aspects of ethical literacy connected to moral imagination. Thus, creativity cannot be disregarded in the process of drafting and implementing sustainability curriculum, especially in higher education where creativity is traditionally omitted (Cheng, 2019; Sandri, 2013).

Second, temporal thinking and interpersonal literacy were prioritized at similar rates, 5.5% and 14% respectively, across all reviewed learning outcomes, thus showing consonance in application. In working to surmise the reasoning behind these results, the authors posit that a potential driver may be the importance of both temporal and interpersonal competencies in terms of economic advancement for businesses and organizations. Both the ability to analyze and predict trends as well as the capacity to foster resilient relationships are critical skills for the sustainability field, but are also applicable on a broader scale regardless of what profession one holds. Furthermore, the results demonstrate the variety of what is prioritized in professionals' abilities in the context of regional economies, as domestic and international institutions found common ground for these meta-competencies.

Moving away from discernable similarities, American and international HEIs tend to emphasize the remaining three meta-competencies differently, as shown in Figs. 2 and 3 respectively. American learning outcomes focused on ethical literacy as a primary meta-competency more than international universities by approximately nine percent. This suggests a slightly elevated emphasis on the desire for a strong understanding of ethical and moral frameworks, the ability to identify and apply moral arguments, and the consideration of human-environment relationships. Next, approximately 46%, of assessed American learning outcomes were classified as prioritizing systems thinking, making it the front runner for most significant domestic meta-competency. Conversely, the assessed international HEIs favored strategic competence as the most significant meta-competency with an, albeit lower, frequency at 38%.

When comparing the frequency of these meta-competencies between domestic and international sets, the difference was striking. For the domestic pool strategic competence was the primary meta-competency for only 14% of outcomes, suggesting that American institutions are less focused on solution development and implementation compared to their global counterparts. Furthermore, expectations surrounding strategic competence's expression vary, as demonstrated by the difference in prioritized parts of speech for keywords. However, when examining the frequency of systems thinking in international HEIs, as depicted in Fig. 3, it was notably more present as it defined 32% of outcomes—only 6% lower than strategic competence outcomes. This signifies that international universities appear to weigh the initiation and pursuit of coordinated actions similarly to the analysis of systems.

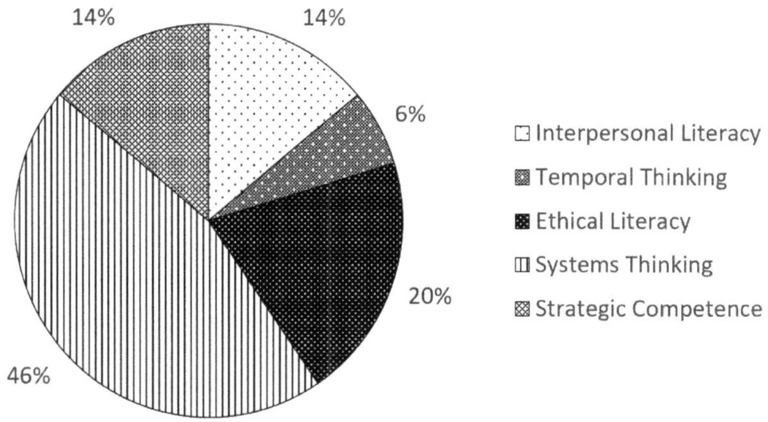

Fig. 2 Frequency of domestic learning outcomes' meta-competency identity, property of the authors

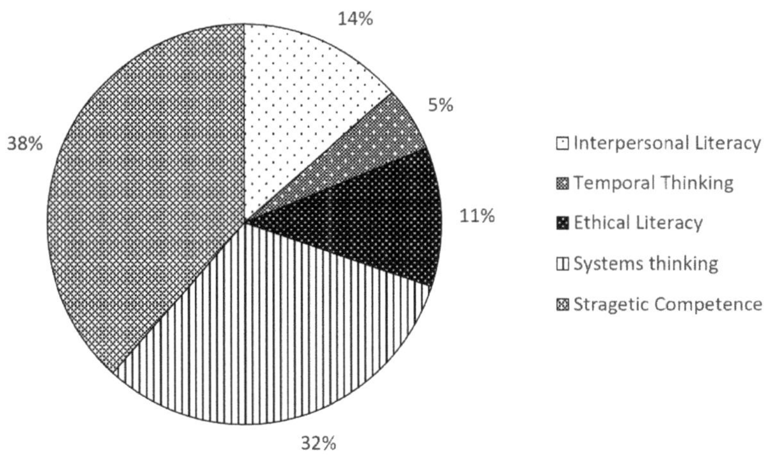

Fig. 3 Frequency of international learning outcomes' meta-competency identity, property of the authors

The authors hypothesize this difference *may* be due to America's superpower status, monopolar/unilateral versus multilateral view of relationships, and cultural eschewing of diplomacy. This hypothesis is untested.

The findings testify to the intricate and integrated nature of sustainability as a discipline, to the point that the discipline itself is a mosaic of countless other disciplines. The prominence of each meta-competency acting as a primary meta-competency within the reviewed learning outcomes cements the interconnected nature required for creating a holistic sustainability curriculum. Thus, students pursuing an education and or career in the field must be adroit in all spaces, which requires a diversified education. To ensure students receive and engage with such an education, the proposed framework can be used to develop an HEI's sustainability curriculum.

However, we suspect that different disciplines and professional fields in different contexts will emphasize different meta-competencies, thereby articulating different learning outcomes. The authors encourage disciplinary educators to assess their academic programs to understand their own praxis and its impacts. Regardless, the authors argue that adoption of a uniform language system, whether the one proposed in this paper or another, can lead to heightened uniformity within the sustainability curriculum sphere while being flexible enough to respond to discipline and depth of exploration. This shared baseline can ensure that citizens, professionals, communities, and networks of practice are equipped with congruent skills and knowledge.

5.1 Further Application

Achieving the primary purpose of the paper, this framework has already been utilized to develop the Pennsylvania State University's learning outcomes, which were formally passed and instituted in April of 2025. Looking toward future research, three potential areas of study arose. As shown, strategic competence is held on par with systems thinking within international HEIs but is significantly less frequently applied in domestic institutions. Thus, an avenue of inquiry is to investigate the potential reasoning behind strategic competence's lower prioritization in American HEIs. Next, as many reviewed outcomes had secondary and tertiary themes present, research can be conducted on which meta-competencies are commonly implemented together. This would continue bridging the understanding of meta-competency preferences via regional learning outcome expression, thus informing best practices. Moving from research to application, institutions continue to encounter changes such as sustainability adjacent issues appearing in diplomacy, policy, and risk literature, the influx of rampant mis—and disinformation, and increasing artificial intelligence access. Are the currently accepted meta-competencies in literature adaptable to learning outcomes related to those issues, and, if not, what changes or additions would be required?

5.2 Limitations

There were four identifiable limitations to this work. First, researchers had to utilize translation software to translate a portion of learning outcomes to English for analysis. The translation of outcomes may have led to the loss of cultural meaning or significance, thus potentially altering the proper sorting of outcomes. Second, the authors' understanding of the literature and meta-competency definitions influenced the sorting process. As a result, cultural or regional understandings of the same material may differ in certain ways, thus again leading to differences in sorting (Svanström et al., 2008). Third, the authors recognize that a binary coding process was utilized with respect to language and culture due to comparing American institutions versus all other international institutions' learning outcomes. Thus, such a binary process inhibits thorough contextualization including the potential overlooking of sub-regional variations, among others. Finally, there was a lack of understanding as to why each meta-competency's prominence was the way it was, both internationally and domestically. This requires a deep dive into concepts such as what Geli de Ciurana and Filho (2006) defined as "considering the cognitive, affective, and action aspects of people." As a result, that topic was suggested as an area of potential future investigation above.

6 Conclusion

Spurred by the international attention sustainability in higher education has received, particularly in the last decade, the authors set out to determine the expression of selected meta-competencies in existing international sustainability curriculum. In the hopes of developing a predictable base framework for sustainability outcome creation, the authors reviewed 400 learning outcomes from over fifty institutions. Using a hermeneutic approach and a deductive, flat coding frame, the learning outcomes, grouped by domestic or international status, were qualitatively sorted by their primary meta-competency. Each meta-competency's code was then developed using identified keywords and themes to create comprehensive banks of information, as featured in the article.

Based on the findings, different priorities and ideals for domestic and international HEIs emerged. Domestic institutions demonstrated a primary focus on systems thinking within their sustainability curriculum, due to comprising 46% of the 158 applicable learning outcomes. Notably, the second most frequently expressed domestic meta-competency was ethical literacy at 20%, which alternatively was the fourth most relevant for international outcomes at 11%. For international institutions, strategic competence was held with a slightly higher priority than systems thinking, at 38% and 32% respectively, thus proposing a focus on action-based learning concepts and experiences. All other meta-competencies were found at similar frequencies. This includes creativity, which was almost non-existent as a primary meta-competency, but was highlighted as being a critical potential secondary or tertiary meta-competency.

Meeting the call for further research and development on sustainability learning outcomes, the implications of the results are twofold. First, the work done here increases the ease of institutions' engagement in learning outcome development. Using these results, individuals or offices at HEIs around the world are better equipped to develop or bolster sustainability learning in the context of their specific institution. Subsequently, the proliferation of well-informed learning outcomes will both increase ESD integration but also has the potential to spur educational policy evolution.

The study was limited by the use of translation software, the authors' cultural and regional understanding, the use of a binary coding process, and the unknown reasoning underlying the prominence of each meta-competency. Looking ahead, future avenues for research include investigating strategic competence's valuation in American HEIs, determining commonly paired meta-competencies, and evaluating the adaptability of commonly accepted meta-competencies to emerging and future global challenges.

Tomorrow's university graduates must fit into the patterns of life on Earth. This proposed framework aims to establish sustainability outcomes that make that more likely. Ultimately, the establishment of such a graduate profile will promote global cooperation for, action on, and engagement with sustainability. In the end, students are the future for the world's climate crisis— both as those being affected by it and those working to solve it. Students are owed a well-grounded, substantial, and integrated sustainability education that will provide them with the competencies to make a difference.

References

Association of University Leaders for a Sustainable Future. (1990). *The Talloires Declaration: 10 Point Action Plan.*

Bartlett, P. W., Popov, M., & Ruppert, J. (2020). Integrating Core sustainability meta-competencies and SDGs across the silos in curriculum and professional development, 71–85. https://doi.org/10.1007/978-3-030-26157-3_6

Brown, R. B., & McCartney, S. (1995). Competence is not enough: Meta-competence and accounting education. *Accounting Education, 4*(1), 43–53. https://doi.org/10.1080/09639289500000006

Brundiers, K., Barth, M., Cebrián, G., Cohen, M., Diaz, L., Doucette-Remington, S., Dripps, W., et al. (2021). Key competencies in sustainability in higher education—Toward an agreed-upon reference framework. *Sustainability Science, Springer Japan, 16*(1), 13–29. https://doi.org/10.1007/s11625-020-00838-2

Brundtland Commission. (1987). *Report of the world commission on environment and development: Our common future.*

Buchanan, P. R. (2022). Communication matters: A study of communication between emergency managers and water systems professionals regarding insufficient access to drinking water [dissertation]. Jacksonville State University.

Builes-Vélez, A. E., Restrepo, J., & Martínez, J. D. D. (2024). The significance of sustainability in higher education: A view to the curricular proposal at a Colombian university. *International Journal of Sustainability in Higher Education, 25*(5), 943–961. https://doi.org/10.1108/IJSHE-09-2023-0423

Cheng, V. M. Y. (2019). Developing individual creativity for environmental sustainability: Using an everyday theme in higher education. *Thinking Skills and Creativity, Elsevier Ltd, 33*. https://doi.org/10.1016/j.tsc.2019.05.001

Earth Charter Commission. (2000). *The earth charter*. The Hague.

Elkington, J. (2008). The triple bottom line: Sustainability's accountants, environmental management: Readings and cases, 2nd edn., SAGE publications, pp. 49–66.

Engle, E. W., Barsom, S. H., Vandenbergh, L., Sterner, G. E., III, & Alter, T. R. (2017). Developing a framework for sustainability meta-competencies. *International Journal of Higher Education and Sustainability*, 1.

EPA. (2021), *Climate Change and Social Vulnerability in the United States: A Focus on Six Impacts*.

Filho, W. L. (2011). About the role of universities and their contribution to sustainable development. *Higher Education Policy, 24*(4), 427–438. https://doi.org/10.1057/hep.2011.16

Geli de Ciurana, A. M., & Filho, W. L. (2006). Education for sustainability in university studies: Experiences from a project involving European and Latin American universities. *International Journal of Sustainability in Higher Education, 7*(1), 81–93.

George, T. (2020). Hermeneutics. In E. N. Zalta (Ed.), *The Standford Encyclopedia of Philosophy (winter 2021)*.

de Haan, G. (2006). The BLK '21' programme in Germany: A 'Gestaltungskompetenz'-based model for education for sustainable development. *Environmental Education Research, 12*(1), 19–32. https://doi.org/10.1080/13504620500526362

Holst, J. (2023). Towards coherence on sustainability in education: A systematic review of whole institution approaches. *Sustainability Science, 18*(2), 1015–1030. https://doi.org/10.1007/s11625-022-01226-8

Intergovernmental panel on climate change (IPCC). (2023). Poverty, livelihoods and sustainable development. In *climate change 2022—Impacts, adaptation and vulnerability* (pp. 1171–1284). Cambridge University Press. https://doi.org/10.1017/9781009325844.010

Jones, P., Trier, C. J., & Richards, J. P. (2008). Embedding education for sustainable development in higher education: A case study examining common challenges and opportunities for undergraduate programmes. *International Journal of Educational Research, 47*(6), 341–350. https://doi.org/10.1016/j.ijer.2008.11.001

Mochizuki, Y., & Fadeeva, Z. (2010). Competences for sustainable development and sustainability: Significance and challenges for ESD. *International Journal of Sustainability in Higher Eduction, 11*(4), 391–403.

Mokski, E., Leal Filho, W., Sehnem, S., & de Andrade Guerra, J. B. S. O. (2023). Education for sustainable development in higher education institutions: An approach for effective interdisciplinarity. *International Journal of Sustainability in Higher Education, 24*(1), 96–117. https://doi.org/10.1108/IJSHE-07-2021-0306

Potter, R., Hiser, K., Evans, T. & Feldman, I. (2023), Key competencies practical approaches to teaching sustainability editing team.

Redman, A., & Wiek, A. (2021). Competencies for advancing transformations towards sustainability. *Frontiers in Education, 6*. https://doi.org/10.3389/feduc.2021.785163

Rieckmann, M. (2012). Future-oriented higher education: Which key competencies should be fostered through university teaching and learning. *Futures, 44*(2), 127–135. https://doi.org/10.1016/j.futures.2011.09.005

Sandri, O. J. (2013). Exploring the role and value of creativity in education for sustainability. *Environmental Education Research, Routledge, 19*(6), 765–778. https://doi.org/10.1080/13504622.2012.749978

Schrand, T., Benton-Short, L., Biggs, L., Bryson, M., & Habron, G. (2013). Teaching sustainability 101: How do we structure an introductory course. *sustainability: The journal of Record, 6*(4), 207–210.

Svanström, M., Lozano-García, F. J., & Rowe, D. (2008). Learning outcomes for sustainable development in higher education. *International Journal of Sustainability in Higher Education, 9*(3), 339–351. https://doi.org/10.1108/14676370810885925

Thomas, K., Hardy, R.D., Lazrus, H., Mendez, M., Orlove, B., Rivera-Collazo, I., Roberts, J.T., et al. (2019) 1 March, Explaining differential vulnerability to climate change: A social science review, *Wiley interdisciplinary reviews: Climate change*, Wiley-Blackwell, https://doi.org/10.1002/wcc.565.

Thomas, L. (2022). *The intersectional environmentalist: How to dismantle Systems of Oppression to protect people + planet*. Voracious.

Tremblay, K., Lalancette, D., & Roseveare, D. (2012). *Assessment of higher education learning*. Feasibility Study Report.
United Nations. (1972), Report of the United Nations conference on the human environment, .
United Nations Commission on (1992), *United Nations Conference on Environment & Development*, Rio de Janerio. *Sustainable Development.*
Wiek, A., Withycombe, L., & Redman, C. L. (2011) July). Key competencies in sustainability: A reference framework for academic program development. *Sustainability Science*. https://doi.org/10.1007/s11625-011-0132-6

Olivia DiPrinzio is the Associate Director of the Pennsylvania Environmental Resource Consortium and an incoming Geosciences doctoral student at the Pennsylvania State University. Her research experience and interests include international climate action planning, sustainable curriculum, international climate policy, and solar energy materials. She has traveled internationally for her research, including attending COP 29 in Baku and working with Māori iwi in Aotearoa New Zealand. Throughout her doctorate, she will be exploring climate adaptation through the lenses of modeling and community engagement for the purpose of just climate policy development.

Dr. Peter Buck is the Director of Education at Penn State Sustainability, Co-Director of Penn State's Local Climate Action Program, and an Associate Teaching Professor in the John and Willie Leone Family Department of Energy and Mineral Engineering. His praxis sits at the nexus of democratic process and ethics, climate change and systemic risk, and educational systems. Dr. Buck has an extensive sustainability-related policy leadership portfolio through elected and appointed offices, testimony to local and state government commissions, non-governmental organization leadership, and public press communication.

Pedagogy for Sustainable Futures: Integrating Culture for Comprehensive Sustainability Learning

Tarsilla Lehmann, Usha Iyer-Raniga, and Kathleen Mahoney

Abstract Education pathways towards quality, inclusive, and culturally informed futures are necessary for developing sustainable futures. Education via culture learning can offer diverse approaches contributing to broadened curriculum development incorporating alternate knowledge systems aiding transformative change to perceptions of 'Self' and 'Other', bridging divides, building solidarity, and deconstructing colonialist global north narratives. Responding to culture deficits in global frameworks (SDGs and ESD), this paper identifies culture as a central and vital pillar for consideration for developing sustainability learning. Culture learning contributes to enhanced reciprocal learning, intercultural knowledge exchange, and leverages learner cultural capital to support equitable, unified, and critical future competencies. Utilising a pedagogical case study, this paper presents the collaborative online international learning sustainability and biculturalism (COIL SB) program and its paralleled sustainability and culture approach, providing implications for developing curricular content. The program divorces from Western-based pedagogy, instead

T. Lehmann (✉) · U. Iyer-Raniga
RMIT University (School of Property Construction and Project Management), Melbourne, Australia
e-mail: tarsilla.lehmann@rmit.edu.au
URL: https://www.linkedin.com/in/tarsilla-lehmann-4766694b/

U. Iyer-Raniga
e-mail: usha.iyer-raniga@rmit.edu.au
URL: https://www.rmit.edu.au/staff/profile?id=usha.iyer-raniga; https://www.linkedin.com/in/usha-iyer-raniga-1b817b16/

T. Lehmann
School of Global, Urban and Social Studies, RMIT University, Melbourne, Australia

U. Iyer-Raniga
Circular Built Environment, Global Alliance for Buildings and Construction, Paris, France

K. Mahoney
SUNY Potsdam, State University of NY (Art Department), Potsdam, USA

Otago Polytechnic, Dunedin, New Zealand

K. Mahoney
e-mail: mahonekm@potsdam.edu

W. Leal Filho et al., *Curriculum Development for Sustainable Development*, World Sustainability Series, https://doi.org/10.1007/978-3-032-22146-9_3

promoting First Knowledges (FKs) acknowledging alternate knowledges as notable and mutually supportive to education including curriculum development. Delivered in two consecutive years: 2021 and 2022, the COIL SB program provides an opportunity to explore how pedagogical approaches inclusive of culture positively impacts learning and sustainable change. Through pedagogical mechanisms drawn from culturally orientated pedagogy (COP) and alternate knowledges, pod groups and learning circles established communities of practice facilitating cross-cultural knowledge exchange and broadened perspective change. The COIL SB program demonstrates that integrating culture learning into education and ESD leads to comprehensive sustainability outcomes including establishing cross-cultural communities of practice; facilitates reciprocal learning and knowledge exchange; enhances learner cultural capital; broadens recognition of alternate knowledge systems; impacts SCs; and change perceptions of 'Self', and 'Other'. Utilising qualitative and quantitative data across the two-year study, the COIL SB case study presents an exemplar of how pedagogy can be designed for transformation and how such intentional pedagogical design is advantageous to sustainable change now and into the future.

Keywords Culturally relevant pedagogy (CRP) · Culturally sustaining pedagogy (CSP) · Culturally orientated pedagogy · Sustainable competencies (SCs) · Education for sustainable development (ESD) · Collaborative learning · First knowledges · Indigenous knowledge

1 Introduction

In a global context where division is rife, promoting pathways for inclusive and culturally informed futures are essential. Education, through culture learning, can offer a way to bridge divides (Holfelder, 2019), rooted in diverse approaches, building awareness, understanding, and solidarity by deconstructing colonialist and global north narratives and systems that maintain divisional conceptualisations. Education and broadened curriculum can play a key role in enhancing understanding of 'Self' and 'Other' via alternate knowledge systems integration contributing to and enhancing reciprocal and intercultural knowledge exchange, and leveraging learner cultural capital to support equitable, unified, and critical future competencies (Cebrián et al., 2021; Demssie et al., 2020; Sposab & Rieckmann, 2024).

At present, broad debates exist regarding the ability of modern education and euro-centric curriculum to adequately develop and endow current and future generations with necessary competencies to navigate complex global challenges. Frameworks have been developed addressing global challenges including the Sustainable Development Goals (SDGs), and 'Education for Sustainability' (ESD) based on economic, social, and environmental assessment and development. While both frameworks provide objectives for mitigating global inequity, the SDGs are criticised for its globalised approach and generalisability in lieu of localised, place-based considerations limiting and excluding fit-for-purpose applications (Corbett & Guilherme, 2021; Hernandez Gonzalez, 2023). Additionally, while the SDGs promote a centralised focus on sustainability based on economic, social and environmental components, consideration of culture and the role of culture within sustainability remains largely absent. ESD, by extension, shares similar critiques of generalisability based on perpetuating Western-based global north perspectives, colonialist alliance, and is further questioned regarding its adequacy for providing comprehensive sustainability learning and outcomes (Holfelder, 2019). Given such critiques, ESD falls short for providing broad cultural understanding for two reasons. First, like its SDG counterpart, ESD disregards the local over the global; and second, culture as a core pillar within the framework is absent, limiting ESD to mere economic, social and environmental considerations. Traditionally, culture acknowledgement has been consolidated within a 'socio-cultural' lens offering understanding and considerations of culture under a generalised social umbrella, leaving cultural acknowledgement as a less than equal partner in such configurations. Calling for pertinent and standalone recognition of culture, this paper posits culture as a vital and independent pillar, acknowledging culture and its contributions separate from blurred social components.

As global challenges exacerbate, how to adequately prepare and equip society and future learners with fundamental competencies for future sustainability is central to education and curriculum development. Competency necessities related to sustainability competencies (SCs), as well as how to transform education, curriculum, and ESD enabling comprehensive learning inclusive of culture are essential to build sustainable futures. ESD must therefore, carefully consider and attend to varied narratives—global and local; Western and Alternate Knowledge (i.e. First Knowledges (FKs))—and reframe current societal ideologies that no longer serve future generations. This paper explores ESD and pedagogy inclusive of culture as a proposed equitable pillar for sustainability along with economic, social and environmental components. Integrating culture into pedagogy and curriculum is posited as essential alongside sustainability learning to convey and shape broadened ideologies, skills, knowledge for future competencies, resulting in enhanced inclusivity, social equity, solidarity, and bridging the human-nature connection (HNC).

Utilising a pedagogical case study established in New Zealand, this paper explores the nexus between sustainability, culture, and pedagogy. The collaborative online international learning sustainability and biculturalism (COIL SB) program demonstrates how an inter-disciplinary program utilising sustainability and culture learning can deliver comprehensive education transforming learning, perceptions of 'Self' and 'Other', and augment current knowledge systems driving forward sustainability bolstering the HNC. Integrating culture as a central pillar of learning, the COIL SB program divorces from traditional pedagogical frameworks, promoting alternate knowledge systems in conjunction with Western-based, Eurocentric sustainability education facilitating a mutually supportive learning environment where learners engage in reciprocal knowledge exchange enabling broadened sustainability and culturalism knowledge. This case study presents how culture can contribute to curriculum demonstrating current gaps within education and provides valuable implications for curriculum development enhancing the learning process.

This paper commences with a review of ESD and its capacity to provide comprehensive approaches for sustainable futures highlighting component deficits of culture. Thereafter, FKs are presented illuminating how alternate knowledge systems contribute to ESD and modern education systems broadening perspective learning for transformative change. SCs are examined regarding questions of what skills and knowledges are needed for sustainable futures, and lastly, culturally orientated pedagogy (COP) is discussed—culturally relevant pedagogy (CRP) and culturally sustaining pedagogy (CSP)—demonstrating interlinkages of culture and education.

2 Sustainable Futures Through Education for Sustainable Development (ESD)

ESD has been posited as an approach and response for achieving sustainable and equitable objectives for society including achieving the SDGs (Holfelder, 2019). ESD, focused on innovative learning driving diverse knowledge acquisition, contributes to societal transformation (Motevalli et al., 2022), however is simultaneously criticised for its confined global north narratives, mono-culture ideology, and narrowed perspectives of sustainable development underpinned by economic-centric directives perpetuating unsustainability (Demssie et al., 2020; Hjorth Warlenius, 2022; Holfelder, 2019; Kopnina, 2020). Such ESD frameworks limit the development of comprehensive approaches in absence of diverse cultural perspectives and integrations including narratives and considerations of marginalised cultures in the global south (Hjorth Warlenius, 2022).

Designed in line with technological, social, and economic perspectives (UNESCO, 2014) it is imperative ESD, now more than ever, aligns to broadened understandings of society—inclusive of culture as a definitive pillar within the framework—forming and enhancing perspectives of environmental and cultural stewardship, global citizenship, and the HNC inclusive of FKs connected to sustainability moving beyond "discursive marginality" (Kahn, 2008, p. 8; Hjorth Warlenius, 2022). Integrating culture into education is seen as fundamental to human and societal sustainable change (Komatsu et al., 2023).

The literature demonstrates an evolution in ESD, expanding its framework to include critique responses to frameworks such as Environmental Education (EE) and Eco-pedagogy approaches (Hjorth Warlenius, 2022). Aligned far more with nature and enviro-centric methodologies, EE and Eco-pedagogy facilitate inclusive consciousness and collectivism inspired by Freire's critical pedagogy (Hjorth Warlenius, 2022; Kopnina, 2020). Current iterations of ESD demonstrate slow yet improved focus on environmental, socio-economic links to sustainability and sustainable futures incorporating enviro-centric considerations borne out of these paralleled frameworks, however, remains limited, if not absent of culture and cultural discourse enabling critical pathways for diverse knowledge transmission and non-hegemonic worldviews (Hjorth Warlenius, 2022; UNESCO, 2020). Such limitations render ESD sufficient for capitalist models of society yet deficient for broader expansive and sustainable realisations in the future (Bedford, 2022). Such deficiencies remain, even in UNESCO's updated roadmap as culture rhetoric remains confined and tokenistic, presenting discourses of "promotion of a culture of peace...global citizenship and appreciation of cultural diversity and of culture's contribution to sustainable development" (UNESCO, 2020, p. 57) positioning culture as a mere afterthought rather than a dominant and centralised component embed it its overarching framework.

Supporters of the cultural approach emphasise that while technological, scientific, and political considerations are important components for creating change, it is culture and how it contributes to the concept of 'Self' and 'Other', combined with 'Nature' and its valuable interlinked applicability for driving forward sustainability change (Komatsu et al., 2023). How one acts, responds, and approaches decision-making is directly congruent with self-perception and the broader perception within the HNC, thus creating vital need for perception change in education and ESD promoting collective engagement rather than current aspirations and rhetoric of selfhood (Komatsu et al., 2023). Changing these perceptions toward sustainable outcomes requires a recalibration in the understanding of where humans stand within the nexus of the environment and incorporating respect and survival awareness of all species and environment facilitated by cultural education (Kopnina, 2020). Culture and its links to FKs and eco-education can be utilised as mechanisms to once more marry humans with the environment (Demssie et al., 2020). It is only through this connectivity and understanding that true sustainability will prevail.

In line with culture integration within ESD, understanding and moving away from Western-based pedagogical approaches towards alternate COP has its merit. UNESCO's emphasis of ESD as a framework that "promotes skills like critical thinking, understanding complex systems, imagining future scenarios and making decisions in a participatory and collaborative way" (UNESCO, 2014, p. 33), provides a directive based on inclusivity, regardless of if such rhetoric is tokenistic or not—true participation and collaboration begins when everyone is included.

Like that of the overarching framework, ESD pedagogy must also aim to incorporate inclusive learning methodologies. Integrating culture perspectives can offer broader and alternative approaches for learning like those presented in eco-pedagogy and critical pedagogy offering quality inclusive education (Corbett & Guilherme, 2021). ESD based on Western ideals positing 'expert', one directional 'passive' learning, can be reframed to become inclusive and collaborative education promoting co-learning, experiential, and place-based learning, and be shaped by both localised and globalised intention (Hernandez Gonzalez, 2023; Hjorth Warlenius, 2022). Integrating culture into ESD as a central pillar can drive forward transformative sustainability pedagogy supporting pedagogical principles such as solidarity where shared ethics and values provide unity within diversity (Bedford, 2022). For Maier (2024), developing solidarity within society is key for societal transformation, therefore providing a renewed ESD framework reflective of all communities, to develop the necessary solidarity needed for change is fundamental. In a time where 'global solidarity…is in retreat' (UN-OCHA, 2025) building solidarity competency is critical, if not vital for future sustainability. The challenge for ESD however, will be providing quality inclusive education removed from limitations of standardisation that hinder components such as quality, sensitivity, and locality (Corbett & Guilherme, 2021).

3 First Knowledges (FKs)

To provide comprehensive ESD—delivering quality, inclusive, and sensitive sustainability education bridging the global with the local—incorporation of FKs is key to addressing current ESD deficits. Given criticisms of education and ESD reflecting global north narratives including colonialist framings and educational imperialism (Demssie et al., 2020; Hjorth Warlenius, 2022; Kopnina, 2020), FKs integration into ESD can address historical inferiority perspectives of Indigenous and First Nations knowledge systems (Santos & May, 2024) and provide an education framework where all learners see themselves in the learning system.

FKs held by First Nations and Indigenous Peoples offer diverse skills, practice, beliefs, and understanding derived from ecological, environmental, medical, technical, cultural, and spiritual knowledge currently absent in Western-based educational systems (Kopnina, 2020). Borne out of naturalistic and cosmological ideologies, FKs present educational and competency opportunities inclusive of HNC that can broaden and redefine ESD and future SCs (Demssie et al., 2020; Kimmerer & Artelle, 2024). The term FKs—borrowed from the First Knowldeges book series (Neale, 2020-2024) recognises and accepts all First Peoples and their knowledge systems. FKs is utilised to encompass all variants of traditional and indigenous knowledge,[1] acknowledging its primary 'first' rightful place in its entirety into education.

Alternate knowledges, gained via place-based pedagogy can inject much needed locality into ESD avoiding limitations of standardisation and current colonial-based deficits (Hernandez Gonzalez, 2023). From an education standpoint, dismantling cognitive imperialism where Western-based education is favoured over indigenous and alternate knowledge systems is vital for moving towards integration efforts (Kimmerer & Artelle, 2024; Kopnina, 2020; Santos & May, 2024). FKs present a breadth and depth to knowledge yet to be fully realised in education and ESD (Latip et al., 2024) highlighting the limiting nature of education projected by the global north perpetuating culture erasure and exclusion. The need for inclusive education therefore is evident as the world strives for a globally unified ideology towards sustainability and SCs. Amalgamated concepts of education are becoming an area of interest as rhetoric and critique regarding modern education highlight ongoing short falls regarding exclusionary approaches, limited historical representation, and

[1] Indigenous Knowledge (IK), Traditional Knowledge (TK), Traditional Cultural Expression (TCE), Traditional Ecological Knowledge (TEK), African Indigenous Knowledge systems (AIKS).

obstructing comprehensive pedagogical approaches needed for culturally diverse learners (Demssie et al., 2020; Santos & May, 2024).

Education should reflect "cultural democracy" (Kopnina, 2020, p. 285) providing a platform where Western-based modern knowledge and FKs converge to provide a model for true education for future sustainability. Marrying FKs with modern education supports concepts of cultural diversity, heralded by Freire as ethically imperative for education empowering and displacing oppression (Torres, 2014). Such cross-cultural approaches are posited as "quality education, configured as a situation in which everyone can learn from everyone else" (Corbett & Guilherme, 2021, p. 450) building necessary solidarity and knowledge exchange.

FKs integration contributes in two ways. First, FKs are deeply place-based and localised promoting inherent critical contextual knowledge of geographical, environmental, and spiritual relevance. Educational pedagogy, therefore, will remain diverse, context-dependent, facilitating fit-for-purpose SCs associated to specific contexts. Second, despite pedagogical diversity for local understanding, FKs provide opportunities for the global, offering new knowledge and practice for global learning and application.

While FKs integration remains limited within education and ESD, scholars increasingly highlight correlation benefits. Santos and May (2024) discuss the incorporation of African Indigenous Knowledge systems into colonial South African education through pedagogy offering learners an enhanced and whole-self experience in university contexts. Pedagogical techniques including movement, music, ceremony, embodiment, and ancestral acknowledgement support learning addressing learner self-esteem and mitigates exclusion. By acknowledging and incorporating FKs from localities, education can provide learners with a broader sense of 'Self' and 'Other' by "acknowledging the wholeness of the student, and the knowledge systems, and techniques of embodiment … deploy[ed], [through which] a more grounded graduate might emerge" (Santos & May, 2024, p. 55). Fully realising *all* alternate knowledge systems in education—moving away from Eurocentric exclusions—can provide enriched localised pedagogy that supports decolonised approaches (Demssie et al., 2020; Santos & May, 2024) leading to fuller educational representation.

In Turtle Island (North America), FKs offer alternate perspectives to support sustainability education. Leonard (2021) and Kayira et al. (2022) offer bridging strategies for FKs and modern education. Leonard (2021) presents Indigenous Higher Education Institutions (IHEIs) focused on Indigenous environmental sustainability education providing a roadmap for program creation in non-indigenous institutions. Kayira et al. (2022) similarly, present a tangible experience of FKs integration and decolonising pedagogy within environmental education. Both offer valuable insight for FKs integration; however, these scholars caution on navigating definitional differences of sustainability within multi-perspective contexts. Sustainability conceptualisation differences have been highlighted between FKs and ESD with the former based on reciprocal partnership and protection of the earth, while the latter remains situated in anthropocentric and economic perspectives (Kayira et al., 2022; Leonard, 2021). Nonetheless, IHEI and FKs integrated pedagogy exemplars provide essential models for furthering FKs integration into education aiding curriculum development, playing a key role to ensuring crucial capacity building, and Indigenous surveillance and safeguarding of knowledge systems enabling current and future usage (Leonard, 2021).

Limited research exists on FKs relating to ESD and SCs development reflecting the inherent lack of culture consideration within these frameworks. Regardless, various scholars support the merit of broader cultural explorations for education and ESD including Demssie et al. (2020) who present research from Ethiopia where traditional knowledge systems exist in parallel to dominant Western-based education. Seen as an imported Eurocentric system, modern education in Ethiopia fails to adequately relate to localised contexts. Moreover, there is a distinct failure to acknowledge and value deep-rooted traditional FKs within this context presenting a "education-reality gap" (Demssie et al., 2020, p. 8). For learners, education is deeply misaligned to the cultural realities and experiences of students, therefore limiting the relevance and effectiveness of education (Demssie et al., 2020).

For Latip et al. (2024), FKs enable and aid regional context understanding, enhance teacher and learner sensitivity and awareness of environmental challenges, and bridges the gap between modern education and FKs via collaborative avenues, knowledge exchange, and building and enhancing SCs. According to Latip et al. (2024), three central challenges prevent FKs integration including modern education not reflecting realities of students 'out-of-class' contexts; FKs misalign or contradict Western-based knowledge; and deficits in teacher knowledge and capacity to utilise FKs. To mitigate ongoing disconnects between FKs and modern education various components to learning are suggested including incorporating localised worldviews; extending learning beyond classrooms; nation orientated and locally relevant curriculum; adopting reciprocal pedagogy (teachers become facilitators not experts); utilising learner cultural capital; utilising fit-for-purpose pedagogy where multiple approaches are used dependent on purpose and need; cooperative learning engagement between diverse students; and incorporating practice-based learning, modelling, and place-based learning enabling transfer of sustainability knowledge, practices and attitudes (Demssie et al., 2020; Hernandez Gonzalez, 2023).

4 Sustainability Competencies (SCs)

Current and future generations face complex global challenges calling upon skills and knowledge—SCs—currently limited or absent in present deliveries of education (Mac Neil & Khare, 2025). Future competencies, arguably different from competencies required for today's neoliberal economic context (Kopnina, 2020; Redman & Wiek, 2021), are being shaped by global frameworks such as the SDGs and ESD promoting universalised sustainability principles in education. Despite directives, these frameworks are criticised for their generalisation and lack of guidance regarding specific learning objectives for SCs development and scholars continue to call for definitive definitions and frameworks aiding clear pathways for development and transformative change (Mac Neil & Khare, 2025; Redman & Wiek, 2021).

SCs literature has steadily increased in notability, however, remains largely ambiguous and varied in definition, development, and utilisation (Mac Neil & Khare, 2025; Redman & Wiek, 2021; Sposab & Rieckmann, 2024). A result of diverse ideological perspectives and frameworks in which SCs are situated (i.e., technological, sociological, economic and environmental), creating clear implementation for SCs in education remains a challenge even within ESD (Brundiers et al., 2021; Mac Neil & Khare, 2025). Contested ideals, varied future competency necessity ideation, and diverse terminologies and definitions in SCs all contribute to ongoing challenges to produce definitive approaches and frameworks for SCs (Brundiers et al., 2021; Cebrián et al., 2021; Mac Neil & Khare, 2025; Redman & Wiek, 2021; Sposab & Rieckmann, 2024). Despite these challenges, Redman and Wiek's (2021) unified framework has been largely recognised in the sustainability domain (Brundiers et al., 2021; Mac Neil & Khare, 2025; Muhonen, Timonen, & Väänänen, 2024), identifying eight key competencies required for future sustainability (see Fig. 1). While this framework provides a broad scope of understanding and development for SCs, scholars highlight the need for enhanced Indigenous and culturally diverse perspective integration (Brundiers et al., 2021). Similarly, the authors of this paper note SCs and ESD frameworks remain largely situated in Western-based domains limiting and obstructing scope, understanding, and comprehensive approaches needed for inclusive futures. Figure 1 therefore, presents an adapted framework with an integrated culture perspective enabling broadened and comprehensive SCs development.

Definitional consensus of SCs are understood as amalgamative skillsets comprising of cognitive knowledge, practical skills and abilities, and personable motives, ethical values, and attitudes that facilitate broad understanding, and interconnectedness utilised in varying contexts (social, private or institutional) to navigate and solve sustainability problems and drive forward sustainable development (Brundiers et al., 2021; Cebrián et al., 2021; Demssie et al., 2020; Mac Neil & Khare, 2025; Sposab & Rieckmann, 2024). This understanding reinforces the need not only for sustainability learning but also for paralleled culture learning contributing to holistic systems-thinking; intergenerational equity understanding; inclusive ethics and justice awareness; comprehensive design, innovation, strategies, and action plans; true cross-cultural engagement and collaboration; reflective capacity awareness of 'Self' and 'Others'; and inclusive solution processes.

There is general agreement that for education to serve future generations adequately and enable transformative change, "a fundamental reorientation of education at all levels and in all areas of education is required to promote sustainable lifestyles, democracy and welfare of human beings" (Cebrián et al., 2021, p. 2). Thus, considering the rise and popularity of ESD, closer examination of how ESD can impact and drive forward SCs through pedagogy is necessary. Links between sustainability education and development of SCs have been established (Mac Neil & Khare, 2025; Wang, Sommier, & Vasques, 2022), however the degree to which ESD and SCs impact real transformative change remains limited (Venn et al., 2022). Moreover, the degree to which culture is considered and discussed within this ESD and SCs nexus is equally limited rendering such frameworks and its learning objectives for SCs incomplete for futures aiming for inclusive sustainable breadth and depth.

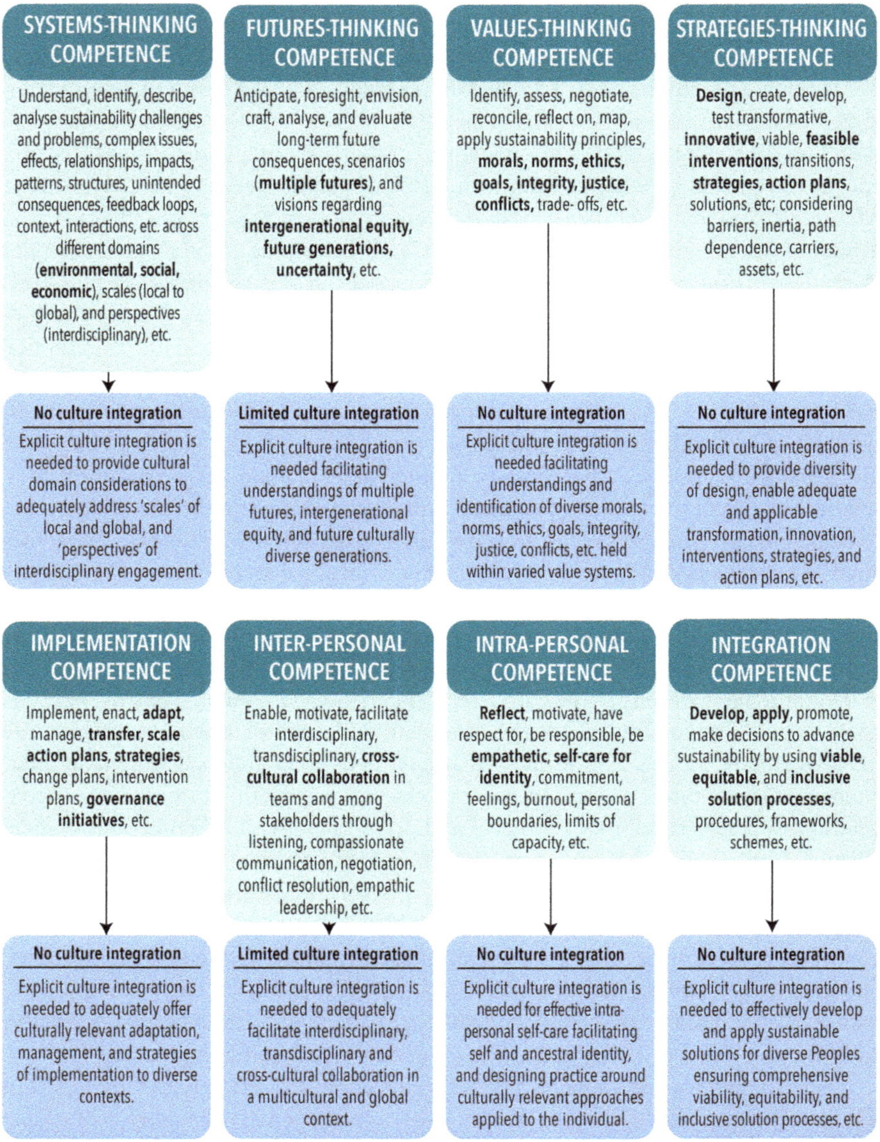

Fig. 1 Key sustainability competencies and culture integration capabilities (linked to bold font). Adapted from: Redman and Wiek (2021)

Acknowledging the broadly accepted ESD and SCs frameworks, it is suggested more comprehensive considerations of ESD and SCs inclusive of culture is explored. Examinations of culture relative to SCs remains limited and continues to be largely situated within professional skill development discourse (Mac Neil & Khare, 2025; Venn et al., 2022) aligned to perceptions of a future connected to business and economic domains rather than facilitating reimagining's of a culturally inclusive society beyond current economic and monetary iterations.

Shaping and perpetuating SCs within education and ESD, Mac Neil and Khare (2025) highlight numerous implementation challenges for SCs within educational institutions citing limitations in organisational structures; teacher knowledge and training; and curriculum appropriateness and redesign. A review of pedagogy that integrates both sustainability and culture perspectives provide an opportunity to both mitigate current limitations and enhance positive impacts for ESD achieving broadened applicable SCs for the future. Championing contextual-based approaches, Sposab and Rieckmann (2024) ensue caution in applying generalised SCs frameworks to all cultural contexts highlighting the need for collaboration to adequately achieve effective localised applicability. Similarly, collaborative mechanisms between educational institutions and regional stakeholders are seen as crucial for knowledge generation and transfer supporting true holistic sustainable development (Muhonen et al., 2024). Collaborative approaches emphasising cross-cultural dialogue and connections can be utilised via COP approaches facilitating diverse perspectives, acceptance of alternate value and knowledge systems helping to expand on limited worldviews and SCs.

5 Culturally Oriented Pedagogy (COP)

To transfer SCs into education, COP supporting localised and globalised culture integration must be realised. COP approaches such as CRP and CSP have emerged focusing on critical thinking, bridging culture and modern subjects for enhanced student cognition, and solution-based skills application to complex global problems.

Developed by Ladson-Billings (2021, 2024), CRP is designed to encourage greater use of culture and cultural references into education providing a pedagogical model for teachers (Cooper & Voigt, 2025; Takona, 2025). Key principles include building cultural competence and developing critique of cultural imbalances (Cooper & Voigt, 2025; Ladson-Billings, 2024; Takona, 2025). CSP expands on CRP supporting thematic areas of resistance, sovereignty, sustaining cultural identity, and focusing on academic achievement via cultural and ethnic identity care, and socio-political awareness (Johnson, 2022; Ladson-Billings, 2021, 2024; Mbah et al., 2024; Moulton, 2024).

As pedagogy expands and reshapes in response to shifting global perspectives, COPs can become central components of pedagogical design addressing criticisms of ESD and the SDGs and their limited recognition of FKs (Ladson-Billings, 2021, 2024; Mbah et al., 2024; Paris, 2012). Offering principal underpinnings of critical pedagogy, multicultural education, critical race theory, and decolonisation, CRP and CSP lay the foundations for collaborative approaches and mechanisms driving forward supportive knowledge systems outside of current modern education. COPs, therefore, are an essential aspect of designing pedagogy and curriculum that integrates FKs and culture as a central pillar, necessary to prepare students as global citizens (Bamber, 2019; Jessen et al., 2020). COPs facilitate cross-cultural communication and engagement enabling learners to develop competencies attuned to local and global contexts, develop confidence, self-actualisation, and identity in education through cultural acknowledgement.

While CRP and CSP have similar drivers, and are similarly educator-initiated, the way in which knowledge is shared differs. The former creates curriculum related to student culture and identity, however, maintains teacher-authority, while the latter, creates a co-constructive environment where knowledge is driven by learners and teachers in a collaborative reciprocal exchange (Jules & Ford, 2023; Momo et al., 2020; Paris, 2012). Given CRP and CSP's development history, both pedagogies share substantial collaborative underpinnings in addition to a central emphasis on ownership of narrative driving forward the importance of FKs as legitimate and necessary for a globalised world. Pedagogy and curriculum, therefore, must mirror learners' history, ancestry, and culture to ensure inclusion and validity (Madiratta, 2024).

Collaboration, therefore, is an essential component for COP and is quickly becoming a focus to drive forward sustainable development seen to foster situated learning and knowledge development in communities of practice (Funk, 2021; Tailby et al., 2020; Venn et al., 2022). Additionally, COP utilising FKs, accommodates and accepts diverse approaches to sustainability (Leonard, 2021), unlike prescriptive Western-based educational framing. Approaches to sustainability, therefore, will be situated in complexity as varied pathways become available dependent on choice providing breadth or limitations subjective to institutions and individual choice. Such choices render sustainability at a crossroad where it is not possible or even desirable for advocating a singular approach. Given the highlighted contextual difference of diverse FKs, place-based learning will therefore be dictated by differing socio-political underpinnings making prescriptions of learning and more specifically applications of comprehensive ESD different.

Despite inherent similarities and objectives for learning, the literature demonstrates little to no explicit linkages between ESD, CRP, and CSP. Zidny (2021) and Kimmerer and Artelle (2024) however, present exemplar integrations of FKs into modern science disciplines offering enriched cultural and place-based learning contributing to amalgamated knowledge. Linking FKs with modern knowledge enables broadened understanding of globally diverse systems which are overlooked via a singular approach (Dorji et al., 2024; Kimmerer & Artelle, 2024; Zidny et al., 2021). In broader conceptualisations of education, CRP and CSP remain separate from ESD, however, it is hoped it does not remain so. While ESD persists with culturally devoid sustainability perspectives, outcomes for sustainable futures and SCs will remain deficient. The absence of culture as a central pillar in education and ESD significantly diminishes the advantages and contribution of FKs toward current and future society therefore recognising FKs and COPs into overarching frameworks is needed for comprehensive application. It is pertinent now more than ever to move forward with pedagogical approaches facilitating broader conceptualisations, consciousness, and contribute to SCs (Wang et al., 2022) reducing the obvious gap in modern education.

6 Method

The COIL SB program is presented as a case study exemplar (Yin, 2017) of how ESD with an integrated culture component can contribute and deliver pedagogical implications such as broader learning, impact cognition and perception change, and contribute to inclusive educational redesign benefiting all learners.

Delivered in two consecutive years (2021 and 2022), the COIL SB program provided an exploratory longitudinal examination of how ESD through parallelled sustainability and culture learning can yield positive outcomes for learners that are broader then current economic, social, and environmental framings. Original research on the program was conducted exploring the transformative learning impacts on student and faculty participants with mixed-method approaches utilised including pre-course and post-course surveys and qualitative focus group interviews to establish baseline and evidence of transformative data. Focus group interviews were carried out weekly throughout the 3-week program together with final post-course interviews to assess transformative learning occurrence as a whole in both student and faculty participants. Outcomes of this transformative learning research were detailed in two prior papers (Lehmann et al., 2025a, 2025b).

While the original research explored transformative learning, the results indicated culture as a vital component in the learning process for transformative learning. The results reflected the need and gap of culture in education and ESD, therefore leading to the recognition and acknowledgement of culture as a key for learning. Culture thus became centralised as a core and vital component for individual and collective sustainability transformation, therefore, this paper draws on the original research, amalgamating the data and recoding all participants across both program years as Learners = L, rather than separate learner cohorts, to demonstrate and support culture integration as an integral pillar for education and ESD.

The COIL SB program, co-taught by self-identifying European and Māori facilitators, emphasised multi-perspective learning via online delivery, enabling flexible engagement and diverse inter-cultural knowledge exchange. Through pedagogical mechanisms drawn from modern and FKs systems, the COIL SB program utilised learning circles and pod groups to establish communities of practice facilitating knowledge exchange and learning. Pod groups consisted of 4-6 participants and larger learning circles engaged whole class participants facilitating relationship building and knowledge sharing. Designed as an interdisciplinary course teaching sustainability and culturalism knowledge to a global cohort, the program saw culture as a key component for perpetuating sustainability bridging the HNC. The COIL SB program offered insight into how parallelled perspective learning designed to establish and evoke cultural understanding of 'Self' and 'Other' within a local and global context encourages the HNC resulting in enhanced sustainability awareness and behaviour. Utilising COP, the program delivered two parallel courses: (1) Sustainability knowledge (concepts and frameworks), and (2) Bicultural nuances of New Zealand (Māori and European), centralising and emphasising culture to develop and

Fig. 2 Program syllabus COIL SB (2021 and 2022)

enhance learning within a sustainability perspective. Figure 2 outlines the COIL SB program's syllabus reflecting indicative content based on ESD and culture to convey multi-perspective learning to learners.

7 Results

The results in this paper are drawn from the longitudinal mixed-method research carried out on the COIL SB program. The original research highlighted cultural integration for education and ESD to be of specific merit for transforming and enhancing pathways for inclusive and sustainable futures. Qualitative and quantitative results from the research are utilised to demonstrate the broad impact of culture integration via COP presented through six observed outcomes of the COIL SB program: (1)

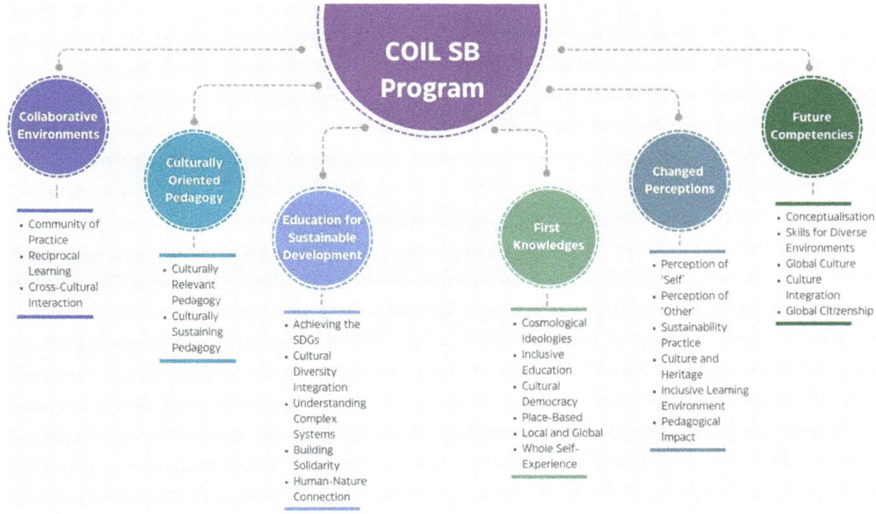

Fig. 3 Observed pedagogical implications for transformative sustainability learning

Collaborative environments; (2) Culturally orientated pedagogy; (3) Education for sustainable development; (4) First Knowledges; (5) Changed perceptions; and (6) Future competencies (see Fig. 3).

7.1 *Collaborative Environments*

The COIL SB program provided a unique platform for delivery and pedagogical approach. Responding to the pandemic, the COIL SB program moved online, enabling flexible international access facilitating collaborative international learning creating a community of practice and simultaneously ensured cultural diversity and inclusion. In conjunction with its delivery, the program also ensured a collaborative COP facilitating diverse knowledge exchange, and specifically emphasised alternate knowledge systems acknowledging its value to modern education.

7.1.1 Community of Practice

Communities of practice were established in the COIL SB program by incorporating student pod groups (4-6 people) and reflective circle interactions (larger class-based groups) drawn from FKs practice providing learners with opportunities to engage with other global participants facilitating cross-cultural discussion, knowledge sharing, and diverse feedback loops. Pedagogical mechanisms inspired by FKs

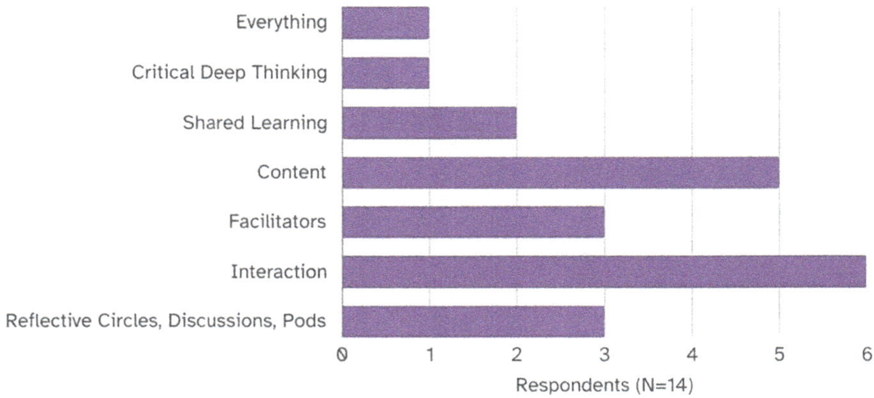

Fig. 4 Program satisfaction areas—post-course survey (2021)

practice facilitated and established relationships and collaboration allowing intercultural exchange leading to broadened scope and understanding of 'Self' and 'Other' cultures. Participants reflected on the communities of practice discussing the positive benefits of exchange and the opportunity to engage within diverse cultural contexts enabling the sharing of personal histories and stories:

> "I did relate a lot of my stories to them,… we kind of like passed the mike, the metaphorical mike to each other, talking our own stories about our own experiences, and so I did tell them about how things are in Guyana, what we have here, what we don't have, and based on what I say like a lot of stuff we kind of relayed back to each other like culturally." (L1)

> "Even though I was hesitant, I'm actually loving it now because I'm getting that gleam and that glimpse into these other cultures from around the world that I never would have got the opportunity to do in class. [I] would have been with a bunch of 16-17 year old kids that I'm like… we're all in the same cultural category here. … I'm loving it… it's just insightful and it's powerful and cool that we have this technology to be, you know, you." (L12)

> "I almost echo everything [participant] is saying, of that excitement, and just the evolution of learning. I found this experience really embodies this idea of learning together, from and with each other, you know, sort of from, with, about each other as well." (L4)

Results from the 2021 post course survey similarly reflected participants value of the COIL SB programs' pedagogical approach to learning, highlighting areas of program satisfaction including interaction, program content, reflective circles and pod groups, and components of shared learning (see Fig. 4).

7.1.2 Reciprocal Learning and Cross-Cultural Interaction

The focus on shared learning and knowledge exchange was a central component of the COIL SB program where reciprocal learning was encouraged, breaking away from Western-based forms of education where educators are situated in constructivist pedagogy as 'experts' in one-directional learning. The COIL SB program established a collaborative learning environment where learners were facilitated by teachers, however, were also simultaneously, autonomously engaged and designing their learning and projects through the acknowledgement of diverse knowledge systems, cultural experience, and learners' prior knowledge. This was aided by learner-led reflective circles and pod groups where learners discussed program content and engaged cross-culturally in a global online context. Participant responses highlighted this opportunity, navigating pod and learning experiences, interacting with each other, and setting their own boundaries through this reciprocal learning approach:

> "I loved our reflective groups as well, because it was different folks and we talked about different subjects. So, and everybody got a chance to be heard, and a good chance to listen as well." (L6)

> "We're able to share information back and forth and kind of just talk about some of the different theories or even what we were thinking of doing for our project, we were bouncing ideas off of each other as to how to improve the project or how to spin it so that it more closely aligned with the course material…. so, yeah it was beneficial, actually it would be nicer if we had more of it." (L2)

> "The best part of the programme was the fact that I was able to interact and engage with people from other parts of the world, that the pods and groups were with different people, and we were given time to discuss some of the emotions we felt as we progressed through the course." (LX)

> "The best parts were the gained knowledge, getting to know the people involved, self-awareness, reflection, self- esteem boost, Māori culture, and relaxed learning." (LX)

> "I think that having the kind of more immersive elements as far as having the discussions and having the opportunity to discuss with different people as far as having the pods, but then also having the larger reflective circles and then the lectures and …. yeah, I think it, it works well. I definitely think it works better to have it in that sort of like round table, like I said, immersive discussive-like format within that period of time. I mean in general it resonates more with me anyway. But especially… if we were to have just the, as you said, just having the more traditional format of westernised learning structures within that period of time, I think that would have been even more challenging for me." (L9)

7.2 Culturally Orientated Pedagogy

The COIL SB program utilised COP including CRP and CSP to acknowledge diverse knowledge systems, establishing communities of practice that promoted culturally inclusive student-led learning. Results reflected the benefit of COP with participants commenting on knowledge sharing, culture acknowledgement, and recognition of ancestral identity, and alternate knowledge systems:

"Mine is just identifying what ability is and is not, and by culturalism and what it is and what it is not. I had an impression of what I thought sustainability was when I took this class.... I didn't identify the depth of what my culturalism was until I took this class. ... It was just merging of cultures. I was like we do that all the time. We have like, I have like 6 or 7 nationalities, you know, we're all what is that 'poi', 'poi' — mixed dogs, whatever. And, you know, sustainability was like, on a whole different level too. So that's where I'm at with that." (L10)

"Appreciated [name] explanations of concepts and ideas... sparked so many thoughts about how landscape influenced identity... I never thought about the impact of landscape and how this impacts our practices, inform our intellectual, emotional, and body intelligence... loved the fingers woven together with the gaps to explain the dynamic interdependence of sustainability, culture, and biculturalism ...[name] discussions and stories made the dichotomies and similarities of culture together with impact on practice very meaningful...Need more of [name] with extended learning opportunities" (LX)

7.3 Education for Sustainable Development (ESD)

The COIL SB program married ESD with culture, enabling learners to achieve an integrated conceptualisation of sustainability and culturalism impacting on sustainability cognition and understanding. Participants reflected on global frameworks including the SDGs, and the need for education reform facilitating broader acknowledgment of culture and broader sustainability conceptualisations:

"I love that whole cultural identity, because the cultural identity, really, I'm getting such a deeper understanding now about this. That identity drives how we feel about sustainability, and the practices we engage in... it's not just recycling, it's about that protection and taking care of what we have, so changing our patterns of behaviours, and identifying where we are engaging in these patterns and finding ways to move it in a different way, so the SDGs I was really ok, now it's deeper, and the cultural piece is so important, one of the things that really stuck out for me, you know, [facilitator] always did this type of thing where [name] do the fingers (interweaves fingers), and the gaps in the middle, that's the gaps to me, I'm focusing on finding those gaps to either influence or interfere with sustainability practices, and recognising biculturalism, and what it really is all about, and recognising the indigenous people." (L4)

"We need to teach both at the same time, teach both from a very young age and I think it would definitely come from the higher-most level.... first of all implementing it from a government level and ensuring that all ministries are definitely looking at sustainability and bi-culturalism... once we have something from a higher level and we teach both sustainability and culturalism to our future generations, I think we can go somewhere in preserving both because we can't afford to lose any." (L1)

"SDGs are a deadening framework, and I find it painful to see them have such a standing in this course and many other spaces. Until they are re-written in a culturally relevant way, I think we should only use these as a reference to contrast against more alive and equally holistic sustainability frameworks." (LX)

7.3.1 Sustainable Futures

The COIL SB program research delved into understanding the potentials for amalgamated education and its impact on the future. Participants shared perceptions of the future (Fig. 5) and perceptions of 'Self' in the future (Fig. 6) reflecting optimistic views regarding broader changing structures of institutions including education; future change relative to global politics and policy on climate change; systems-thinking changes requiring an incorporation of sustainability and cultural considerations; changes in sustainability and practice; culture integration; and enhanced linkage between sustainability and culturalism.

Participants similarly shared responses regarding perspectives of 'Self' in the future presenting global futures that were dynamic and bolstered perspectives of critical change including (1) changed systems, concepts and ideas, not just of sustainability but of broader societal structures; (2) changes in the professional and work landscape regarding enhanced global experience, engagement, and transformation; and (3) changes in knowledge and approaches to cultural retention (see Fig. 6). Results demonstrate participant future perceptions are disconnected from current societal iterations with most participants indicating a future world different than current exclusionary neo-liberal structures.

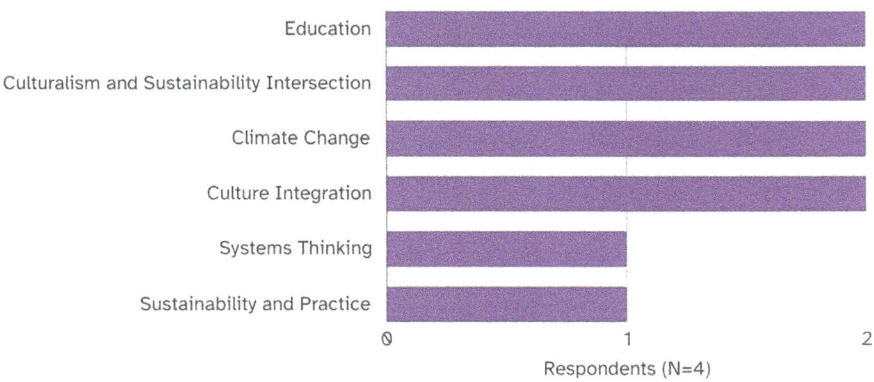

Fig. 5 Future perspectives of the world in 5-10 years (2021)

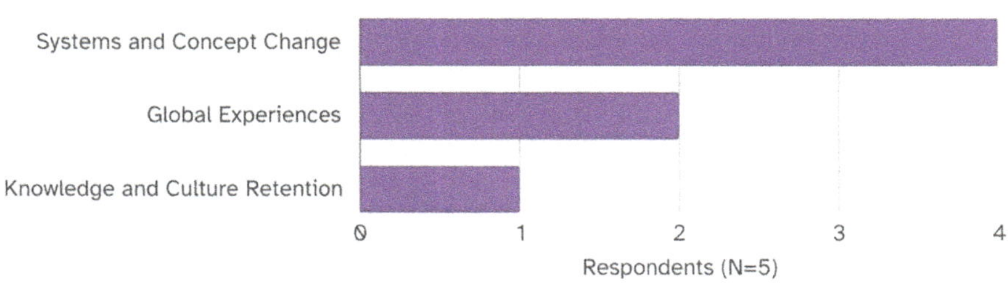

Fig. 6 Future perceptions of 'Self' and the world in 5-10 years (2021)

One participant particularly highlights current societal structures and the need for change:

> "I guess identifying broken systems and then re-evaluating that against through this new lens and saying actually…That's what it's like with society, though, and culture, right? Once these things have been put in place, nobody wants you to challenge it. Nobody wants you to question it. It's like, you know….. but we're here to make those ripples. We have to be that disruption." (L12)

7.4 *First Knowledges (FKs)*

FKs was a central component throughout the COIL SB research with participants highlighting how FK integration and intercultural interactions prompted self-reflective thinking about culture, identity consciousness and acknowledgement of 'Self', transformation, and broader scopes for pedagogical design:

> "So that reminded me when [participant] talked about that, I thought, somehow sharing everyone creates this pattern, and that is how I begin to change and transform, and that is a missing piece, the indigenous, and this is what we have done, again in parallel, when the institution says, 'we are going to indigenise our-;' and I have to tell you, I so don't like that term, 'indigenise curriculum,' it just sounds artificial; it sounds inauthentic, it sounds like token to me, so I prefer the words of inclusion or being pro-something, as opposed to talking about colonising or decolonising, I think it's about reflecting, it's about metacognition, it's about transforming, it's about understanding our past so we don't repeat it, and being active and changing the future." (L4)

Another participant highlighted how the absence of acknowledging culture not only in education but in overall society deeply impacts self-identity, playing a part in a capitalist, neo-liberal societal structures:

> "I think it really played a big part in my life. Growing up and sort of sweeping that under the rug and not identifying with that took a huge part of my identity away. Like, I didn't really have an identity as a youth. It was sort of lost in that that Western culture doesn't really foster identities. It fosters like employees, I guess…. go to school, learn the skill set, come out, go get a job. As opposed to like, who are you in this community? And yeah, so I've been finding it, really, very interesting."(L12)

Other participants discussed their perception of culture and indigeneity, its varied conceptualisation, impact on the HNC, and positive connection to frameworks:

> "I emphasise that indigenous people are all over the world, that when I speak of my Celtic roots, we're indigenous, we're a people who were in love with the land, in tune with nature and the seasons." (L5)

> "I have two grandsons. And I didn't really speak about them in my introductions yet, but they're [a] huge part of why I'm out here, because I came out here to meet my second grandson and my daughter lives out here. And when I met him, I was like, I can't go back home. I need to stay here because passing on that knowledge, that [participant] was talking about and It's so important because I didn't get it. My granny was in residential school, so [was] my grandmother on my dad's side and then my dad didn't go, but he went to Indian

> day school …so from my grandmother, the language was gone, the culture was taken away. She didn't practise medicine, she didn't, she knew all the medicines she was being taught at a very young age to what the medicines were. So, we lost that, we lost the language, we lost the practises like Indigenous, our, our Ojibwe, Anishinaabe ways of saying all those things." (L13)

Another participant discussed culture integration and COP yet cautioned limiting or tokenistic pathways, and highlight opportunities to engage better and broader:

> "So, I've always understood that when it comes to curriculum, that when it comes to this movement forward, it's a responsibility we all share, and so as an indigenous person I can't take all of the burden of that responsibility, and I think sometimes a lot of that responsibility is placed on us as indigenous people to lead, and of course we do lead, but we also need others leading with us." (L8)

> "I'm the one who created it at [institution], I'm the one that really shifted the process at [institution], of the ways in which we're doing land acknowledgements, moving it from a place where I as an indigenous person comes in and delivers the land acknowledgements and nobody else does that, for me. I've been really dissatisfied with the idea of land acknowledgements, with the idea that we're reading from this paper that means absolutely nothing… the idea that we need to personalise it, every day is an opportunity to connect with land, and to connect with the Peoples who have come into being in those places, and that includes indigenous as well as non-indigenous people." (L8)

> "I do think that there are limits to what a non-indigenous person [can do]…,because I think that there's a level of empowerment that happens in the context of an indigenous person telling their own stories, and so that's always the stance that I've taken, and sometimes I've gotten surprise from, particularly from non-indigenous faculty when I say like, as an indigenous person, I'm not going to talk about all indigenous people's experiences, and even as an Anishinaabe person I can't talk about every Anishinaabe's experience. I can talk about my family, my community, and the limited knowledge that I have, and recognise it as that." (L8)

7.5 Changed Perceptions

The COIL SB program facilitated inclusive learning environments where a globally diverse learner cohort were able to integrate culture and heritage into the learning process. As a result, participants reflected on changed perceptions of 'Self', changed sustainability practice, and new ways of conceptualising:

> "This is problematic. But it's an exciting problematic because we've got the opportunity to try to create, change and originate new ideas or initiatives and whatnot instead of just being like, OK, let's be part of this broken system. It's like….actually let's rethink and re-examine if this is even correct." (L12)

> "This was life changing for me. I will not be the same because of this course for the better. I will continue to be the change to influence others for the betterment of my community and the world." (LX)

"From the beginning, I've always wanted to [pay] homage [to] my Māori history …put that into my designs. The first part of the year we did a wee dive into biculturalism which sparked my curiosity because you can probably see, I don't look Māori and I'm not…. I'm a 1/6th …. I'm only part Māori, but that's enough to be Whakapapa here in iwi, but growing up I never really affiliated with my Māori side, and it was never really encouraged so it was sort of just swept under the rug cause I'm from a very white family and I never really explored it. So now, for the first time I'm like delving into exploring it. And I'm seeing all the depth of culture and everything." (L12)

"Great course—it was important to advance my personal journey and thinking of my 'self' in the places that I exist. I think it worked very well as an on-line course but of course the experience and opportunity to develop relationships and learn from each other in-person would have been even more powerful. But at the same time this approach was probably more inclusive for some who otherwise may not have been able to participate. It would have been a very difficult time for many 'Canadian' Indigenous people to be away from home and family." (LX)

"So, as I go into new projects, there'll be things I try to do, to be sustainable or have that bicultural aspect, but I don't know.… In terms of daily life, I think it's just been more of a thinking thing for me, like reconsidering standard practises, just rethinking everything. It's been really challenging to be honest. It's sort of like one of those things that I'm like, wait we live [in a] fundamentally wrong society that's based upon these worldviews that aren't bi-culturally acceptable or sustainable and I'm like, how do we undo this? Or like, you know, how do you go up to a society 150 years in and say you guys are doing it wrong and we need to change it. … I'm finding myself challenged at almost every turn, … like from, the trees we used to build stuff, like, is that sustainable? It's like, oh, this is a really big, seems like really big subject…. but it's definitely challenging me and teaching me, like I'm learning. It's just that practicality of how do I incorporate that …into my daily life, you know? So, I guess that's the journey I'm on for the week." (L12)

"My biggest take away was learning yourself. So, you gotta know yourself and then you can know the other person. Like I said, the hierarchy of needs, self-realization's on the top and that's what we've been taught. That …you do all this other stuff, then you can get to know yourself and first nation's knowledge, self-realisation's first and then you can do the other stuff. So that was huge for me. Know yourself, then you can know the other person. If you don't know who you are, which is one of my biggest issues because I am also a mutt, then that's unsustainable in itself, and you're not gonna be effective and you know it. It's there in the science part about it which is a lot of the sustainability that I think of, and we it got in there…. I learned quite a bit about that but that personal self-realisation….. Yeah, that was the big point for me. The biculturalism. I always thought it was two….. it's whatever, it's all."(L11)

7.5.1 Pedagogical Impact

The results from the COIL SB program demonstrated cultural integration impacted pedagogy. Learners who were also faulty indicated the program effected changes in pedagogical thinking early into the program timeline establishing a pattern of change (see Fig. 7).

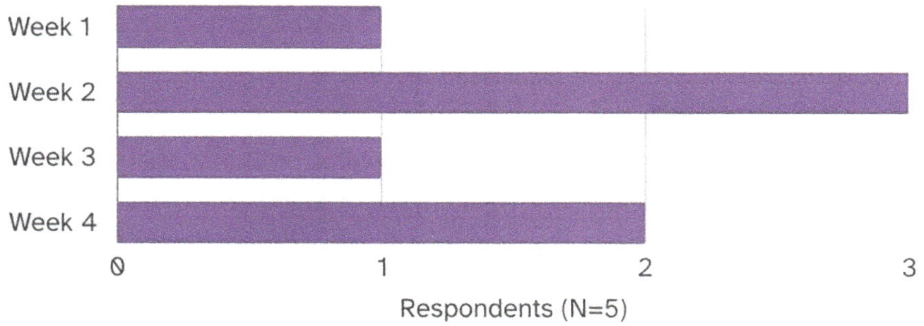

Fig. 7 Pedagogical changes in faculty (2021)

Participants reported transformations attributed to culture integration regarding cross-cultural interaction and knowledge sharing experiences:

> "It just changed my whole outlook on how education can work in, and we're seeing that again in this course with everybody, just the sharing of ideas and learning from one another." (L5)

> "I was starting to think in more depth than I have not taught before, and especially when it came to culture, cultural practices woven in with sustainability, I found that was the missing piece that I feel I've found this week…. when I explore sustainability …I felt it was all parallel, everything moving in straight lines, but not woven, and I found for the first time I'm actually seeing the missing piece is the culture." (L4)

> "So, what I see is helping me redefine and define the concepts that I'm bringing to the classroom already, I'm getting to hear a lot more examples from other cultures, and what their experience is, and those are the kinds of things that I can bring into my classrooms." (L6)

Delving deeper into pedagogical change and transformation, results demonstrated a broad spectrum of change for faculty participants. Participants not only presented a change regarding how they viewed sustainability and culturalism concepts, but also identified further changes regarding identity, indigenous contexts and perspectives, SDGs, environmental stewardship, educational course design, and broader considerations of institutional systems thinking (see Fig. 8).

7.6 Future Competencies

The COIL SB program research also highlighted conceptualisation and perceptions of future competencies for future students. Participants identified the need to build skills for a diverse and dynamic working environment, focused on global culture and becoming global citizens.

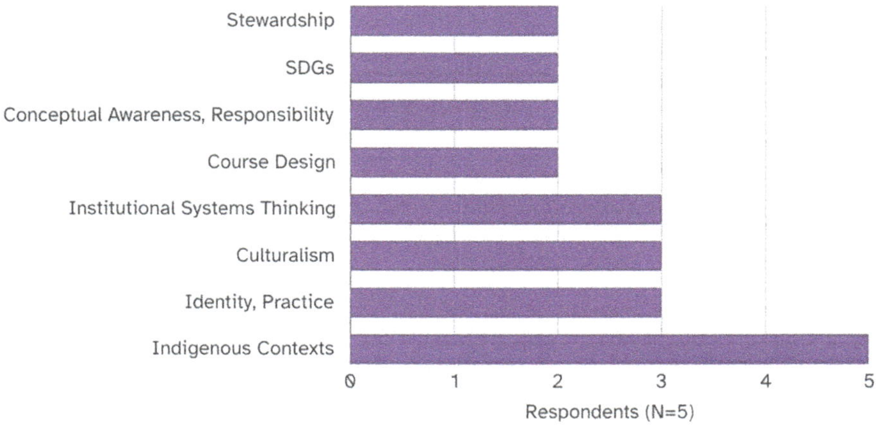

Fig. 8 Types of pedagogical change for faculty (2021)

Participants indicated the need for culture integration providing the base for responsibility for sustainability and providing flexible transformation for future professional work:

> "There was lots of talk in the educational institutions of globalisation and the impact it has on teaching and learning, and I hear about preparing students to be global citizens, I would love to see these global citizens become responsible people who carry the torch for sustainability and culturalism, because globalisation means then, will we have a global culture and what would that culture look like, and what would the practice be for that global culture." (L4)

> "To me I'm thinking ahead, of our future students, so I always connect it to the global citizen, what does that look like; students today and in the future to me is not the student of the past like myself, we sort of had a career sort of pathway planned out and you kind of just, you became something, and then you're that, and I think today young people are constantly transforming." (L4)

> "What they're great at is not being stuck in one sort of career pathway, you're constantly evolving, and that new student is going to look different, they're going to craft their own learning, they're going to, perhaps when they graduate, their diploma's going to be a sort of multi-sort of, diploma, it won't be what we offer them, but it's what they take away from us." (L4)

> "I think as we go forward, like we're all gonna be slash people, 'I'm a this and a this', 'I'm a this and a this and a this,' so yeah, I think, the days of having …just one job you do is probably on its way out …My parents' generation was probably the last generation that even maybe would have had that happen, even then, both my parents were born in the '50s and they had several different companies that they worked for, not as many as now, but yeah,…. I think their parents' generation was the last one where it was like you work for one company for your whole life, then you retire and that's it, or look at having one job that you do, or one type of job being the same thing." (L2)

8 Culture, Sustainability, and Future Curricula

Current modern education throughout the world falls short of providing comprehensive approaches for culture integration including acknowledging FKs as alternate and notable knowledge systems perpetuating erasure and inadequate iterations of society and reality that should reflect diversity and inclusion (Demssie et al., 2020; Santos & May, 2024). Similarly, global frameworks such as ESD and the SDGs—that exist to enhance social justice, sustainability, and global solidarity—remain largely culture deficient. Given these deficits, this paper provided an exploratory exemplar from the COIL SB program centralising and emphasising culture to aid and advance sustainability learning. The results demonstrate culture integration not only is possible for ESD and education programming, but that culture is a necessary pillar for consideration for producing sustainable futures enhancing pedagogical outcomes such as cross-cultural engagement and knowledge sharing; stimulating reciprocal learning; and contributes to learner cultural capital, identity, and awareness of 'Self' and 'Other'. Therefore, curriculum needs to explicitly engage with cultural attributes.

As demonstrated by the literature, ESD and SDGs remain situated in social, economic and environmental domains, with limited, if not absent recognition of culture (Demssie et al., 2020; Hjorth Warlenius, 2022; Holfelder, 2019; Kopnina, 2020). Culture within these frameworks continues to be recognised as an amalgamated domain, where culture is melded with the social, significantly limiting and minimising culture in its actualisation and as an independent and standalone pillar for assessment and learning. Continued use of amalgamated 'socio-cultural' domains will severely compromise legitimate pathways for culture outcomes, rendering FKs and traditional practices mute, and obscure all-encompassing and comprehensive approaches for sustainability in a globalised society. Given these deficits, it is imperative education, ESD, and the SGDs are reconfigured to integrate culture enabling broadened perspectives based on quality, inclusivity, and sensitivity, and that new framework iterations are cognisant of culture reflecting inclusion and global diversity.

Highlighting these gaps, this paper suggests more can be done to address culture deficits and the education-reality gap (Demssie et al., 2020). Achieving sustainable futures will be reliant on inclusive frameworks and mechanisms that incorporate all narratives not just those of the global north. Such narrowed perspectives—for society and sustainability—severely hinder adequate creation, design, and implementation of solutions both locally and globally. Integrating culture into education and ESD requires primary acknowledgement of culture as a standalone pillar, and second a recognition of FKs as notable alternate knowledge systems. The COIL SB program did just that. Through amalgamated education and drawing on melded pedagogical mechanisms (modern and FKs), the COIL SB program provided an exemplar of culture integration and its transformative impact on learners. Culture integrated education with comprehensive approaches aids the deconstruction of colonialism and educational imperialism (Hjorth Warlenius, 2022; Leonard, 2021; Zalli, 2024) producing education models that are reflective of all communities. The results confirm comprehensive approaches led to improved cultural self-reflection and identity of 'Self', as well as enhanced conceptualisation of indigeneity, and HNC.

Recognition of FKs also bridges divergence between the local and global. Global frameworks perpetuate standardisation whilst FKs prioritise place-based, fit-for-purpose approaches emphasising diversity. The challenge therefore for education, ESD, and the SDGs will be to navigate and diversify standardised applications suitable for a sustainable future recognising all knowledge systems not just those in the global north. Similarly, conceptualisations of sustainability will also require diversification regarding definition, approach, and implementation marrying FKs and modern education to incorporate global narratives. The results reveal culture integration played a part in sustainability perception changes. Participants commented on new ways of thinking about sustainability through stewardship and caring of the land, and enhanced systems-thinking regarding current 'broken' systems no longer adequately serving what is needed.

From a pedagogical standpoint, culture integration provides critical contributions for "cultural democracy" (Kopnina, 2020, p. 285). Amalgamated education provides empowerment for oppressed communities; therefore, reshaped education can act as a tool for solidarity and identity rather than selfhood (Komatsu et al., 2023). The results buttress this point with several participants reflecting on improved solidarity and cultural identity cognition via personal recognition of ancestral identity, and broadening perspectives of 'Self' and 'Other' through intercultural exchange. Similarly, culture and FKs integration provides avenues for broadened pedagogical design and curriculum, providing diverse alternate mechanisms for learning currently absent in modern education. FKs sharing is based on reciprocal and intergenerational exchange rooted in naturalistic and cosmological ideologies (Kimmerer, 2013) and is comparatively different to modern education mechanisms. The results reflect beneficial outcomes from alternate pedagogical mechanisms such as pod groups and learning circles which emphasised and enhanced inter-cultural knowledge sharing and influenced the teacher-learner process facilitating learner cultural capital and place-based, fit-for-purpose pedagogy contributing to local and global understandings of sustainability (Hernandez Gonzalez, 2023). The results also demonstrate positive impacts for faculty participants. These participants reported newfound ways of teaching previously absent in modern education. Participants highlighted new conceptualisation of sustainability with the missing piece of culture completing this concept.

Culture integration in education and ESD can also contribute positively to broadened SCs. SCs frameworks are currently shaped to respond to today's societal necessities, however competencies for sustainable futures are yet to be determined. Striving for comprehensive frameworks addressing all domains is essential for inclusive sustainable futures. As demonstrated via an adapted SCs framework from Redman and Wiek (2021) culture integration remains largely absent in SCs rhetoric, however, can and should be integrated. The results reflect that while future SCs remain ambiguous, participants highlighted globalisation, changed systems and concepts, and greater global exchanges for future iterations of society. SCs must therefore, be tailored to support inclusivity, diversity, and diverse knowledge systems that contribute to globalised conceptualisation and facilitate global citizenship.

To achieve comprehensive SCs, ESD must equally be comprehensive, and COP presents an achievable approach for integrating culture and FKs pedagogical mechanisms into education. CRP and CSP provide diverse alternate mechanisms that incorporate cultural references highlighting cultural imbalances, power dynamics, and emphasises resistance, sovereignty and identity (Ladson-Billings, 2021, 2024). By emphasising critical pedagogy, multiculturalism and decolonialisation COPs enables a broad platform for accommodating diversity. Unlike current prescriptive ESD approaches, COPs offer collaborative avenues for sustainability learning supporting diverse concepts of sustainability. The degree to which ESD will enable such collaboration and diversity is yet to be determined, however remains pertinent to this paper. The results reveal that COPs delivered through ESD can contribute to positive engagement on a global and local scale. Utilising COPs, the COIL SB program did indeed facilitate reciprocity, collaboration, and relationships as key tenets of the learning experience.

The COIL SB program and its exploratory research provides a palpable case for culture integration. While the original research did not examine culture integration specifically, the research results revealed culture to be a central and vital component for ESD and sustainable futures. Implications for this research are widespread including impacts on furthering literature, responding to current conceptual, literary, and institutional gaps, and calls for reconfiguration of ESD and the SDGs in future framework iterations inclusive of culture integration, as well as education and curriculum design.

While this research is an asset for supporting culture integration in curricula, various limitations are acknowledged regarding sample size, study design, and an absence of longitudinal results providing long term impact. Furthermore, research constraints are identified with the dataset related to quality and collection. Initial research into the COIL SB program aimed to understand education and transformative learning via paralleled perspectives of sustainability and culturalism, therefore results orientate towards the initial aims of the research and provide limited opportunity for comprehensive data supporting examinations of culture integration specifically. The initial dataset, targeting paralleled perspectives, rather demonstrated the links between curriculum and education design together with culture, leading to enhanced learning. While the data is identified as limited for quality assessments for culture integration, the findings provide an exemplary implication for education and curriculum design as outlined in this paper speaking to broader awareness, consciousness, cross cultural engagement, and solidarity.

Data collection limitations are also highlighted regarding the data set for 2021 and 2022 relating to feedback limitations to aid specific curriculum development. While such drawbacks are identified, the authors maintain the relevance and importance of the pedagogical implications of this paper and research. Further research is recommended to explore ongoing benefits of culture integration including exploring the curriculum underpinnings for cross cultural engagement, acknowledgement of place-based culture, and integration of learner cultural capital for personalised education providing broadened awareness of 'Self' and 'Other', supporting inclusivity and diversity, and to ensure global frameworks support global narratives and all cultures accordingly.

9 Conclusion

At present modern education and curriculum reflect limited ability to develop comprehensive learning for future learners inhibiting development of necessary SCs needed for sustainable futures. Similarly, global frameworks such as the SDGs and ESD are found deficient in their components based on economic, environmental, and social considerations. This paper discusses culture integration as a central component for sustainability learning for sustainable futures and highlights culture deficiencies in ESD and the SDGs, as well as modern education calling for acknowledgement and recognition of culture as an independent and standalone pillar for assessment.

The COIL SB program, originally researched to assess transformative learning in course participants via paralleled perspective learning, is utilised as a case study demonstrating the beneficial impact of culture integration into curriculum and pedagogy. The results, drawn from a mixed method approach of pre-course and post-course surveys and focus group interviews illustrates culture and FKs integration can indeed contribute to enhanced and comprehensive learning and therefore, has a place in curricula. Results indicate culture integration in education can establish cross-cultural communities of practice; facilitate reciprocal learning and knowledge exchange; enhance learner cultural capital; broaden recognition of alternate knowledge systems; positivity impact SCs; and change perceptions of 'Self', 'Other' and current societal structures.

In an increasingly globalised world, culture integration is central and vital to building sustainable futures in full recognition of all knowledge systems. Without such recognition, global frameworks and approaches will remain deficient and risk inadequacy; consequently, culture integration is imperative to support the future pathways for sustainability that are framed for quality, inclusivity, and diversity. As such, this research and paper provides a basis and exemplar for culture integration conceptualisation, advocating and encouraging pedagogical and curriculum design change, and promoting inclusivity not only within education but also throughout current social structures.

References

Bamber, P. (2019). *Teacher education for sustainable development and global citizenship: Critical perspectives on values, curriculum and assessment* (1st ed.). Routledge. https://doi.org/10.4324/9780429427053

Bedford, T. (2022). Constructing a transformative sustainability pedagogy: Teacher empowerment for a sustainable future. *Journal of Teacher Education for Sustainability, 24*(1), 5–18. https://doi.org/10.2478/jtes-2022-0002

Brundiers, K., Barth, M., Cebrián, G., Cohen, M., Diaz, L., Doucette-Remington, S., Dripps, W., Habron, G., Harré, N., Jarchow, M., Losch, K., Michel, J., Mochizuki, Y., Rieckmann, M., Parnell, R., Walker, P., & Zint, M. (2021). Key competencies in sustainability in higher education—Toward an agreed-upon reference framework. *Sustainability Science, 16*(1), 13–29. https://doi.org/10.1007/s11625-020-00838-2

Cebrián, G., Junyent, M., & Mulà, I. (2021). Current practices and future pathways towards competencies in education for sustainable development. *Sustainability, 13*(16), 8733. https://doi.org/10.3390/su13168733

Corbett, J., & Guilherme, M. (2021). Critical pedagogy and quality education (UNESCO SDG-4): The legacy of Paulo Freire for language and intercultural communication. *Language and Intercultural Communication, 21*(4), 447–454. https://doi.org/10.1080/14708477.2021.1962900

Cooper, D. J., & Voigt, M. (2025). Advancing culturally relevant pedagogy in college chemistry. *Chemistry Teacher International, 7*(2), 259–270. https://doi.org/10.1515/cti-2024-0086

Demssie, Y. N., Biemans, H. J. A., Wesselink, R., & Mulder, M. (2020). Combining indigenous knowledge and modern education to foster sustainability competencies: Towards a set of learning design principles. *Sustainability, 12*(17), 6823. https://doi.org/10.3390/su12176823

Dorji, T., Rinchen, K., Morrison-Saunders, A., Blake, D., Banham, V., & Pelden, S. (2024). Understanding how indigenous knowledge contributes to climate change adaptation and resilience: A systematic literature review. *Environmental Management, 74*(6), 1101–1123. https://doi.org/10.1007/s00267-024-02032-x

Funk, J. (2021). Caring in practice, caring for knowledge. *Journal of Interactive Media in Education, 2021*(1), 1–14. Article 11. https://doi.org/10.5334/jime.648.

Hernandez Gonzalez, F. (2023). Exploring the affordances of place-based education for advancing sustainability education: The role of cognitive, socio-emotional and Behavioural learning. *Education Sciences, 13*(7), 676. https://doi.org/10.3390/educsci13070676

Hjorth Warlenius, R. (2022). Learning for life: ESD, ecopedagogy and the new spirit of capitalism. *The Journal of Environmental Education, 53*(3), 141–153. https://doi.org/10.1080/00958964.2022.2070102

Holfelder, A. K. (2019). Towards a sustainable future with education? *Sustainability Science, 14*(4), 943–952. https://doi.org/10.1007/s11625-019-00682-z

Jessen, T., Ban, N., Claxton, N., & Darimont, C. (2020). Contributions of indigenous knowledge to ecological and evolutionary understanding. *Frontiers in Ecology and the Environment, 20*(2), 93–101. https://doi.org/10.1002/fee.2435

Johnson, A. P. (2022). Culturally responsive teaching in higher education. *The International Journal of Equity and Social Justice in Higher Education, 1*(1), 9. https://doi.org/10.56816/2771-1803.1008

Jules, T. D., & Ford, D. Y. (2023). Decolonizing higher education syllabi: Beyond the aesthetics of the syllabus. *Diverse Issues in Higher Education, 40*(14), 8–12. https://link.gale.com/apps/doc/A772956748/AONEu=anon~b4e21d2f&sid=googleScholar&xid=acf95b95

Kahn, R. (2008). From education for sustainable development to ecopedagogy: Sustaining capitalism or sustaining life. *Green theory & praxis: The journal of ecopedagogy, 4*(1). https://doi.org/10.3903/gtp.2008.1.2

Kayira, J., Lobdell, S., Gagnon, N., Healy, J., Hertz, S., McHone, E., & Schuttenberg, E. (2022). Responsibilities to decolonize environmental education: A co-learning journey for graduate students and instructors. *Societies (Basel, Switzerland), 12*(4), 96. https://doi.org/10.3390/soc12040096

Kimmerer, R. W. (2013). *Braiding sweetgrass: Indigenous wisdom, scientific knowledge and the teachings of plants*. Milkweed Editions.

Kimmerer, R. W., & Artelle, K. A. (2024). Time to support indigenous science. *Science, 383*(6680), 243–243. https://doi.org/10.1126/science.ado0684

Komatsu, H., Silova, I., & Rappleye, J. (2023). Education and environmental sustainability: Culture matters. *Journal of International Cooperation in Education, 25*(1), 108–123. https://doi.org/10.1108/JICE-04-2022-0006

Kopnina, H. (2020). Education for the future? Critical evaluation of education for sustainable development goals. *The Journal of Environmental Education, 51*(4), 280–291. https://doi.org/10.1080/00958964.2019.1710444

Ladson-Billings, G. (2021). Three decades of culturally relevant, responsive, & sustaining pedagogy: What lies ahead? *The Educational Forum (West Lafayette Ind.), 85*(4), 351–354. https://doi.org/10.1080/00131725.2021.1957632

Ladson-Billings, G. (2024). How pedagogy makes the difference in U.S. schools. *Daedalus (Cambridge, Mass.), 153*(4), 96–110. https://doi.org/10.1162/daed_a_02106

Latip, A., Hernani, & Kadarohman, A. (2024). Local and indigenous knowledge (LIK) in science learning: A systematic literature review. *Journal of Turkish Science Education, 21*(4), 651–667. https://doi.org/10.36681/tused.2024.035

Lehmann, T., Iyer-Raniga, U., & Mahoney, K. (2025a). Learning for sustainability: Adult transformative learning through sustainability and culturalism perspectives. *Social Sciences & Humanities Open, 11*, 101523. https://doi.org/10.1016/j.ssaho.2025.101523

Lehmann, T., Iyer-Raniga, U., & Mahoney, K. (2025b). The COIL classroom: Utilising collaborative multi-stakeholder learning for sustainable futures. (under review).

Leonard, K. (2021). Turtle Island (North America) indigenous higher education institutions and environmental sustainability education. *Journal of Comparative and International Higher Education, 13*(3), 90–133. https://doi.org/10.32674/jcihe.v13iSummer.3279

Mac Neil, P., & Khare, A. (2025). Adopting sustainability competencies in management education—A scoping review of progress. *Standards, 5*(2), 13. https://doi.org/10.3390/standards5020013

Maier, A. (2024). Solidarity—An aim of education?: Reflections on a vital topic of education for sustainable development (ESD) from the perspective of a responsive pedagogy. *Hungarian Educational Research Journal, 14*(4), 364–382. https://doi.org/10.1556/063.2024.00265

Madiratta, J. (2024). Theoretical foundations of culturally responsive teaching and connections to Saskatchewan curriculum and indigenous education. *in education, 29*(1), 35–49. https://doi.org/10.37119/ojs2024.v29i1.759

Mbah, M. F., Clifton, N., & Kushnir, I. (2024). Internationalising higher education curricula for sustainable development: Considerations for indigeneity and (Inter)culturality. *Discover Sustainability, 5*, 307. https://doi.org/10.1007/s43621-024-00520-y

Motevalli, S., Saffari, N., Michael, M. T. A., & Abadi, F. H. (2022). Enculturation, education and sustainable development: Understanding the impact of culture and education on climate change. *International Education Studies, 15*(4), 31. https://doi.org/10.5539/ies.v15n4p31

Momo, B., Hoople, G. D., Chen, D. A., Mejia, J. A., & Lord, S. M. (2020). Broadening the engineering canon: How culturally responsive pedagogies can help educate the engineers of the future. *Murmurations, 2*(1), 6–21. https://doi.org/10.31946/meee.v2i1.32

Moulton, H. (2024). Using culturally sustaining and revitalizing pedagogy in the composition classroom to engage students in critiquing supremacy culture. *Journal of Pedagogical Sociology and Psychology, 6*(3), 79–94. https://doi.org/10.33902/jpsp.202424973

Muhonen, T., Timonen, L., & Väänänen, K. (2024). Fostering education for sustainable development in higher education: A case study on sustainability competences in research, development and innovation (RDI). *Sustainability, 16*(24), 11134. https://doi.org/10.3390/su162411134

Neale, M (ed.) (2020-2024) First Knowledges Series (Songlines: The Power And Promise (2020); Design: Building On Country (2021); Country: Future Fire, Future Farming (2021); Plants: Past,

Present And Future (2022); Law: The Way Of The Ancestors (2023); Innovation: Knowledge And Ingenuity (2023); Astronomy: Sky Country (2022); Health: Spirit, Country And Culture (2024). Thames and Hudson Australia.

Paris, D. (2012). Culturally sustaining pedagogy: A needed change in stance, terminology, and practice. *Educational Researcher*, 93–97. https://doi.org/10.3102/0013189X12441244

Redman, A., & Wiek, A. (2021). Competencies for advancing transformations towards sustainability. *Frontiers in Education (Lausanne), 6*, 785163. https://doi.org/10.3389/feduc.2021.785163

Santos, D., & May, T. (2024). The academic space is not a safe space to be an indigenous person; responding to the trauma of the settler-colonial university through african indigenous knowledge system (AIKS)-informed pedagogy. *Teaching Anthropology, 13*(2), 53–62. https://doi.org/10.22582/ta.v13i2.724

Sposab, K., & Rieckmann, M. (2024). Development of sustainability competencies in secondary school education: A scoping literature review. *Sustainability, 16*(23), 10228. https://doi.org/10.3390/su162310228

Takona, J. (2025). Harmonizing STEM rigor and cultural responsiveness in college-level statistics: A review. *American Journal of STEM Education, 8*, 109–124. https://doi.org/10.32674/phxvdp70

Tailby, E., Whatman, S., & Sammel, A. (2020). Community and school collaboration: Initiatives that enable primary students to embed indigenous knowledges. In A. Sammel, S. Whatman, & L. Blue (Eds.), *Indigenizing Education*. Springer. https://doi.org/10.1007/978-981-15-4835-2_4

Torres, C. A. (2014). *First Freire: Early writings in social justice education*. Teachers College Press. ISBN: 9780807755341.

UN-OCHA United Nations Office for the Coordination of Humanitarian Affairs (2025). Mr. tom fletcher, under-secretary-general for humanitarian affairs and emergency relief coordinator-remarks at ECOSOC humanitarian affairs segment opening session (18 June 2025, Geneva). Retrieved June 20 2025. https://www.unocha.org/publications/report/world/mr-tom-fletcher-under-secretary-general-humanitarian-affairs-and-emergency-relief-coordinator-remarks-ecosoc-humanitarian-affairs-segment-opening-session-18-june-2025-geneva#:~:text=We're%20in%20a%20moment,funding%20is%20also%20in%20retreat.

UNESCO United Nations Educational, Scientific and Cultural Organization. (2014). UNESCO road map for implementing the global action Programme on education for sustainable development. Retrieved June 17 2025. https://sustainabledevelopment.un.org/content/documents/1674unescoroadmap.pdf

UNESCO United Nations Educational, Scientific and Cultural Organization. (2020). *Education for sustainable development: A roadmap*. 7, place de Fontenoy, 75352 Paris 07 SP. https://doi.org/10.54675/YFRE1448

Venn, R., Perez, P., & Vandenbussche, V. (2022). Competencies of sustainability professionals: An empirical study on key competencies for sustainability. *Sustainability, 14*(9), 4916. https://doi.org/10.3390/su14094916

Wang, Y., Sommier, M., & Vasques, A. (2022). Sustainability education at higher education institutions: Pedagogies and students' competences. *International Journal of Sustainability in Higher Education, 23*(8), 174–193. https://doi.org/10.1108/IJSHE-11-2021-0465

Yin, R. K. (2017). *Case study research: Design and methods* (5th ed.). SAGE.

Zalli, E. (2024). Globalization and education: Exploring the exchange of ideas, values, and traditions in promoting cultural understanding and global citizenship. *Interdisciplinary Journal of Research and Development, 11*(S1), 55. https://doi.org/10.56345/ijrdv11n1s109

Zidny, R., Solfarina, S., Aisyah, R. S. S., & Eilks, I. (2021). Exploring indigenous science to identify contents and contexts for science learning in order to promote education for sustainable development. *Education Sciences, 11*(3), 114. https://doi.org/10.3390/educsci11030114

Tarsilla Lehmann is a researcher at RMIT University, currently at the School of Property and Construction Management. She holds a Masters in International Urban and Environmental Management and has been engaged in various projects and roles across areas of climate change adaptation and resilience, urban development, disaster response, and socio-cultural sustainability research. Tarsilla has served as guest reviewer and speaker to undergraduate and post graduate students advising and advancing students' understanding of socio-cultural perspectives and considerations including presenting at architecture design research studios and international development seminars. Deeply passionate about diverse cultures, Tarsilla strives to encourage the mutually beneficial link between culture, humanity, and sustainability.

Usha Iyer-Raniga is Professor of Sustainability in the Built Environment at RMIT University. She is Co-lead of Circular Built Environment (CBE) under the GlobalABC/One Planet Network's Materials Hub hosted by United Nations Environment Programme. The circularity research and engagement associated with CBE responds to Sustainable Development Goals 4, 6, 7, 8, 9, 12, 13, 15 and 17. As a practitioner-academic, Usha's teaching and research has been focused on accelerating the transition to sustainable and circular futures across the Global North and Global South. Usha has been invited as key-note speaker and invited speaker at national and international conferences including COPs 26–29, seminars, and workshops. She is reviewer for the Australian Research Council and for national and international referred journals and refereed conferences. She is on the editorial board for refereed journals and has led Special Issues. Usha serves on the Boards of various not for profit organisations and has also served as panel judge for state, national and international sustainability and circular economy awards.

Kathleen Mahoney is an Associate Professor of Graphic Design and New Media at SUNY Potsdam and an adjunct Professor of Art and Art History at St. Lawrence University in Canton, NY. Kathleen was an award-winning Art Director in and around New York City for 10 years before she began her teaching career. She has shown her fine art in New York at galleries in Manhattan, on Long Island, and in Ithaca, NY. She has shown internationally in Belgium and Barcelona, Spain. Her joint student project with New Zealand Fulbright scholar in residence Dr. Caroline McCaw, *Anei Au: Here I am Am* was presented in LA and Kolding Denmark. Kathleen expects to complete her research for her DPP in 2025.

Artificial Intelligence and Sociodigital Futures: Curricular Experiences for Sustainable and Immersive Education

Tiago Negrão de Andrade and Maria Cristina Gobbi

Abstract This study critically examines the integration of Artificial Intelligence (AI) into educational curricula within the framework of sociodigital futures, analyzing its epistemological, ethical, and political implications. While AI, combined with immersive technologies like augmented and virtual reality, offers transformative potential for personalized and interactive learning, it also introduces significant challenges, including data colonialism, algorithmic bias, and the privatization of education. Drawing on critical sociotechnical, decolonial, and digital governance perspectives, the research highlights tensions between innovation and equity, emphasizing how AI-mediated curricula can either reinforce hegemonic knowledge systems or foster pluralistic, justice-oriented pedagogies. Through a systematic literature review (PRISMA-guided) of interdisciplinary sources (2010–2025), the study reveals that AI's role in education extends beyond technical implementation, shaping conceptions of authorship, knowledge legitimacy, and democratic futures. Key findings underscore the need for critical algorithmic literacy, epistemic sovereignty, and ethical frameworks to ensure AI serves sustainable, inclusive, and socially just education. The study concludes by advocating for empirical research on localized AI practices and interdisciplinary collaborations to address gaps in teacher preparation and digital equity.

Keywords Artificial intelligence · Sociodigital futures · Critical pedagogy · Decoloniality · Algorithmic governance · Immersive learning

---

T. N. de Andrade (✉) · M. C. Gobbi
Faculty of Architecture, Arts and Communication, Graduate Program in Media and Technology, UNESP, São Paulo, Brazil
e-mail: tiago.negrao@unesp.br

M. C. Gobbi
e-mail: cristina.gobbi@unesp.br

W. Leal Filho et al., *Curriculum Development for Sustainable Development*, World Sustainability Series, https://doi.org/10.1007/978-3-032-22146-9_4

1 Introduction

The incorporation of Artificial Intelligence (AI) into educational processes represents one of the most significant transformations in the contemporary educational scenario. Far from being merely a technical innovation, this integration reveals a field of tensions, possibilities, and epistemic disputes, especially when analyzed from the perspective of sociodigital futures (Di Felice & Schlemmer, 2022). The exponential advancement of AI, associated with technologies such as augmented reality (AR), virtual reality (VR), and remote laboratories, projects scenarios in which learning environments become increasingly sensorial, immersive, and mediated by algorithms (Lampropoulos, 2025a, 2025b). Market projections indicate that, by 2030, AI-based solutions in education are expected to generate billions of dollars globally, radically transforming the ways of teaching, learning, assessing, and producing knowledge (Williamson & Hogan, 2020).

However, this trend is not without controversy. The field of critical technology studies has shown that digital futures are neither neutral nor inevitable, but rather social constructions, deeply influenced by political, cultural and economic choices (Bell, 2003; Facer, 2011; Selwyn, 2016). Over the past few decades, research on educational technology has evolved from instrumentalist approaches, focused on the effectiveness of tools, to sociotechnical and critical approaches, which question the impacts of technologies on the production of subjectivities, epistemologies and structural inequalities (Couldry & Mejias, 2019; Mignolo & Walsh, 2018). Currently, the debate is shifting to understanding how AI not only changes teaching methods, but redefines the very concepts of curriculum, authorship, knowledge and citizenship.

The central objective of this study is to analyze the integration of AI in the construction of curricula in the context of sociodigital futures, understanding the curriculum as a sociotechnical and epistemic device. This device is not limited to the organization of formal knowledge, but functions as a field of negotiation and dispute over which knowledge is legitimate, which epistemologies are recognized, and which futures are possible (Williamson et al., 2020). Technically, AI applied to education operates based on statistical models, neural networks, natural language processing, and machine learning algorithms, capable of performing tasks ranging from personalizing learning paths to generating content and automating assessment processes (Knox, 2020; Retbi & El Faddouli, 2024).

Despite these potentialities, AI presents structural challenges when compared to traditional pedagogical approaches. Unlike models centered on human interaction, dialogicity, and the collective construction of knowledge (Laurillard, 2012), AI tends to reduce educational processes to predictive and correlational patterns. This generates tensions between the promise of personalization and the risk of standardization, surveillance, and performativity (Dou, 2025; Williamson & Hogan, 2020). Furthermore, AI applied to education does not escape ethical and epistemic criticisms related to the opacity of algorithms, data governance, and the reproduction of structural biases (Benjamin, 2019; Noble, 2018).

One of the most acute challenges lies in the coloniality of data and the reproduction of hegemonic epistemologies. As Couldry and Mejias (2019) and Mignolo and Walsh (2018) warn, AI systems, when fed mainly by data produced in the Global North, tend to make local, indigenous, peripheral and Global South knowledge invisible, reinforcing historical asymmetries under the guise of technological innovation. This phenomenon imposes the urgent need to rethink data sovereignty, algorithmic justice and epistemic plurality in the educational field.

Given this scenario, this study aims to critically analyze how the integration of AI into educational curricula reconfigures pedagogical practices, epistemologies and power relations. The research is guided by questions such as: How can AI-mediated curricula be built in an ethical, sustainable and socially just way? What are the epistemic, ethical and political risks associated with the platformization and coloniality of data? And what possibilities emerge for building more plural, inclusive and democratic educational futures?

By mobilizing theoretical frameworks from sociotechnical studies, decoloniality, and AI ethics, this article seeks to contribute to a broader understanding of the role of technology in education, overcoming instrumentalist readings and proposing paths for the construction of pedagogical practices guided by social justice, sustainability, and digital sovereignty. It is therefore a question of understanding AI not as an end in itself, but as a means that, when critically appropriated, can be at the service of the formation of subjects capable of inhabiting, disputing, and reinventing futures.

2 Methodology

The research adopts a qualitative, theoretical, documentary and critical approach, combined with a systematic literature review, conducted according to the PRISMA protocol guidelines, with the aim of ensuring rigor, transparency and reproducibility in the process of selecting and analyzing the materials. The study is based on epistemic references from the fields of Sociotechnical Studies, Digital Governmentality, Critical Sociology of Education and Decolonial Studies, allowing an in-depth analysis of the intersections between artificial intelligence, curriculum and sociodigital futures. The data sources included scientific articles, books, institutional reports, public policy documents and regulatory frameworks produced between 2010 and 2025, collected in the Scopus, Web of Science, ERIC, Redalyc and SciELO databases, as well as gray literature from international organizations and official documents. The definition of the materials considered descriptors such as "artificial intelligence in education", "digital curriculum", "AI ethics", "algorithmic governance", "digital futures" and "coloniality of data", applying inclusion criteria that favored studies aligned with critical, sociotechnical and decolonial approaches, and exclusion criteria that discarded publications with a purely technical, instrumental or commercial bias. The analytical process was conducted through the articulation of critical discourse analysis and thematic analysis, with manual coding, seeking to map, problematize and tension the discourses and practices that configure the processes of integrating artificial intelligence into curricula, their epistemic, ethical and political implications, as well as the challenges and possibilities for the construction of educational projects committed to sustainability, epistemic plurality and social justice.

3 Discussion and Results

The analysis shows that the integration of artificial intelligence into educational processes, in the context of sociodigital futures, does not occur in a neutral or conflict-free manner, but operates as a sociotechnical device permeated by epistemic, ethical and political disputes. The findings reveal that, although immersive technologies associated with AI expand the possibilities for building sensory, interactive and potentially more inclusive learning environments, these same infrastructures carry logics that can deepen processes of data coloniality, platformization and reproduction of historical asymmetries. The tensions emerge from a structural paradox: on the one hand, AI enables the creation of innovative pedagogical practices, centered on playfulness, digital aesthetics and the personalization of learning; on the other, it reinforces dynamics of performativity, algorithmic surveillance and the commodification of education. By problematizing these contradictions, the analysis shows that AI-mediated curricula are not just spaces for the circulation of content, but arenas of dispute over which epistemologies, knowledge, and ways of existing are legitimized or silenced in digital ecosystems. This dynamic implies challenges that go beyond the technical dimension, requiring a critical reflection on the processes of authorship, algorithmic governance, and epistemic sovereignty. At the same time, experiences emerge that point to possibilities for subverting and reinventing these models, when technology is appropriated in a situated, ethical way, oriented toward social justice, sustainability, and the construction of more plural, inclusive, and democratically contested educational futures.

Figure 1 presents a conceptual framework that articulates sociodigital futures in education mediated by artificial intelligence, centered on the Sociotechnical Curricular Framework. This model integrates theoretical foundations, pedagogical potentials, and critical challenges, which unfold methodologically into fundamental structural tensions. Based on this analysis, recommendations are proposed aimed at building sustainable futures, with a focus on critical algorithmic literacy, epistemic sovereignty, and democratic governance.

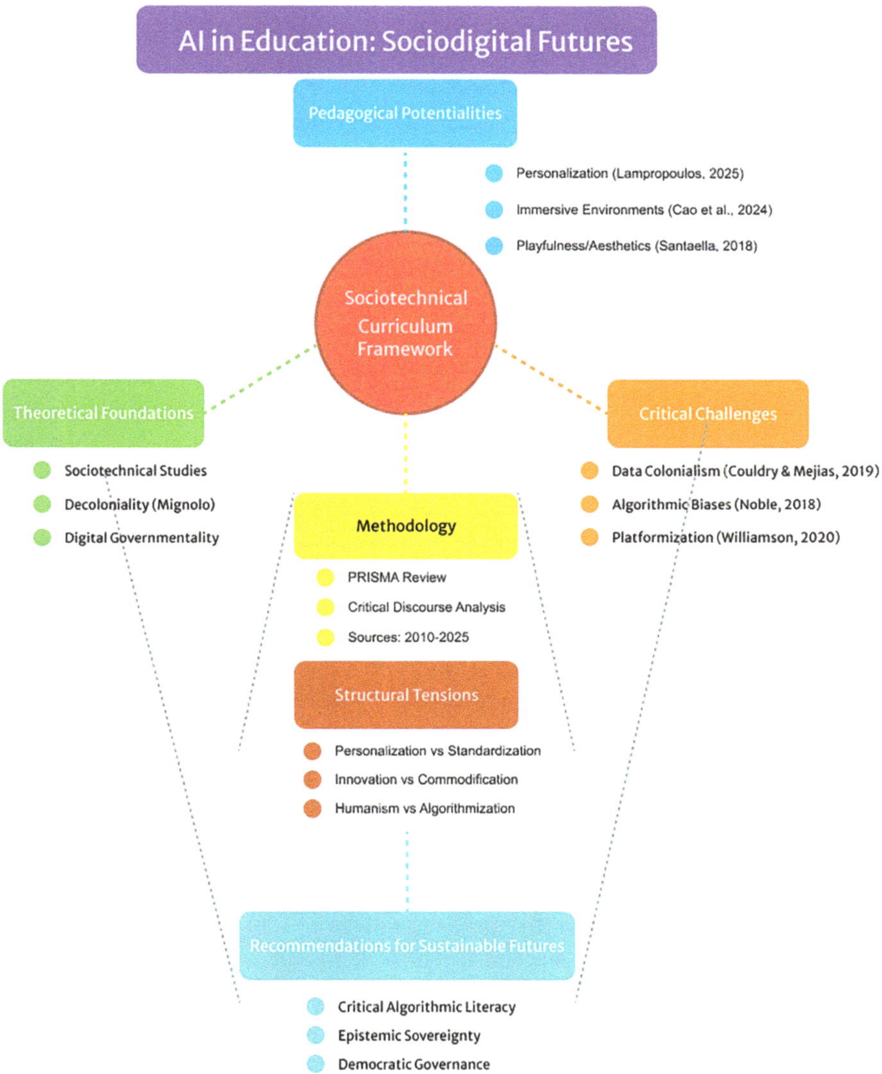

Fig. 1 Conceptual framework of sociodigital futures in education based on a sociotechnical curricular framework

4 Futures Sociodigits and Curriculum Construction—Between Possibilities and Disputes

The incorporation of Artificial Intelligence (AI) into educational processes, especially when analyzed from the perspective of **Sociodigital Futures**, reveals a scenario deeply marked by tensions, paradoxes and epistemic disputes. As highlighted by Di Felice and Schlemmer (2022), thinking about the future from a sociotechnical lens requires understanding it not as a deterministic line, but as a contingent space, crossed by power relations, technological choices and pedagogical decisions that shape, from the present, the possibilities of the future.

This understanding directly dialogues with the contributions of authors such as **Bell** (2003) and **Selwyn** (2016), who point out that digital technologies in education are not neutral, but carry within themselves values, ideologies and world models that can both open up emancipatory possibilities and reinforce historical asymmetries. In fact, as argued **Make**, thinking about educational futures mediated by AI forces us to abandon linear conceptions of technological progress and adopt a critical perspective, capable of understanding that futures are socially constructed, disputed and, often, unequally distributed.

In this scenario, the curriculum ceases to be just a repository of formal knowledge and becomes a **sociotechnical device**—a space where values, epistemologies, languages and ways of existing in the digital world are negotiated. The work of Santaella and Policarpo (2018) already warned that digital networks produce a new aesthetics of knowledge, based on interactive, collaborative and hypertextual processes. This view is expanded by Lampropoulos (2025b), who shows how the combination of AI, augmented reality and virtual reality redefines learning environments, making them immersive, sensorial and deeply mediated by algorithms.

On the other hand, this same algorithmic mediation is not without risks. The work of Retbi and El Faddouli (2024) warns of the implications of generative AI in education, from the automation of assessment practices to the ethical challenges associated with authorship, privacy and reproduction of algorithmic biases. This point is crucial and is echoed in the analyses of Noble (2018) and Benjamin (2019), which demonstrate how algorithms, when not critically analyzed, can reproduce and amplify structures of oppression, racism, sexism and social exclusion.

At the same time, pedagogical experiences that critically incorporate AI, such as those reported by Luz et al. (2024), reveal the potential of these technologies for building more inclusive, creative, and sustainable educational practices. Projects that use AI to create digital museums, remote laboratories, or gamified learning environments (Cao et al., 2024; Schulz et al., 2023) demonstrate that technology can be an ally in building more open and plural educational futures—as long as they are anchored in critical and socially committed epistemologies.

This discussion is also linked to debates on decolonization of digital knowledge, as they defend Pinky and Walsh and Couldry and Mejias (2019). For these authors, any curriculum project in the context of sociotechnological futures necessarily needs to confront the logics of coloniality of data and algorithms, which tend to universalize Western epistemologies, making local, indigenous and Global South knowledge invisible.

Therefore, the epistemological, ethical, and political challenges in building AI-mediated curricula are not peripheral—they are central. They require educators, technology developers, and policymakers to recognize the curriculum as a field of sociotechnical dispute. In this sense, building sustainable and democratic educational futures demands pedagogical practices that articulate technological innovation with social justice, training for digital citizenship, and commitment to the Sustainable Development Goals, especially SDG 4 (Quality Education).

This perspective is also supported by the work of Williamson et al. (2020), who advocate a critical and reflective approach to the use of AI in education, emphasizing that the design of algorithmic systems must be guided by ethical, inclusive and transparent principles.

In short, the construction of curricula for Sociodigital Futures is, above all, a political-pedagogical project. In it, AI and other digital technologies are not ends in themselves, but means that can—and should—be put to the service of educational practices that form critical, reflective subjects, capable of inhabiting, disputing and reinventing futures.

5 Pedagogical Potential of Artificial Intelligence for Sustainable and Immersive Education

The growing integration of artificial intelligence (AI) and immersive technologies—such as augmented reality (AR), virtual reality (VR), remote labs, and automated storytelling—opens up a field of pedagogical possibilities that redefine teaching and learning practices in the twenty-first century. This technological convergence is not just a technical issue, but also profoundly epistemological, ethical, and political, as it affects the ways in which we construct, access, and share knowledge.

Authors such as Lampropoulos (2025a, 2025b) highlight that the combination of AI, AR and VR significantly expands learning environments, making them more sensorial, interactive and personalized. This immersion allows subjects to learn in digital ecosystems that simulate complex phenomena, providing experiences previously restricted to physical laboratories or specific contexts. This is particularly visible in the studies by Cao et al. (2024), which demonstrate how the integration of AI and VR in sustainable architecture training not only enhances technical understanding, but also stimulates critical thinking about sustainability and environmental impact.

However, these potentialities are not free from tensions. As Dou (2025) warns, immersive environments mediated by AI, although rich in possibilities, also carry risks associated with capturing attention, standardizing experiences, and commodifying the educational process. This criticism directly dialogues with the reflections of Gee (2007), who points out that the use of playful and interactive elements—such as games, simulations, and gamified narratives—can be both a powerful catalyst for complex learning and a vehicle for distraction and superficiality, when not guided by critical pedagogical intentionality.

In this sense, it is essential to understand that AI is not a neutral tool. It embeds data models, algorithmic logics, and biases that, if not critically addressed in the training process, can reproduce epistemic and cultural inequalities. Noble (2018) and Benjamin (2019) show, in a compelling manner, how algorithmic systems often reinforce racial, gender, and class discrimination by automating search, classification, and recommendation criteria. Translating this analysis to the educational context, it becomes clear that a curriculum that incorporates AI must necessarily include critical training in algorithmic literacy and digital ethics.

On the other hand, pedagogical experiences such as those reported by Luz et al. (2024) demonstrate that it is possible to use AI to promote more democratic, creative and collaborative educational processes. In this study, elementary school children used programming languages, associated with AI, to create digital narratives, games and simulations that significantly expanded their ability to critically read the digital world.

Digital aesthetics, discussed by Policarpo and Santaella (2018), also emerges as a central axis in the pedagogical transformation operated by AI. The construction of immersive environments is not only a functional issue, but also an aesthetic and sensitive one. The possibility of creating educational experiences that combine sounds, images, AI-generated narratives and interactivity expands the modes of perception and production of meaning, making learning not only a cognitive practice, but also a sensory and affective one.

These processes, however, can only be considered sustainable if they are accompanied by robust public policies, inclusive digital infrastructure, and continuous and critical teacher training. Williamson and Hogan (2020) warn that the increasing privatization of digital education and dependence on commercial platforms may compromise pedagogical autonomy, shifting curricular decisions to agents external to the educational space, such as big techs and data companies.

Therefore, the contemporary challenge is not only to integrate AI and immersive technologies into educational practices, but to do so in an ethically oriented, pedagogically critical and socially committed way. Building sustainable and immersive education requires overcoming instructionist and commodified models, putting technology at the service of building critical, creative subjects capable of facing the complex challenges of the contemporary world.

In this sense, thinking of AI as an ally of education requires that it be understood as a sociotechnical artifact, whose implications go beyond the instrumental dimension. AI pedagogy, therefore, is not limited to the use of intelligent tools, but involves the construction of educational practices that articulate playfulness, aesthetics, ethics, sustainability and social justice.

6 Playfulness, Aesthetics and Interactivity as Drivers of Learning—The Mediation of Artificial Intelligence in the Construction of Knowledge

The emergence of educational technologies that combine artificial intelligence (AI), augmented reality (AR), virtual reality (VR) and immersive environments ushers in a significant transformation in the ways of teaching and learning. The triad of playfulness, aesthetics and interactivity, when mediated by AI, profoundly reconfigures pedagogical practices, both in their epistemological and political and sensory dimensions.

As regards the playfulness, authors such as Gee (2007) argue that digital games are not mere entertainment devices, but complex contexts of situated learning, where subjects face challenges, solve problems and construct meanings in interactive environments. This potential is amplified by AI, which allows the dynamic personalization of scenarios, challenges and feedback, adjusting the pace and difficulty of learning to the characteristics of each student, as highlighted by Lampropoulos (2025b) and Schulz et al. (2023). Such possibilities materialize, for example, in platforms that use AI to create adaptive educational games or virtual laboratories where students can experiment, simulate and interact in sensory environments.

However, this appropriation of playfulness carries contradictions. When decontextualized from a critical pedagogical project, gamification can transform learning into a logic of extrinsic reward, reinforcing competitiveness, performativity and symbolic consumption, as Selwyn (2016) and Williamson and Hogan (2020) warn. This forces us to reflect on whether gamified practices are at the service of critical and emancipatory education or whether they reproduce neoliberal logics of constant evaluation, surveillance and productivity.

The dimension of the digital aesthetics appears as a fundamental vector in the construction of immersive educational environments. Policarpo and Santaella (2018) argue that, in digital networks, knowledge is permeated by an aesthetic that is not merely decorative, but constitutive of the process of constructing meaning itself. This aesthetic manifests itself in the interfaces, sounds, visual textures, movements and interactive dynamics of AI-mediated environments. Murray (2012) complements this view by arguing that the aesthetics of digital immersion generates cognitive and emotional experiences that are different from traditional ones, expanding the frontiers of learning.

The integration of aesthetics and interactivity, in turn, reconfigures the very notion of authorship and knowledge construction. Laurillard (2012) already argued that the transmissive model of education gives way, in the digital environment, to a dialogical, interactive and co-constructive pedagogy. With AI, this interactivity intensifies, allowing subjects not only to consume information, but to become co-creators of narratives, experiences and environments. This is evident in the educational metaverse environments analyzed by Gosling et al. (2024) or in the experiences with Scratch reported by Luz et al. (2024), where students are led to program, build and interact in digital ecosystems that demand systemic thinking, creativity and collaboration.

However, one cannot ignore the risks associated with the uncritical aestheticization of education. Noble (2018) and Benjamin (2019) warn that the algorithms that mediate digital experiences—including aesthetically sophisticated ones—carry structural biases that reproduce exclusions, erasures, and sociotechnical hierarchies. Thus, aestheticization, when not accompanied by an epistemological and ethical critique, can transform the educational environment into a space of seductive simulations, but devoid of critical density and commitment to social transformation.

Therefore, the potential of AI in mediating playfulness, aesthetics and interactivity is undeniable. It expands the boundaries of learning, making it more sensitive, affective, sensorial and, when well-guided, deeply meaningful. However, this power is only fully realized when articulated with pedagogical projects guided by principles of social justice, inclusion, sustainability and critical training.

AI, in this context, is not an end, but a means that must be at the service of educational practices capable of forming subjects who not only inhabit sociotechnological futures, but who are capable of disputing them, reinventing them and reconstructing them based on an ethics of care, collectivity and social transformation.

7 Epistemological, Ethical and Political Challenges in Incorporating Artificial Intelligence into Sustainable Curricular Proposals

The emergence of artificial intelligence (AI) in education is not a mere technological phenomenon, but a process deeply intertwined with epistemological, ethical and political dynamics. Incorporating AI into the curriculum necessarily means facing challenges that transcend the instrumental dimension, requiring a critical review of how we conceive of knowledge, authorship, learning and, above all, the social function of education itself in the twenty-first century.

From the point of view epistemological, AI shifts fundamental paradigms. By operating through probabilistic models, predictions, and pattern recognition, it introduces a logic that strains traditional concepts of knowledge and authorship. Retbi and El Faddouli (2024) point out that generative AI challenges the boundaries between human authorship and automated production, calling into question the notion of originality, especially in assessment processes and in the production of educational content. This issue is explored in depth by Knox (2020), who argues that AI translates knowledge into statistical models, which, on the one hand, offers new paths for the personalization of learning, but, on the other, tends to empty the qualitative, contextual, and critical dimensions of knowledge.

You ethical dilemmas arising from this scenario are equally unavoidable. AI brings with it problems of algorithmic opacity, data governance, privacy and reproduction of structural biases. Researchers such as Noble (2018) and Benjamin (2019) highlight how algorithmic systems perpetuate racial, gender and class discrimination, often naturalized under the apparent neutrality of codes. In education, this risk materializes in automated assessment systems, predictive performance platforms and digital environments that operate without transparency regarding the criteria they use to classify, recommend or punish subjects.

Not flat political, AI fits into a broader context of platformization, privatization, and data coloniality. Williamson and Hogan (2020) denounce that the advance of large technology corporations over public education is not just an economic issue, but an epistemological and governmental one. These platforms not only provide infrastructure, but also define what counts as knowledge, how one learns, and how one measures performance. Couldry and Mejias (2019) expand on this critique by introducing the concept of*coloniality of data*, revealing how the massive extraction of information by AI reproduces colonial structures, concentrating power in the Global North, while exploiting and making invisible local, community and Global South knowledge.

Authors such as Mignolo and Walsh (2018) emphasize that any AI project in education must necessarily confront the legacies of epistemological coloniality. Without this perspective, the use of AI runs the risk of being yet another layer of silencing of indigenous, Afro-diasporic, peripheral and popular knowledge.

However, this is not a scenario of technological determinism. The experiences reported by Luz et al. (2024) and the reflections of Di Felice and Schlemmer (2022) show that AI, when appropriated in a critical, situated and ethical way, can be a powerful instrument for the construction of inclusive, collaborative educational practices aligned with the principles of sustainability. The key lies in the construction of curricula that not only incorporate AI, but that critically address it—forming subjects capable of understanding, questioning and transforming the algorithmic infrastructures that permeate their lives.

Facer (2011) and Selwyn (2016) insist that thinking about the digital futures of education is not just about preparing students to operate technologies, but about enabling them to imagine, dispute and build more just, democratic and plural sociotechnical futures. This involves including discussions on algorithmic justice, data sovereignty, plural epistemologies and AI ethics in the curriculum.

So the debate is not whether or not we should adopt AI in education—that is already a fact. The central question isas, for whom and for what purposes this technology will be incorporated. This is a dispute that is not technical, but deeply political, ethical and epistemic.

Therefore, building sustainable curricula in the context of AI requires facing three fundamental imperatives: (1) epistemological—pluralizing the ways of knowing and resisting the universalization of algorithmic models; (2) ethical—ensuring transparency, justice and democratic governance of data; and (3) political—breaking with the logic of data coloniality and the privatization of education, putting technology at the service of emancipation, social justice and sustainability.

Analytical Synthesis: The emergence of Artificial Intelligence (AI) in educational processes, especially when situated on the horizon of sociodigital futures, represents not only a technical innovation, but a profound epistemological, ethical and political transformation (Di Felice & Schlemmer, 2022). As Bell (2003) and Selwyn (2016) warn, the incorporation of digital technologies into the curriculum is not neutral; it carries with it sociotechnical projects, hegemonic epistemologies and disputes over the definition of what counts as knowledge in the twenty-first century.

Thinking about sociodigital futures in education, therefore, requires breaking with the linearity of the technocentric discourse that naturalizes AI as inevitable and devoid of intentionality. In this sense, Facer (2011) proposes that futures are not given, but socially constructed, and, therefore, permeated by power relations, cultural choices and pedagogical decisions that shape not only the present, but also the possibilities of the future. The curriculum, in this sense, becomes a sociotechnical device and, simultaneously, a field of epistemic and political dispute (Selwyn, 2016; Williamson et al., 2020).

On the one hand, the pedagogical potential of AI associated with immersive technologies—such as augmented reality (AR), virtual reality (VR), and remote laboratories—is widely recognized. Studies such as those by Lampropoulos (2025a, 2025b) and Cao et al. (2024) demonstrate that these technologies expand the ability of individuals to experiment, simulate, and construct knowledge in sensory, interactive, and collaborative environments. The possibility of personalizing learning, mediated by AI, favors training processes that are more adherent to the specificities of learners and to contemporary challenges, such as sustainability and training for digital citizenship.

However, the adoption of these technologies is not free from tensions and risks. Dou (2025) draws attention to the effects of attention capture, standardization of experiences and commodification of education promoted by immersive environments mediated by AI. The criticism is reinforced by Gee (2007), who problematizes the uncritical use of gamification, capable of converting learning into a mechanism of performativity and reward, disconnected from critical formative processes.

The aesthetic dimension of digital, often neglected in traditional analyses, emerges as a central vector in the configuration of learning environments. Santaella and Policarpo (2018) argue that, in digital culture, aesthetics is not an adornment, but an epistemic operator that conditions ways of perceiving, understanding and constructing meaning. Murray (2012) complements this view by indicating that the aesthetics of immersion profoundly transforms the cognitive and affective experiences of subjects, shifting learning from a purely rational practice to a sensorial, affective and situated experience.

This scenario, however, also highlights the challenges related to algorithmic governance and epistemic justice. Researchers such as Noble (2018) and Benjamin (2019) demonstrate how algorithmic systems reproduce and amplify social, racial, and gender hierarchies, naturalizing discrimination and erasure under the guise of technological neutrality. In the educational field, these dynamics materialize in automated assessment systems, prediction-based learning platforms, and personalization models that often reinforce structural inequalities.

On an epistemological level, Knox (2020) warns that AI operates through statistical models that tend to reduce the complexity of knowledge to probabilistic patterns, weakening the qualitative, contextual and relational dimensions of knowledge. This generates a structural tension in the curriculum, which begins to oscillate between the promise of algorithmic personalization and the risk of emptying educational practices of a dialogical, situated and critical nature.

From a political perspective, Williamson and Hogan's (2020) analysis highlights how the platformization of education, associated with the privatization of data and dependence on corporate infrastructures, threatens pedagogical autonomy and subordinates curricular decisions to the interests of large technology companies. This criticism is expanded by Couldry and Mejias (2019), who denounce the coloniality of data as a new regime of exploitation, in which epistemologies from the Global South are systematically made invisible and subordinated to the logics of extraction, accumulation and control promoted by the Global North.

In this context, the construction of AI-mediated curricula cannot be dissociated from a decolonial perspective, as proposed by Mignolo and Walsh (2018). For these authors, confronting the coloniality of digital knowledge is a sine qua non condition for the construction of educational projects committed to social justice, epistemic plurality and sustainability.

Despite the challenges, experiences such as those reported by Luz et al. (2024) demonstrate that it is possible to mobilize AI in a critical and situated way, promoting pedagogical practices based on collaboration, creativity, and the problematization of the algorithmic systems that structure contemporary social life. This requires not only the adoption of technologies, but, above all, the incorporation into the curriculum of critical training in algorithmic literacy, digital ethics, and data justice.

Therefore, the question is not whether or not AI should be incorporated into education—this is already an irreversible reality—but how, under what conditions, and for what purposes. The construction of curricula for sociodigital futures is, therefore, above all an ethical, epistemic, and political dispute. It requires educational practices that articulate playfulness, aesthetics, and interactivity, without giving up the radical commitment to the formation of critical, reflective subjects capable of inhabiting, contesting, and reinventing futures.

8 Conclusion

This study contributes to broadening the critical debate on the incorporation of Artificial Intelligence (AI) into educational processes, analyzing how it reconfigures epistemologies, pedagogical practices, and curricular projects in the context of sociodigital futures. Based on an approach based on critical, decolonial, and sociotechnical perspectives, it was shown that AI, far from being a neutral tool, operates as a sociopolitical artifact that can both expand emancipatory possibilities and reproduce historical and epistemological asymmetries.

Among the main contributions, the articulation between the debates on digital aesthetics, playfulness, interactivity and the ethical, epistemic and political challenges of AI in education stands out. The study also offers a theoretical systematization that integrates the epistemological, aesthetic and political dimensions, proposing that the construction of curricula for sociodigital futures must be anchored in principles of social justice, epistemic plurality, sustainability and digital sovereignty. In addition, the work reinforces the urgency of including critical training in algorithmic literacy, digital ethics and data governance in the curriculum, contributing to the formation of subjects capable of inhabiting, disputing and reinventing futures mediated by technologies.

However, this study has some limitations. It is a theoretical-conceptual analysis that, although critically dense, does not incorporate direct empirical research on educational practices that already use AI in different contexts. The lack of empirical data limits the possibility of evaluating, in practice, how these sociotechnical dynamics materialize in concrete educational environments, especially in realities marked by technological, economic, and cultural inequalities. In addition, the scope of this study prioritizes an analysis focused on critical references from the sociology of education, science, technology, and society (STS) studies, and decolonial studies, without covering, in more depth, psychological, neuroscientific, or applied technological perspectives.

In view of this, it is recommended that future studies advance in at least three directions. First, empirical research should be conducted to analyze how pedagogical practices mediated by AI, AR, and VR develop in different educational contexts—both in the Global South and the Global North—considering their potential and contradictions. Second, research should be conducted to further analyze the impacts of AI on teacher training, especially with regard to the ethical, critical, and technological competencies required for transformative action. Finally, it is suggested that interdisciplinary studies be expanded to articulate education, AI ethics, data governance, and digital sovereignty, in order to propose models of technological development aligned with the principles of social justice, equity, and sustainability.

Thus, it is expected that this study can not only contribute to theoretical advancement in the field of education and sociotechnical studies, but also provoke reflections and pedagogical practices committed to building more fair, inclusive and democratically contested educational futures.

References

Bell, W. (2003). *Foundations of futures studies: Human Science for a new era* (Vol. 1). Transaction Publishers.

Benjamin, R. (2019). *Race after technology: Abolitionist tools for the new Jim code.* Polity Press.

Cao, Y., Gao, X., Yin, H., Yu, K., & Zhou, D. (2024). Reimagining tradition: A comparative study of artificial intelligence and virtual reality in sustainable architecture education. *Sustainability, 16*(24), 11135. https://doi.org/10.3390/su162411135

Couldry, N., & Mejias, U. A. (2019). *The costs of connection: How data is colonizing human life and appropriating it for capitalism.* Stanford University Press.

Di Felice, M., & Schlemmer, E. (2022) As ecologias dos metaversos e formas comunicativas do habitar, uma oportunidade para repensar a educação [The ecologies of metaverses and communicative forms of inhabiting: an opportunity to rethink education]. *Revista e-Curriculum, 20*(4), 1799–1825. https://doi.org/10.23925/1809-3876.2022v20i4p1799-1825

Dou, X. (2025). Research on the development of immersive communication industry driven by AIGC in metaverse scenarios. *SHS Web of Conferences, 213*, 02037. https://doi.org/10.1051/shsconf/202521302037

Facer, K. (2011). *Learning futures: Education, technology and social change.* Routledge.

Floridi, L., Cowls, J., Beltrametti, M., Chatila, R., Chazerand, P., Dignum, V., et al. (2018). AI4People—An ethical framework for a good AI society: Opportunities, risks, principles, and recommendations. *Minds and Machines, 28*(4), 689–707. https://doi.org/10.1007/s11023-018-9482-5

Gee, J. P. (2007). *What video games have to teach us about learning and literacy* (2nd ed.). Palgrave Macmillan.
Gosling, M. S., Campos, A. F. D., Otoni dos Santos, R. F., & Pereira, F. C. M. (2024). Metaverso: Uma análise de mapeamento científico [Metaverse: A scientific mapping analysis]. *Informação & Informação, 29*(2), 24–47. https://doi.org/10.5433/1981-8920.2024v29n2p24
Knox, J. (2020). Artificial intelligence and education in China. *Learning, Media and Technology, 45*(3), 298–311. https://doi.org/10.1080/17439884.2020.1745403
Lampropoulos, G. (2025a). Intelligent virtual reality and augmented reality technologies: An overview. *Future Internet, 17*(2), 58. https://doi.org/10.3390/fi17020058
Lampropoulos, G. (2025b). Combining artificial intelligence with augmented reality and virtual reality in education: Current trends and future perspectives. *Multimodal Technologies and Interaction, 9*(2), 11. https://doi.org/10.3390/mti9020011
Laurillard, D. (2012). *Teaching as a design science: Building pedagogical patterns for learning and technology*. Routledge. https://doi.org/10.4324/9780203125080
Luz, S. D., Niero, L. T. L., Borges, M. K., & Lucas, E. R. O. (2024) Formação de leitores ubíquo com Scratch: Experiência com estudantes do quarto ano do Ensino Fundamental na Rede Pública [Ubiquitous reader training with Scratch: Experience with fourth-year elementary students in public schools]. In *Anais do WIE 2024. Sociedade Brasileira de Computação*. https://doi.org/10.5753/wie.2024.242565
Mignolo, W., & Walsh, C. (2018). *On decoloniality: Concepts, analytics, praxis*. Duke University Press.
Murray, J. (2012). *Inventing the medium: Principles of interaction design as a cultural practice*. MIT Press.
Noble, S. U. (2018) Algorithms of oppression: How Search engines reinforce racism. NYU Press. https://doi.org/10.18574/nyu/9781479837243.001.0001
Policarpo, C., & Santaella, L. (2018). A estética do conhecimento nas redes digitais [The aesthetics of knowledge in digital networks]. *Dialogia, 28*, 157–169. https://doi.org/10.5585/dialogia.n28.8455
Retbi, A., & El Faddouli, N.-E. (2024) The impact of generative artificial intelligence on education: A comparative study. *European Conference on e-Learning*, 1–9. https://doi.org/10.34190/ecel.23.1.2975
Schulz, T., Liu, S., Peukert, C., Toreini, P., Maedche, A., & Weinhardt, C. (2023). An immersive "Proto AI Learning Factory" concept for teaching AI competencies. *SSRN Electronic Journal*. https://doi.org/10.2139/ssrn.4470430
Selwyn, N. (2016). *Education and technology: Key issues and debates*. Bloomsbury Publishing.
Williamson, B., & Hogan, A. (2020) Commercialisation and privatisation in/of education in the context of Covid-19. *Education International Research*, 1–52.
Williamson, B., Eynon, R., & Potter, J. (2020). Pandemic politics, pedagogical practices: Digital technologies and distance education during the coronavirus emergency. *Learning, Media and Technology, 45*(2), 107–114. https://doi.org/10.1080/17439884.2020.1761641

Tiago Negrão de Andrade Ph.D. in Media and Technology from FAAC—Faculty of Architecture, Arts and Communication—Bauru Campus, São Paulo State University "Júlio de Mesquita Filho". Graduated in Social Communication—Specialization in Public Relations from the University of Sorocaba (UNISO).

Maria Cristina Gobbi CNPq Research Productivity Scholarship. Associate Professor in History of Communication and Media Culture in Latin America (2014). He completed his Post-doctorate (2008) in the Graduate Program in Integration of Latin America at the University of São Paulo—(PROLAM / USP), a Ph.D. in Social Communication from the Methodist University of São Paulo (2002). Head of the Social Communication Department of FAAC-Unesp (2017–2019). He was a Fellow of the Institute of Applied Research (IPEA)—year 2010–2013. Professor of the Graduate

Program in Communication, professor and was vice-coordinator (2009–2017) of the Postgraduate Program in Media and Technology, both of the Universidade Estadual Paulista (UNESP). Administrative Director (2013–2016 management) of SOCICOM (Brazilian Federation of Scientific and Academic Communication Associations). She was Director-secretary of the Folkcom Network. In the 2008–2011 period she was Director of Documentation for INTERCOM (Brazilian Society of Interdisciplinary Communication Studies) and currently is part of the management (2017–2019) in the Fiscal Council. Ad-hoc Consultant (Institutional and Communication Courses) of the National Institute of Educational Studies and Research of the Ministry of Education (INEP/MEC). Coordinator of the Research Group on Latin American Communications Thinking (PCLA) of CNPq. She was Alternate Director of the Unesco Chair of Communication (1998–2010). Assistant Editor—Latin American Association of Communication Researchers—ALAIC. Winner of the Luiz Beltrão Award—Category: Academic Maturity of the Brazilian Society of Interdisciplinary Communication Studies (2014). He works in the area of Communication, with an emphasis on History of Communication and Media Culture, History and Memory (Communication), Latin American Communication, Youth Cultures, Popular Culture, Cultural Diversity, Media-cultural Profiles (Institutions and Personalities), Theories of Communication and Culture in Latin America, Journalism and its social and technological nuances and Digital Technologies in Latin America, Gender and Human and Social Development.

Embedding Sustainability: Case Studies of Successful Curriculum Models for a Greener Future

Kudzanayi Chiteka and C. C. Enweremadu

Abstract This chapter critically examines curriculum development for sustainable development, exploring how educational frameworks and pedagogical innovations can advance the integration of sustainability principles across diverse educational contexts. Anchored in the paradigm of Education for Sustainable Development and aligned with the Sustainable Development Goals, particularly SDG 4, it investigates the conditions under which interdisciplinary, experiential, and participatory teaching practices can foster sustainability mindsets, systems thinking, and transformative learning. Drawing on global case studies, the chapter analyses curriculum models that link core disciplinary knowledge to real-world sustainability challenges, emphasizing design thinking, reflective practice, and collaborative problem-solving. It also considers the role of curriculum designers in developing integrative, competency-based frameworks, and the importance of assessment tools that capture critical and systems thinking. The chapter highlights the vital role of policymakers in enabling curricular transformation through inclusive policies, sustainable funding mechanisms, and national strategies aligned with the SDGs. Finally, it underscores the need for cross-sector collaboration and investment in teacher capacity as foundational to building adaptive and future-ready education systems capable of equipping learners to navigate complex sustainability challenges.

Keywords Sustainability education · Curriculum models · Interdisciplinary learning · Environmental stewardship · Global case studies

K. Chiteka · C. C. Enweremadu (✉)
Department of Mechanical, Bioresources and Biomedical Engineering, University of South Africa, Science Campus, Florida, South Africa
e-mail: enwercc@unisa.ac.za

K. Chiteka
e-mail: echitekk@unisa.ac.za; kchiteka@gmail.com

W. Leal Filho et al., *Curriculum Development for Sustainable Development*, World Sustainability Series, https://doi.org/10.1007/978-3-032-22146-9_5

1 Introduction

Sustainability has become critical in the 21st century due to growing environmental, social, and economic challenges. Issues like climate change, resource depletion, and pollution threaten both the planet and human well-being, exacerbated by inequalities, unsustainable consumption and weak governance. The United Nations' Sustainable Development Goals (SDGs), adopted in 2015, offer a global framework for addressing these challenges, with SDG 4 focusing on education's transformative role in promoting inclusivity, empowerment, and resilience (Kushwah et al., 2024). Education for Sustainable Development (ESD) plays a central role in this effort, equipping individuals with the knowledge, skills and values necessary to foster a just, inclusive, and environmentally responsible world. ESD integrates social, environmental, and economic aspects of sustainability, promoting critical thinking and systems analysis, essential for addressing global challenges. Educational institutions must align curricula, pedagogy, governance and community engagement with sustainability principles, fostering learners who lead transformative change (Beasy et al., 2023). ESD supports multiple SDGs, including climate action (SDG 13), gender equality (SDG 5), and responsible consumption (SDG 12), and helps societies adapt to technological, demographic, and ecological shifts.

This chapter aims to critically examine how sustainability principles are embedded in curriculum design and delivery across different educational levels. It focuses on identifying effective models and strategies that support transformative learning and align with the SDGs, especially SDG 4 on quality education. The core objectives of the chapter are:

1. To analyze innovative curriculum models that integrate sustainability thinking across disciplines and institutional levels.
2. To assess pedagogical approaches and institutional strategies that foster sustainability competencies, systems thinking, and student agency.
3. To showcase global case studies that exemplify successful implementation of sustainability education in diverse contexts.
4. To derive actionable insights for curriculum designers, educators, and policymakers aiming to embed sustainability in formal education systems.

By addressing these objectives, the chapter contributes to the global discourse on curriculum development for sustainable development, offering both theoretical frameworks and practical guidance for systemic educational transformation.

2 Methodology

This chapter employs a qualitative, interpretive methodology based on document and case study analysis to examine how sustainability is embedded in curriculum development across educational systems. The study draws from a diverse range of sources, including peer-reviewed journal articles, university publications, national policy frameworks, and grey literature. A structured literature search was conducted across major academic databases, including Scopus, Web of Science, and the Directory of Open Access Journals, using search terms such as "Education for Sustainable Development," "sustainability curriculum models," "interdisciplinary learning," and "transformative pedagogy." Except for a few cases, materials published between 2015 and 2024 were prioritised to ensure conceptual and contextual relevance. The incorporated insights enabled a comprehensive understanding of both theoretical frameworks and practical implementation. Case studies were purposively selected to reflect diverse educational levels and geographic contexts, with an emphasis on documented curricular integration, pedagogical innovation and institutional transformation.

While this chapter offers a comprehensive synthesis of curriculum models and sustainability education strategies, some methodological limitations were noted. The analysis was based exclusively on secondary data, which may exclude undocumented or emerging practices, particularly in underrepresented regions. The purposive selection of case studies, while aimed at ensuring contextual and thematic diversity, introduces potential selection bias and may limit the generalisability of the findings. Additionally, language barriers and disparities in publication accessibility may have constrained the inclusion of relevant non-English sources. The interpretive and qualitative nature of the synthesis, while suitable for capturing conceptual trends and cross-cutting themes, does not permit empirical validation or assessment of implementation fidelity. These limitations underscore the need for complementary primary research to deepen understanding of how sustainability is operationalized in diverse educational settings.

3 Results and Discussion

This section presents the key findings of the chapter, derived from a comprehensive analysis of case studies, institutional documents, and scholarly literature on sustainability curriculum models. The results are organized thematically into four core areas: curriculum models for sustainability, strategies for embedding sustainability, illustrative international case studies, and the broader impacts and outcomes of these initiatives. Each subsection integrates results with critical discussion to highlight emerging patterns, successful practices, and ongoing challenges.

3.1 Curriculum Models for Sustainability

Curriculum models for sustainability aim to equip learners with the competencies needed to address complex global challenges such as climate change, biodiversity loss, and social inequality (Hays & Reinders, 2020). These models share core features including interdisciplinarity, critical thinking, experiential learning and systems thinking, with a growing emphasis on transformative, participatory, and context-sensitive approaches. They seek not only to transfer knowledge but to shape ethically grounded, action-oriented global citizens.

Interdisciplinary and Transdisciplinary Curriculum Models focus on breaking down disciplinary silos to encourage a more integrated understanding of sustainability. Thematic-integrative and interdisciplinary approaches link scientific, economic, and sociocultural dimensions, helping learners analyze challenges through multiple lenses. Transdisciplinary models extend this integration by involving external stakeholders, such as communities, industry and policymakers in the educational process. These collaborative learning environments foster systems thinking and equip learners to co-create solutions grounded in real-world complexity (Holst, 2023).

Competency-Based and Practice-Oriented Models emphasize the acquisition of sustainability-specific competencies that prepare students for real-world action. Competency-based education focuses on learning outcomes such as systems thinking, strategic problem-solving, and change agency (Evans, 2019). These models are often complemented by experiential learning strategies, including fieldwork, service learning and internships, which provide authentic settings for students to apply and refine their skills.

Critical Pedagogy and Transformative Learning Models promote reflection and critical analysis of the structural causes of unsustainability. Rooted in Freirean educational theory, these models challenge learners to question dominant paradigms, power structures and cultural norms that perpetuate environmental and social injustice (Thomas, 2009). Transformative learning emphasizes emotional and ethical development alongside intellectual growth, cultivating learners who are prepared to advocate for systemic change.

Agile and Problem-Based Curriculum Models are designed to remain responsive in the face of evolving sustainability challenges. Agile curriculum models emphasize adaptability, allowing educational content and methods to evolve with societal needs, technological innovations and student interests (López-Alcarria et al., 2019). Problem-Based Learning (PBL) engages students in solving complex, real-life sustainability problems through inquiry-driven collaboration.

The I3E Model (Inform, Engage, Empower) offers a strategic institutional approach to embedding sustainability across university curricula. It begins by informing the university community through awareness campaigns, training programmes, and communications. Engagement follows, with stakeholders, including faculty, students, and staff, actively participating in curriculum development and sustainability initiatives (Cebrián, 2018). Empowerment is achieved by building sustainability competencies and leadership skills through professional development and integrated learning experiences.

The CoDesignS ESD Framework is a design-centered model that emphasizes sustainability integration through co-design processes and life-cycle thinking (Ahmad et al., 2023). It incorporates cognitive, socio-emotional and behavioural learning domains to foster transformative pedagogy. The framework promotes early sustainability considerations in design projects and involves diverse stakeholders, such as designers, engineers, policymakers and communities, in collaborative solution development. It integrates Life Cycle Sustainability Assessment and Environmental, Social, and Governance (ESG) indicators to evaluate long-term impacts. Its modular nature offers flexibility, allowing users to tailor sustainable design strategies to specific project needs.

Together, these curriculum models; ranging from interdisciplinary and competency-based frameworks to strategic institutional and co-design approaches, offer diverse pathways for embedding sustainability in education. By fostering holistic thinking, real-world engagement, and institutional transformation, they help prepare learners to become effective agents of change in pursuit of a more sustainable future.

3.2 Strategies for Embedding Sustainability

3.2.1 Experiential and Project-Based Learning

Experiential and project-based learning (PBL) are powerful strategies in sustainability education, offering students opportunities to apply theoretical knowledge in real-world contexts. Experiential learning, which includes activities such as outdoor education, sustainability audits and eco-clubs, enhances both cognitive and affective learning by promoting active participation and reflection.

PBL, rooted in experiential learning, involves students in complex, authentic projects that connect academic content with practical applications. This method encourages self-regulated learning, collaboration, and accountability, as students take ownership of their learning, solve open-ended problems and present their findings to relevant stakeholders (Goldman et al., 2020). Both experiential learning and PBL emphasize critical thinking, creativity, and practical problem-solving, benefiting disciplines like engineering, vocational studies, and social sciences by linking classroom knowledge with real-life challenges and enhancing teamwork and communication skills (Habib & Nagata, 2020).

Experiential and project-based learning therefore enrich sustainability education by making learning relevant, engaging, and skill-oriented. These methods foster real-world competencies like collaboration, critical thinking, and problem-solving, preparing students for professional environments.

3.2.2 Community Engagement and Partnerships

Community engagement and partnerships are central to embedding sustainability in education by linking classroom learning with real-world challenges. Universities and schools increasingly collaborate with local governments, NGOs, and businesses to create authentic learning experiences that address environmental and social issues. These partnerships enrich curricula and highlight the societal relevance of sustainability education.

Local-global linkages are key to sustainability innovation, especially within multinational corporations, where subsidiaries drive local adaptation and global innovation. By customizing solutions, developing technologies, and creating scalable strategies, subsidiaries enable knowledge exchange across global networks, enhancing learning and cross-border collaboration (Tippmann et al., 2012). Addressing sustainability challenges often requires joint efforts from local and global actors, with effective solutions emerging from shared strategies adaptable to various contexts. This empowers local stakeholders to shape broader outcomes (Brown et al., 2021). In product development, strong local-global ties foster innovation as subsidiaries balance local insights with global ecosystems, sustaining innovation. By combining local knowledge with global frameworks, institutions and communities can co-create solutions to complex sustainability issues.

3.2.3 Institutional Commitment and Whole-School Approaches (WSAs)

Institutional commitment is essential for embedding sustainability in education, and WSAs integrate it across governance, curriculum, operations, and culture. This systemic method ensures lasting impact by aligning formal and informal learning environments. WSAs are grounded in five core principles including coherence, continuous learning, participation, responsibility, and long-term commitment, that drive actions in governance, curriculum, campus operations, and community engagement, all supported by a sustainability-focused culture (Holst, 2023).

Institutional leadership plays a critical role, with strategic tools like sustainability charters linked to more effective implementation. Clear short-, medium-, and long-term plans are needed for system-wide integration (Lozano et al., 2015). Yet, policy gaps often obstruct progress, calling for systems thinking and active stakeholder engagement. Professional development is vital, as educators in committed institutions are more likely to undergo training and model sustainable behaviours (Sasaki et al., 2024).

Long-term success depends on continuous learning, stakeholder involvement, and cultural change. Embedding sustainability requires firm institutional commitment and WSAs with integrated governance, learning environments, frameworks, and engagement strategies.

3.3 *International Case Studies*

3.3.1 Eco-Schools Programme

The Eco-Schools Programme is a globally recognized initiative advancing whole-school sustainability through a structured, participatory framework centred on student leadership, environmental literacy, and institutional transformation. It was developed by the Foundation for Environmental Education (FEE) and integrates sustainability into school operations, curricula and community engagement by empowering students to lead environmental action and decision-making (Schröder et al., 2020). Central to the programme is the "Seven Steps Methodology," a cyclical model beginning with the formation of an Eco-Committee to coordinate activities, followed by an Environmental Review to assess the school's ecological footprint. This informs an Action Plan, which is then implemented, monitored, and evaluated. Sustainability education is further embedded through curriculum integration, community outreach, and adoption of an Eco-Code to guide long-term commitment. The programme fosters shared responsibility by involving students, teachers, staff, and external stakeholders, encouraging school-wide reform and systemic thinking (Andreou, 2020) (Fig. 1).

In Ireland, the Eco-Schools initiative evolved into the Green-Campus Programme (GCP), tailored for higher education and large institutions. The GCP has been adopted by all Irish universities and over half of the country's Institutes of Technology. These

Fig. 1 The seven steps methodology of the eco-schools programme

institutions commit to ongoing environmental improvement via student-led committees, audits, and community engagement. Notably, University College Cork and Cork University Hospital have embedded the programme into their strategic frameworks, serving as benchmarks for sustainable campus operations. Success is attributed to collaboration with government agencies, NGOs, and a structured verification process that ensures accountability and measurable impact (Ryan-Fogarty et al., 2016).

Portugal also offers strong examples of Eco-Schools integration in higher education. At the Polytechnic Institute of Coimbra (IPC), the programme involved ecological footprint assessments and behaviour change campaigns, becoming central to IPC's culture despite ongoing challenges in ecological literacy (Leal et al., 2021). Similarly, the Instituto Politécnico de Beja (IPBeja) adopted environmental audits and a campus-wide sustainability policy via its Eco Green Campus initiative, engaging all sectors in sustainability (Pardal et al., 2020). A broader review of Portuguese HEIs found that even newly participating institutions achieved meaningful outcomes (Sousa, 2022). These cases show that with structured frameworks and strong stakeholder engagement, the Eco-Schools model can drive transformative sustainability in higher education. However, sustaining momentum remains difficult due to limited resources, staff turnover, and a tendency toward superficial engagement with tasks like recycling, rather than addressing deeper issues such as consumerism or ecological justice.

3.3.2 Green Schools Initiative (GSI)

The GSI, originating in the United States, offers a holistic model for embedding sustainability in education by integrating health, environmental education, and resource efficiency into school operations, curricula, and community engagement. It promotes healthy, sustainable, and environmentally responsible schools through its "Four Principles" framework (Fig. 2), which guides greening efforts. The integration of traditional knowledge and local governance, as seen in Fiji's Green School programme, has expanded GSI's global impact, addressing food insecurity and indoor air pollution while preserving indigenous heritage and empowering communities, especially women (Katz et al., 2024).

Environmental education remains central to GSI's transformative approach. Hands-on, project-based learning including recycling, composting, and gardening, develops critical thinking and deepens students' understanding of sustainability. Khofi (2024) highlights how such experiential education fosters environmental literacy and pro-environmental behaviour from an early age. International case studies reinforce this. In Bulgaria, Afrikanov (2024) shows how primary teachers embedded sustainability themes through activities that increased engagement and ecological awareness. Similarly, a large-scale evaluation of Beijing's Green School Programme emphasized stakeholder collaboration, policy alignment, and innovative assessments as key to scaling sustainability education across 98 schools (Yuan et al., 2024).

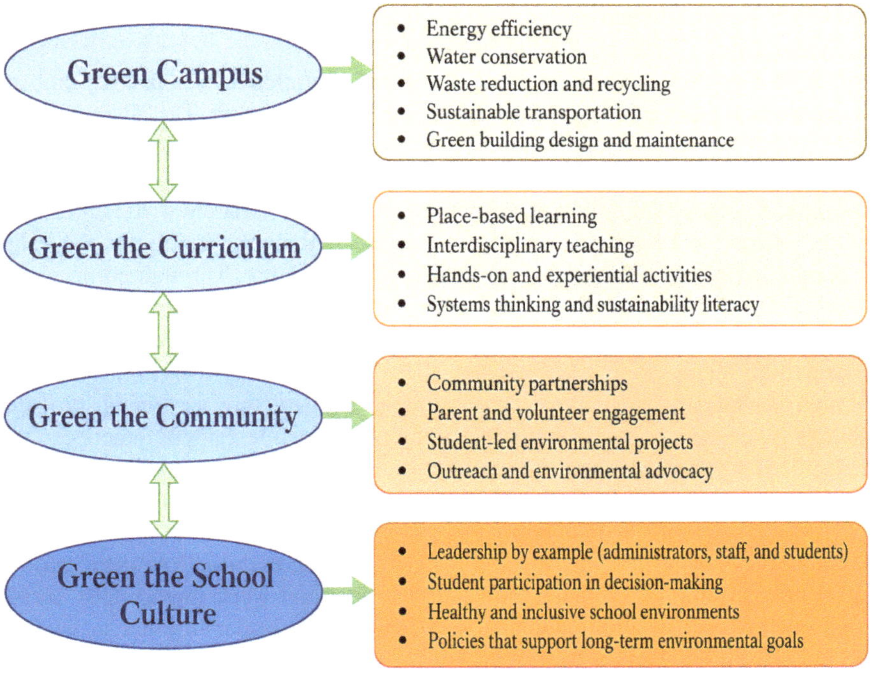

Fig. 2 Major components of the green schools initiative

Resource efficiency is another cornerstone. Schools aim to reduce their ecological footprint via sustainable materials, energy and water conservation, and waste reduction. These efforts not only improve operations but also offer rich, real-world learning opportunities. Denan et al., (2017) show that in Malaysia, Indonesia, and Thailand, successful green school transformations require institutional support, investment, and active participation from teachers and communities.

These examples show that the Green Schools Initiative is a scalable, impactful model for sustainability in education. Anchored in health, environmental literacy, and operational reform, it prepares students to be informed, responsible, and proactive global citizens.

3.3.3 University of British Columbia (UBC)

UBC leads in advancing sustainability in higher education through its comprehensive University Sustainability Initiative (USI). UBC's strategy goes beyond environmental efforts, integrating interdisciplinary academics, applied research, and faculty-student partnerships to foster institutional and cultural change. Signature initiatives like SEEDS, the Sustainability Programme, Sustainability Ambassadors, and Campus as a Living Lab (CLL) actively engage students, staff, and stakeholders in real-world problem-solving (Teslenko, 2019; Pilon et al., 2020).

UBC's distinctive integration of academic and operational sustainability embeds environmental goals into institutional planning and campus life. This dual approach reflects broader trends, such as the University of Pennsylvania's ISAC Programme, which encourages faculty-student collaboration to develop sustainability-focused courses and a shared commitment to environmental action (Dmochowski et al., 2016).

The Master of Engineering in Clean Energy Engineering highlights the value of combining academic learning with professional training. Through co-op placements with BC Hydro, FortisBC, and others, students gain hands-on experience in emissions reduction and energy efficiency, boosting employability while advancing sustainability goals (Reilly & Teslenko, 2015). Sustainability learning also occurs through the "hidden curriculum," including green infrastructure, sustainable procurement (e.g., eco-conscious promotional items), and visible campus practices (Wu et al., 2015). UBC also embraces digital innovation, incorporating gamified learning, data-driven tools, and Education 4.0 to enhance interactivity and equip students for complex sustainability challenges (Shenkoya & Kim, 2023). Its model illustrates how sustainability can be embedded across university structures through strategic coordination, interdisciplinary collaboration, and a focus on real-world impact. UBC serves as a benchmark for institutions preparing graduates for leadership in a sustainable and just global society.

3.3.4 Green School Bali

Green School Bali offers a radical reimagining of education, where learning occurs in open-air bamboo classrooms surrounded by nature. The model integrates environmental science, entrepreneurship, and cultural studies, creating a living laboratory for sustainability. It embodies a holistic approach that merges environmental awareness, sustainable architecture, and community engagement into a learning experience deeply rooted in sustainability.

As a non-governmental organization, Green School Bali not only educates the next generation but also raises global awareness of environmental issues. It engages local and international communities by fostering an educational setting that prioritizes sustainability. Through environmental stewardship, counselling, and partnerships with various organizations, the school promotes sustainable lifestyles and amplifies global awareness of sustainability systems (Utama, 2020).

A defining aspect of the school is its commitment to sustainable architecture. Buildings are primarily constructed from bamboo, a renewable resource, and include energy-efficient features such as natural lighting and ventilation. This approach reduces the environmental footprint and enhances learning by aligning the built environment with nature. The campus meets green building standards, emphasizing energy and water conservation, sustainable materials, and high indoor air quality (Astuti & Shania, 2024).

The curriculum blends environmental science, entrepreneurship, and cultural studies. Students participate in hands-on, real-world projects that connect academic learning with practical applications. Globally recognized as a model of Education for Sustainable Development (ESD), the school encourages permaculture practices and fosters deep connections to nature through its innovative design (Iwan et al., 2018).

Green School Bali also pioneers sustainable energy management, operating on a hybrid of grid and photovoltaic power supported by an IoT-based monitoring system. This system optimizes energy use and serves as a learning tool, allowing students to engage with real-time data and build skills in sustainable technology. By integrating green building practices, a nature-immersive curriculum, and active community engagement, the school equips students with the knowledge and skills to shape a sustainable future. Its innovative model continues to inspire institutions seeking to embed sustainability into education.

3.3.5 University of Vermont (UVM) Case Study

UVM exemplifies a strategic, holistic approach to campus sustainability, integrating curriculum development, faculty engagement, and operational transformation. A major milestone was the adoption of a university-wide sustainability general education requirement, placing UVM among the few research universities with such a mandate (Hill & Wang, 2018). This was supported by the Sustainability Faculty Fellows programme, which advanced interdisciplinary teaching and faculty leadership in sustainability. The Office of Sustainability also engaged students in applied research and surveys, revealing strong environmental values but underscoring the need for deeper institutional commitment beyond promotional messaging (Conner et al., 2018).

This whole-institution approach embeds sustainability not only in academics but also across operations spanning from energy use to waste management, ensuring consistency and relevance for the campus community (Weiss et al., 2021). A key to UVM's success has been its collaborative ethos, involving faculty, students, administrators, and community members in shaping goals. This inclusive process fostered shared ownership and accountability, critical for durable curriculum reform.

Nonetheless, challenges remain. Institutional inertia and the lack of dedicated policies continue to impede curricular progress. UVM's case shows that overcoming such barriers requires strong leadership, strategic alignment, and long-term policy integration. While UVM offers a replicable model for embedding sustainability through multi-stakeholder engagement, lasting change ultimately hinges on sustained institutional commitment and policy support.

3.3.6 Comparative Analysis

Table 1 presents a comparative analysis of five leading sustainability education programmes. The Eco-Schools Programme and GSI focus on K–12 education, with

Eco-Schools expanding into tertiary settings via participatory audits and student-led efforts, while GSI emphasizes health, traditional knowledge, and community involvement. In contrast, UBC and UVM operate in higher education. UBC integrates sustainability into research, operations, and digital tools, while UVM focuses on curriculum reform and faculty development. Green School Bali offers an alternative K–12 model, immersing students in sustainability through architecture, permaculture, and IoT-based learning. Although all programmes promote environmental responsibility, their approaches vary from institutional governance to hands-on learning, underscoring the need for contextual adaptation and systemic integration in sustainability education.

3.4 Impacts and Outcomes

3.4.1 Student Learning and Behaviour Change

Sustainability curricula are essential for building environmental literacy, critical thinking, and civic responsibility in students. These programmes embed environmental issues into education, enhancing knowledge while developing skills like problem-solving, ethical reasoning, and analysis of complex challenges. By combining theory with real-world experience, sustainability education deepens understanding of ecological systems and their societal impacts. Experiential learning has been shown to increase students' willingness to act sustainably and strengthens their belief in making a meaningful difference (Colding, 2020). Critical thinking empowers students to assess environmental issues, consider diverse perspectives, and make informed, sustainability-oriented decisions.

Civic responsibility is another core outcome, fostering active engagement in community issues. By embedding social responsibility into curricula, these programmes cultivate the skills and mindsets needed for environmental and social action. Such involvement enhances students' commitment to collective well-being and their desire to address environmental and social challenges (Ardoin et al., 2023). Thus, sustainability curricula that emphasize literacy, critical thinking, and civic duty prepare students to become informed, responsible advocates for a more sustainable and equitable future.

3.4.2 Institutional Transformation

Adopting sustainability as a core educational value drives institutional transformation, marked by shifts in culture, policy, and stakeholder engagement. This begins with a cultural shift where sustainability becomes central to the institution's mission, emphasizing social justice, environmental stewardship, and critical thinking. Faculty commitment and student involvement are key to fostering this change, encouraging

Table 1 Comparative analysis of international sustainability education programmes by institutional level, focus, engagement strategies, and implementation features

Theme	Eco-schools	Green schools initiative	UBC	UVM	Green school Bali
Level	K–12 & tertiary	Primarily K–12	Higher Education	Higher Education	K–12 (alternative)
Core focus	Whole-school reform	Health & environment	Interdisciplinary academia and operations	Curriculum and governance	Holistic, place-based learning
Student role	Leadership in audits and planning	Project-based learning	Co-researchers and implementers	Research and feedback agents	Sustainability as lifestyle
Technology integration	Moderate	Low–Moderate	High (IoT, gamification)	Moderate	High (energy monitors)
Strengths	Scalable, participatory	Culturally adaptable, health-centred	Research-practice integration	Governance-driven reform	Architecture-education synergy
Challenges	Surface-level engagement risk	Policy dependency	Maintaining cross-departmental cohesion	Institutional inertia	Resource intensity, niche scope

collaboration and reflection on shared values. The result is an institutional ethos that supports sustainable practices on campus and in the broader community.

Policy evolution is also essential for embedding sustainability into governance and operations. Principles of sustainability should guide decisions across curriculum design, campus management, and community outreach. Strong leadership and effective change management ensure policies are forward-looking, adaptable, and resilient to global challenges. Broad stakeholder involvement, from faculty and students to administrators and external partners, enhances the process by bringing diverse perspectives, accelerating progress, and promoting inclusivity. Institutions that integrate sustainability successfully become living models, practicing and promoting sustainable values while contributing to global goals like the SDGs.

3.4.3 Societal and Environmental Contributions

Integrating sustainability principles into curricula and institutional strategies equips students to tackle global challenges and advance the SDGs. By aligning education with values like resilience, alternative economies, and collaboration, institutions foster the skills and mindsets needed for long-term societal transformation, empowering students to become proactive agents of change (Kioupi & Voulvoulis, 2020).

Beyond theory, many institutions implement practical frameworks that engage students in environmental projects such as renewable energy, recycling, and local restoration. These activities build environmental awareness and equip students with the skills to lead sustainability initiatives (Altassan, 2023). Environmental education that emphasizes civic leadership and community engagement also enhances social-ecological resilience, preparing students to address local challenges and instilling a sense of shared responsibility (Maharramli et al., 2021). By fostering sustainability leadership, institutions influence both campus and community practices, embedding sustainability as a core societal value.

4 Challenges and Limitations

A major barrier to educational innovation is institutional resistance, often stemming from rigid administrative structures and unsupportive leadership. This resistance impedes the adoption of new practices and hinders the development of an innovative culture in schools and universities. The lack of clear policies or frameworks further complicates efforts, leaving educators uncertain about how to implement and sustain new approaches (Schophuizen & Kalz, 2020; Valdés Sánchez & Gutiérrez-Esteban, 2023).

Resource constraints including limited funding, materials, and technology, also challenge innovation. With reduced government and research funding, many higher education institutions struggle to invest in innovative solutions or staff development

(Sharma & Sharma, 2021). Teacher preparedness is another key issue, as many educators feel unready to adopt new pedagogical roles or technologies. Inadequate training and support often leave them without the skills or confidence to implement innovations effectively (Kunnari & Ilomäki, 2016).

Cultural and contextual factors add to these challenges. Traditional teaching norms and entrenched assessment systems can generate resistance among both educators and students. Misalignment between institutional culture and the collaborative mindset needed for innovation further inhibits progress. Addressing these barriers requires learner-centered approaches, supportive governance, and systemic efforts that promote experimentation and collaboration.

Sustained innovation depends on systemic support, continuous professional development, and aligned policies. Strong leadership and coordinated top-down strategies are essential for building institution-wide systems that foster innovation. By promoting a culture of learning and adaptability, institutions can more effectively navigate complexity and achieve meaningful change (Meng, 2024).

5 Recommendations

To foster sustainability mindsets, educators should use experiential and participatory pedagogies such as design thinking, real-world problem-solving, and contemplative art projects, engaging students with sustainability challenges and building creative confidence and literacy. Combining transformative learning with design thinking promotes collaboration and prepares students to become changemakers. Reflective practices like self-awareness and critical reflection help students internalize sustainability values, strengthen leadership, and encourage pro-environmental behaviours (Yang et al., 2021).

Integrating sustainability themes across disciplines is vital for developing systems thinking and a holistic understanding of the SDGs. Interdisciplinary pedagogies enhance problem-solving and encourage student ownership of learning, nurturing new mindsets and behaviours (Alm et al., 2021). Universal pedagogies, including collaborative projects, community-based learning, and agile methods, support adaptability, creativity, and systemic thinking, motivating pro-environmental actions (Tran, 2024). Mapping pedagogical methods to sustainability competences ensures students gain future leadership skills.

Curriculum designers should create integrative, thematic frameworks linking core subjects through real-world themes, improving critical thinking, creativity, collaboration, and communication, while boosting motivation (Hakim, 2017). Assessment of integrative and systems thinking remains a challenge, but tools like the Interdisciplinary Writing Assessment Profiles and the Integrative Learning VALUE (Valid Assessment of Learning in Undergraduate Education) rubric help evaluate critical argumentation and reflective self-assessment, ensuring that students develop critical and systems thinking skills (Ortega et al., 2017). Student-centered approaches, including digital simulations and peer learning, promote engagement and conceptual change, while professional development resources for educators ensure effective curriculum implementation (Santos et al., 2019).

Policymakers play a crucial role by creating supportive policies and national strategies aligned with the SDGs, particularly SDG 4, to ensure inclusivity and long-term educational success, especially for marginalized communities. Collaborative funding mechanisms and technology-driven resource management can address funding inequities and ensure financial sustainability for teacher capacity building and institutional development (Festus et al., 2019). Investment in teacher training and professional development is essential for achieving quality education and sustainable development. Policymakers should prioritize cross-sector collaboration to promote innovation, drive system-wide reform, and develop alternative funding models that support ESD.

6 Conclusion

This chapter explored how sustainability can be meaningfully embedded in curriculum design and educational practice within the broader context of the United Nations Sustainable Development Goals, particularly SDG 4 on quality education. Through a qualitative, interpretive methodology based on literature and global case study analysis, the study identified a diverse range of curriculum models, pedagogical strategies, and institutional frameworks that support sustainability integration.

The findings reveal that effective sustainability education requires interdisciplinary and practice-oriented curriculum models, supported by experiential learning, community engagement, and whole-institution approaches. International case studies demonstrate the transformative potential of initiatives like Eco-Schools, the Green Schools Initiative, and university-led programmes at UBC and UVM, each offering context-specific pathways to embed sustainability across levels and regions.

These insights have important implications for curriculum designers, educators, and policymakers. A systemic and collaborative approach which is grounded in adaptive governance, inclusive pedagogies, and strategic planning, is essential for scaling sustainability education. Moreover, strong institutional leadership, professional development, and coherent policy frameworks are critical enablers of sustainable education reform.

Looking ahead, future efforts should focus on longitudinal research to evaluate implementation impacts, especially in underrepresented regions, and develop robust assessment tools that capture sustainability competencies. There is also a pressing need for increased investment in educator training and for strengthening global-local partnerships that can foster innovation and equity in education for sustainable development.

References

Afrikanov, L. (2024). Teaching for a sustainable future: Case studies from the green schools project in Bulgaria. *Education in Natural Sciences and Advanced Technologies, 33*(3–4), 219–226.

Ahmad, N., Toro-Troconis, M., Ibahrine, M., Armour, R., Tait, V., Reedy, K., Malevicius, R., Dale, V., Tasler, N., & Inzolia, Y. (2023). CoDesignS Education for Sustainable Development: A Framework for Embedding Education for Sustainable Development in Curriculum Design. *Sustainability, 15*(23), 16460.

Alm, K., Melén, M., & Aggestam-Pontoppidan, C. (2021). Advancing SDG competencies in higher education: exploring an interdisciplinary pedagogical approach. *International Journal of Sustainability in Higher Education, 22*(6), 1450–1466.

Altassan, A. (2023). Sustainable Integration of Solar Energy Behavior Change, and Recycling Practices in Educational Institutions: A Holistic Framework for Environmental Conservation and Quality Education. *Sustainability, 15*(20), 15157.

Andreou, N. (2020). Towards a Generation of Sustainability Leaders: Eco-Schools as a Global Green Schools Movement for Transformative Education. In *Green Schools Globally* (pp. 31–45). Springer.

Ardoin, N. M., Bowers, A. W., & Gaillard, E. (2023). A systematic mixed studies review of civic engagement outcomes in environmental education. *Environ. Educ. Res.*. Accessed 5 May 2025. https://doi.org/10.1080/13504622.2022.2135688

Astuti, F., & Shania, N. (2024). Penerapan Prinsip Arsitektur Berkelanjutan Pada green school bali: studi kasus penggunaan material alami [application of sustainable architecture principles at green school bali: a case study of the use of natural materials]. *Venus: Jurnal Publikasi Rumpun Ilmu Teknik, 2*(6), 01–17.

Beasy, K., Smith, C., & Watson, J. (Eds.). (2023). *Education and the UN Sustainable Development Goals: Praxis Within and Beyond the Classroom.* Springer Nature.

Brown, R., Rossi, F., Solovey, K., Tsao, M., Wolf, M. T., & Pavone, M. (2021). On Local Computation for Network-Structured Convex Optimization in Multiagent Systems. *IEEE Transactions on Control of Network Systems, 8*(2), 542–554.

Cebrián, G. (2018). The I3E model for embedding education for sustainability within higher education institutions. *Environmental Education Research, 24*(2), 153–171.

Colding, J. (2020). Emerging ideas in Sustainability Science: Critical thinking and lifelong environmental learning, report. *Stockholm University.* https://doi.org/10.17045/sthlmuni.11590929.v1

Conner, D., Falkner, A., Lantieri, N., McGavisk, B., & McShea, B. (2018). Stakeholder Perceptions of Campus Sustainability Efforts: Lessons from Vermont. *Sustainability, 10*(11), 3849.

Denan, Z., Awang, A. H., Mazlan, M. A. H., Majid, N. H. A., Rahim, Z. A., & Sanusi, N. A. Z. (2017). The Implementation of Environmental Education and Green Programmes in Schools to Achieve Sustainability. *Advanced Science Letters, 23*(7), 6261–6265.

Dmochowski, J. E., Garofalo, D., Fisher, S., Greene, A., & Gambogi, D. (2016). Integrating sustainability across the university curriculum. *International Journal of Sustainability in Higher Education, 17*(5), 652–670.

Evans, T. L. (2019). Competencies and Pedagogies for Sustainability Education: A Roadmap for Sustainability Studies Programme Development in Colleges and Universities. *Sustainability, 11*(19), 5526.

Festus, A. B., Ifeoma, U. K., Orobosa, O. S., & Nancy, O. O. (2019). Sustainable Policy Options for Funding Capacity Building of Teachers for the Achievement of Education 2030 in Nigeria. *Journal of Economics and Sustainable Development, 10*(4), 80.

Goldman, J., Kuper, A., Baker, G. R., Bulmer, B., Coffey, M., Jeffs, L., Shea, C., Whitehead, C., Shojania, K. G., & Wong, B. (2020). Experiential Learning in Project-Based Quality Improvement Education: Questioning Assumptions and Identifying Future Directions. *Academic Medicine, 95*(11), 1745.

Habib M K & Nagata F (2020). Mechatronics: Experiential Education and Project Based Learning, *In 2020 21st International Conference on Research and Education in Mechatronics (REM)*, (pp. 1–6). https://doi.org/10.1109/REM49740.2020.9313073.

Hakim, S. (2017). Implementation Impact of Integrative Thematic Learning for Developing the Student and Teacher's Competencies in The Islamic Elementary School (Madrasah Ibtidaiyah) in Yogyakarta Special Region., Sunan Kalijaga. *International Journal on Islamic Educational Research, 1*(1), 83–109.

Hays, J., & Reinders, H. (2020). Sustainable learning and education: A curriculum for the future. *International Review of Education, 66*(1), 29–52.

Hill, L. M., & Wang, D. (2018). Integrating sustainability learning outcomes into a university curriculum: A case study of institutional dynamics. *International Journal of Sustainability in Higher Education, 19*(4), 699–720.

Holst, J. (2023). Towards coherence on sustainability in education: a systematic review of Whole Institution Approaches. *Sustainability Science, 18*(2), 1015–1030.

Iwan, A., Rao, N., & Poon, K. K. Y. (2018). Characteristics of Green Schools: Observations of Award-winning Green Preschools in Bali, Berkeley and Hong Kong. *Journal of Education for Sustainable Development, 12*(2), 140–159.

Katz, S., Matairakula, U., Cinavilakeba, J., Dradra, T., Carter, R., Tikoibua, T., & Mitchell, B. G. (2024). Cultivating Wellbeing: Traditional Wisdom and Sustainability in Fiji's Green Schools. *Proceedings of the Nutrition Society, 83*(OCE1), E7.

Khofi, M. B. (2024). The Green School Concept in Elementary Schools as an Effort to Form Sustainable Behavior and Environmental Awareness, Al-Adzka : Jurnal Ilmiah Pendidikan Guru Madrasah Ibtidaiyah, *14*(2), 206–225.

Kioupi, V., & Voulvoulis, N. (2020). Sustainable Development Goals (SDGs): Assessing the Contribution of Higher Education Programmes. *Sustainability, 12*(17), 6701.

Kunnari, I., & Ilomäki, L. (2016). Reframing teachers' work for educational innovation. *Innovations in Education and Teaching International, 53*(2), 167–178.

Kushwah, L., Navrouzoglou, P., & Cheng, W. (2024). Embedding Sustainability in Existing Curriculum by Using Authentic Assessments, Rubrics, and Formative Assessments: A Case Study. In T. Wall et al. (Eds.), *Sustainability in Business Education, Research and Practices* (pp. 241–254). Cham.

Leal, Á., Simões, H., Figueiredo, J., Loureiro, A., Seco, S., & Ferreira, A. (2021). Implementation of the Eco-Schools Programme in Higher Education - A Path to Environmental Sustainability at IPC. In *In Proceedings of ICoWEFS 2021. Cham: Springer International Publishing* (pp. 516–525). https://doi.org/10.1007/978-3-030-75315-3_56

López-Alcarria, A., Olivares-Vicente, A., & Poza-Vilches, F. (2019). A Systematic Review of the Use of Agile Methodologies in Education to Foster Sustainability Competencies. *Sustainability, 11*(10), 2915.

Lozano, R., Ceulemans, K., Alonso-Almeida, M., Huisingh, D., Lozano, F. J., Waas, T., Lambrechts, W., Lukman, R., & Hugé, J. (2015). A review of commitment and implementation of sustainable development in higher education: results from a worldwide survey. *Journal of Cleaner Production, 108*, 1–18.

Maharramli, B., Bredow, V. L., & Goodwin, L. (2021). Using civic ecology education to foster social-ecological resilience: A case study from Southern California. *The Journal of Environmental Education, 52*(6), 445–462.

Meng, L. (2024). The Changes and Challenges of Educational Technology Innovation on the Role of Teachers. *Advances in Educational Technology and Psychology, 8*(5), 180–186.

Ortega P E, Lagoudas M Z and Froyd J E (2017). Overview and Comparison of Assessment Tools for Integrative Thinking, In *2017 ASEE Annual Conference & Exposition*. https://peer.asee.org/overview-and-comparison-of-assessment-tools-for-integrative-thinking (Accessed 5 May 2025).

Pardal A, Romeira T and Durão A (2020). Eco Green Campus: Challenges and Opportunities. The Study Case of Polytechnic Institute of Beja, E3S Web of Conferences, 171, p. 01010.

Pilon, A., Madden, J., Tansey, J., & Metras, J. (2020). Campus as a Living Lab: Creating a Culture of Research and Learning in Sustainable Development. In *Teaching and Learning Strategies for Sustainable Development* (pp. 213–227). Emerald Publishing Limited.

Reilly, J., & Teslenko, T. (2015). Enhancing ESD Through the Master of Clean Energy Engineering Co-op Programme: A Canadian Case-Study. In W. Leal Filho et al. (Eds.), *Integrative Approaches to Sustainable Development at University Level* (pp. 511–523). Springer.

Ryan-Fogarty, Y., O'Carroll, D., O'Mahony, M. J., & O'Regan, B. (2016). Development of the Green-Campus Programme in Ireland: Ensuring Continuity of Environmental Education and Action for Sustainable Development. In W. Leal Filho & P. Pace (Eds.), *Teaching Education for Sustainable Development at University Level* (pp. 269–284). Springer.

Santos, J., Figueiredo, A. S., & Vieira, M. (2019). Innovative pedagogical practices in higher education: An integrative literature review. *Nurse Education Today, 72*, 12–17.

Sasaki, O., Yonehara, A., & Kitamura, Y. (2024). The influence of the whole school approach on implementing education for sustainable development in Japan. *PROSPECTS, 54*(1), 203–220.

Schophuizen, M., & Kalz, M. (2020). Educational innovation projects in Dutch higher education: bottom-up contextual coping to deal with organizational challenges. *International Journal of Educational Technology in Higher Education, 17*(1), 36.

Schröder L-M U, Wals A E J and van Koppen C S A (Kris) (2020). Analysing the state of student participation in two Eco-Schools using Engeström's Second Generation Activity Systems Model, *Environmental Education Research, 26*(8), 1088–1111.

Sharma, M. K., & Sharma, R. C. (2021). Innovation Framework for Excellence in Higher Education Institutions. *Global Journal of Flexible Systems Management, 22*(2), 141–155.

Shenkoya, T., & Kim, E. (2023). Sustainability in Higher Education: Digital Transformation of the Fourth Industrial Revolution and Its Impact on Open Knowledge. *Sustainability, 15*(3), 2473.

Sousa, S. (2022). Some Reflections on the Role of the Eco-Schools Programme in the Promotion of Sustainable HEIs: A Case Study in Portugal. *Administrative Sciences, 12*(4), 149.

Teslenko, T. (2019). Engaging Students and Campus Community in Sustainability Activities in a Major Canadian University. In W. Leal Filho & U. Bardi (Eds.), *Sustainability on University Campuses: Learning, Skills Building and Best Practices* (pp. 3–20). Springer International Publishing.

Thomas, I. (2009). Critical Thinking, Transformative Learning, Sustainable Education, and Problem-Based Learning in Universities. *Journal of Transformative Education, 7*(3), 245–264.

Tippmann, E., Scott, P. S., & Mangematin, V. (2012). Problem solving in MNCs: How local and global solutions are (and are not) created. *Journal of International Business Studies, 43*(8), 746–771.

Tran, M. T. (2024). Fostering sustainable mindsets: A critical exploration of educational psychology in business education. *The International Journal of Management Education, 22*(3), 101054.

Utama, D. N. (2020). The role of non-governmental organization in creating awareness of global environmental issues: discourse analysis of the 2016/2017 green school annual report. *JARES (Journal of Academic Research and Sciences), 5*(2), 9–20.

Valdés Sánchez, V., & Gutiérrez-Esteban, P. (2023). Challenges and enablers in the advancement of educational innovation. The forces at work in the transformation of education. *Teaching and Teacher Education, 135*, 104359.

Weiss, M., Barth, M., & von Wehrden, H. (2021). The patterns of curriculum change processes that embed sustainability in higher education institutions. *Sustainability Science, 16*(5), 1579–1593.

Wu S, Liu M, Kung A, Naghavi P and Lee J (2015). *An investigation into UBC sustainable swag.* https://open.library.ubc.ca/collections/18861/items/1.0108868.

Yang, C., Ivanova, E., & Hufnagel, J. (2021). Using contemplative photography in transformative sustainability management education: Pedagogical applications in the United States Russia, and Germany. *The International Journal of Management Education, 19*(3), 100568.

Yuan X, Yu L, Li X, Luo J, She H, Yang J, Lin Z, Zhao J, Zhao F, Meng R, Zhang Y & Zhao Y (2024). Promoting education for sustainable development through the green school Programme to achieve SDGs: Insights from a case study of Beijing's primary and secondary schools, *International Journal ofChinese Education, 13*(2). 2212585X241259192

Kudzanayi Chiteka is a researcher and educator with a strong background in renewable energy, sustainability, and engineering education. He holds a PhD in Engineering specialising in solar energy, an MSc in Renewable Energy, an MTech in Production Engineering, and a BTech in Industrial and Manufacturing Engineering. Dr Chiteka's research interests focus on solar energy systems and sustainable development. He has served in various academic and leadership roles in university settings. Currently he is a Post-Doctoral Research Fellow at the University of South Africa, and he actively contributes to advancing renewable energy solutions and fostering capacity building in higher education. He is passionate about promoting green technologies, interdisciplinary collaboration, and practical pathways for sustainable development in Africa and beyond.

C.C Enweremadu is a distinguished scholar in the Department of Mechanical, Bioresources and Biomedical Engineering at the University of South Africa (UNISA). He has a strong academic and research footprint, authoring over 100 peer-reviewed journal articles focusing on renewable and sustainable energy. He leads cutting-edge studies in renewable fuels, and in his teaching and leadership roles within UNISA's School of Engineering and Built Environment, he mentors postgraduate students and post-doctoral research fellows and fosters collaborations that bridge theoretical innovation and industrial applications. His work contributes significantly to Africa's sustainable energy transformation, marrying rigorous research with practical impacts in bioresource engineering.

Integrating Sustainability Teaching in Higher Education in South Africa: Critical Reflections

Florah Sewela Modiba

Abstract Educating university students about sustainable development ensures they understand how to support efforts to meet Sustainable Development Goals (SDGs) while becoming environmental stewards. However, integrating the SDGs into a curriculum that has not outlined these goals is a challenge due to the lack of resources to implement programmes and planning, which is compounded by the academic pressures educators face. This chapter aims to explore how to incorporate SDGs in a preexisting curriculum in efforts to equip students with sustainability skills. A scoping review, combined with critical reflective narratives, is used to examine sustainability teaching from an educator's perspective in the Development Studies discipline, guided by transformative learning theory. The challenges and opportunities are presented on how amending lesson outcomes can be used to ensure sustainable teaching. This approach demonstrates how educators across various disciplines can adapt these integration strategies to their specific contexts, enabling interdisciplinary collaboration in curriculum redesign and sustainability education. Reflexive narratives demonstrate that incorporating sustainability goals into the assessment process enhances understanding of these goals and their potential implementation across various aspects of people's lives. The results reveal limited studies on sustainability integration in South African higher education, with reported implementations concentrated mainly in Geography, Environmental Management, and Computing disciplines. The guidelines provided demonstrate transferable principles that can be applied across various disciplines, illustrating how interdisciplinary approaches can enhance sustainability teaching without redesigning the curriculum. It is also noted that course outcomes can be used to include SDGs in the learning outcomes. These findings highlight the need for increased reporting on sustainability teaching in South African universities and for policymakers to ensure adequate institutional capacity for SDGs integration in higher education.

Keywords Sustainability teaching · Higher education · Sustainable Development Goals · Curriculum · South Africa

F. S. Modiba (✉)
Department of Development Studies, Faculty of Business and Economic Sciences, Nelson Mandela University, Gqeberha, South Africa
e-mail: Florah.Modiba@mandela.ac.za

W. Leal Filho et al., *Curriculum Development for Sustainable Development*, World Sustainability Series, https://doi.org/10.1007/978-3-032-22146-9_6

1 Introduction

The integration of sustainable development goals (SDGs) in the curriculum stems from the need to ensure broad participation in meeting the 2030 targets set by the United Nations (2025). Higher education institutions HEIs) have responded by designing curricula that support quality education (SDG4) and sustainable development objectives (Pham Xuan & Håkansson Lindqvist, 2025; Price et al., 2024). Within the existing measurement frameworks, education for sustainability is used in the university rankings as part of the criterion to assess universities' efforts, and Africa is reported to have low participation in the rankings (Leal Filho et al., 2024). Locally, the South African policy framework, the National Development Plan of 2012, has aligned its 2030 policy implementation plan with the UN targets (Republic of South Africa, 2012). As part of reporting, HEIs are required to demonstrate alignment with policy targets in their operations, making sustainability teaching crucial for achieving some targets (Anstey, 2024). Institutions of higher learning's participation in the global SDG ranking and in research that focuses on SDGs shows the commitment of higher education institutions to sustainability practices (Anstey, 2024; Nhamo & Chapungu, 2024; QS World University, 2025). Therefore, embracing sustainability teaching could help improve the reach of some targets.

According to Boarin and Martinez-Molina (2022) there has been a change in the sustainability teaching, having moved from what must be taught to how it should be done, thus requiring a structured approach to implementing sustainability teaching. Nevertheless, teaching students about sustainable development goals (SDGs) in courses such as development studies presents unique challenges. There is often an assumption that the discipline is at the core of addressing certain SDGs, and, therefore, educators and students are generally aware of the implications of not understanding SDGs, particularly regarding climate issues (Tshabalala et al., 2025; United Nations, 2025). This reflective narrative explores the intentional reflexive approach to teaching SDGs in undergraduate classes, examining ways to help students understand the scope of each SDG and how to contribute to national targets within their communities.

The use of relational pedagogies has been cited as instrumental in understanding the intricate dependencies between ecological systems and humans, which can be achieved by including affective approaches (Jones, 2023). The use of affective strategies can be beneficial in understanding the interconnectedness of human behavioural practices that influence climate issues and that of SDGs (Edwards & Küpers, 2024; Price et al., 2024). As argued by Rusaanes et al. (2024), this radical pedagogic approach illustrates the interlinkages between climate change and social, economic and environmental issues. Thus, the approach allows for educators and students to engage deeper with SDGs and practices that are detrimental to the environment within the curriculum framework.

The benefit of sustainability teaching is that it prepares students to understand global challenges and how these concepts affect them at a local level. According to Tshabalala et al. (2025), education for sustainable development (ESD) helps prepare African youth to understand challenges related to climate change and foster sustainability. The authors further emphasised the importance of integrating knowledge, competencies and values into the educational system. This integration can be achieved by understanding climate science, recognising social and economic aspects, and connecting them to broader sustainability goals, thereby promoting climate literacy and justice. Such use of sustainability concepts helps students think of mitigation and adaptation strategies using both local and globally informed strategies. Scholars such as Leal Filho et al. (2018) have argued for the importance of integrating sustainable development into the university curriculum, emphasising the need for educators and students to commit to achieving this goal. Such integration helps to prepare students for global challenges and opportunities.

The teaching of SDGs to support sustainable development practices has been published in various fields. For example, studies have used game learning as a method, which has proved to help students understand SDGs better and improve their problem-solving skills as they attempted simulated games linked to sustainability themes (Ahmadov et al., 2025). Similarly, Rusaanes et al. (2024) have used it in health sciences to explore how nursing science students can use lessons learned in rural communities. Scholars from disciplines such as mathematics have argued the importance of including sustainable development principles in these fields, and ensuring that students gain skills that will enable a sustainable future (Maqoqa, 2025). However, this type of teaching in the development studies discipline is limited in the literature corpus. Buerkle et al. (2023) argue that while HEIs are prioritising inclusion of SDGs in their curricula, there is less attention on the teaching efforts. This chapter aims to showcase a case study of how the curriculum has been redesigned to incorporate sustainability teaching that fosters transformative learning (Boarin & Martinez-Molina, 2022). The use of a transformative approach to teaching and learning helps unpack how educators in other disciplines can embed SDGs within a pre-existing curriculum. Thus, contributing to reducing the gap in sustainability teaching in the African context. This approach can guide educators across disciplines in incorporating sustainable development principles and practices into their classrooms, demonstrating interdisciplinary collaboration opportunities that could enhance curriculum effectiveness even when current course outcomes do not explicitly address sustainability. In doing so, it contributes to the body of knowledge in education for sustainable development and the development discipline by demonstrating how critical reflection on teaching practices and intentional pedagogical approaches can benefit not only SDG knowledge retention but also the development of scholars who will embody this knowledge and apply it in their careers and communities (Edwards & Küpers, 2024; Pham Xuan & Håkansson Lindqvist, 2025). The study further contributes to policy formulation, especially in higher education, so that policymakers can create a supportive environment to ensure that the curriculum accommodates important actions that foster sustainable practices. Therefore, the purpose of this study is to explore how to embed SDGs in an existing curriculum to complement the learning outcomes of a module. In doing so, the following research questions are posed to achieve the aim:

1. How can SDGs be integrated into a pre-existing curriculum?
2. Which approaches can be used to enable educators to incorporate SDGs within the original course objectives?
3. How can guidelines be designed to integrate SDGs into existing module outcomes?

2 Theoretical Framework

The theoretical framework used to underpin this study is based on transformative learning, which Mezirow (1997) defines as a process of effecting change in a frame of reference. According to Kitchenham (2008), transformative learning is at the centre of adult learning, providing a frame of reference for self-reflection and encouraging autonomy and critical thinking. It taps into the core and can potentially lead to thinking that is socially responsible. It further allows students to think of alternatives to problems and to work with others to solve problems (Mezirow, 1997). This quality is particularly relevant in teaching sustainability to young adults at universities, as they are sensitised to the immediate actions required during their degree qualification and the subsequent steps they will take upon entering the business world after graduation. It strengthens their graduate attributes and provides an opportunity to continue finding ways to implement change that contributes to sustainability (Price et al., 2024).

Cranton (2009) takes transformative learning further to include the educators and asserts that, for transformative education to be realised, there is a need to be cognisant of students, content and context of learning. This view corroborates Taylor's (2007) assertion that more studies that focus on how transformation can be fostered are needed. Nevertheless, transformative learning has been criticised by several scholars. Some argue that the term is too bold and refers to change, which might not be the case for students completing a degree (Newman, 2012). However, this current study only views transformation as a process that students and their educators might not observe publicly, but as a change that might be linked to specific changes, be it retention of concept or better application of such (Boarin & Martinez-Molina, 2022). It can also involve making decisions that promote the learning acquired at a certain level, as explained in the results section. While Newman (2012) posits that observed/non-observable changes should rather be termed "good learning", the gravity of the term "transformation" is understood and considered relevant in this study. Figure 1 illustrates the transformation journey, and as it can be assessed drawing from Mezirow (1997), Kitchenham (2008) and Cranton (2009).

Taylor (2007) further argues that transformative learning can be fostered by providing supportive environments in higher education and offering direct and engaging learning environments, which, of course, must recognise the diversity of students. Critical reflexivity, a tiered process of self-reflection aimed at gaining insight to inspire future action, is also adopted in this study to explore how the

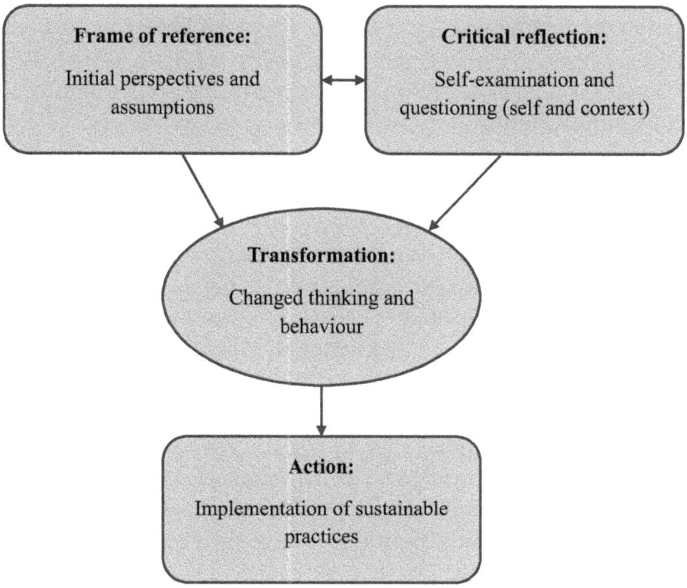

Fig. 1 Transformative learning process (researcher)

teaching of SDGs can be strengthened. This study employed transformative learning to analyse how the teaching of SDGs can be integrated into an existing curriculum and how that transformed students' engagement with issues of sustainable development.

3 Sustainability Education

Sustainable development goals offer a framework of 17 goals that the United Nations (2025) uses to measure progress in various sectors, including social environment, business and governance. This global framework is not only important for helping countries to address their socio-economic challenges, but also inequalities that continue to burden the progress of development, especially in developing countries such as South Africa (Nhamo & Chapungu, 2024). This framework offers an opportunity for alternative pedagogical approaches to help students develop their global citizenship by exploring how local and international behaviours impact the environment, ultimately leading to climate events. Sustainability education stems from SDG 4 on quality education (Pham Xuan & Håkansson Lindqvist, 2025). This goal is further supported by targets that outline the type of education required for students to attain relevant education. Target 4.7 specifies the importance of acquiring the knowledge and skills necessary to build sustainable development in different facets of their lives, thus strengthening the transformative agenda inherent in SDGs (Velempini, 2025). These skills, in the education sector, can be achieved in various ways, including digital literacy. According to Maqoqa (2025), digital competencies are required to capacitate students and educators to navigate challenges related to the fourth industrial revolution, thus contributing to the integration of sustainable practices in higher education.

The study of Leal Filho et al. (2025) has demonstrated how sustainability teaching has evolved, with clusters such as environmental education, SDGs and sustainability competencies, to mention a few. Their findings show varying teaching methods where case studies, research and project-based learning are primarily used as didactic tools for learning. Embedding sustainable development in higher education can, therefore, be achieved in various ways. It could mean adopting practices that help universities or institutions of higher learning align with the SDGs (Buerkle et al., 2023). It could also be achieved by ensuring that the curriculum incorporates sustainable development principles as part of its outcomes. As can be noted in the study of Price et al. (2024), issues like carbon literacy can be tracked institutionally, and embedding sustainable education in the curriculum can help students gain knowledge in sustainable development and acquire relevant skills. Therefore, institutional strategy is key in promoting sustainability education. It is also argued that framing sustainability teaching, particularly issues related to climate change, in a way that enables students in African countries to understand the intergenerational effects of such events and resonate with their context is necessary (Tshabalala et al., 2025). As such, educational strategies are crucial to informing climate resilient strategies (Chineka & Yasukawa, 2020; Tshabalala et al., 2025).

Earlier studies have argued for the need to re-evaluate and redirect sustainability teaching from a dominant Western approach to African-centric methods that are ethical and encompass epistemic justice (Kumalo, 2017). This view proposes an engagement with the environment from a position of understanding that humans are interconnected to the environment, thus mirroring Jones' (2023) narrative when calling for a unified planet that acknowledges collectiveness, opposing Freire's view on the separateness between human and non-human life. Similarly, Kumalo (2017) advocates for decolonial and relational approaches to development cognisant of the existing interconnectedness and not undermining sustainability efforts. Thus, the SDGs framework offers pathways for different countries to work towards aligning targets that will essentially harmonise human development and the planet for mutual thriving (Jones, 2023).

Other curriculum-related studies have focused on the skills that graduates should have. For example, the study conducted by Neboh et al. (2023) highlights the importance of incorporating new required skills into the supply chain management curriculum, including the need for sustainability teaching. However, it also noted that the scope of the study provided limited details on this topic. In disciplines like information systems, limited skill competencies in sustainability were cited as a challenge (Hamadi et al., 2024). While there is appreciation of sustainability-related skills, a gap still exists in the South African context to showcase how SDGs are addressed in classrooms, supporting institutional reporting requirements for the country's SDGs target contributions.

Jones (2023) argues that critical self-reflection and storytelling are core tools in critical pedagogy, given that they help in addressing both social and environmental issues. Critical pedagogies offer alternative pedagogical strategies that meet the complex demands of sustainability education while enhancing critical thinking skills (Hamadi et al., 2024; Serafini et al., 2022). It also helps students to adopt concepts and understand how they resonate with their lived realities, particularly in their local contexts. A study by Volkmann and Fraunhofer (2023) revealed how students were able to connect the sustainability-related content learned to their major modules, and they could align the sustainability principles to courses such as fashion design and nursing. Their findings further highlighted the importance of collaborative learning and allowing more time for students, especially in diverse groups, to discuss more cross-cultural perspectives and how this can transform potential barriers into pedagogical assets, especially in understanding concepts related to climate change.

4 Methodology

This qualitative reflective narrative was combined with a systematic scoping review to explore how other scholars in the South African context approach sustainability teaching. The scoping review is designed to map clusters of studies and summarise current studies on a specific topic (Kitchenham et al., 2012; Segooa et al., 2025b), thus making it suitable for this research. The first phase focused on literature studies guided by the Preferred Reporting Items for Systematic Reviews and Meta-Analysis (PRISMA) to identify relevant papers (Page et al., 2021). A search was conducted using the following keywords: "sustainability teaching," "empirical studies," and "South African universities" in the ScienceDirect, Scopus and Google Scholar databases. Boolean operators "Or" and "And" were applied in the database search to test the search returns to include terms such as sustainable development goals and SDGs in assessing the records returned while searching. However, the inclusion of the terms led to a low return in studies. The use of three databases was to ensure that more studies on this topic can be accessed. The inclusion and exclusion criteria were established to guide paper selection and quality review, as not all articles meeting the inclusion criteria are necessarily high-quality studies for the researched topic. Table 1 outlines the inclusion and exclusion criteria followed.

The search records from ScienceDirect, Google Scholar and Scopus returned 23, 121, and 1 articles, respectively, as shown in Fig. 2. Of the 144 records screened from the two databases, no duplicates were found. The screening led to 11 studies that were eligible for inclusion. However, although the studies focused on South Africa and the topics aligned with the research questions, only five could be reviewed. The exclusion of the seven records that were assessed for certainty (Page et al., 2021) was based on the articles not entirely meeting the inclusion criteria after they were reviewed in detail to extract data. The first set of three records was not peer reviewed, as they were submitted for qualification purposes. The second set consisted of two book chapters published by Springer, which were not accessible due to the costs attached to the

Table 1 Inclusion and exclusion criteria (researcher)

Criterion	Inclusion	Exclusion
Geographical scope	South African studies	Non-South African studies
Language	English language	Non-English language
Sector focus	Higher education/University	K12 or secondary schools
Publication type	Published peer-reviewed journal articles and book chapters	Dissertations, theses, conference proceedings, pre-prints, grey literature
Publication period	Publications between 2015 and 2025	Before 2015
Institutional context	South African universities and higher education institutions	Other types of educational institutions and non-educational contexts

publications. In the last set, one article did not discuss how sustainability teaching was approached, and the other was a literature study that included international primary studies in its review to inform the findings, which would not provide results reflecting experiences in South Africa. The importance of delimiting the studies to the South is due to the institutional, cultural and policy directives that influence the higher educational sector. Other exclusion justifications are illustrated in Fig. 1. Artificial Intelligence tools, such as Claude, were used to assist with screening and selecting studies, guided by the inclusion and exclusion criteria, to minimise bias. It is argued that generative AI tools help support research activities, particularly systematic literature review studies (Schryen et al., 2025; Segooa et al., 2025a). Therefore, the records pulled from Google Scholar were reviewed by both the author and Claude AI to monitor inconsistencies in paper selections. This process helped to identify articles that were included but were not supposed to be part of the reviewed papers.

The second phase of data collection involved presenting the auto-ethnography shared by the lecturer teaching a course in development studies. Auto-ethnography is described as an account of an individual's own life used to explore the extent of their social understanding. According to Sanjakdar and Premier (2023), this qualitative research method can be used in presenting essays or stories to reveal the discovery and recount an epiphany in a character's life. It can, therefore, be used to showcase the culture of higher education in each context. It is, thus, a powerful tool that can be used to further understand the intertwined nature of being and knowing, especially in the process of teaching sustainability in higher education. The study, therefore, uses this method to provide a reflective account of how sustainable development teaching was experienced by the educator, reflecting on the teaching period from 2021 to 2025, for a module adopted in 2020. It highlights how this process can be improved through a critical reflection of what is being taught in class and how to change the teaching practice within a predefined curriculum to embed SDGs. As noted by Bofylatos and Perez (2024), auto-ethnography allows for rigorous introspection, thus making it a practical reflective method. This approach challenges educators to identify gaps in

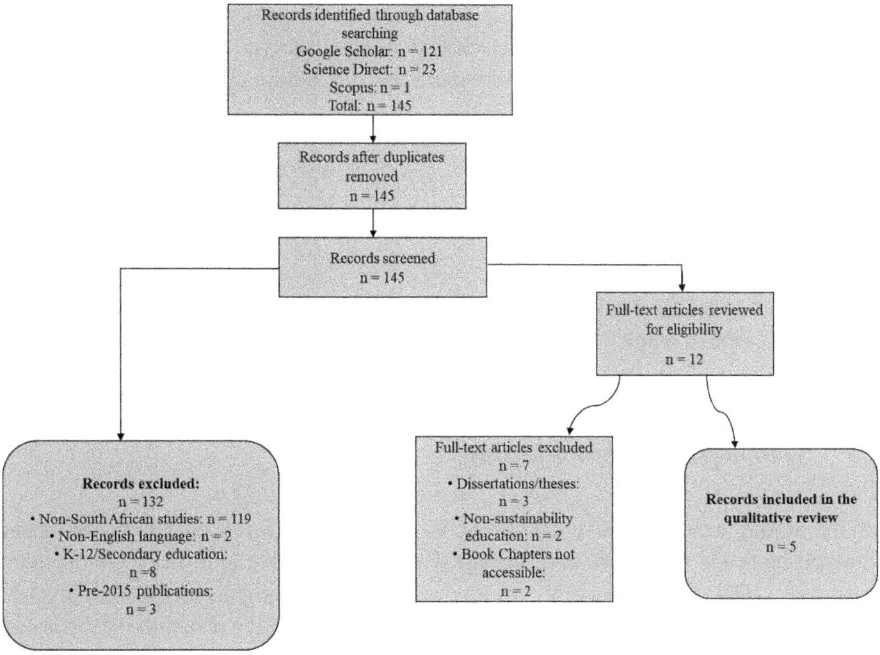

Fig. 2 Adapted PRISMA flow (Page et al., 2021)

supporting students' understanding of sustainability and its application. Therefore, the narrative helps to observe the social impact of the gaps in sustainability teaching and assists the author to reflect deeper on the issues in their communities and future careers. The data was analysed using content and narrative analysis. As illustrated in Fig. 3, the scoping review and reflective narratives were used to integrate findings emerging from the content analysis.

The scoping review provides a structured overview of existing research, while the reflective narratives give depth through lived experiences. The insights from

Fig. 3 Integrative analysis (researcher)

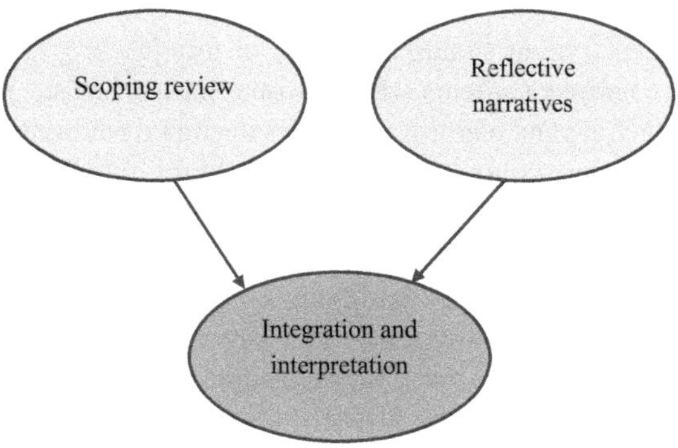

the review shape the focus of reflections, offering new perspectives to be explored in the literature. These methods converge in the integration and interpretation phase, producing a multilayered understanding of teaching sustainability in higher education.

5 Results

This section presents the findings from the systematic review and the account of auto-ethnography on integrating sustainable development goals in teaching.

5.1 Systematic Review Findings

Of the studies reviewed, five employed a qualitative research approach, one used a mixed methods approach, and the last used a quantitative approach. Goldschagg and Long (2021) and Mawonde and Togo (2021) used a mixed methods approach, suggesting that the combination of qualitative and quantitative research methods can be effective when integrating SDGs in the classroom. The studies further employed different theories. Technological studies utilised the technological acceptance model, while those focusing on environmental management employed the new ecological paradigm and grounded theory. In contrast, the other study used structure-agency theory.

At an institutional level, in one university, it was found that students understood the sustainability discourse better than administrators, who often spoke only to broader policy issues without linking them to the SDGs framework (Uleanya et al., 2025). Students were citing sustainable development goals, such as education and climate action. Their exposure was also cited to have been informed by online platforms and peers. The study cited a lack of structured programmes as a challenge, particularly to administrative staff, in supporting broader sustainability efforts within the institution. Table 1 provides a summary of the findings from the reviewed studies.

The results from Raath and Hay (2019) indicate that education for sustainable development should form part of training programmes, particularly for pre-service geography students. They also emphasise the use of systems thinking and cooperative teaching and learning strategies to embed environmentally sustainable principles into their curriculum. The study also highlighted the importance of connecting social, economic, and political dimensions with biophysical environment aspects to frame ESD within the broader sustainability context. Goldschagg and Long (2021) suggest that environmental worldviews of pre-service teachers can be engaged and possibly shifted through targeted environmental education modules, although the impact on concern may be low if the engagement is limited, which means that more engagement is required from the students; also, institutional efforts might be helpful to spike this interest in sustainability education in relation to environmental management. On the other hand, Mawonde and Togo (2021) highlight the role of institutional support in integrating SDGs through work-integrated learning and campus-based SDGs-related practices. The authors propose developing sustainability centres and incentivising student participation through rewards, certifications and committed competitions, which implies curriculum integration that involves off-campus and practical learning components. Such practices may increase participation and interest of students in sustainable development education. Furthermore, Mawonde and Togo (2021) argue that students who are not on campus, especially those in an open distance learning context, face various factors, such as physical distance, financing and misconceptions about SDGs, and this leads to a general lack of interest in sustainability practices. They further assert the need for curriculum designs that consider delivery context and student engagement barriers, which could be an indirect integration strategy.

Gumbi et al. (2024) highlight how digitalisation, specifically through digital game-based (DGBL) learning, can improve engagement and offer contextualised learning, and help students acquire 21st-century digital skills. These digital tools are central to achieving SDGs like quality education (SDG 4) and innovation (SDG 9). They further suggest that the incorporation of technology-mediated pedagogies can be a powerful mode for embedding sustainability concepts within the traditional STEM curricula. The study also reported an interest from the pre-service teachers who were engaged in this programme, who found the digital tool handy and enjoyed skills related to critical thinking, problem solving, and sustainability awareness embedded in the DGBL.

5.2 Reflective Account on Teaching SDGs

Teaching development studies, especially at the undergraduate level, comes with different responsibilities for an educator. The learning facilitation requires outcomes that are not only curriculum-aligned but also within the course's content set, enabling students to acquire knowledge and skills aligned with the graduate attributes. At a university where the educator is based, students must achieve graduate attributes that help them to become socially conscious and responsible citizens. They are also required to have adaptive expertise as part of their life and career skills. These attributes are also linked to other core areas, which include foundational knowledge and literacies, as well as learning and innovation competencies (Nelson Mandela University, 2023). Therefore, in a classroom context, while teaching development planning, the educator must integrate opportunities for students to practice and build skills such as critical thinking, creativity, communication and collaboration.

These skill sets are linked to the knowledge component of the module, which primarily focuses on local development planning, assessing the impact of economic development and case studies that showcase equity, participation, conflict and cooperation in development planning. The scope allows for the introduction of SDGs; however, given that the module was designed prior to the SDGs era, it was not necessarily aligned with the Millennium Development Goals either. I have spent over

5 years teaching the module, which I adopted in 2020. At the time, due to COVID-19 disruptions and the need to switch from contact classes to virtual ones, both educators and students had to re-adjust to new forms of learning. From 2022, efforts were made as part of practical examples shared during the learning facilitation process to introduce SDGs into the module gradually. For example, I decided to ensure that, as part of the formative assessments, questions relating to the SDGs form part of the class presentations to enable students to work on these topics extensively, to encourage increased knowledge. However, bringing in SDGs as examples during lectures, particularly those relating to socio-economic issues and climate change, as they had a direct impact on our geographical position, was not sufficient. This approach was insufficient because not all topics selected by students, as constructed by their educator, included the SDGs component. As a result, there would be either one or two assessments that would focus on SDGs as part of the class presentations. The class would get an opportunity to engage with the content, but this was limiting for the groups that were not tasked to research such topics. It also provided the benefit of enhancing an understanding of the SDGs for the groups allocated topics, which might have enabled them to explore sustainability issues and disadvantaged those who did not get an opportunity in class to research sustainability-related topics (Table 2).

As part of critical reflections on how this course was taught and the growing importance and need for universities to embrace sustainability practices, the educator saw it fit to redesign the module outline to teach and assess sustainability-related issues intentionally. The broad nature of the module outcomes enabled the incorporation of various SDGs into the teaching units without deviating from the core outcomes. Additionally, the educator ensured that students were actively engaged in learning about SDGs through peer-to-peer learning by allowing each student to select at least two SDGs at the beginning of the semester. SDGs were allocated to groups through a draw system, with each group receiving a specific set of SDGs to focus on. Each group then researched a topic and linked it to their assigned SDGs. This collaborative approach enabled students to learn about their own SDGs in depth while also gaining knowledge about other SDGs through their peers' group presentations.

Therefore, at the beginning of the semester, as part of introducing development planning and the core curriculum content, SDGs were also introduced and discussed. Students were informed that their presentations should demonstrate understanding of their allocated SDGs while explaining interconnections with other SDGs. Assessments were designed to encourage students to learn not only their allocated SDGs but also to develop knowledge of the broader framework of all 17 SDGs as outlined by the United Nations. This approach, compared to how the SDGs were approached in my class in the previous years, was more productive in ensuring that students know more about sustainable development, mainly because formative and summative assessments included testing that knowledge. Class discussions now involve digging deeper to understand how each SDG is experienced in various communities, as represented in the classroom and globally. The global context was introduced using visuals and case studies from research that has shown the impact of various SDGs and how they are being addressed in different regions. Such engagements enabled

Table 2 Summary of results

Source	Study focus	Methodology	Theory	Main findings
Goldschagg and Long (2021)	Teachers' environmental attitudes	Mixed methods approach	New Ecological Paradigm (NEP) and grounded theory	• Limited changes in students' worldview • Increased environmental concerns and awareness were noted • Students indicated the impact on future practices as they would share what they had learned • There is a need for ongoing research on local contexts to identify factors influencing positive environmental worldviews
Gumbi et al. (2024)	Teachers' views on the use of digital game-based learning for sustainable STEM education	Quantitative approach	Technological Acceptance Model (TAM)	• The teachers accepted the digital game-based learning as an effective tool for sustainable STEM education • The tool was perceived to be valuable and easy to use • It supported the sustainable development of diverse skills and lifelong learning • Offers contextualised and self-directed learning, contributing to quality education
Mawonde and Togo (2021)	SDG-related practices for students in Environmental management	Mixed methods approach	Structure-agency theory	• Students not involved in campus or off-campus SDG projects • Misunderstanding and limited interest in SDG education • Sustained comprehensive environmental education is required • Financial challenges affecting students' participation in campus SDG initiatives

(continued)

Table 2 (continued)

Source	Study focus	Methodology	Theory	Main findings
Raath and Hay (2019)	Teaching sustainability for pre-service geography students	Qualitative approach	O'Donoghue's model for the environment	• The use of systems thinking helped students to have a contextual understanding of sustainability concepts related to Geography • Learning included processes and patterns • Systems thinking enabled understanding of complex interactions and interrelationships in sustainability issues
Uleanya et al. (2025)	Sustainability in higher education	Qualitative	Co-evolution and theory of change	• Students showed high levels of sustainability awareness and engagement • Exposure to sustainability through course content, as there were mentions of SDGs 4 and 13, which the administrative staff lacked

Source Researcher's compilation

students to develop a practical understanding of the effects of various SDGs, particularly those linked to industry innovation and infrastructure (SDG 9), climate action (SDG 13), life below water (SDG 14), peace, justice and strong institutions (SDG 16), as well as sustainable cities and communities. Examples from their communities were also shared by students, thus enabling them to compare local and global issues.

The redesign of the course necessitates the educator's understanding of the injustice, or rather the indirect injustice, that has been occurring in the course over the previous years due to a lack of intentionality in integrating SDGs into development planning. It also helped me gain perspective on how easy it is to assume that by practising in the discipline, we are indirectly fitting into the sustainability practice, which is not always the case, especially when not all students engage deeply with the content and its practical implications. This sustainability content misalignment has led to the perpetuation of injustices through the learning and teaching processes for both educators and students, as the university environment provides students with a platform to learn as much as they can. Therefore, when there is an opportunity to bring in SDGs in the classroom, it not only helps students to gain more knowledge as they link that to the concept learned in class, but it also prepares them for the global world and helps them to acquire attributes related to being global citizens.

One of the benefits of this approach was that it enabled me, as an educator, to identify and share programmes linked to sustainability in education with my students. As a result, their confidence in the content allowed them to attempt those opportunities. For instance, at the end of the semester, I learned that some of the students from my class have been appointed as SDGs campus ambassadors. Going forward, this means that I need to strengthen the engagement in SDGs in my teaching assessments and create communities of practice so that, at the departmental level, the integration of SDGs can then be applied across all modules and levels of study. Being conscious about sustainability has led me to discover several sustainability education opportunities, which I now share with postgraduate students. Notably, some courses are free for doctoral students, preparing them for a potential academic career.

6 Discussions

This section presents the discussion of results emanating from section five and the guidelines that can be used to integrate SDGs into the course.

6.1 Integrating SDGs into the Curriculum

Sustainable development goals can be effectively embedded in course content by establishing a plan and structure that aligns curriculum requirements with these objectives. The need for structured programmes, as cited by Uleanya et al. (2025), aligns with reflective findings, indicating that the need to be intentional about teaching SDGs must be embraced at an institutional level. This institutional approach will also expose disciplines that do not interact with the SDGs and encourage them to make efforts to align them to their disciplinary focus. Maqoqa's (2025) emphasis on the need to acquire digital competencies for effective SDGs incorporation in learning and teaching underscores the potential gap in sustainability education competencies. This finding was highlighted in the author's reflective account, suggesting that SDGs may not be incorporated due to a lack of 'know-how' to integrate it into the existing curriculum. This integration becomes even more challenging in higher education settings where there are no existing strategies institutionally to promote sustainable development practices, as seen in the research of Price et al. (2024).

In cases where sustainability education had been explored, uptake and engagement were a problem. The challenges related to low interest in SDGs and the misunderstanding of the goals, cited by Mawonde and Togo (2021), are not uncommon. There is limited engagement with sustainability content. This finding aligns with my class experience, which led me to link them to learning outcomes and assessments. Lack of participation in the project will equally be limited if both learners and educators do not prioritise sustainability education. I have observed that institutions of higher learning circulate several calls on sustainability to staff and students; however, unless we are attuned to such content, these calls will not have any meaning. Therefore, educators need to encourage students to participate by sharing available information. It is through educators' interest and intentionality that students will have more information about SDGs. It is also a challenge for educators to find platforms that can empower them in teaching the SDGs (Boarin & Martinez-Molina, 2022). Institutional strategy for SDGs is another important factor that will set the foundation for incorporating SDGs or sustainability teaching.

6.2 Approaches for Teaching SDGs

Goldschagg and Long (2021) study reported limited change in how the students viewed the world in line with what was learned in class. The testing results showed no increased level of their pro-environmental attitudes. However, awareness levels increased, as did concerns, when noting their contributions to environmental harm. The author further correctly points out the need to have more engagement with sustainability education beyond one module. As posited by Leal Filho et al. (2018), faculties should encourage departments to adopt sustainability practices, enabling them to align with the goals across programmes and modules, and ensuring the success of sustainability education. The continued engagement with SDGs will boost the students' confidence in the topic and encourage them to participate in other programmes that deal with sustainability practices (Pham Xuan & Håkansson Lindqvist, 2025; Price et al., 2024).

The aspect of low interest was similar for Goldschagg and Long (2021) and Mawonde and Togo (2021), signalling a lack of readiness from the students. Nevertheless, there are some positives because the concerns noted and interest in SDGs-related projects by students in the reflective account show the need for educators to continue discussing such topics. Alternative methods of teaching such topics, as proposed by Leal Filho et al. (2025) and Ahmadov et al. (2025), could also be tested to see if it might change students' reactions. Mawonde's (2021) study also shared concerns regarding a lack of communication about SDGs or dedicated webpages on institutional sustainability. They further felt that this institutional gap relating to SDGs could be linked to students' lack of interest in the matter.

6.3 SDGs Integration Guidelines

Considering the findings, this section presents the Gradual SDG integration framework, depicted in Fig. 4, developed to guide educators through the systematic embedding of sustainability goals with guidelines. The framework proposes a gradual integration of SDGs in the classroom. The first step involves understanding the institution's graduate attributes and learning outcomes related to the course where SDGs should be embedded. This alignment can be achieved by linking SDGs to existing institutional outcomes, such as "socially conscious and responsible citizens" in my case. The second step involves determining which SDGs are relevant to each unit taught in the module. When mapping these SDGs to the course material, ensure that examples are available to link back to learning attributes. This stage enables educators to experiment with what works for students. It also helps the educator feel confident in their engagement with SDG-related material. This level is also about assessing students' knowledge or prior knowledge of sustainability issues. Educators can begin by incorporating examples of sustainable practices into their teaching and then transition to a comprehensive redesign.

The next step involves redesigning the course to ensure that the materials taught in class or for module lessons can then be assessed using both formative and summative assessments. Integrating sustainability into assessment practice is important because it helps students understand the importance of the content, given that they are assessed based on it. Therefore, educators should ensure both formative and summative assessments to test SDGs knowledge. Approaches that can be used include participatory learning, which allows students to learn in groups, potentially enhancing their continued engagement with SGDs beyond the classroom. Additionally, peer learning mechanisms can help encourage a broad interest in SDGs content. Therefore, the use of group work and presentations is important to enhance engagement. The educator needs to decide on the pace that they will take prior to redesigning the module

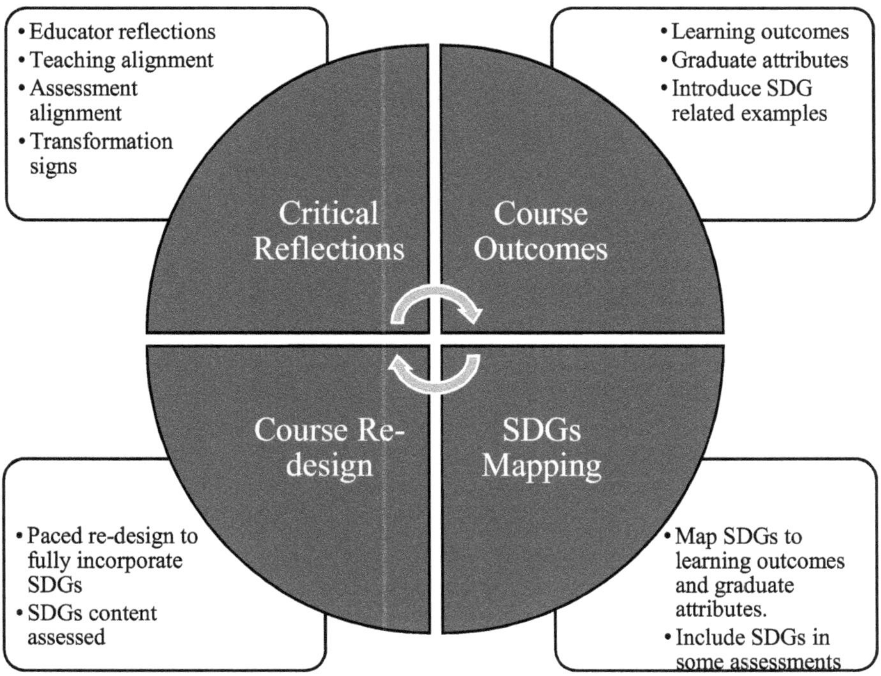

Fig. 4 Gradual SDG integration framework (researcher)

guide to include SDGs. This step is similar to the previous one, where the educator must assess the pace and the students' readiness. Therefore, in some courses, there might be a need to introduce a few SDGs at that time and re-evaluate them in the following teaching cycle. The last step involves reflecting on what has worked from both the educators' and students' perspectives regarding the integration of the SDGs. A continuous reflection and regular evaluation of integration effectiveness is important for educators to identify areas that still require improvement. The formation of a community of practice is also advisable because it encourages academics to work towards departmental-level integration of SDGs across modules and promotes faculty-wide engagement.

7 Conclusion

This chapter aimed to explore how the SDGs can be embedded in an existing curriculum to ensure that a cohort of students is introduced to sustainability teaching and practices. A combination of systematic scoping review and reflexive autoethnography was used to understand sustainability teaching in South African HEIs and how to integrate SDGs in a development studies discipline module. The results showed that there is an attempt, within the South African context, to integrate sustainable teaching, such as the teaching of SDGs. However, few empirical studies were available, providing a limited perspective on how academics are approaching sustainability teaching in disciplines like Geography and Environmental Management. The findings further indicate that there is interest from students in learning about sustainable development goals, and others understand the impact they will have in their future careers; however, in specific contexts, students are unable to participate in sustainable development goals initiatives due to financial challenges. Innovative studies have also been pursued, utilising digital game-based teaching for sustainable development education. Such advances are important not only to inspire lifelong learning for students but also to make teaching about SDGs more relatable for students, particularly young students. One of the challenges that was mentioned in the studies was that some students have no interest in sustainable development education or content that, with the gaming approach, could address such challenges.

In efforts to support sustainability teaching, the chapter proposed a framework and guidelines on how to include SDGs content in a pre-set curriculum, with an understanding that it is not easy to change the curriculum. Thus, educators can still use teaching units and align them to sustainable development. The SDGs framework has been proven to be broad enough, with the 17 goals to be incorporated into different disciplines, including mathematics, as was observed in the study of Maqoqa (2025). Practical implications of this research include guidance on how to incorporate SDGs in an existing curriculum. The results are also important for policymakers as they indicate the importance of supporting educators so that they can integrate sustainability teaching, which will benefit students in strengthening their global citizenship, and help institutions of higher learning and the national government towards meeting their sustainability targets.

The limitations of this study are primarily methodological, as the chapter only provides the ethno-autography and literature to report on teaching sustainability. The reflective account is based on one educator who might have overlooked other significant discoveries in teaching the SDGs. Nevertheless, the results showed similarities in the importance of incorporating SDGs into education. Future studies that include in-depth discussions through interviews and focus groups with students and educators are required to guide academics in incorporating sustainability practices into their classrooms. Other studies may also consider using collaborative, multidisciplinary approaches that can mitigate these limitations while drawing perspectives from various disciplines.

Acknowledgements The author acknowledges Mr. Tshifhiwa Netshiukwi for editing this work and Generative Artificial Intelligence tools like ChatGPT, Claude, and Grammarly for language support and brainstorming during manuscript preparation.

References

Ahmadov, T., Karimov, A., Durst, S., Saarela, M., Gerstlberger, W., Wahl, M. F., & Karkkainen, T. (2025). A two-phase systematic literature review on the use of serious games for sustainable environmental education. In *Interactive learning environments* (Vol. 33, Issue 3, pp. 1945–1966). Routledge. https://doi.org/10.1080/10494820.2024.2414429

Anstey, G. (2024). *South African universities' contribution to Sustainable Development Goals unveiled at the Higher Education conference.* Universities South Africa. Retrieved September 8, 2025, from https://usaf.ac.za/sa-universities-contribution-to-sustainable-development-goals-unveiled-at-the-he-conference-2024/

Boarin, P., & Martinez-Molina, A. (2022). Integration of environmental sustainability considerations within architectural programmes in higher education: A review of teaching and implementation approaches. *Journal of Cleaner Production, 342.* Elsevier Ltd. https://doi.org/10.1016/j.jclepro.2022.130989

Bofylatos, S., & Perez, D. (2024). *Reflective narratives in social design: Towards an autoethnographic approach.* DRS2024. https://doi.org/10.21606/drs.2024.1694

Buerkle, A., O'Dell, A., Matharu, H., Buerkle, L., & Ferreira, P. (2023). Recommendations to align higher education teaching with the UN sustainability goals—A scoping survey. *International Journal of Educational Research Open, 5.* https://doi.org/10.1016/j.ijedro.2023.100280

Chineka, R., & Yasukawa, K. (2020). Intergenerational learning in climate change adaptations; limitations and affordances. *Environmental Education Research, 26*(4), 577–593. https://doi.org/10.1080/13504622.2020.1733494

Cranton, P. (2009). Understanding and promoting transformative Learning: A guide for educators of adults. *Canadian Journal of University Continuing Education, 35*(1).

Edwards, M. G., & Küpers, W. (2024). Feelings for the planet: An alternative vocabulary for incorporating biospherefocused emotions into management learning and education. *Academy of Management Learning and Education, 23*(4), 600–625. https://doi.org/10.5465/amle.2023.0180

Goldschagg, P., & Long, D. (2021). Influencing generations: Pre-service teachers' environmental worldviews at a South African University. *Southern African Journal of Environmental Education, 37*(1). https://doi.org/10.4314/sajee.v37i3.8

Gumbi, N. M., Sibaya, D., & Chibisa, A. (2024). Exploring pre-service teachers' perspectives on the integration of digital game-based learning for sustainable STEM education. *Sustainability (Switzerland), 16*(3). https://doi.org/10.3390/su16031314

Hamadi, M., Imtinan, U., & Namisango, F. (2024). Sustainability education in information systems' curricula: A conceptual research framework. *Education and Information Technologies, 29*(12), 14769–14787. https://doi.org/10.1007/s10639-023-12409-w

Jones, N. (2023). Reimagining Freire: Beyond human relations. *Cultural Studies of Science Education, 18*(1), 205–216. https://doi.org/10.1007/s11422-023-10154-7

Kitchenham, A. (2008). The evolution of John Mezirow's transformative learning theory. *Journal of Transformative Education, 6*(2), 104–123. https://doi.org/10.1177/1541344608322678

Kitchenham, B., Brereton, P., & Budgen, D. (2012). Mapping study completeness and reliability—A case study. In *16th international conference on evaluation & assessment in software engineering (EASE 2012)* (pp. 126–135).

Kumalo, S. H. (2017). Problematising development in sustainability: Epistemic justice through an African ethic. *Southern African Journal of Environmental Education, 33*(1), 14. https://doi.org/10.4314/sajee.v33i1.2

Leal Filho, W., Raath, S., Lazzarini, B., Vargas, V. R., de Souza, L., et al. (2018). The role of transformation in learning and education for sustainability. *Journal of Cleaner Production, 199*, 286–295. https://doi.org/10.1016/j.jclepro.2018.07.017

Leal Filho, W., Viera Trevisan, L., Sivapalan, S., Mazhar, M., Kounani, A., et al. (2025). Assessing the impacts of sustainability teaching at higher education institutions. *Discover Sustainability, 6*(1). https://doi.org/10.1007/s43621-025-01024-z

Leal Filho, W., Viera Trevisan, L., Wahaj, Z., Déda Araújo Nunes, D., et al. (2024). University rankings and sustainable development: The state of the art. *International Journal of Sustainability in Higher Education.* https://doi.org/10.1108/IJSHE-11-2023-0530

Maqoqa, T. (2025). Transforming and sustaining digital mathematics teaching and learning in higher education institutions in South Africa: A literature review. *TWIST*, *20*(3), 142–156. https://doi.org/10.5281/twist.10049652#468

Mawonde, A., & Togo, M. (2021). Challenges of involving students in campus SDGs-related practices in an ODeL context: The case of the University of South Africa (UNISA). *International Journal of Sustainability in Higher Education, 22*(7), 1487–1502. https://doi.org/10.1108/IJSHE-05-2020-0160

Mezirow, J. (1997). Transformative learning: Theory to practice. *New Directions for Adult and Continuing Education, 1997*(74), 5–12. https://doi.org/10.1002/ace.7401

Neboh, N. D., Nkeleme, E. I., & Shakantu, W. (2023). Skills for the post-COVID era of supply chain management education in South Africa: A review. *Journal of Contemporary Management, 20*(1), 136–152. https://doi.org/10.35683/jcm22006.189

Nelson Mandela University. (2023). *Vision 2030 Strategy.* https://publications.mandela.ac.za/publications/media/Store/documents/Institutional/Vision-2030-Pamphlet-Feb-2023.pdf

Newman, M. (2012). Calling transformative learning into question: Some mutinous thoughts. *Adult Education Quarterly, 62*(1), 36–55. https://doi.org/10.1177/0741713610392768

Nhamo, G., & Chapungu, L. (2024). Seven years of embracing the sustainable development goals: Perspectives from University of South Africa's academic staff. *Frontiers in Education, 9.* https://doi.org/10.3389/feduc.2024.1354916

Page, M. J., McKenzie, J. E., Bossuyt, P. M., Boutron, I., & Hoffmann, T. C., et al. (2021). The PRISMA 2020 statement: An updated guideline for reporting systematic reviews. *Systematic Reviews, 10*(1). https://doi.org/10.1186/s13643-021-01626-4

Pham Xuan, R., & Håkansson Lindqvist, M. (2025). Exploring sustainable development goals and curriculum adoption: A scoping review from 2020–2025. *Societies, 15*(8), 212. https://doi.org/10.3390/soc15080212

Price, E., Nicholson, D. T., Dunk, R., Lawler, C., Carney, M., et al. (2024). Enabling change agents for sustainable development: A whole-institution approach to embedding the UN Sustainable Development Goals in higher education. *International Journal of Sustainability in Higher Education, 25*(7), 1333–1350. https://doi.org/10.1108/IJSHE-02-2024-0112
QS World University. (2025). *QS world university rankings: Sustainability 2025*. Retrieved September 8, 2025, from https://www.topuniversities.com/sustainability-rankings
Raath, S., & Hay, A. (2019). Preservice geography students' exposure to systems thinking and cooperative learning in environmental education. *Journal of Geography, 118*(2), 66–76. https://doi.org/10.1080/00221341.2018.1516231
Republic of South Africa. (2012). *National Development Plan: Our future, make it work*. Retrieved August 14, 2025, from https://www.gov.za/sites/default/files/gcis_document/201409/ndp-2030-our-future-make-it-workr.pdf
Rusaanes, V., Eide, H., Brembo, E. A., Gladhus, L., & Oswald, B. M., et al. (2024). Educating nursing students for sustainable future rural health-care services: An umbrella review. *International Journal of Nursing Studies, 151*, 104688. https://doi.org/10.1016/J.IJNURSTU.2023.104688
Sanjakdar, F., & Premier, J. (2023). Teaching for social justice in higher education: Reflexive and critical auto-ethnographic narratives of hope, resilience, and change. *Teaching and Teacher Education, 127*. https://doi.org/10.1016/j.tate.2023.104114
Schryen, G., Marrone, M., & Yang, J. (2025). Exploring the scope of generative AI in literature review development. *Electronic Markets, 35*(1). https://doi.org/10.1007/s12525-025-00754-2
Segooa, M. A., Modiba, F. S., & Motjolopane, I. (2025a). Generative artificial intelligence tools to augment teaching scientific research in postgraduate studies. *South African Journal of Higher Education, 39*(1). https://doi.org/10.20853/39-1-6275
Segooa, M. A., Modiba, F. S., & Motjolopane, I. (2025b). Integrating generative artificial intelligence into higher education: A framework for different types of reviews. *International Journal of Science Annals, 8*(1), 2025. https://doi.org/10.26697/ijsa
Serafini, P. G., de Moura, J. M., de Almeida, M. R., & de Rezende, J. F. D. (2022). Sustainable development goals in higher education institutions: A systematic literature review. *Journal of Cleaner Production, 370*. https://doi.org/10.1016/j.jclepro.2022.133473
Taylor, E. W. (2007). An update of transformative learning theory: A critical review of the empirical research (1999–2005). *International Journal of Lifelong Education, 26*(2), 173–191. https://doi.org/10.1080/02601370701219475
Tshabalala, S., Lunga, W., & Baloyi, C. (2025). Climate change education: Preparing African youth for future challenges. *Sustainability (Switzerland), 17*(10). https://doi.org/10.3390/su17104267
Uleanya, C., Adamu, C. D., & Yassim, K. (2025). Prioritizing sustainability consciousness among university stakeholders: Solution for educational leaders. *Cogent Education, 12*(1). https://doi.org/10.1080/2331186X.2025.2534153
United Nations. (2025). *The Sustainable Development Goals Report 2025*. Retrieved August 14, 2025, from https://unstats.un.org/sdgs/report/2025/The-Sustainable-Development-Goals-Report-2025.pdf
Velempini, K. (2025). Assessing the role of environmental education practices towards the attainment of the 2030 Sustainable Development Goals. *Sustainability (Switzerland), 17*(5). https://doi.org/10.3390/su17052043
Volkmann, L., & Fraunhofer, H. (2023). Environmental literacy, sustainable education and posthumanist pedagogy: Teaching the climate crisis in a global, transatlantic online setting. *International Journal of Development Education and Global Learning, 15*(2). https://doi.org/10.14324/ijdegl.15.2.02

Florah Sewela Modiba is a Doctor of Development Studies and is employed as a senior lecturer and an academic programme leader of postgraduate studies in the Department of Development Studies at Nelson Mandela University. She also teaches at the undergraduate and postgraduate levels. Her postgraduate supervision for MA and Ph.D. students is in community development, entrepreneurship, and rural and township economies. Her research interests are in public policy, sustainable rural livelihoods, rural and township economies, entrepreneurship, digital transformation and academic mentorship.

Scaffolding Sustainability Curriculum Within and Across Institutions: A Cautionary Tale

Shannon Donovan, Dorn Van Dommelen, and Jamie Trammell

Abstract The Environment & Society program, founded in the spring of 2010 at the University of Alaska Anchorage, was an interdisciplinary program created by faculty in the Department of Geography and Environmental Studies. Program development sought to explicitly scaffold sustainability theories and practices across the program's curriculum. This was operationalized by creating a foundation of sustainability knowledge in lower-division courses and moving students towards more complex layers of education and experience throughout their time in the program. Coursework was grounded in critical thinking and reflection and included strands of consistency anchored by professional development, service learning, and capstone experiences. While individual courses had student learning outcomes, all courses focused on programmatic outcomes which were ultimately assessed through student ePortfolios. The final assignment in the major charged students with developing a portfolio that linked their learning experiences across the curriculum to demonstrate student growth in sustainability, skill development, and hands-on, professional experience gained in the program. The Environment & Society faculty integrated best practices used by other programs and received awards for their innovative teaching techniques. Ten years later, the Environment & Society program at the University of Alaska Anchorage was dismantled due to budget cuts. One faculty member stayed in Alaska and wove sustainability curriculum into other programs, outside of the Environment & Society program. Other faculty members left the state and brought their expertise to different universities. This included one faculty member who joined a similar program focused on sustainability and has been able to bring pieces of his

S. Donovan (✉)
Biology Department and Conservation Planning Solutions LLC, Eastern Oregon University, La Grande, Oregon, USA
e-mail: smdonovan5678@gmail.com

D. Van Dommelen
Department of Anthropology and Geography, University of Alaska Anchorage, Anchorage, Alaska, USA
e-mail: dvandommelen@alaska.edu

J. Trammell
Environmental Science, Southern Oregon University, Ashland, Oregon, USA
e-mail: trammelle@sou.edu

W. Leal Filho et al., *Curriculum Development for Sustainable Development*, World Sustainability Series, https://doi.org/10.1007/978-3-032-22146-9_7

former program to grow a new program. Another faculty member is working outside of a sustainability program but works to integrate sustainability practices and experiences into their university's curriculum. The goal of our chapter is twofold: (1) to share how sustainability education can be programmatically scaffolded so that the whole of the knowledge gained is larger than the individual pieces taught within a given course, and (2) to promote the scaffolding of sustainability education both within sustainability-focused programs as well as across other disciplines. We discuss the methods used to scaffold sustainability learning across disciplines and within the major, the challenges and opportunities we faced, and how we work to continue to integrate sustainability lessons and practices across the curriculum.

Keywords Sustainability · Experiential education · Community engagement · Higher education

1 History and Introduction

In the Spring of 2010, the University of Alaska Anchorage, College of Arts and Sciences, created a new, and innovative program, offering a B.A. and B.S., called Environment & Society (Environment & Society). The program had its origins in the early 1990s with the development of an Environmental Studies Minor that was fully in place in 1998. This minor was offered, along with several core courses, as an interdisciplinary program. Still, it became clear after a few years of tepid enrolments that a full-fledged program was the only way that environmental studies would succeed in the university. This led to the development of a new department, the Department of Geography and Environmental Studies, in the Fall of 2006, which drew from select faculty in the disciplines of geography, biological sciences, and economics. Once this new department was formed, it was charged with developing a new bachelor's program as well as hiring faculty members with a dedicated disciplinary focus in environmental studies and sciences.

In time, the new department developed a program it styled "Environment & Society", though it took a painfully long time to gain approval of the program, and it was not finally passed by the University of Alaska Board of Regents until the Spring of 2010. The program was built around two basic principles, which were considered innovative within the fairly traditional College of Arts and Sciences. In the first place, the program leveraged local professionals working in positions of environmental planning, policy and regulation to develop a curriculum that would prepare students for immediate engagement in the workforce after graduation. This included the development of a "studio" style course that required students to work on hands-on, problem-solving environmental and climate-focused projects, writing actual policy documents, like categorical exclusions, or holding charrettes on planned development with close proximity to the university in the city's "UMED District".

The curriculum in the program was decidedly focused on problem-solving, in part because environmental policy work is so important in the State of Alaska. The federal government owns over 60% of the state's lands, and 10% of the state's lands are owned by Alaska Native corporations. This means that federal environmental policy is crucial in all aspects of the state's land use and in development efforts. Students graduating in any natural resource management program, or cultural resource management program, like the university's applied anthropology programs, need to know how to navigate the challenges of federal policy and "solve problems" regularly. Hence, the department embraced the idea of an engaged, pragmatic program. For many in the college, this approach was unusual and led to difficult issues in constructing faculty workloads and finding suitable times to offer courses. Yet, it matched well the university's broader mission. While the University of Alaska Fairbanks is the more traditional institution with aspirations to be a recognized tier-one research university, the University of Alaska Anchorage (UAA) is an open-enrollment university that aims to meet workforce demands and offer pragmatic programs. Therefore, the needs, interests, and abilities of students at UAA varied greatly. We realized early on that some of our students were well equipped to apply sustainability-related content quickly, while others needed more time and assistance to synthesize the information. We quickly realized that many of our students needed reinforcement of sustainability takeaways throughout their program tenure. Similar to the best practices outlined by Brundiers et al. (2010), we address this challenge by creating more hands-on, community-based opportunities and introducing strands of consistency within the Environment & Society program.

The approach taken by the department was not at all outside of the scope of work done by departments in other colleges at the University of Alaska Anchorage. The university was a fairly early adopter in developing community engagement within the institution and gained recognition as a Community Engagement classified university in 2006, the first cycle of such classification by the Carnegie Foundation. Developing an engaged curriculum that was responsive to community needs was an established practice in the College of Health and the university's Community and Technical College. Arts and Sciences, however, was a relative newcomer to engaged curriculum, though it was embracing, at this time, a range of high-impact practices. Faculty members in the Department of Geography and Environmental Studies were quite committed to both HIPs and engagement and recognized the importance of promoting interdisciplinary and high-impact curriculum within the context of sustainability education (Hecht et al., 2019).

The other principle that gained much more traction in the college and across the university was the idea of intentionally working to reinforce, or scaffold, and build upon a set of concepts by using consistent language, outcomes, and methods across coursework and to build outcomes, progressively, in courses throughout the program. The term "scaffolding" is used here with a bit of reservation, because the geographers, environmental scientists, and biologists who largely carried the development of the program didn't really recognize, at first, that their efforts part of a broader movement in the academy with a bonafide literature to back up what we were implementing (Gunawardena & Wilson, 2021). As development and then redevelopment of the curriculum took place, we increasingly understood that what we were doing was a best practice, and created a transformative learning experience (Sterling, 2011). We therefore refined our efforts based on what we learned from our own experiences and experts in pedagogical development to transform and scaffold our curriculum.

Much of what we implemented reached its apogee with the development of two professional development courses, which will be explained in some detail later. Students completed, during these courses, an ePortfolio, which could be used by them professionally, and was also used to measure program outcomes. Our work represented what, for our campus, was a fairly innovative melding of professional development, which is generally praised in our dual-mission university, and academic assessment, which is mostly disdained by teaching faculty.

The department also changed in its composition during this process, which is important to understand. While we started as more of an interdisciplinary grouping than a true department, faculty were eventually forced to choose between their traditional academic departments and membership in the Department of Geography and Environmental Studies. This shaped the department into a small but enthusiastic team, completely dedicated to the program, especially as we saw the departure of several department members as they took new appointments at other universities. By the time this gleaning process ended, the department had four full-time tenure track faculty members and a small group of committed adjuncts. While the core faculty directed the department's work, a loose alliance of faculty members from other departments acted in an advisory role. This played an important role as the department considered the program's learning outcomes and how best to reach these. And, as noted before, this ultimately turned into an interdisciplinary movement to "scaffold" learning outcomes throughout the program's courses, including those not directly offered by the department.

2 Methods

The development of the Environment & Society program was guided by a clear objective: to prepare students to address Alaska's pressing environmental and sustainability challenges through a curriculum that was rigorous, experiential, and professionally relevant. This goal required a strategic and collaborative design process that engaged environmental professionals, integrated high-impact educational practices, and scaffolded learning outcomes across the curriculum.

To ensure alignment with workforce needs, program faculty consulted with a panel of experts representing government agencies, nonprofit organizations, private firms, and tribal entities engaged in environmental planning, regulation, and sustainability. These consultations informed the selection of key knowledge domains, professional competencies, and pedagogical approaches. The feedback emphasized the importance of interdisciplinary knowledge, applied skills, and familiarity with regulatory and planning processes, particularly within the context of Alaska's unique environmental and political landscape (Fig. 1).

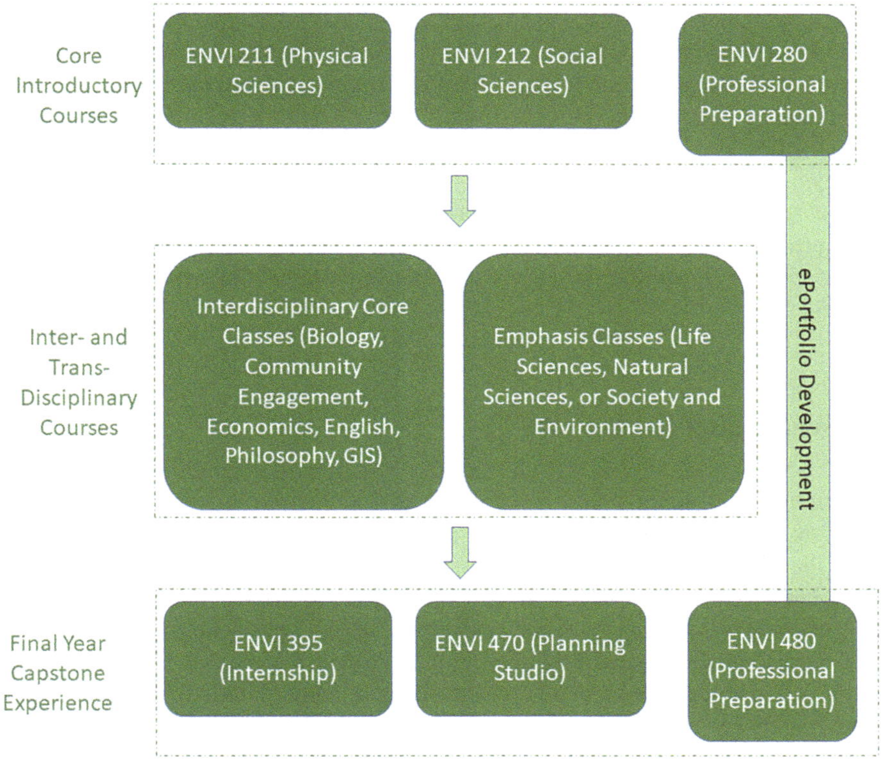

Fig. 1 Scaffolded environment and Society curriculum co-produced with experts in the workforce

The curriculum was designed to scaffold knowledge and skills progressively. Foundational courses in environmental science and studies were supported by prerequisites in math, chemistry, statistics, and communication. As students advanced through the program, they engaged with increasingly complex content in field methods, GIS, environmental ethics, and public science writing. These courses provided both technical training and critical reflection, with each course contributing to a program-wide set of learning outcomes.

A central feature of the program was the integration of experiential learning. Students participated in hands-on projects through a required internship and an advanced studio course, "Environmental Planning and Problem Solving," where they tackled real-world challenges, conducted stakeholder engagement, and developed professional deliverables such as NEPA documents, GIS maps, and policy analyses.

Faculty ensured consistency and coherence across the curriculum by convening annual retreats. These retreats allowed instructors to align on shared definitions, learning outcomes, and assignment design, thereby reducing gaps in student learning. The use of shared artifacts and reflection prompts across courses further supported curricular integration.

To assess student progress and program effectiveness, students curated ePortfolios throughout their academic journey. These portfolios contained artifacts from multiple courses and included reflective narratives that demonstrated how each artifact aligned with specific program learning outcomes. The ePortfolios served dual purposes: they functioned as professional tools for students and as assessment instruments for faculty.

This methodical, collaborative approach (rooted in scaffolding, engagement with professionals, and continuous assessment) allowed the program to provide a comprehensive, workforce-ready education in sustainability and environmental studies.

3 Scaffolding Review

Curriculum scaffolding is an instructional approach that provides temporary support structures to help students bridge the gap between what they currently know and what they need to learn (Van de Pol et al., 2010). Like physical scaffolds used in construction, educational scaffolds offer strategic assistance that gradually decreases as learners develop competence and confidence. The ultimate goal of scaffolding is to cultivate independent learners who can think critically (Browne et al., 2009) and eventually tackle bigger and more complex challenges, a particularly important competency for those studying the environment (Acosta-Gonzaga & Ramierz-Arellano, 2022).

In the case of our program, students, as described below, were on a curricular path that began with basic environmental and sustainability concepts and progressed toward a comprehensive, "studio" course that revolved around problem solving. An important intermediate step was the development of a stacked professional development course (described below). Enabling scaffolding in the curriculum allowed the faculty to incrementally build upon core competencies as we prepared students to address complex environmental issues.

Scaffolding is traditionally defined as mapping core competencies and learning outcomes across the entire program, and ensuring there is a clear progression of knowledge and skills from introductory to advanced courses to meet those competencies (Kruiper et al., 2022). We know from Bloom's taxonomy that students must master subject-matter content before moving on to something more complex. This is why critical thinking is typically scaffolded into curricula, as the creativity and flexibility required come through and after a certain level of subject matter knowledge (Gunawardena & Wilson, 2021). Scaffolding significantly improves learning outcomes according to meta-analytic research, with especially powerful benefits when it supports students' ability to monitor and reflect on their own thinking (Doo et al., 2020). Research by Coulson and Harvey (2013) demonstrates this effect clearly in experience-based learning contexts, where strategically scaffolded reflection activities throughout the learning process markedly enhanced each phase of student development.

Thus, the curriculum we developed sought to build these skills and the ability of students to reflect over the course of their entire academic career with a mind to helping them apply what they were learning, increasingly, in complicated professional contexts. This meant structured reflections at all levels of the curriculum were included to ensure students could connect the semester-level learning to broader program learning outcomes, all with a professional development face.

4 Curriculum Structure

The curriculum that the department developed was not dramatically outside of the norm for an environmental studies program. Students were required to take a lower-division, sophomore-level environmental science course and another in environmental studies, where they built the core competencies. Both of the courses required that students had already completed several core courses in math, statistics, natural, and social sciences. Later, as the program developed, the faculty shifted the prerequisite courses in the program to require physical geography and chemistry, in addition to introductory statistics, applied calculus, public speaking, and technical writing, which were usually taken concurrently with the core environmental courses. Notably, at this early stage in the curriculum, faculty required students to produce a common "artifact" that was required for inclusion in an electronic portfolio (see below) and served as an early learning element through which faculty members could encourage reflection on learning outcomes success, the early phase of our curricular scaffolding.

After their introductory courses, students completed both a field method course, offered by the department, and the department's GIS course, both at the junior level. In addition, students usually completed environmental ethics (offered by the philosophy department) and public science writing (English) in their junior years. In each of these courses, students, again, produced learning artifacts, this time much more meaningful and technically demanding.

In their senior year, students completed conservation biology, offered by the department, a special readings seminar, the "studio" course noted above, and a concentration of six courses in "Spatial Science", "Social Science", or "Biology". In addition, every student was required to be professionally placed in an internship, even though it was greatly resisted by the college's administration. Requiring methods classes, along with a specific concentration and the studio class, were explicit requests made by the professionals with whom we consulted when developing the program. They were specifically concerned that students should have some sort of specialization that would allow them to work in interdisciplinary teams (as required by NEPA,) and then the studio class was conceived so that students would have the chance to put that interdisciplinary work into practice (Fig. 2).

The course was titled Environmental Planning and Problem Solving, and it met for a regular lecture class each week, but on Fridays, the entire day was taken up in a studio setting. Students were given real or "life-like" environmental problems to solve, requiring them to develop field skills, organize community charrettes, write

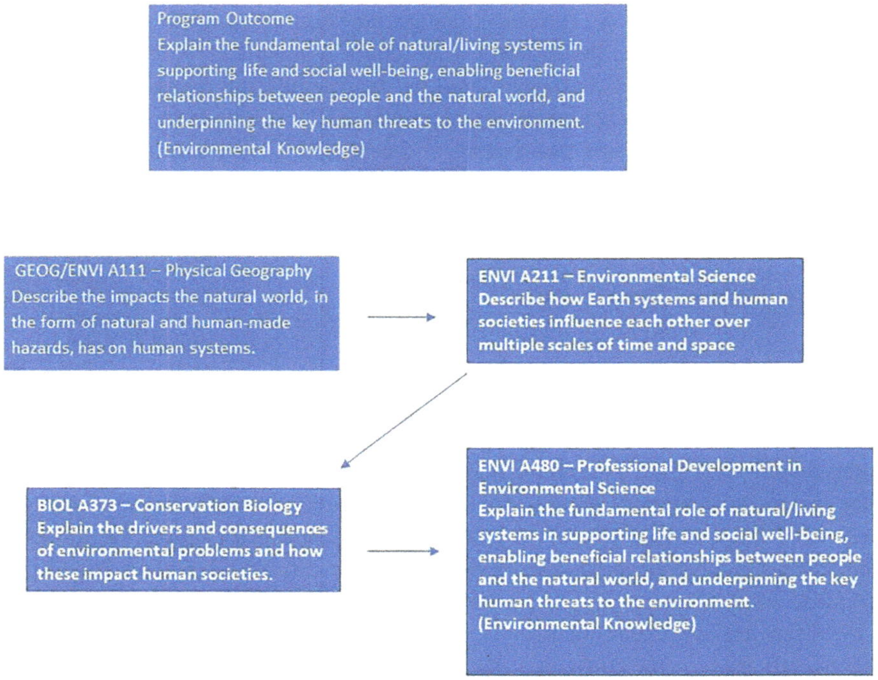

Fig. 2 Example of outcomes scaffolding in one outcome

documents like categorical exclusions and portions of environmental assessments, and engage in climate action planning. Students visited a wide range of sites and worked on projects such as a planned road extension through a wetland owned by the university, a faux project expanding the municipality's wastewater treatment plant to meet EPA standards, a highway development project with significant 4(f) implications in a major part of Anchorage's greenbelt, and partnered with scientists, policymakers, and rightsholders to create Anchorage's Climate Action Plan.

However, what was particularly unique about the curriculum was the department's concerted efforts to scaffold learning outcomes and measure progress in those outcomes through two 1-credit professional development courses. These two courses were "stacked", meaning that the first credit was earned during a student's sophomore year and the other credit was earned during the senior year, with students attending the course together. Thus, two cohorts of students were mixed, the idea being that the slightly more seasoned students could help to mentor the newer students. All students shared lessons and discussions on building a resume, either finding or reporting on an internship, graduate school prospects, and many other professional topics. But the key product of these courses, at least to faculty members, was the construction of an ePortfolio.

In all of the classes taught by the program and several taught in allied disciplines that were required by the program, students were required to produce prescribed artifacts, as described above, that could be included in their electronic portfolios. Each of these artifacts was designed to demonstrate achievement in certain programmatic student learning outcomes. Beginning with the sophomore-level professional development course, these artifacts were placed in the student portfolios. Students were required to reflect on their artifacts and the degree to which they demonstrated, or not, the desired program outcome. The idea was to encourage them to think about what they were learning in the program, what they needed to learn, and how they could grow in their knowledge during the remainder of their career.

At the same time that students were collecting artifacts and reflecting on their progress, faculty members were, of course, using the artifacts and reflection on the artifacts to gauge our own successes and failures as teaching faculty members. This informed the ongoing progress of students who were at the beginning of our program, and any need to redirect teaching efforts. But for students who were in the ultimate year of their program, student development of the ePortfolio was used to assess programmatic outcomes and success.

5 Outcomes and Hands-On Experiences

As described above, the Environment & Society program was intentionally developed in consultation with environmental professionals and regularly refined based on ongoing feedback from the field. From the program's inception, faculty remained in close conversation with professionals across government, nonprofit, and private sectors to ensure curricular relevance and career preparedness for our students. This long-standing collaboration shaped not only the content of our courses but also the program's structural components, including its culminating experiences.

One of the key ways we maintained our connection to professional practice was through the internship requirement. Every student was placed with an external organization, agency, or business, allowing them to apply classroom learning in real-world contexts. Faculty worked closely with site supervisors to coordinate placements, align learning objectives, and ensure that internships supported both student development and organizational needs. These placements became important feedback loops for the program, allowing us to track shifts in workforce expectations and student preparedness.

Several of our courses were designed with direct input from practitioners, especially our Environmental Planning and Problem-Solving course and our Field Methods course. These courses emphasized applied, interdisciplinary learning and reflected the complex, multi-scalar nature of environmental work in Alaska. The studio-format planning course, in particular, asked students to address real or simulated projects involving NEPA compliance, community engagement, and regulatory analysis. Students visited project sites, met with stakeholders, and developed policy documents under realistic constraints—experiences that mirrored the work they would soon encounter as professionals. They developed maps, education and outreach materials, surveys, and draft sections of a climate action plan.

The two professional development courses—taken at the sophomore and senior levels—were central to the program's scaffolding of career readiness. These courses were intentionally "stacked" so that early- and late-career students could learn from one another, modeling mentoring relationships and helping demystify the path from undergraduate study to post-graduate life. Through these courses, students developed resumes, practiced networking, explored graduate and career opportunities, and most significantly, constructed ePortfolios that served both as assessment tools and professional artifacts. These ePortfolios contained curated "artifacts" from across the curriculum—field reports, GIS projects, reflective essays, technical documents—each aligned to program-level learning outcomes. Students were required to reflect on how each artifact demonstrated (or failed to demonstrate) progress toward these outcomes, encouraging metacognition and self-directed learning.

The capstone experience built on this foundation. In their final year, students participated in a structured studio course or a major independent project, synthesizing their learning in an applied context. Capstone topics were often derived from community or agency partnerships, ensuring both relevance and reciprocal value. These projects required students to integrate technical skills, policy knowledge, and ethical considerations while engaging with real environmental challenges. For some, this meant drafting sections of an environmental impact statement; for others, it meant designing outreach materials for public participation in planning processes. Faculty served as mentors and evaluators, and capstone work was also incorporated into the final phase of ePortfolio development and program assessment.

Together, these outcomes and capstone structures exemplified the program's commitment to scaffolded, experiential, and professionally grounded education. Students did not simply graduate with transcripts—they left with tangible demonstrations of their competencies, tested in the kinds of complex settings they would soon navigate professionally. Faculty, in turn, were able to use these cumulative experiences to continuously assess and evolve the program, strengthening its identity and deepening its alignment with the evolving field of environmental work.

6 University Cuts

Increasing stress on higher education resulting from declining enrollments, reduced state funding, and (often misplaced) interests in eliminating redundant programs at different educational institutions within the same state affects university systems across the country (Mitchell et al., 2019). At the University of Alaska Anchorage, the effects resulted in the loss of several small programs, including the Environment & Society program. In February 2020, the faculty of the small but mighty program was told that the Environment & Society program would be eliminated, and plans were made to "teach out" our remaining students. This plan was then implemented within a single semester, leaving the faculty little time to craft a thoughtful exit strategy. The faculty pressed for answers regarding how a decision to eliminate an environmental program focused on Alaskan issues could be justified; however, they were met by police officers at listening sessions and superficial answers at meetings. Interestingly, studies have shown that sustainability-focused college programs were the fastest growing green program between 2016 and 2021 (U.S. Department of Education NCES, 2023). While Union representatives, community members, other faculty members, and students both from within and outside of Environment & Society rallied in support of the program, the University of Alaska Anchorage administration elected to focus its attention on growing the health sciences programs and the Environment & Society program was eliminated.

The faculty came together and collectively went through the range of emotions related to the loss of the program and its significance to the region and communities, the impact the dismantling of the Environment & Society program would have on our students, and the long-term effects of the deletion of the program would have on our careers as professors in sustainability fields. We ultimately decided to push ahead in ways that would allow us to continue supporting sustainability despite the loss of our program.

We channeled our energies to (1) help Environment & Society students find pathways that would allow them to graduate with their chosen Environment & Society degree. Given state and federal guidelines (Higher Learning Commission, n.d.), universities need to provide reasonable opportunities to complete the degree they are pursuing, even if this is cut. Helping students figure out pathways included negotiating for some subset of Environment & Society courses be taught for a defined period, identifying courses outside of the program (and in some cases outside of the University of Alaska Anchorage) that could act as substitutes for required courses, and help students build revised timelines to ensure the opportunity to graduate with an Environment & Society degree; (2) embed sustainability coursework into other programs within the University of Alaska Anchorage. This included modifying a subset of courses and adding them to other existing programs including Anthropology and Biology. This also provided justification to retain one Environment & Society faculty member who could teach the remaining Environment & Society courses; and (3) explore and apply for jobs at other universities with sustainability programs. Doing this allowed us to bring and adapt pieces of our innovative programming to universities outside of Alaska.

While the loss of the Environment & Society program was incredibly difficult, it provided us with the challenge of continuing to scaffold sustainability across the curriculum in various ways and with varying degrees of success. To date, three of the Environment & Society faculty continue to work within the higher education sector. Below we share details on how transitions happened, to what degree they have been successful, and lessons learned that could be applied to others looking to embed sustainability curriculum both within and outside of sustainability programs.

7 Rebirth and Future

The demise of the Geography and Environmental Studies department was very difficult for all persons involved. However, the lessons learned and skills developed through the Environment & Society program, although fragmented, have taken root in other academic departments and institutions. Of the four faculty members who were housed in the Environment & Society program, three remained in academia.

The founding member of the Environment & Society program remained at the University of Alaska Anchorage and moved out of the Geography and Environmental Studies department into the Anthropology department. With their shift into a new department came a shift in curriculum and the opportunity to adapt and teach a handful of existing Environment & Society courses to meet the newly assigned department's needs. Opportunities also opened to create new educational materials and to scaffold sustainability content into the Anthropology program. Most notably, the original capstone program from the Environment & Society program was retained and is now a capstone course in a Cultural and Natural Resource Management Minor, which is popular with students in applied anthropology and the Natural Science, B.S. While the course has, through necessity, shifted from a course focused on complex environmental problem solving in a studio environment to a more traditional lecture-based course, it still retains much of its "hands-on" flavor. Students continue to learn about and work on real-life projects related to environmental regulation and planning.

The retention of this course has had an important impact on the anthropology program by exposing students to the realms of environmental policy beyond those that usually are studied by students of Cultural Resource Management (for example, Section 106 of the National Historic Preservation Act) and places these sorts of laws and regulations in the context of the National Environmental Policy Act and the many other environmental laws that are key to understanding development, especially in an Alaskan context. Another positive impact of integrating environmental and sustainability curriculum in an anthropology program has been the "cross-fertilization" of the curriculum. Students who, in the past, focused on environmental policy, now understand its deep ties to cultural management issues. And students in the cultural/archaeological interest area are learning about environmental law and also being exposed to key techniques in the field, especially geographic information systems.

Of course, it would be naive to believe that as much had been gained as was lost. The university retains a very limited environmental B.S., but it is really just a smattering of courses and makes no systematic effort to scaffold learning outcomes in a sustainable curriculum program. Students take almost no courses that could be called "environmental studies," and they can graduate from the program without ever having formally studied seminal writings, such as Rachel Carson or Aldo Leopold. It has only been through curricular wrangling that environmental law and regulation is required of students, and geographic information systems, sadly, remain simply an option in the program.

Another faculty member from the Environment & Society program ended up at a small regional public university, teaching and now chairing an Environmental

Science, Policy, and Sustainability program. Bringing much of what that faculty member learned at the University of Alaska Anchorage, the department utilizes ePortfolios for their programmatic assessment, scaffolds their curriculum to build field, technical, and writing skills, and embraces service-learning through traditional lecture classes, lab classes, and field experience courses. The department has not yet adopted the stacked sophomore and senior professional development course, but they have adapted their first quarter, junior year cohort class to bring in many of the same professional development assignments. Several of the same challenges exist, especially when considering cross-department alignment and producing ePortfolio artifacts across all core classes.

Moving to a new institution and adapting content has its challenges, but moving from one environmental program to another allowed the faculty member to pull from their existing knowledge base and experiences to create meaningful learning journeys for students. Challenges included understanding the environmental and political landscape regarding sustainability education, adapting course content that was developed within the 15-week semester system to the 10-week quarter system (especially challenging for field-based and civic engagement projects), and starting over in creating deep and meaningful relationships with new partners and figuring out community needs. However, the program was established in 2008, and the list of partners and alumni in the region greatly enhanced the faculty members' ability to bring community-based learning to the institution.

The third faculty member who elected to stay in academia also moved to a small regional university. However, this faculty member's charge was to start a new sustainability program during the height of COVID. While there was no doubt a need for a sustainability-focused program, decreasing enrollments at universities at both national and regional levels made recruitment very difficult. Providing students with hands-on experiences was challenging. Meeting new community partners, creating new relationships, and developing mutually beneficial projects during times of social distancing was very limiting. While successful in developing long-term and meaningful projects, placing students in internships, and creating lasting community relationships, the new program was dismantled, and the faculty member was absorbed into another department. Some of the courses offered through the newly dismantled program were carried over. However, they are mostly considered electives and are not part of the required curriculum. There was some movement to start a Master's degree focused on sustainability. While the faculty member prepared a Master's degree prospectus, there has been no known forward progression with the idea.

In addition to the COVID and enrollment challenges, given that the university is in a rural area, sustainability programs are often perceived as being in conflict with extractive land uses, ranching, and farming in the region. Even including the words sustainability and environment in a program title is construed as political. However, not being an established faculty member at the university and not having a sustainability-focused program required the faculty member to scaffold sustainability education in new and different ways than before. Opportunities have been sought to integrate sustainability education in new and unique ways. This means the faculty member is often coming in through the side door rather than the front door to deliver sustainability education. One benefit is that students who might not otherwise take sustainability courses or learn sustainability content receive it—sometimes piecemeal and sometimes as whole projects. This creates an opportunity to weave sustainability curriculum in ways that programs have never considered before—providing students with new and interesting perspectives while casting a wider net.

8 Lessons Learned

While the Environment & Society journey formally ended after 10 years, the faculty continues to create sustainability education that ripples across universities in different ways. The life cycle of the Environment & Society program at the University of Alaska Anchorage offers a valuable case study in both the promise and vulnerability of interdisciplinary, sustainability-focused education in public universities. Over its relatively short existence, the program has illuminated several lessons that may inform similar efforts elsewhere—whether embedded within existing departments, newly launched interdisciplinary initiatives, or full-scale degree programs. During a time when universities are facing even deeper cuts and programs similar to those continue to have their resilience tested, we have the following lessons learned to offer.

- Innovation Must Be Matched by Institutional Alignment

One of the core innovations of the Environment & Society program—its applied, community-engaged curriculum and use of scaffolded, reflective ePortfolios—was both its strength and a source of tension. While the program aligned with the University of Alaska Anchorage's mission as an open-enrollment, workforce-oriented university and complemented the institution's Carnegie Community Engagement classification, it was often at odds with the more traditional structures and expectations within the College of Arts and Sciences. Institutional alignment—across colleges, departments, and administrative leadership—is crucial to sustain innovative programs over time, especially in times of budgetary or enrollment strain.

- Community-Engaged, Problem-Solving Curricula Are Valuable and Viable

The program's success in preparing students for professional roles in environmental planning, policy, and regulation was rooted in its use of real-world problems, local partnerships, and applied field experiences. The studio course, interdisciplinary methods requirements, and hands-on policy simulations produced graduates who could immediately contribute in complex environmental contexts. These approaches resonated with community and professional stakeholders, even if they required extra work and unconventional scheduling. The lesson: applied, civic-engaged learning can thrive in interdisciplinary environmental programs and should be central to their mission.

- Scaffolding and Assessment Through ePortfolios Strengthen Learning and Program Identity

The structured use of scaffolded learning outcomes—culminating in ePortfolios that were both reflective and accessible—helped anchor the program's identity and track student development over time. This model not only facilitated meaningful assessment but also helped students articulate their competencies in professional settings. The stacked sophomore/senior professional development courses created valuable mentoring opportunities and reinforced the program's learning culture. The lesson here is that clear scaffolding, combined with integrated reflection and assessment, can yield significant benefits for both students and faculty.

- Faculty Commitment Is Critical, but Institutional Support Is Non-Negotiable

The faculty's dedication to the program was evident in their curricular innovations, student mentorship, and efforts to keep the program alive even after its closure was announced. However, institutional decisions—often opaque and driven by larger economic or political forces—can override even the most committed internal efforts. Faculty cohesion and innovation can carry a program far, but without broader institutional investment and recognition, sustainability remains fragile.

- Closure Is Not the End: Curricular Resilience and Adaptation Matter

Although the Environment & Society program was officially closed, its pedagogical DNA has lived on in new contexts. Faculty members embedded sustainability content in other programs at the University of Alaska Anchorage and brought their approaches to new institutions, adapting to different academic calendars, political climates, and student populations. These efforts underscore the importance of designing curricula that can be modular, transferable, and adaptable. Sustainability education, in particular, benefits from being woven into broader academic offerings rather than confined to isolated programs.

- Systemic Vulnerabilities Affect Even Well-Designed Programs

The program's closure, despite rising national interest in environmental and sustainability education, reflects a systemic challenge in higher education: small, interdisciplinary programs are particularly vulnerable in periods of fiscal austerity or shifting political priorities. The Environment & Society program's elimination, despite growing demand for such education and strong student and community support, points to a need for broader advocacy, data-informed decision-making, and protective policies for programs that serve regional and global needs.

9 Limitations

While the Environment & Society program achieved many of its goals and developed a strong identity rooted in experiential and scaffolded learning, several limitations emerged over the life of the program and in the aftermath of its closure.

Institutional Misalignment and Vulnerability: One of the most significant challenges was the lack of sustained institutional alignment. Although the program's values and methods were consistent with the university's mission and community engagement classification, its interdisciplinary structure and nontraditional pedagogy were sometimes viewed as outliers within the College of Arts and Sciences. This disconnect limited long-term support and ultimately contributed to the program's vulnerability during budget cuts.

Scalability and Transferability: The program's intensive, faculty-driven approach required significant time, coordination, and commitment. Annual retreats, shared curriculum design, and close student mentorship were resource-intensive, making the model difficult to scale or replicate without robust institutional support. When faculty transitioned to new institutions, they found that some of the program's most innovative elements—like stacked professional development courses or comprehensive ePortfolios—were challenging to implement in systems with different calendars, structures, or administrative priorities.

Context-Specific Design: The program was deeply shaped by Alaska's environmental and regulatory context. While this regional focus provided authenticity and relevance, it also limited direct applicability in different geographic and political settings. Faculty who brought elements of the program to new institutions often had to adapt significantly to fit local needs and perceptions, particularly in rural or politically conservative regions where sustainability language and goals were met with skepticism.

Assessment Challenges: Although the ePortfolio was a powerful tool for reflection and assessment, it placed a high demand on both students and faculty. Some students struggled with the reflective components or failed to see the value of the portfolio until the end of the program. Additionally, integrating artifacts across multiple departments required ongoing coordination that was not always sustainable.

Impact of Program Closure: The abrupt termination of the program, with minimal transition planning, undermined the continuity of student learning and faculty work. While efforts were made to support students and embed sustainability in other areas, the loss of a centralized program structure meant that much of the scaffolding and coherence were lost. Fragmentation has made it more difficult to track long-term student outcomes or maintain community partnerships at the same level of depth and consistency.

Despite these limitations, the core lessons and methods of the Environment & Society program continue to inform sustainability education in new contexts. However, they also underscore the importance of systemic support, adaptable design, and institutional commitment for the long-term success of innovative, interdisciplinary programs.

10 Conclusions

The Environment & Society program at the University of Alaska Anchorage demonstrates both the promise and the precariousness of designing and delivering scaffolded sustainability education within higher education. From its inception, the program was developed through a collaborative process that consulted with professionals in the field, embedded high-impact practices, and intentionally scaffolded learning outcomes across the curriculum. These methods created a coherent learning pathway that allowed students to progress from foundational knowledge to advanced, applied sustainability work in professional contexts.

Central to the program's design was the understanding that sustainability education must be both interdisciplinary and experiential. Through internships, community-based projects, and the integration of professional development and reflection via ePortfolios, the program created opportunities for students to internalize and demonstrate sustainability competencies. The use of a stacked professional development course and artifacts across disciplines provided a powerful vehicle for students to connect theory to practice and cultivate their capacity to navigate complex, real-world environmental challenges.

However, the limitations faced by the program point to critical lessons for institutions and educators alike. Despite its innovative design and strong alignment with sustainability workforce needs, the program's intensive scaffolding approach required sustained coordination and institutional investment. Its closure highlighted the vulnerability of small, interdisciplinary programs to shifting political and economic priorities, even when they are delivering demonstrable value to students and communities. Moreover, as faculty worked to translate and adapt the program's core principles to other institutions, it became clear that local context, institutional culture, and structural support all shape the success and sustainability of curricular innovation.

These dual insights—into both method and constraint—underscore an urgent takeaway: sustainability education is not only about content, but about the systems that support how it is taught, assessed, and valued. Scaffolded approaches that include structured reflection, experiential learning, and cross-disciplinary integration offer students the best chance to develop the knowledge and tools necessary for creating a sustainable future. Yet such approaches must be matched by administrative vision, institutional flexibility, and long-term investment.

The Environment & Society program may have been formally dissolved, but its pedagogical contributions live on, reflected in the courses, programs, and faculty who continue to adapt and expand its principles. This chapter affirms that sustainability education can be both rigorous and responsive, and that scaffolded, real-world learning is essential to preparing students to meet the urgent challenges of our time. For institutions and educators committed to sustainability, the task ahead is clear: build resilient, adaptable structures that support deep learning across disciplines, ensure alignment between innovation and institutional priorities, and embed sustainability education in ways that will endure, even in uncertain times.

References

Acosta-Gonzaga, E., & Ramirez-Arellano, A. (2022). Scaffolding matters? Investigating its role in motivation, engagement and learning achievements in higher education. *Sustainability, 14*(20), 13419. https://doi.org/10.3390/su142013419

Browne, L., Hough, M., & Schwab, K. (2009). Scaffolding: A promising approach to fostering critical thinking. *SCHOLE: A Journal of Leisure Studies and Recreation Education, 24*(1), 114–119. https://doi.org/10.1080/1937156X.2009.11949644

Brundiers, K., Wiek, A., & Redman, C. L. (2010). Real-world learning opportunities in sustainability: From classroom into the real world. *International Journal of Sustainability in Higher Education, 11*(4), 308–324. https://doi.org/10.1108/14676371011077540

Coulson, D., & Harvey, M. (2013). Scaffolding student reflection for experience-based learning: A framework. *Teaching in Higher Education, 18*(4), 401–413. https://doi.org/10.1080/13562517.2012.752726

Doo, M. Y., Bonk, C., & Heo, H. (2020). A meta-analysis of scaffolding effects in online learning in higher education. *International Review of Research in Open and Distributed Learning, 21*(3), 60–80. https://doi.org/10.19173/irrodl.v21i3.4638

Gunawardena, M., & Wilson, K. (2021). Scaffolding students' critical thinking: A process not an end game. *Thinking Skills and Creativity, 41*, Article 100848. https://doi.org/10.1016/j.tsc.2021.100848

Hecht, A., Crowley, E., Andrews, J., & Buczynski, S. (2019). Sustainability education: Perspectives and practices across higher education. *Environmental Management, 63*, 1–13. https://doi.org/10.4324/9781849776516

Higher Learning Commission. (n.d.). *Substantive change: Teach-out arrangements.* https://www.hlcommission.org/accreditation/policies/teach-out-arrangements/

Kruiper, S. M., Leenknecht, M. J., & Slof, B. (2022). Using scaffolding strategies to improve formative assessment practice in higher education. *Assessment & Evaluation in Higher Education, 47*(3), 458–476. https://doi.org/10.1080/02602938.2021.1958767

Mitchell, M., Leachman, M., & Saenz, M. (2019). State higher education funding cuts have pushed costs to students, worsened inequality. *Center on Budget and Policy Priorities, 24*, 9–15.

Sterling, S. (2011). Transformative learning and sustainability: Sketching the conceptual ground. *Learning and Teaching in Higher Education, 5*, 17–33.

U.S. Department of Education, National Center for Education Statistics. (2023). *Integrated Postsecondary Education Data System (IPEDS), Completions Survey: 2016–2021.* Sustainability Studies (CIP Code 30.3301). https://nces.ed.gov/ipeds/use-the-data

Van de Pol, J., Volman, M., & Beishuizen, J. (2010). Scaffolding in teacher–student interaction: A decade of research. *Educational Psychology Review, 22*, 271–296. https://doi.org/10.1007/s10648-010-9127-6

Dr. Shannon Donovan is an Associate Professor of Sustainable Communities at Eastern Oregon University and the founder of Conservation Planning Solutions LLC. Shannon has been working across disciplines to address sustainability issues for over 25 years. Her work bridges gaps between policy, management, and education.

Dr. Dorn Van Dommelen is Professor of Geography at the University of Alaska Anchorage where he has been a faculty member for over 30 years. He currently serves as the Chair of the Department of Anthropology and Geography. Dorn's disciplinary interests include landscape and settlement history and population geography.

Dr. Jamie Trammell is an Associate Professor of Environmental Science, Policy, and Sustainability at Southern Oregon University. He is a landscape ecologist interested in modeling landscape change with an explicit focus on the integration of socioeconomic and biophysical drivers.

Walking the Sustainability Talk: The Case of a Higher Education Institution

Yuan Zheng Li

Abstract The purpose of this chapter is to find out what and how higher education institutions address global warming, provide evidence to reflect on climate change curriculum development in universities, and contribute to advancing sustainable development education. By analyzing sustainability initiatives at a Canadian university, we investigated our research question about institutional responses to climate change, more specifically those related to the transmission of climate change knowledge. Drawing on sociological perspectives on social change and environmental practices, we intended to examine sustainable development initiatives, reflect on the university's sustainability journey, and look ahead to foster understanding and changes in sustainability education. Three major findings emerged from our analysis. First, the university has taken sustainability initiatives. Secondly, climate change teaching was connected to research activities. Our third finding revealed that climate change responses have been implemented through a multi-stakeholder process that have led to positive environmental and social impacts. The contribution of this paper is threefold. Taking stock of existing sustainability strategies and activities could inform educators, policymakers, and practitioners about educating and mobilizing people to take climate action. Furthermore, the study provided evidence to reflect upon climate change curriculum development, shedding some light on a multi-pronged approach to advance sustainable development practices that would contribute to mapping pathways towards a more sustainable future.

Keywords Climate change education · Curriculum development · Case study · Change agent · Implementation

Y. Z. Li (✉)
Michael D. Penner Institute on ESG and Scotiabank Sustainable Development Innovations Observatory, Université de Montréal, Montreal, Canada
e-mail: yuan.zheng.li@umontreal.ca

W. Leal Filho et al., *Curriculum Development for Sustainable Development*, World Sustainability Series, https://doi.org/10.1007/978-3-032-22146-9_8

1 Introduction

The concept of sustainable development first gained traction in the 1980s as a means of achieving economic development, social inclusion, and environmental sustainability. Some people have expressed their concerns about the feasibility of sustainable development, given that many sustainability issues such as poverty, income inequality, and climate change are complex—the current climate response appears to support their point of view. In spite of the existential threat to human and environmental well-being posed by global warming,[1] societies are not on track to cut emissions.[2] Although there are many hurdles facing sustainable development, we argue that higher education institutions (HEIs) can play a key role in addressing complex issues like climate change and identifying solutions to improve sustainability outcomes. HEIs have been taking steps to tackle sustainability issues through education, research, and operational practices. However, much more work is required when it comes to rising to the challenge that is global warming at 2.6 °C and beyond.[3]

Given that university campuses have large ecological footprints through their operations, infrastructure, and resource consumption, we propose examining university activities and practices to determine whether and how they respond to sustainability, more particularly global warming and climate change education (CCE). Drawing on environmental sociology, our study hypothesized that universities became change agents for sustainability by developing interdisciplinary research and education and fostering collaboration among stakeholders to achieve their goals. Université de Montréal's (UdeM's) significant campus footprint made it a compelling case for studying sustainability. Also, its position as one of Canada's leading research universities would increase our chances for obtaining information that would enable us to identify recent CCE-related practices. By tracing the university's sustainability trajectory, the aim of our study was to find out what and how it has addressed global warming to date, provide evidence for reflection on climate change curriculum development in universities, and contribute to advancing sustainable development education.

This chapter is organized in four parts. The first part discusses the role of HEIs in addressing climate change. The second explains in detail our research methodology via a flowchart, followed by a results and discussion section. Using tables and figures, the third part of the chapter reports the data on sustainability initiatives as regards activities and operational practices, institutional approaches (including governance and administration, research, teaching, operations, and collaboration with communities), as well as the outcomes achieved. In relation to our research question, hypothesis, and existing literature, we analyze the possibilities for and the challenges around curriculum development and redevelopment for CEE, in addition

[1] IPCC (2022).

[2] UNEP (2024).

[3] Ibid.

to education for sustainable development (ESD). Finally, the last part explores the implications of our findings, presents research limitations, and offers suggestions for future studies.

2 Climate Change and HEIs

2.1 Climate Change and Education

Education is key to climate change mitigation and adaptation efforts (UNESCO & MECCE, 2024). For instance, it can prepare people to make complex decisions (Feinstein & Mach, 2020) and mobilize communities to reduce vulnerability (Pal et al., 2023). However, recent studies have shown that CCE efforts have been insufficient (Kwauk, 2022; UNESCO & MECCE, 2023, 2024). For example, countries have fallen short in terms of integration, stakeholder inclusion, education quality, and climate justice (Kwauk, 2022), perhaps stemming from implementation issues such as a lack of support (European Commission/EACEA/Eurydice, 2024) and funding (UNESCO & MECCE, 2023).

Despite increasing involvement in climate change worldwide at the higher education level, challenges related to CCE approaches, methods, and tools remain. Empirical curricular development has illustrated that the lack of comprehensive framework for content and multi-year climate change curricula implementation process posed considerable issues for universities (Sullivan et al., 2022; Takinana & Baars, 2022). With respect to these challenges, HEIs could be an important actor in closing the implementation gap (Molthan-Hill et al., 2019). In fact, various CCE approaches have already been deployed by universities to develop their curricula (Molthan-Hill et al., 2019). University researchers and students have been advocating for climate inclusion curricula (Hampshire et al., 2022; Omrani et al., 2020), developing collaborative, experiential, and problem-based learning strategies (Filho et al., 2021; O'Dwyer et al., 2023) while incorporating climate change into current curricular activities (Sullivan et al., 2022).

2.2 Universities as Change Agents for Sustainability

To understand the roles HEIs have played in implementing sustainability initiatives such as CCE efforts, we drew on ecological modernization—a theoretical notion of environmental sociology (Hajer, 1995; Huber, 1991, as cited in Mol & Jänicke, 2009; Spaargaren & Mol, 1992) that claims a transformation of key institutions (Huber, 1982, as cited in Mol, 2000) could solve environmental problems within existing structures (Mol, 1995; Spaargaren, 1997). Applying sociological perspectives directs attention toward whether and how changes in sustainability practices have occurred in systems where actors and stakeholders interact. The research question is, then, whether and how universities have responded to sustainability, and more particularly climate change and CCE. We hypothesized that universities became change agents for sustainability by developing interdisciplinary research and education, as well as fostering collaboration among stakeholders to achieve their goals.

Academic CCE literature—including studies related to curriculum development in higher education from various disciplines—helped us understand the current state of practice in the field. However, little research on CCE at HEIs exists (Molthan-Hill et al., 2019). Focusing on the national and transnational levels, previous studies on the topic have provided a wide breadth of insights. Contrasted with a broader overview, our research aims at contributing to the literature by taking a granular look and uncovering patterns that are not apparent, without delving into specific details at the institutional level.

3 Methodology

3.1 Data Collection

We employed a case study-based methodology at Université de Montréal, the third largest Canadian public research university in terms of research volume. UdeM comprises 13 faculties and schools, as well as 465 research units. Located on Mount Royal, the 65-hectare main campus features 40 buildings, in addition to being home to the majority of the university's facilities and its affiliated schools: HEC Montréal and Polytechnique Montréal.[4] Beyond the main campus, UdeM opened the new MIL campus science complex, which houses four departments[5] and hosts various facilities and satellite campuses across the province of Quebec.

The university was selected as a laboratory for understanding HEIs' sustainability practices for two reasons. First, Université de Montréal has already made some progress in terms of ESD. Second, the school offers over 600 undergraduate and graduate programs and more than 12,500 courses.[6] Studying an institution at the forefront of ESD implementation that also provides a wide array of courses across disciplines serves to increase the chances of identifying recent practices, including teaching and learning activities related to climate change.

Several search methods were used to find institutional sustainability and climate change policy documents and information. Using the terms (in French) "sustainable development" OR "environment" OR "greening" OR "climate change" OR "carbon" OR "greenhouse gas emissions" OR "Indigenous", we reviewed institutional webpages devoted to sustainability- and climate-related issues, during which

---

[4] Université de Montréal (2025a).

[5] Université de Montréal (2025b).

[6] Université de Montréal. (n.d.-e).

we obtained a number of policy documents, including a sustainable development policy, sustainability strategic plans, action plans, a responsible investment policy, a progress assessment report, a verified carbon footprint report, a sustainability course list,[7] and a list of research experts on the United Nations Sustainable Development Goals (SDGs).

In order to uncover sustainability efforts that encompass a wide range of operations and facilities, we checked experts' faculty webpages, the online university repository (Papyrus),[8] and the latest university news and updates. We did so by looking for the above-mentioned terms, with a focus on climate change in said experts' research projects, publications, courses they teach, theses and dissertations they supervise,[9] as well as any community activities and services in which they get involved. The examples of sustainability initiatives were chosen in line with the literature reviewed and the diversity represented in intervention types. Finally, we created a data collection inventory to prepare the data for analysis.

3.2 *Data Analysis*

To conduct an analysis of sustainability practices, we combined the information gathered from multiple sources as categorized in our inventory and transferred it to a Microsoft Excel spreadsheet to create a new, consolidated data set. When the information was transferred from each source to the Excel file, we verified it item-by-item, checked for obvious inconsistencies, and corrected errors to ensure data transfer accuracy.

Based on the new spreadsheet, we first looked for what has been done in terms of sustainability, utilizing a matrix table that categorized each activity type (governance, teaching, research, outreach, and campus operations), with a focus on teaching and training around climate change issues. The second phase of analysis involved identifying whether and how climate change was reflected in the consolidated data. For example, the table enabled us to pinpoint information about climate-related courses—in terms of title, detailed description, faculty, department, level of study, training and research activities on climate change—as well as uncover any relationships between them.

An analytical schema (Fig. 1) was drawn up to illustrate how the institution has created change and achieved its goals through an implementation process in which stakeholders were involved.

The schema would guide us in (1) identifying key elements of the process by which a targeted outcome was achieved, (2) explaining how climate change-related ideas and themes manifested themselves in existing curricula, and (3) reflecting upon the particular challenges faced by faculty members, students, and stakeholders around the implementation of sustainability practices.

---

[7] Université de Montréal (2022).

[8] https://umontreal.scholaris.ca/home.

[9] Université de Montréal (2025c).

Fig. 1 Analytical schema

4 Results and Discussion

4.1 Sustainability Initiatives

Université de Montréal has implemented sustainability initiatives across all five types of activities (governance, teaching, research, outreach, and campus operations), as reflected in Fig. 2.

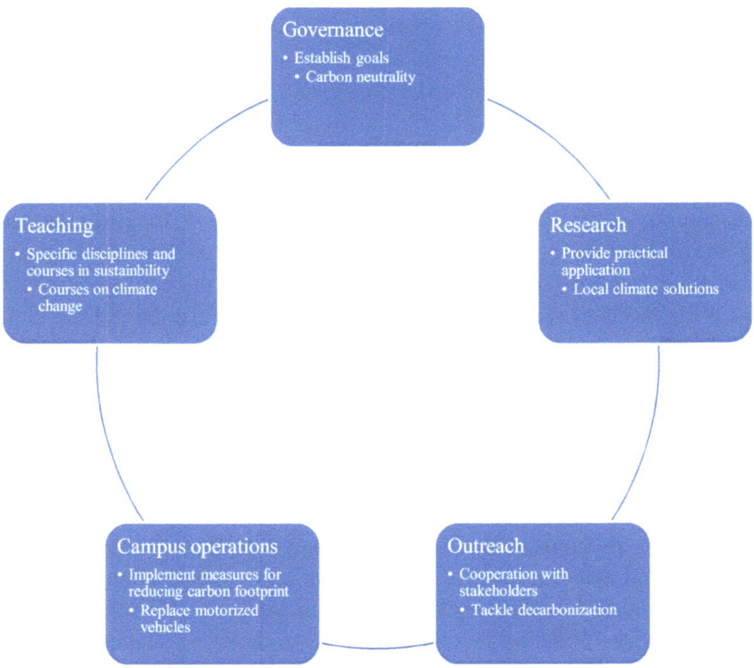

Fig. 2 Sustainability initiatives—type and example of climate action

We analyzed categorized data by activity type through paying attention to institutional responses to climate change, more specifically those related to teaching and training.

4.1.1 Teaching

At Université de Montréal, nearly 80% of departments offer sustainable development courses, representing over 20% of its total courses. For example, the bachelor's degree in environmental geography curriculum allows students to develop a transdisciplinary approach to space and territories and integrate the human and physical dimensions of socio-environmental challenges, tackling issues such as natural resource management, climate change, global urbanization, biodiversity, and migration.[10] In 2009, Université de Montréal launched a graduate diploma—referred to as a DESS in Quebec—in environment and sustainable development and a master's degree in sustainable development in 2019 (Fig. 3),[11] both of which combine theory and practice covering an array of disciplines.

In undergraduate and graduate curricula, we found climate change content in 11 credit courses spanning governance, climate justice, impacts on health in biological and health sciences, earth sciences, as well as the humanities and social sciences:

Undergraduate courses (6)

- Atmospheric Chemistry and Climate Change
- Climate Change
- Climate Change and Justice
- Ecosystem Ecology
- Geomorphology
- History of the Environment in Canada.

Graduate courses (5)

- Biodiversity: Role, Threats, Solutions
- Climate Change: Health Impacts
- Environmental Biogeochemistry
- Politics and Climate Change
- Social Issues and Governance.

Covering a wide range of climate-related topics:

- Biogeochemistry of nutrients and response to environmental disturbances and changes
- Chemical reactions of the atmosphere and disturbances resulting from human activities

[10] Université de Montréal (n.d.-f).

[11] Ibid., 6.

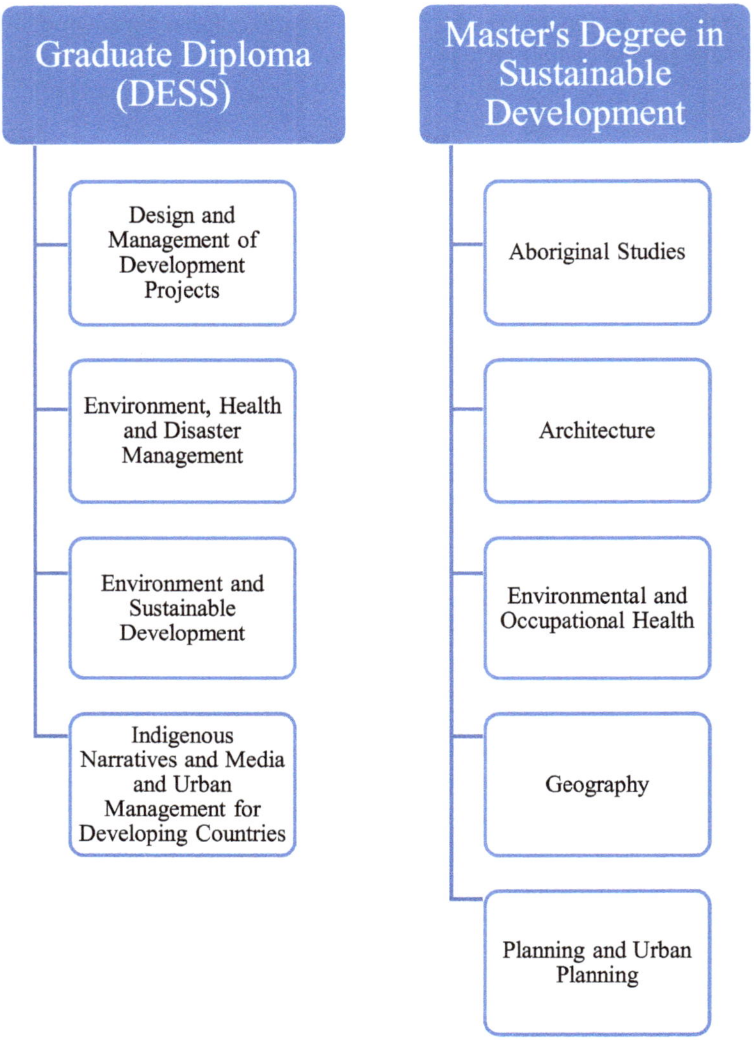

Fig. 3 Graduate diploma in sustainable development

- Democracy, governance, and social issues: resource management, population growth and consumption
- Ecological imperialism and occupation of territory
- Gaps between scientific knowledge and response of public policies
- Geomorphology and land use: role of the climate in the differentiation of forms
- Importance of biodiversity, climate change, habitat destruction, and conservation
- Justice, inequality, and equity
- Mobilization of actors, transnational projects, and legal actions
- Principles, global and local solutions, and equitable measures

- Response of ecosystems and geosystems to climate variations and improvement of climate models
- Role of living organisms in modern flows of carbon, nitrogen, and phosphorus.

Teaching from historical, Canadian, and international comparative perspectives, course instructors used multiscale and interdisciplinary analysis to develop knowledge of climate change impacts, improve understanding of biophysical and socio-economic risks, and produce climate models. Students were able to learn how to analyze large-scale data, link Earth's physical and biological processes to make predictions, develop arguments from evidence, apply interdisciplinary approaches, and propose solutions to solve complex problems at multiple scales.

4.1.2 Postgraduate Training, Teaching, and Research

11 Ph.D. theses and 29 Master dissertations on climate change have appeared in the online university repository over the past decade (2015–2025). The number of these within the last five years has quadrupled, with dissertations increasing by 40%, as shown in the Table 1.

At Université de Montréal, 32 out of 304 SDG experts worked on SDG 13 and hailed from six faculties and 15 departments, as follows: Faculty of Arts and Sciences (14), Faculty of Law (1), Faculty of Medicine (1), Faculty of Planning (1), Faculty of Veterinary Medicine (1), and Faculty of Public Health (1). The production of theses and dissertations appears to associate positively with the number of courses offered and climate change projects, as shown in Table 2. For instance, the department that offered more courses and in which researchers engaged in more research projects produced nearly 40% of theses and dissertations.

It seems that climate change teaching is closely connected to research activities in the way that it integrates knowledge, applications, and opportunities into the student educational experience. Some postgraduates and their supervisors shared similar research topics, such as contaminants, permafrost, and the impacts of climate change in Arctic region or boreal forests, as illustrated in Table 3. Several of these researchers published multi-authored articles on climate change in peer-reviewed journals.

Table 1 Doctoral Theses and Master's Dissertations

Year of submission	2015–2019	2020–May 2025	2015–May 2025
Doctoral theses	2	9	11
		17	
		17–29	
		26–40	
Master's dissertations	12	17	29
		26	
Total	**14**	**26**	**40**

Table 2 Researchers, courses, and projects

Faculty of	Department/ School of	SDG 13 researchers	Courses in climate change	Theses/ Dissertations	Research projects
Arts and Sciences		***14***	***10***	***6/23***	
	Biological Sciences	3	2	1/3	3
	Geography	**6**	**3**	**3/8**	**6**
	History	1	1		
	Physics	1	0		1
	Political Science	3	1	1/8	3
	Continuing Education	0	**1**		
	Chemistry	0	**1**	0/1	
	Unspecified Department	0	**1**		
	Anthropology	0	**0**	1/0	
	Criminology	0	**0**	0/1	
	Sociology	0	**0**	0/1	
	Translation	0	**0**	0/1	
Law		***1***	*0*	*½*	
Medicine		***4***	***0***	***0***	
	Medicine	2	0		
	Pediatrics	1	0		
	Kinesiology and Exercise Science	1	0		
	Rehabilitation	1	0		
Planning		***4***	***0***	***0/2***	1
	Management	1	0		
	Urban Planning and Landscape Architecture	3	0		
Veterinary medicine		***2***	***0***	***2/0***	
	Veterinary Biomedicine	1	0		
	Clinical Sciences	1			
Public health		***6***	***1***	***2/2***	**6**
	Social and Preventive Medicine	4	0		

(continued)

Table 2 (continued)

Faculty of	Department/ School of	SDG 13 researchers	Courses in climate change	Theses/ Dissertations	Research projects
	Environmental and Occupational Health	2	1		
	Total	**32**	**11**	**11/29**	

Table 3 Climate change: researchers' projects, thesis and dissertation

Action	Research project	Thesis and dissertation
Improving understanding	**Permafrost** dynamics, migration and health, social and political responses, **impacts of climate change**: **contaminants**, diseases, health, migration	Naskapi knowledge in Kawawachikamach, role of innovation in the transition to health-promoting food systems, **impacts of climate change**: application of the Convention on the Law of the Sea, characterization of mineral **dust,** thawing **permafrost** of **boreal forest** wetlands, degrading **permafrost** bog, climate variability in **boreal forests**, climate election, climate skepticism
Monitoring	Multidisciplinary Observatory for **Arctic** Climate Change and Extreme Events, monitoring program for a **boreal forest** watershed with **permafrost**	N/A
Mitigation	Carbon footprint reduction	N/A
Adaptation	Climate Resilience, heat in schools, local food systems, adapting workplaces	Cryostratigraphy of **permafrost**: to adapt the built environment (Nunavik), adaptation policies
Solution	Model projections, nature-based climate solutions, integrate equity into the design of green spaces, substitution input materials: use less toxic materials or lean into renewable ones	Modelling avian species distribution and fire in **boreal forests**
Improving justice, equity, & inclusivity	Equitable access to health services, vulnerability of population and climate change	Protecting marine mammals in the **Arctic** region, tropical forests, and preservation

4.1.3 Operational Practices and Outreach

Université de Montréal has generated significant greenhouse (GHG) emissions from its fleet of vehicles, food services, commuter- and work-related travel (Université de

Montréal, n.d.-d). UdeM has implemented a series of measures to reduce its GHG emissions, from replacing motorized vehicles and gas-powered equipment (Université de Montréal, n.d.-d) to changing energy consumption behaviours. It conducted a pilot project that enabled 100 people to measure and reduce their carbon footprint using a newly developed GHG emissions calculator. Since then, the Sustainable Development Unit (SDU) has developed a free mobile app in French (*Votre empreinte*), which is available on its website. Users can calculate their carbon footprint by adding up GHG emissions from commuting, professional travel, and food consumption.[12] To foster climate resilience in the community, the institution participated in various biodiversity conservation efforts (Université de Montréal, n.d.-a), ranging from creating a Biodiversity Centre and adopting a Biodiversity Plan, to joining the Nature Positive Universities Alliance,[13] which unites over 500 HEIs across the world around addressing both climate and biodiversity issues in a bid to build more resilient ecosystems, help nature recover, and limit climate change.[14]

Université de Montréal has engaged in active shareholding on climate issues (engagement and voting) and invested in low-carbon assets. Besides setting reduction targets, it has demonstrated its commitment to regularly measuring the carbon footprint of its investment portfolio and integrating environmental, social, and governance (ESG) factors into the decision-making process. Furthermore, Université de Montréal has teamed up with other Canadian institutions to leverage shareholder power. For instance, after signing the Canadian Universities Charter for Environmentally Responsible Investing in a Time of Climate Change (2020) (Université de Montréal, n.d.-c), Université de Montréal began participating in a shareholder engagement initiative on climate issues that brings together 15 Canadian university endowments and pension plans: the University Network for Shareholder Engagement (UNIE).[15] As one of UNIE's founding members, it has contributed to developing the program[16] and building bridges between universities and businesses to activate change.

[12] Université de Montréal. (n.d.-g).

[13] Université de Montréal. (n.d.-h).

[14] Nature Positive Universities (n.d.).

[15] A shareholder engagement program for university endowments and pension plans, leveraging their power as institutional investors to meaningfully address climate change-related risks (SHARE, 2025) https://share.ca/initiatives/unie/.

[16] Université de Montréal. (n.d.-i).

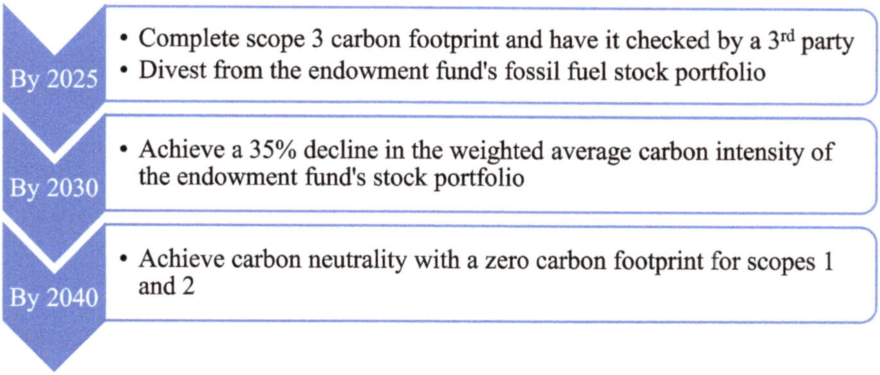

Fig. 4 Goal timeline for carbon neutrality

4.2 Multi-actor Change Process

4.2.1 Institutionalizing Sustainable Development

Since the adoption of UdeM's Sustainable Development Policy (2014), climate change responses have been gradually embedded in both institutional policy and practices. "We have an obligation to step up the pace in the fight against climate change, both in our own practices and in the essential contribution of our researchers to solutions to the problems posed by this major generational challenge," stated Daniel Jutras, Rector of Université de Montréal in the report on the Sustainable Development Action Plan 2021–2023.[17] The university has set objectives and targets to achieve carbon neutrality by 2040—and reduce the carbon intensity of the endowment fund's portfolio (Fig. 4).

To translate goals into changes, UdeM has planned implementation processes, adopted control measures, checked against commitments, and reported results. More specifically, incremental targets have been set for scope 1 and 2 emission reductions to achieve an overall decrease of 20% below 2004–2005 levels by 2025, and 40% by 2030 (Université de Montréal, 2023). In addition, the management committee has approved a comprehensive reduction plan of GHG emissions for each scope (referring to scopes 1–3 emissions)[18] (Université de Montréal, n.d.-a, n.d.-d) and a climate change resilience and adaptation plan (Université de Montréal, n.d.-b). To support its efforts, UdeM established a carbon fund to offset the GHG emissions generated by employee professional trips (Université de Montréal, 2023) and integrated

[17] Our translation from French.

[18] Université de Montréal (2023). Scope 1: Direct emissions sources including heating buildings with natural gas, refrigerant system leaks, and emissions from service vehicles; Scope 2: Purchased electricity consists of hydroelectric power provided by Hydro-Québec; Scope 3: Indirect emissions includes procurement of goods and services, followed by daily commuting by students and employees, professional travel, and management of leased space.

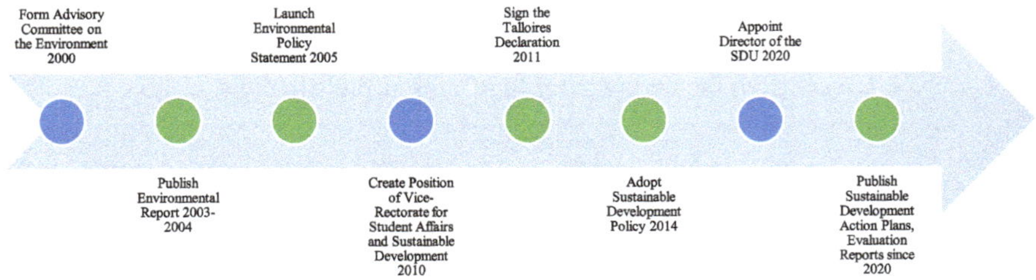

Fig. 5 Sustainability timeline

fossil fuel divestment into its overall strategy. The publication of the university's first verified carbon footprint report in 2023 has not only helped UdeM monitor its emission reduction efforts to achieve its targets, it has also provided a solid basis for measuring future progress and refining institutional emission reduction strategies—all this according to SDU's sustainable development coordinator and climate change advisor.[19]

At Université de Montréal, sustainability considerations have been integrated into governance structures and operations, from strategic planning and decision-making to training. Today, sustainability practices have moved beyond short-term isolated initiatives towards a long-term commitment that aligns with its vision. The sustainability timeline below shows key moments and the evolution of sustainable development at UdeM, starting in the year 2000 (Fig. 5).

Sustainable development has been institutionalized by defining roles and responsibilities, as well as establishing mechanisms for measuring, reporting, monitoring, and evaluating. In so doing, people, services, and units have been made accountable in terms of performance and improvement. For example, the guidelines for the implementation of the sustainable development policy outline have created accountability for specific goals and actions. The Vice-Rector in charge of sustainable development is responsible for its implementation, which is also overseen by a sustainable development committee. New positions (equity, diversity, and inclusion advisors, climate change and energy management) have been created to support the integration of sustainable development in departments and faculties (Université de Montréal, n.d.-d). In addition, the latter may obtain services offered by the SDU to fulfill their objectives—such as GHG emissions reduction—which could help them quantify GHG emissions and implement their reduction plans (Université de Montréal, 2023).

4.2.2 Inter-institutional Collaboration

To mobilize research and training activities, Université de Montréal and its two affiliated schools (Polytechnique Montréal and HEC Montréal) created an online platform (LA REF) providing university community members with resources to learn about sustainability teaching and training methods, as well as approaches

[19] LaSalle (2024).

across disciplines. They built a multi-disciplinary project (CLAD) that offers a forum where people can discuss environmental and societal issues.[20] Furthermore, an online toolbox was conceptualized to facilitate the integration of sustainable development in research practices, from responsible procurement, management (laboratory residual materials, digital tools), and professional mobility to the pooling of equipment and infrastructure.[21]

Interdisciplinary and multi-institutional projects such as Chemins de transition and One Health Initiative[22] have been established to rally researchers, experts, practitioners, and decision-makers around climate change adaptation, decarbonization, and health issues at the human-animal-environment interface. Chemins de transition collaborates with Ouranos, a consortium on regional climatology and adaptation to climate change that deals with decarbonization challenges.[23] One Health Initiative, on its end, adopts cross-sectoral teaching and co-construction of knowledge practices, including the participation of Indigenous peoples and marginalized communities.[24] Recently, a philanthropic chair was created for transitioning to carbon neutrality. Aiming to identify, understand, and anticipate problems before they arise, the chair will develop methods that accelerate the transition to carbon neutrality and mitigate its impacts from social, environmental, political, economic, and technical perspectives.[25]

4.2.3 Community of Change Agents

The SDU deployed community mobilization by creating a student led eco-team. The team was tasked with helping launch an eco-responsible newsletter and conducting awareness campaigns around reducing single-use containers and improving residual materials management. The unit also designed educational programs such as the SDG competition, in addition to holding workshops and training courses on eco-responsibility and sustainability integration. Conferences and workshops were held to stimulate cross-faculty collaboration in sustainability research. A competition awarding $100,000 was launched by Innovation Laboratory to advance new, eco-responsible solutions.[26] Supported by their institutions, community members could actively work to improve their local areas, addressing issues and fostering positive

[20] Université de Montréal (n.d.-j).

[21] Université de Montréal. (n.d.-k).

[22] A transdisciplinary approach that aims to balance and optimize the health of people, animals, and ecosystems, responding collectively, responsibly, and sustainably to a range of global issues from a common good perspective (CDC, 2025). Retrieved from https://www.cdc.gov/one-health/about/index.html#:~:text=One%20Health%20is%20a%20collaborative%2C%20multisectoral%2C%20and%20transdisciplinary,between%20people%2C%20animals%2C%20plants%2C%20and%20their%20shared%20environment.

[23] https://cheminsdetransition.org/en/.

[24] https://uneseulesante.umontreal.ca/accueil/.

[25] UdeMNouvelles (2025).

[26] Université de Montréal. (n.d.-l).

transformation. A podcast series on eco-anxiety was created by and for students,[27] enabling participants to share views, learnings, expertise, and resources on a phenomenon that is on the rise among young people.

In parallel with teaching and research activities—and by collaborating with their colleagues and civil society—researchers equipped people with the skills to make environmentally sustainable choices beyond knowledge transfer. Research team leaders obtained financial support from private and public funding organizations to help create new projects on climate change and train professionals and students in field research. Various efforts aligned with collective goals. For instance, the mathematics research centre published articles covering topics related to the SDGs, highlighting the role of mathematics in mapping poverty, measuring the gender gap, modelling a pandemic, predicting climate change, and measuring biodiversity.[28]

Prior to the 26th United Nations Climate Change Conference, ten researchers made themselves available to discuss themes that were addressed with the media at COP26.[29] A political scientist contributed to creating an interactive, free, downloadable map illustrating how Canadians view climate change.[30] Two public health researchers participated in producing a podcast on climate displacement, sharing experiences of people in vulnerable situations.[31] Using global warming data collected by a Université de Montréal geomorphologist's team, a professor from the department of medicine, an Inuit composer, and pop-rock and throat singers presented their co-production of songs and a documentary to the public at a museum, inspiring people in the fight against global warming.[32]

4.3 Goal Achievement

An internal indicators report demonstrated progress towards meeting the university's goals and targets. Université de Montréal has completed or is on the way to accomplishing all items (46) set out in the Sustainable Development Action Plan 2021–2023. Nearly $18 million has been committed to infrastructure-dedicated funding for green energy projects and those linked to the net zero carbon transition. It has also addressed items stipulated (68) in the Sustainable Development Action Plan 2024–2029. UdeM has completed carbon footprint scopes 1 and 2 and had them verified by a third party (Université de Montréal, n.d.-d, 2023). As of December 31, 2023, all external equity and bond managers had identified and analyzed the short-, medium-, and long-term climate risks and opportunities for the university's portfolio,

[27] https://laboinnovation.umontreal.ca/index.php?id=33935.

[28] St-Cyr-Leroux (2022).

[29] UdeMNouvelles (2021).

[30] Heinrich (2021).

[31] Nouvelles (2024).

[32] Letarte (2025).

integrated climate change risks and opportunities into the investment process, and measured the portfolio's carbon footprint (Université de Montréal, 2023).

International performance metrics such as rankings have been used by UdeM's administrators as quantitative and qualitative measures that assess sustainability performance as compared with other institutions. Université de Montréal moved up 14 and 16 places to 28th place in the Times Higher Education (THE) 2024 Impact Ratings[33] (25th for action on climate change SDG 13) and came in 65th place in the QS World University Rankings: Sustainability 2025.[34] Specifically, sustainable development teaching and its influence climbed 78 places to rank 68th in the teaching category, representing a rise of 48 (85th). These rises in rankings indicate that the university has made progress in the UN SDGs and teaching.

Being recognized for its success, the senior management has been motivated to engage in a process that helps track progress against goals, identify areas for improvement, and make informed decisions about policy design. To fight against global warming, Université de Montréal has been taking the necessary steps to reduce its emissions on the road towards carbon neutrality—from stating goals, adapting strategies, fixing timetables, implementing action plans, and measuring outcomes to forging a continuing feedback loop. The university has played a key role in advancing knowledge and identifying evidence-based solutions that inform policy. By collaborating with a wide range of stakeholders and sharing resources, it has engaged in the management of its own operations and infrastructure to build local resilience and deliver positive, tangible, and intangible environmental and social impacts.

4.4 Discussion

The findings have confirmed our hypothesis. The university has tackled sustainability issues through interdisciplinarity and engaged communities to achieve desirable outcomes around sustainable development that encompass economic prosperity, environmental conservation, social equity, and beyond. To address climate-related issues, Université de Montréal aligned GHG emissions reduction with institutional sustainability strategies. Collaboration and partnership have amplified their efforts, from reducing the impact of resources and infrastructure to promoting changes in attitude and behaviour when it comes to climate action.

At Université de Montréal, available courses on climate remained piecemeal, despite its growing emphasis on climate change. There is room for climate inclusion in classes. For instance, faculty members could help pinpoint opportunities within their schools and departments to draw up a plan that advances climate education. The university could then develop a climate change education strategy that aligns climate change research agendas and related capacities, as our analysis revealed that climate change teaching was connected to research activities. On the contrary, studies

33 THE (2024).

34 Topuniversities.com (2025).

(Liu et al., 2022; Navarrete-Welton et al., 2022) have shown that a lack of research on climate change and its impacts make it difficult to cultivate student interest and engagement both inside and outside the classroom.

The treatment of climate change content across disciplines at Université de Montréal is within sustainability education curricula, with input from socioeconomic and environmental considerations that emphasize interconnectedness with science and the humanities. The credit courses offered appear to be taught within SDG 13's target (13.3), which aims at "improving education, awareness-raising, and human and institutional capacity on climate change mitigation, adaptation, impact reduction, and early warning" (United Nations, n.d.). To prevent fragmentation and dispersal across climate change curricula (Eilam, 2024), it may behoove UdeM to develop a coherent vision that enables the systematic and organized construction of climate change knowledge.

Furthermore, the university would be well positioned to implement curriculum changes by drawing on the expertise of sustainability leaders—as well as Indigenous perspectives, knowledge, and practices—to view newly created analysis models through a different, enriched lens. To build a dynamic learning environment and foster community-based initiatives, institutions could create living labs and incorporate research design alongside Indigenous communities. Instructors could apply problem-based learning methods to engage learners with real-world sustainability challenges and provide them opportunities to implement knowledge and applications. Such initiatives would enrich their educational experience and empower students to make behavioural and attitude changes. According to a review of CCE literature (Monroe et al., 2019), activities, deliberative discussions, and school or community projects have been proven to be effective.

At a micro level, using a multi-stakeholder approach to weave meaningful climate change information into current curricula (Monroe et al., 2019; Sullivan et al., 2022) may boost learner interest and engagement. At an organizational level, building a bridge between faculty members from different schools and departments could maximize synergies. Given that research grants on climate change have so far been predominantly provided to research in physical sciences, health, and technological innovation, universities could offer additional funding support and encourage multi- and transdisciplinary research projects in social sciences and humanities. Doing so would create new knowledge and support teachers to build capability, effectively breaking silos. They could promote bottom–up initiatives led by professors committed to bridging social sciences, the humanities, and natural and applied sciences for CCE.

5 Conclusion

Focusing on measurable goals and objectives, Université de Montréal has integrated sustainable development into its decision-making and strategies. Strong leadership and a clear vision have strengthened its short-term and long-term commitments.

Integrating sustainability into all aspects of the organization is complex, which requires good governance practices from planning steps to tracking, monitoring, evaluating, and adjusting priorities. Engaging with multiple stakeholders is essential in the implementation process. Significant efforts have been made to mobilize people, including designated individuals within the university, employees, professors, students, community members, partners, and collaborators from other institutions. They have acted as transmitters and implementers who advance change and turn obstacles into possibilities to achieve their goals. They have managed to build more sustainable communities and promote social changes via thinking and acting sustainably.

Interdisciplinary sustainability education is at the vanguard of the transition to a more sustainable society. Integrating climate change into curricula across schools and departments can address the multi-dimensional aspect of climate change challenges. Climate change curriculum development is a process involving not only scientific research, but also diverse knowledge systems and perspectives. The university could implement a framework to bridge these systems and perspectives for improving responses to complex challenges. HEIs can play their part in changing social practices and building a sustainable future. To achieve sustainable development, sustainability practices must extend beyond university walls to span participatory research, training students to work with communities and help develop capacities, forming knowledge networks, and mapping pathways to long-term success.

The contribution of this paper is threefold. Taking stock of the existing sustainability strategies and activities can serve to inform educators, policymakers, and practitioners. Such a study would provide insight into using a multipronged approach to implement sustainable development practices and contribute to finding pathways towards a more sustainable future. However, our study was constrained by the exact search terms we employed, which may not capture certain pertinent articles in the literature review. Moreover, the single institution study made it difficult to generalize findings. Despite its limitations, the lessons learned stemming from our case study examples—such as interventions in climate change—provided opportunity for reflection about how an HEI has integrated sustainability into its systems and achieved its goals. Additionally, evidence uncovered by the study, such as close relations between climate change teaching and research activities, may lead to insights relevant for climate change curriculum development. We suggest identifying key actors to participate in future study and deepen this research. In light of new insights into CCE, further study could analyze climate change curriculum implementation challenges and facilitators in an organizational context.

References

Centers for Disease Control and Prevention (CDC). (2025). *One health.* Retrieved from https://www.cdc.gov/one-health/about/index.html

Eilam, E. (2024). Interrogating climate change education epistemology: Identifying hindrances to curriculum development. *ECNU Review of Education, 8*(1), 112–143. https://doi.org/10.1177/20965311241240491

European Commission/EACEA/Eurydice. (2024). *Learning for sustainability in Europe: Building competences and supporting teachers and schools [Eurydice report].* Publications Office of the European Union.

Feinstein, N. W., & Mach, K. J. (2020). Three roles for education in climate change adaptation. *Climate Policy, 20*(3), 317–322. https://doi.org/10.1080/14693062.2019.1701975

Filho, W. L., Sima, M., Sharifi, A., Luetz, J. M., Salvia, A. L., et al. (2021). Handling climate change education at universities: An overview. *Environmental Sciences Europe, 33*(1), 109. https://doi.org/10.1186/s12302-021-00552-5

Hajer, M. A. (1995). *The politics of environmental discourse: Ecological modernization and the policy process.* Clarendon Press.

Hampshire, K., Islam, N., Kissel, B., Chase, H., & Gundling, K. (2022). The planetary health report card: A student-led initiative to inspire planetary health in medical schools. *The Lancet Planetary Health, 6*(5), e449–e454. https://doi.org/10.1016/s2542-5196(22)00045-6

Heinrich, J. 2021. *Réchauffer l'opinion publique* [Warm up public opinion]. Retrieved from https://nouvelles.umontreal.ca/article/2021/11/10/rechauffer-l-opinion-publique/

IPCC. (2022). Climate Change 2022: Impacts, adaptation, and vulnerability. In *Contribution of Working Group II to the sixth assessment report of the intergovernmental panel on climate change.* Cambridge University Press.

Kwauk, C. (2022). *The climate change education ambition report card: An analysis of updated Nationally Determined Contributions submitted to the UNFCCC and National Climate Change Learning Strategies.* Education International. https://www.ei-ie.org/en/item/25344:the-climate-change-education-ambition-report-card

LaSalle, M. (2024). *Université de Montréal Releases Its—First carbon footprint report.* Retrieved from. https://nouvelles.umontreal.ca/en/article/2024/12/12/universite-de-montreal-releases-its-first-carbon-footprint-report/

Letarte, M. (2025). *L'art inuit pour sensibiliser au-réchauffement climatique* [Inuit art to raise awareness of global warming]. Retrieved from https://nouvelles.umontreal.ca/article/2025/06/18/l-art-inuit-pour-sensibiliser-au-rechauffement-climatique/

Liu, I., Rabin, B., Manivannan, M., Laney, E., & Philipsborn, R. (2022). Evaluating strengths and opportunities for a co-created climate change curriculum: Medical student perspectives. *Frontiers in Public Health, 10*, 1021125. https://doi.org/10.3389/fpubh.2022.1021125

Mol, A. P. J. (1995). *The refinement of production modernization theory and the chemical industry.* Jan van Arkel/International Books.

Mol, A. P. J. (2000). The environmental movement in an era of ecological modernisation. *Geoforum, 31*(1), 45–56.

Mol, A. P. J., & Jänicke, M. (2009). The origins and theoretical foundations of ecological modernisation theory. In A. P. J. Mol, D. A. Sonnenfeld & G. Spaargaren (Ed.), *The ecological modernisation reader* (pp. 17–27).

Molthan-Hill, P., Worsfold, N., Nagy, G. J., Filho, W. L., & Mifsud, M. (2019). Climate change education for universities: A conceptual framework from an international study. *Journal of Cleaner Production, 226*, 1092–1101. https://doi.org/10.1016/j.jclepro.2019.04.053

Monroe, M. C., Plate, R. R., Oxarart, A., Bowers, A., & Chaves, W. A. (2019). Identifying effective climate change education strategies: A systematic review of the research. *Environmental Education Research, 25*(6), 791–812. https://doi.org/10.1080/13504622.2017.1360842

Nature Positive Universities. (n.d.). Our Purpose. Retrieved from https://www.naturepositiveuniversities.net/

Navarrete-Welton, A., Chen, J. J., Byg, B., Malani, K., Li, M. L., et al. (2022). A grassroots approach for greener education: An example of a medical student-driven planetary health curriculum. *Frontiers in Public Health, 10*, 1013880. https://doi.org/10.3389/fpubh.2022.1013880

Nouvelles. 2024. *Le projet "Balado sur les déplacements climatiques: Récits des personnes en situation de vulnérabilité" des professeures Muriel Mac-Seing et Mylène Ratelle reçoit un financement du RRQSP dans le cadre de leur appel à projet Climigrant* [The project "Podcast on climate displacement: Stories of people in vulnerable situations" by professors Muriel Mac-Seing and Mylène Ratelle receives funding from the RRQSP as part of their Climigrant call for projects]. Retrieved from https://espum.umontreal.ca/lespum/nouvelles/nouvelle/news/detail/News/le-projet-balado-sur-les-deplacements-climatiques-recits-des-personnes-en-situation-de-vulnerabil/

O'Dwyer, S., Geoghegan, E., Nisonen, E., Rosa, R. C.-D. L., Pelsmakers, S., et al. (2023). Architectural education: Methods for integrating climate change design (CCD) in the curriculum. In *AMPS: A focus on pedagogy: Teaching, learning and research in the modern academy* (Vol. 28, pp. 167–189). https://orca.cardiff.ac.uk/id/eprint/157096/

Omrani, O. E., Dafallah, A., Castillo, B. P., Amaro, B. Q. R. C., Taneja, S., et al. (2020). Envisioning planetary health in every medical curriculum: An international medical student organization's perspective. *Medical Teacher, 42*(10), 1107–1111. https://doi.org/10.1080/0142159x.2020.1796949

Pal, A., Tsusaka, T. W., Nguyen, T. P. L., & Ahmad, M. M. (2023). Assessment of vulnerability and resilience of school education to climate-induced hazards: A review. *Development Studies Research, 10*(1), 2202826. https://doi.org/10.1080/21665095.2023.2202826

SHARE. (2025). *University Network for Shareholder Engagement (UNIE)*. Retrieved from https://share.ca/initiatives/unie/

Spaargaren, G. (1997). *The ecological modernization of production and consumption essays in environmental sociology*. Wageningen University and Research.

Spaargaren, G., & Mol, A. P. J. (1992). Sociology, environment, and modernity: Ecological modernization as a theory of social change. *Society & Natural Resources, 5*(4), 323–344.

St-Cyr-Leroux, B. (2022). *ODD: une nouvelle ressource pour soutenir la prise de décision fondée sur la science* [SDGs: a new resource to support science-based decision-making] Retrieved from https://nouvelles.umontreal.ca/article/2022/03/15/odd-une-nouvelle-ressource-pour-soutenir-la-prise-de-decision-fondee-sur-la-science/

Sullivan, J. K., Lowe, K. E., Gordon, I. O., Colbert, C. Y., Salas, R. N., et al. (2022). Climate change and medical education: An integrative model. *Academic Medicine, 97*(2), 188–192. https://doi.org/10.1097/ACM.0000000000004376

Takinana, A., & Baars, R. C. (2022). Climate change education in the south pacific: resilience for whom? *Asia Pacific Viewpoint, 64*(1), 72–84. https://doi.org/10.1111/apv.12358

THE. (2024). The ranking measures the actions taken by universities to advance the United Nations (UN) Sustainable Development Goals (SDGs), both on their campuses and beyond. The Impact 2024 ranking assessed over 1960 academic institutions worldwide. Retrieved from https://www.timeshighereducation.com/

Topuniversities.com. (2025). Using a methodology comprised of indicators designed to measure an institution's ability to tackle the world's environmental, social, and governance (ESG) challenges, QS' 2025 sustainability rankings feature nearly 1,800 universities from around the world. Retrieved from https://www.topuniversities.com/sustainability-rankings?tab=indicators&search=universit%C3%A9%20de%20montreal&sort_by=indicator_3824995&order_by=asc

UdeMNouvelles. (2025). *Une nouvelle chaire philanthropique pour faciliter la transition vers la carboneutralité* [A new philanthropic chair to facilitate the transition to carbon neutrality]. Retrieved from https://nouvelles.umontreal.ca/article/2025/04/09/une-nouvelle-chaire-philanthropique-pour-faciliter-la-transition-vers-la-carboneutralite/

UdeMNouvelles. (2021). *COP 26: des experts de l'UdeM prêts à éclairer les débats COP 26* [COP26: UdeM experts ready to shed light on the debates]. Retrieved from https://nouvelles.umontreal.ca/article/2021/10/26/cop-26-des-experts-de-l-udem-prets-a-eclairer-les-debats/

UNEP. (2024). Emissions Gap Report 2024. Nairobi: United Nations. Retrieved from https://doi.org/10.59117/20.500.11822/46404

UNESCO and MECCE. (2023). *Climate change communication and education country profiles: Approaches to greening education around the world*. Global Education Monitoring Report.

UNESCO and MECCE. (2024). *Education and climate change: Learning to act for people and planet*. UNESCO.

United Nations. (n.d.). *Goal 13*. Retrieved March 22, 2025, from https://sdgs.un.org/goals/goal13

Université de Montréal. (n.d.-a). *Développement durable à l' Université de Montréal – Planification stratégique 2021–2023* [Sustainable Development at Université de Montréal—Strategic Plan 2021–2023].

Université de Montréal. (n.d.-b). *Plan d'action en développement durable 2024–2029* [Sustainable Development Action Plan 2024–2029].

Université de Montréal. (n.d.-c). *Rapport d'investissement responsable 2023* [Responsible Investment Report 2023].

Université de Montréal. (n.d.-d). *Transition vers des campus durables Rapport-bilan du Plan d'action en développement durable 2021–2023* [Transition to Sustainable Campuses Report on the Sustainable Development Action Plan 2021–2023].

Université de Montréal. (n.d.-e). Enseignement [Education]. Retrieved from https://durable.umontreal.ca/enseignement-recherche/enseignement/

Université de Montréal. (n.d.-f). Programmes de 1er cycle en environnement et développement durable [Undergraduate programs in environment and sustainable development]. Retrieved from https://admission.umontreal.ca/programmes-de-1er-cycle/environnement-et-developpement-durable/

Université de Montréal. (n.d.-g). *Application mobile Votre Empreinte* [Mobile app Your Footprint]. Retrieved from https://durable.umontreal.ca/empreinte-carbone/application-mobile-votre-empreinte/

Université de Montréal. (n.d.-h). Historique en biodiversité [Biodiversity history]. Retrieved from https://durable.umontreal.ca/en/amenagement-biodiversite/flore/historique-en-biodiversite/

Université de Montréal. (n.d.-i). Program managed by Shareholder Association for Research and Education (SHARE).

Université de Montréal. (n.d.-j). *LA REF Ressources écoresponsables pour la formation* [Eco-friendly resources for training]. Retrieved from. https://durable.umontreal.ca/enseignement-recherche/la-ref/

Université de Montréal. (n.d.-k). *Écoresponsabilité en recherche* [Eco-responsibility in research]. https://durable.umontreal.ca/enseignement-recherche/ecoresponsabilite-en-recherche/

Université de Montréal. (n.d.-l). *4e édition de l'appel à projets - Écoresponsabilité de la recherche* [4th edition of the call for projects - Environmental responsibility in research]. Retrieved from https://laboinnovation.umontreal.ca/quatre-grands-projets/construire-lavenir-durablement/nouvelles-du-projet/nouvelle/news/detail/News/4e-edition-de-lappel-a-projets-ecoresponsabilite-de-la-recherche/

Université de Montréal. (2014). *Sustainable development policy*. Montréal, Canada.

Université de Montréal. (2022). *Document about sustainability course list*. Retrieved from https://durable.umontreal.ca/fileadmin/durable/documents/UdeM_STARS_Sustainability_Course_List_A2022_final_en-US.pdf

Université de Montréal. (2023). *Empreinte carbone: Rapport des émissions de GES sur la période du 1er mai 2022 au 30 avril 2023* [Carbon footprint Report on GHG emissions for the period from May 1, 2022, to April 30, 2023].

Université de Montréal. (2025a). *En chiffres* [In figures]. Retrieved from https://www.umontreal.ca/l-udem/en-chiffres/

Université de Montréal. (2025b). *The MIL Campus*. Retrieved from https://www.umontreal.ca/en/our-campuses/the-mil-campus/

Université de Montréal. (2025c). *Parcourir par Direction de recherche* [Browse by research direction]. https://umontreal.scholaris.ca/browse/advisor

Yuan Zheng Li holds a Ph.D. in sociology with a specialization in environment. She has been engaging in and promoting interdisciplinary collaborations that help advance HEIs' sustainability, climate, and wellbeing commitments. Her research focuses on sustainability implementation in the fields of sociology, environment and development.

Policy Innovations for Sustainability Education: Challenges, Opportunities and Lessons from Namibia and Ghana in Green Curriculum Development

Philip Mensah, Eric Yankson, and Derrick Frimpong Buabeng

Abstract As global environmental and socio-economic challenges intensify, integrating sustainability into education has become essential for achieving sustainable development. This chapter explores how curriculum development processes in sub-Saharan Africa—specifically in Namibia and Ghana—are evolving to embed sustainability principles in national education systems. The chapter aims to assess the extent and effectiveness of green curriculum integration, identify enabling policy innovations, and examine the pedagogical and institutional frameworks supporting sustainability education. Using a comparative case study approach, the study draws on policy analysis, curriculum reviews, and interviews with policymakers, educators, and curriculum developers, complemented by secondary data from government and international frameworks. Findings reveal commendable progress in both countries through cross-sectoral collaboration, competency-based reforms, and localized curriculum models. However, challenges such as fragmented policies, inadequate teacher training, and limited resources persist. The chapter concludes by emphasizing the need for coherent, context-responsive, and transformative curriculum strategies that position sustainability as a foundational element of education reform, offering critical lessons for developing countries aligning their education systems with the sustainable development agenda.

Keywords Curriculum development · Education policy · Green curriculum · Policy innovations · Sustainability education

P. Mensah (✉) · D. F. Buabeng
Department of Land and Spatial Sciences, Namibia University of Science and Technology, Windhoek, Namibia
e-mail: pmensah.st.nust.na@gmail.com; pm319834@gmail.com

E. Yankson
Department of Architecture, Planning and Construction, Namibia University of Science and Technology, Windhoek, Namibia
e-mail: eyankson@nust.na; ericyankson22@gmail.com

W. Leal Filho et al., *Curriculum Development for Sustainable Development*, World Sustainability Series, https://doi.org/10.1007/978-3-032-22146-9_9

1 Introduction

Quality education plays a pivotal role in promoting sustainable development, environmental consciousness, and long-term economic resilience. Integrating sustainability into educational policy and curriculum development is therefore essential, especially for countries aiming to align their systems with the 2030 Agenda for Sustainable Development, particularly SDG 4 (quality education) and SDG 13 (climate action) (Agbedahin, 2019; Shulla et al., 2020). In this regard, policy innovations in sustainability education are critical. Namibia and Ghana, two sub-Saharan African countries with distinct post-independence education paths, provide important comparative cases for examining the challenges, opportunities, and lessons involved in embedding sustainability into national curricula.

Since Namibia's independence in 1990, the government has aimed to transform the apartheid-inherited education system into one that promotes equity, inclusivity, and relevance to global challenges. The Ministry of Education, Arts and Culture has undertaken several curriculum reforms to embed Environmental and Sustainability Education (ESE) within subjects such as Natural Science and Social Studies. Despite this progress, the translation of these policies into effective classroom practices remains inconsistent due to challenges such as limited teacher training, inadequate materials, and weak institutional coordination (Anyolo, 2015; Phuthi, 2021). Although the Education and Training Sector Improvement Program (ETSIP) introduced in 2006 sought to reform the system, alignment between curriculum content and sustainability principles remains incomplete.

Similarly, Ghana has made efforts to integrate sustainability into its national education framework, particularly after the 2007 education reform. Through the Ghana Education Service (GES) and the National Council for Curriculum and Assessment (NaCCA), the government revised the curricula for both basic and senior high school education. The 2019 introduction of the Standard-Based Curriculum (SBC) emphasized critical thinking, ecological awareness, and problem-solving (Ahovi, 2025; Kwarteng, 2024; Setordzi, 2023). Nonetheless, implementation challenges persist, including weak institutional support, gaps in teacher readiness, and a lack of systems to monitor the effectiveness of green curriculum delivery.

At the global level, United Nations Educational, Scientific and Cultural Organization's—Education for Sustainable Development *(UNESCO's—ESD) for 2030 Roadmap* emphasizes curriculum innovation as key to cultivating environmentally responsible citizenship (Chasokela, 2024; Leicht & Byun, 2021; UNESCO, 2020). Yet, translating such frameworks into national strategies requires adaptive approaches that consider specific socio-economic conditions such as poverty, inequality, and infrastructural deficits. In many developing contexts like Namibia and Ghana, ESE implementation is constrained by limited ministerial capacity, overreliance on donor-driven projects, and conflicting national development priorities.

Regionally, the African Union's *Continental Education Strategy for Africa (CESA 2016–2025)* calls on member states to mainstream ESD in education systems. Both Namibia and Ghana have committed to this agenda, though practical outcomes vary. Namibia has collaborated with non-governmental organizations (NGOs) and international agencies to pilot environment-based curricula, but such initiatives are often fragmented and lack national coherence (African Union, 2015; Tubaundule, 2014). Ghana's Environmental Protection Agency (EPA), through school outreach programmes, has promoted local ESE awareness, though these efforts are limited by inconsistent funding and weak institutionalization (Klawinski, 2022).

At the local level, school communities in Namibia—particularly in the arid Kunene and Omaheke regions—face harsh environmental conditions that make ESE crucial (Hailombe, 2011). Yet, many rural schools lack basic resources such as water, electricity, and trained personnel, resulting in a largely theoretical treatment of sustainability issues. In Ghana, similar disparities exist. Rural schools in the Northern and Upper West regions lack infrastructure, while urban schools often struggle to connect sustainability content to learners' daily realities, reducing curriculum impact (Anlimachie, 2022). A key gap in both countries is the limited incorporation of sustainability principles into teacher education. Pre-service and in-service training rarely include structured modules on ESD, leaving many educators ill-equipped to implement green curricula. This underscores the need for more innovative and context-sensitive curriculum frameworks that are aligned with broader sustainability goals.

This study is therefore motivated by the need to examine how policy innovations in Namibia and Ghana support the integration of sustainability education into national curricula. It explores how both countries interpret and implement sustainability education, identifies the challenges faced, and investigates policy opportunities that can inform more effective curriculum reform. The research also aims to generate empirical insights into how existing policies are being translated into educational practice and to assess institutional readiness for promoting ESE. While policy frameworks in both countries reflect a strong rhetorical commitment to sustainability, implementation remains uneven. This study seeks to bridge that gap by examining how curriculum development for sustainable development can serve as both a transformative tool and a policy instrument for long-term ecological and social resilience.

In doing so, the research contributes to the global discourse on curriculum innovation for sustainable development. It recognizes that green curriculum development is not only a technical undertaking but also a political, pedagogical, and ethical process. For countries like Namibia and Ghana, where development pressures and climate risks intersect with educational equity issues, ESE offers a critical pathway for cultivating the knowledge, skills, and values needed for sustainable futures.

2 Theoretical Review

The imperative for sustainability education has gained prominence globally as education systems seek to integrate environmental stewardship, social equity, and economic viability into their curricula. This transformation, often referred to as "green curriculum development," is underpinned by international frameworks such as the United Nations Sustainable Development Goals (SDGs), particularly SDG 4.7, which emphasizes education for sustainable development and global citizenship (Dutra, 2025; UNESCO, 2020). The push for education that transcends traditional pedagogical boundaries and inculcates environmental consciousness, critical thinking, and civic engagement is no longer a marginal concern but a central tenet in educational reform discourses.

In the context of Namibia and Ghana, where education systems are intricately tied to developmental trajectories, policy innovations around green curriculum development are both a necessity and a challenge. The conceptual framework of sustainability education is deeply rooted in the paradigm of transformative learning, which posits that education should not only transmit knowledge but also transform the way individuals perceive and interact with their socio-ecological realities. This aligns with Mezirow's theory of transformative learning, which emphasizes critical reflection and consciousness-raising as mechanisms for meaningful change in attitudes and behavior (Mezirow, 2018). Within sustainability education, this transformation encompasses ecological literacy, systems thinking, and an ethical orientation toward future generations (Warren et al., 2014).

The infusion of sustainability into educational policies requires a shift from a transmissive to a transformative model, calling for curriculum structures that integrate environmental ethics, indigenous knowledge systems, climate literacy, and community engagement. Policy innovation, in this regard, must be understood not merely as new instruments or mechanisms, but as a fundamental rethinking of the purposes and outcomes of education. Theoretical insights from institutional theory suggest that educational policies are shaped by institutional logics, normative frameworks, and actor interests (Hanson, 2001). In contexts like Namibia and Ghana, where postcolonial legacies and contemporary development agendas intersect, policy innovation is contingent upon navigating bureaucratic inertia, resource constraints, and socio-political complexities.

Theories of policy change, such as Kingdon's multiple streams framework, elucidate how windows of opportunity emerge when problems, policies, and politics align (Shephard et al., 2021). For green curriculum development, such alignment may be catalyzed by environmental crises, donor support, or shifts in national development priorities. Education for Sustainable Development (ESD) has also been framed through critical pedagogy, as advocated by Paulo Freire, which calls for an emancipatory approach that empowers learners to question dominant narratives and become agents of change (Dutra, 2025). This perspective is relevant in postcolonial African contexts where curricula often reflect Eurocentric knowledge systems, marginalizing indigenous ecological knowledge and local experiences (Morrow, 2008).

Integrating green principles into curricula must, therefore, involve more than environmental topics; it necessitates a reorientation of pedagogical methods to foster participatory learning, problem-solving, and ethical reasoning. Systems theory further supports this, positing that education systems function within broader ecological, economic, and social systems, requiring holistic approaches to curriculum design (Porter et al., 2009). Such systems thinking is central to sustainability education, which aims to illuminate the interconnectedness of human and natural systems. Practically, this involves embedding sustainability across disciplines rather than treating it as a standalone subject. Achieving this requires curricular coherence, teacher training, and institutional commitment. Namibia's Integrated Environmental Education Policy and Ghana's Environmental Education Strategy reflect attempts to mainstream sustainability, though implementation gaps remain, particularly in aligning national curricula with intended sustainability outcomes.

Curriculum theory also informs green education. Pinar's reconceptualist approach views curriculum as dynamic and reflective rather than fixed content (Pinar, 2013). This supports curriculum innovation that is responsive to environmental and social change and values learners' lived experiences. Experiential learning theories, such as those by Abdallah (2025), reinforce the value of hands-on, place-based learning in fostering sustainability competencies. Policy transfer and borrowing frameworks also offer insight. Phillips and Ochs (2004) caution that models from the Global North may not align with African socio-cultural realities. In Namibia and Ghana, policy innovations must be contextually relevant, participatory, and localized. Balancing global imperatives with national strategies such as Namibia's Vision 2030 and Ghana's Education Strategic Plan is key.

Policy network theory sheds light on the multiplicity of actors involved, including government ministries, NGOs, donors, and communities (Stone, 2004). This interplay can create opportunities or bottlenecks. Donor-driven efforts can support reform but may impose external priorities. Local innovations may offer contextual solutions but require institutional support. Effective policy innovation thus requires inclusive governance, cross-sectoral coordination, and capacity building. The capabilities approach by Sen and Nussbaum emphasizes expanding freedoms and opportunities for learners (Buzzelli, 2015; Rajapakse, 2016). In education, this means cultivating capabilities for environmental citizenship and ethical reasoning, aligned with UNESCO's sustainability competencies (Parra et al., 2020).

Rogers' diffusion of innovation theory helps explain how green curricula spread. Adoption depends on perceived advantages, compatibility, and ease of implementation (Menzli et al., 2022). In Namibia and Ghana, factors such as exam-driven teaching, lack of resources, and teacher training affect uptake. Political economy theories also show that green curriculum development challenges dominant ideologies of growth and consumption. In Namibia and Ghana, education is often viewed through the lens of employability and growth, requiring a rethinking of its broader societal purposes (Henry, 2020). Theories of curriculum alignment stress coherence between objectives, instruction, and assessment (Jin et al., 2019). Misalignment leads to superficial engagement. Decentralization and teacher autonomy can enhance contextualization.

Finally, critical curriculum theorists advocate for ecological democracy and learner agency. A pedagogy of hope (Bourn, 2021) is essential in African contexts, promoting justice, resilience, and sustainability. Green curriculum development, then, must be participatory, justice-oriented, and grounded in local knowledge. Figure 1 illustrates the integrated conceptual framework guiding this study. At the

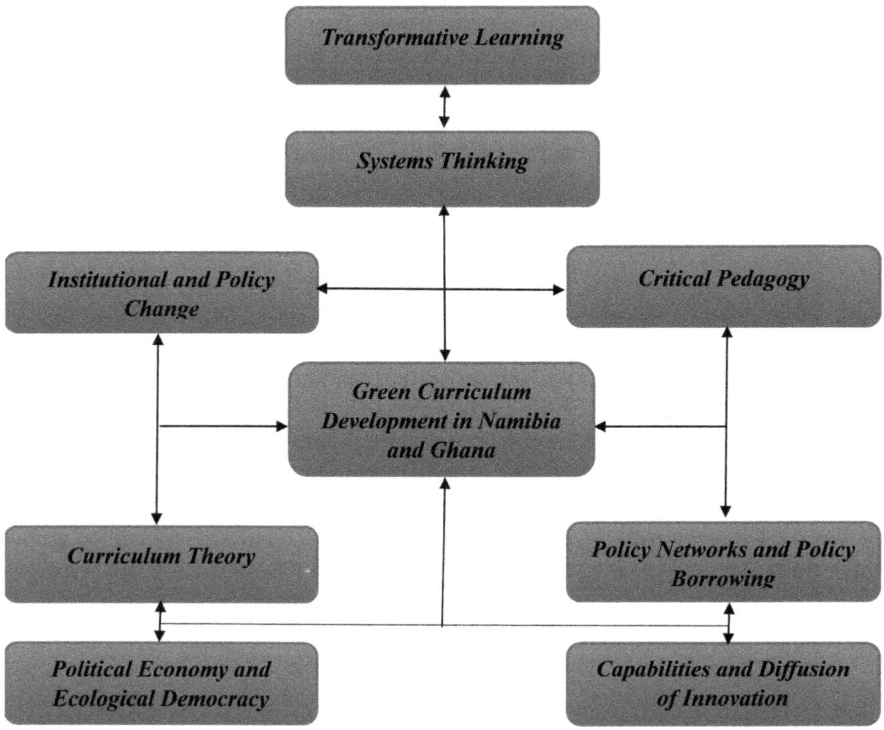

Fig. 1 Theoretical foundations guiding green curriculum development. *Source* Authors' own research (2025)

core lies the goal of green curriculum development in Namibia and Ghana, influenced by interrelated theories and perspectives. Transformative learning and systems thinking underpin the foundation, from which institutional change and critical pedagogy flow into curriculum development. Supporting frameworks such as curriculum theory, policy networks, political economy, diffusion of innovation, and the capabilities approach demonstrate the complex interplay between policy, pedagogy, and context that shape sustainability education in diverse settings.

3 Methodology

This qualitative study adopts a comparative interpretivist approach, suitable for examining how sustainability education policies are constructed, interpreted, and implemented by different actors within their local contexts (Duggan et al., 2015; Ishtiaq, 2019). The interpretivist paradigm supports exploration of how policymakers, educators, and civil society actors make sense of green curriculum development. The comparative lens provides analytical leverage to uncover cross-contextual lessons while acknowledging socio-political uniqueness. The research locale selection was informed by purposive sampling to reflect regional diversity, governance models, and reform trajectories. Namibia's centralized education policy contrasts with Ghana's decentralized system. These institutional differences are useful for examining how policy innovations are conceptualized and embedded in green curriculum processes. Both countries also participate in regional and global ESD platforms, allowing insights into how global frameworks are translated nationally (Oe et al., 2022).

The study relied primarily on document analysis and key informant interviews (KIIs). Document analysis included national education policies, curriculum frameworks, strategy papers, legislative texts, and sustainability reports. This allowed interrogation of how sustainability is framed, which tools are employed, and how institutional roles are assigned. To enhance depth, a critical policy analysis lens (Young & Diem, 2018) was employed to go beyond the text and examine how policies are formulated, by whom, and with what implications. Semi-structured interviews complemented the documentary analysis by capturing experiential and process-oriented insights. A total of 12 participants were deem appropriate and selected using purposive and snowball sampling, targeting officials from education ministries, curriculum developers, NGO representatives, teacher educators, and international development partners. Moreover, the study adheres to British Educational Research Association ethical guidelines such as informed consent, anonymity, and confidentiality assurance (Ramrathan et al., 2017).

Interview data were transcribed and analyzed using NVivo software, with codes derived both inductively (grounded theory) and deductively (guided by ESD principles). This allowed themes to emerge organically while ensuring alignment with values-based and systems-thinking criteria (Furst-Bowe, 2011). Thematic coding help reveal patterns of discourse, policy assumptions, and implementation logics across both countries. The temporal scope spans from 2000 to 2025 to capture policy shifts, reforms, and institutional learning processes. This historical approach supports understanding of how green curriculum efforts have evolved, institutional mandates have shifted, and how policy coherence (or incoherence) has affected progress. In other words, the desktop document analysis adhered to Maguire and Delahunt's (2017) six-phase framework for thematic content analysis (TCA), including data familiarization, initial coding, collating codes into potential themes, reviewing and refining themes, interpretation and synthesis of findings, and triangulation and validation.

An important methodological feature is the recognition of policy borrowing and translation (Burdett & O'Donnell, 2016; Phillips & Ochs, 2004). This perspective helps reveal how global sustainability norms are localized, reinterpreted, or resisted within specific national contexts. It also accounts for the influence of donor agencies, regional frameworks, and NGOs in shaping ESD trajectories (Burdett & O'Donnell, 2016; Lotz-Sisitka, 2016). The researchers maintained a reflective journal throughout data collection and analysis to manage positionality and interpretive bias. Regular peer debriefings, triangulation of data sources, and prolonged engagement with participants and documents contributed to credibility and rigor (McLeod,

2024). While no quantitative metrics were used, analytical precision was ensured through methodical coding, clear audit trails, and transparent documentation of interpretations.

Drawing on Uljens and Ylimaki (2017), the study treats curriculum not as a neutral output but a site of political and cultural negotiation. The analysis attends to the epistemological tensions in incorporating indigenous knowledge systems, local environmental contexts, and culturally grounded pedagogies. The decolonial education perspective (Zavala, 2016) informed analysis of whether green curricula in Namibia and Ghana reflect local realities or replicate externally imposed models. Thus, the integrated framework in Fig. 2 illustrates the methodological structure guiding the study's design, tools, analytical approach, and expected outcomes followed by a results and discussion section.

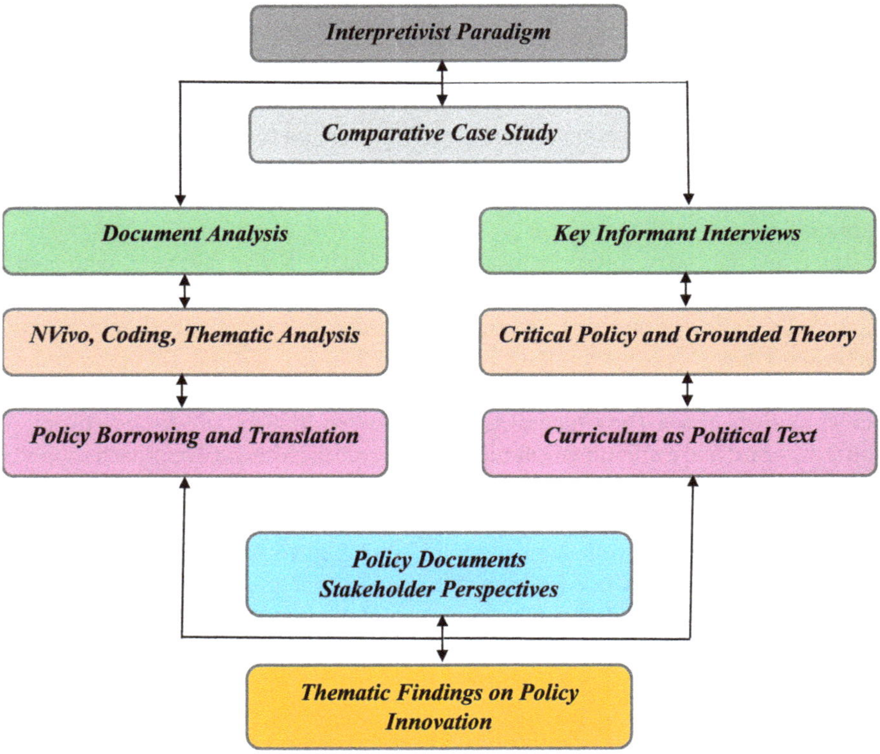

Fig. 2 Methodological framework for analyzing policy innovations in sustainability education. *Source* Authors' own research (2025)

4 Results and Discussion

Having discussed the emerging theories and analytical frameworks guiding this study, this section now presents and critically analyzes the success stories, patterns, and limitations associated with policy innovations for sustainability education in Namibia and Ghana. Through a synthesis of qualitative insights, policy review, and curriculum content analysis, the section unveils how green curriculum development has been shaped, contested, and operationalized within both national contexts. The discussion is informed by the comparative application of the methodological tools outlined earlier, emphasizing the intersections of policy formulation, pedagogical design, and institutional implementation across varying educational tiers.

4.1 Institutionalization of Sustainability Education

The first step in tackling the integration of sustainability into national education systems is to address the research question: *How are sustainability education principles integrated within the national education policy frameworks of Namibia and Ghana?* The empirical findings indicate that the integration of sustainability education principles into national policy frameworks has gained strategic momentum in both Namibia and Ghana, catalyzed by commitments to SDG 4 (quality education) and SDG 13 (climate action). National policy documents, curriculum frameworks, and interviews with education officers and curriculum developers demonstrate that while both countries are embedding sustainability education, the depth, scope, and consistency of implementation remain uneven.

In Namibia, sustainability principles are anchored in the 2019 National Environmental Education and Education for Sustainable Development (EE/ESD) Policy. This policy promotes inclusive, lifelong learning for sustainability across education levels. Interviewees at the Ministry of Environment, Forestry and Tourism (MEFT) confirmed that the policy was formulated through extensive stakeholder engagement, involving educators, NGOs, curriculum developers, and environmental policymakers. A curriculum officer from the National Institute for Educational Development (NIED) affirmed, "*The policy framework now gives us a mandate to align textbooks and teacher guides with the principles of sustainable living. The implementation is happening progressively, and by 2025/2026, we aim to have ESD fully mainstreamed.*"

Namibia's National Curriculum for Basic Education (2016) designates sustainability as a cross-cutting themes, with environmental topics introduced in Natural Sciences and Environmental Studies from Grades 1 to 7 (i.e. Junior and Senior Primary) and further expanded in secondary education. However, interviews with subject teachers, such as one in the Otjozondjupa region, revealed implementation bottlenecks: "*We teach about ecosystems and climate change, but we lack context-specific content and updated teaching resources. Sometimes the syllabus is ahead of what the materials allow us to deliver.*" These insights underscore the disconnect between curriculum policy and classroom reality, exposing systemic resource and capacity gaps.

In Ghana, the 2016 Climate Change and Green Economy Learning Strategy (CCGELS), developed by the Ministry of Environment, Science, Technology, and Innovation (MESTI), laid the foundation for sustainability education. This strategy influenced the 2018 National Pre-Tertiary Education Curriculum Framework, which mandates environmental education and green economy integration across all subjects. According to curriculum developers at the Ghana Education Service (GES), this framework targets transformative learning by embedding green themes across Science, Our World Our People, and Religious and Moral Education. A curriculum officer from the National Council for Curriculum and Assessment (NaCCA) noted, "*We were very deliberate about embedding climate change content in the new curriculum, and we ensured that all subjects, from Social Studies to Mathematics, include green thinking where appropriate.*"

Nevertheless, interviews with teachers revealed implementation gaps. A science teacher in the Ashanti region stated, "*We now have climate change in the syllabus, but no textbooks or experiments to demonstrate these concepts. We are improvising most of the time.*" The lack of resources and professional training hinders teachers' ability to deliver green education effectively, revealing a critical misalignment between policy ambition and implementation capacity. In other words, such testimonies underline the gap between curriculum policy and pedagogical practice, highlighting the need for coherent resource alignment and teacher capacity building.

Comparatively, Namibia offers a more structured and institutionally embedded Education for Sustainable Development (ESD) policy framework, supported by regional Environmental Education Centers and UNESCO alignment. Ghana, while demonstrating rapid curriculum innovation and global integration, struggles with cross-sector coordination and resource deficits. This comparison aligns with Salmen's (2018) and Mokski et al.'s (2023) argument that successful ESD requires both policy coherence and institutional readiness.

An additional finding is the role of international partnerships in strengthening national sustainability agendas. Namibia's collaboration with UNESCO and Southern African Development Community Education for Sustainable Development (SADC ESD) has helped competency-based curricula emphasizing systems thinking and participatory pedagogy. Ghana has received support from the United Nations Development Program (UNDP) and the Global Green Growth Institute (GGGI) to train teachers and pilot climate-focused modules. A Ghanaian stakeholder explained, "*These partnerships enabled us to test new ideas and build confidence in green curriculum delivery.*"

Table 1 synthesizes key policy and implementation variables across the two countries, offering a comparative lens through which policy innovations can be better understood. In line with comparative ESD studies in southern Africa (Phuthi, 2021), Namibia's collaborative, phased integration model reflects the value of participatory reform. Ghana's strategy illustrates how alignment with global policy trends

Table 1 Policy frameworks and curriculum integration comparatively

Dimension	Namibia	Ghana
Key policy frameworks	National EE/ESD Policy (2019), National Curriculum for Basic Education (2016)	Climate Change and Green Economy Learning Strategy (2016), National Curriculum Framework (2018)
Focus areas	Biodiversity, climate resilience, water management, energy literacy	Climate change, green economy, waste management, environmental responsibility
Curriculum integration approach	Cross-cutting themes in subjects; progressive integration	Embedded learning outcomes; subject-specific content and interdisciplinary emphasis
Stakeholder participation	Participatory curriculum development; involvement of civil society and environmental agencies	Multi-stakeholder consultations; inter-ministerial coordination
Resource availability	Moderate; environmental education centers and UNESCO support	Limited; reliance on donor support, textbook gaps, limited practical teaching aids
Implementation challenges	Resource gaps, misalignment between policy and practice	Inadequate teacher training, lack of textbooks, monitoring gaps
Strengths	Policy coherence, EE/ESD institutionalization, regional alignment	Strategic national positioning, green economy mainstreaming, policy innovation
Opportunities	Strengthened EE Centers, digital learning platforms, community involvement	International partnerships, curriculum review processes, green entrepreneurship modules

Source Authors' own research (2025)

can mobilize internal reform, albeit with risks of fragmentation. These variations affirm the importance of context-specific approaches in shaping green curriculum outcomes.

In summary, the first empirical objective demonstrates that Namibia and Ghana have established policy frameworks for sustainability education, yet challenges in implementation, resource alignment, and teacher preparation persist. By identifying these gaps, the findings reinforce the study's central aim: to explore how policy innovations can advance sustainability education through systemic, context-aware, and inclusive curriculum development.

4.2 *Barriers to Green Curriculum Implementation*

Having analyzed the extent to which green curricula are embedded within national education frameworks in Namibia and Ghana, the chapter now proceeds to address the question: *What are the major challenges facing the implementation of sustainability education in the curricula of Namibia and Ghana?* Findings reveal a range of systemic and institutional challenges that continue to impede the effective operationalization of sustainability education across both countries. These challenges—emerging from a qualitative thematic analysis of policy documents, curriculum reviews, and key informant interviews—highlight structural inconsistencies, capacity deficits, resource limitations, and pedagogical barriers.

A valid outcome of this study is that both Namibia and Ghana, despite their national commitments to ESD and green policy mandates, face persistent challenges stemming largely from implementation-related weaknesses. The uniqueness of this study lies in its comparative qualitative lens which explores cross-country lessons in green curriculum delivery—a gap that remains under-explored in existing sub-Saharan African literature (Lotz-Sisitka & Lupele, 2016; Zickafoose et al., 2024). These include: limited teacher capacity and professional development; inadequate curriculum integration; resource constraints; overcrowded curriculum and time constraints; absence of stakeholder engagement and collaboration; cultural and societal attitudes; policy and institutional challenges.

A significant challenge identified is the limited capacity of teachers to effectively deliver sustainability education. In Namibia, many educators reported a lack of adequate training and professional development opportunities related to environmental and sustainability topics. This deficiency hampers their ability to integrate sustainability concepts into their teaching practices (Wicke et al., 2024). Similarly, in Ghana, teachers expressed concerns about insufficient training and resources to effectively teach sustainability topics, often relying on outdated materials and lacking access to current information on environmental issues (Atuguba, 2019; Witoszek, 2018). The absence of sustainability education in pre-service teacher curricula in both countries further highlights this gap.

> As a curriculum developer in Namibia noted, "*We have the curriculum content, but many of our teachers were never trained on what ESD actually means or how to teach it. That is a huge problem in rural areas.*" A Ghanaian science teacher echoed this: "*Most of us did not study these topics in college, and now we have to teach them i.e. learners or student ESD related topics without support. It is overwhelming.*"

Both Namibia and Ghana have made efforts to incorporate sustainability education into their national curricula. However, these efforts have often been fragmented and lack coherence. In Namibia, sustainability topics are sometimes treated as add-ons rather than being fully integrated into the curriculum, leading to superficial coverage and limited impact (Kanyimba & Coetzer, 2011; Malua & Olvitt, 2024). Similarly, in Ghana, sustainability education is inconsistently embedded across subjects, resulting in a lack of continuity and depth in students' learning experiences (Agyemang, 2020; Darkwa, 2016).

> An interview with a Namibian headteacher revealed: "*Our Environmental Studies curriculum briefly mentions biodiversity conservation and water scarcity, but it is not embedded throughout. Learners learn facts, not systems thinking.*" This echoes broader critiques from ESD frameworks that integration should go beyond token mentions to truly transform learning (Griffiths, 2021; Kioupi & Voulvoulis, 2022).

Resource limitations also pose a significant barrier to the effective implementation of sustainability education. In both Namibia and Ghana, schools often lack the necessary materials, infrastructure, and financial support. The shortage of teaching aids, laboratory equipment, and digital resources limits the ability of educators to provide hands-on learning experiences (Pacadaljen, 2024; Suleiman, 2014). Moreover, budgetary constraints hinder the development and dissemination of sustainability education materials, affecting not only the procurement but also teacher training.

Educators in both countries have expressed concerns about the overcrowded nature of the curriculum, which leaves little room for the inclusion of additional topics such as sustainability education. Teachers find it challenging to allocate sufficient time to explore environmental issues in depth, resulting in superficial coverage and limited engagement (Gyimah & Ayinselya, 2022; Munasi, 2024). This challenge is compounded by the rigid subject separation that prevents interdisciplinary approaches essential to ESD.

Effective sustainability education requires the involvement of diverse actors. However, interviews revealed that educators are often excluded from curriculum planning processes. In Namibia, a curriculum officer observed: "Consultations are top-down, which results in poor ownership of content." Similarly in Ghana, one teacher shared: "*We only get the syllabus after it is approved. Nobody asks for teacher input before that.*" Local community engagement, particularly the inclusion of indigenous knowledge, remains underutilized despite being vital to contextual relevance (Adom et al., 2024; Mohammad & Adam, 2022).

In some Namibian and Ghanaian communities, environmental issues are perceived as low priority. Misconceptions exist about the practicality and value of sustainability education, especially in regions where immediate economic needs take precedence over environmental stewardship (Kopnina, 2014; Kioupi & Voulvoulis, 2022). A Ghanaian parent noted during a school forum: "*Children should focus on passing exams and getting jobs. Talking about trees and pollution is not useful here.*" This indicates the need for awareness-building and cultural reorientation to support sustainability learning.

In both countries, policy inconsistencies and unclear mandates were cited as recurring issues. Ambiguities in sustainability policy objectives lead to fragmented implementation at the school level. Interviews highlighted bureaucratic inefficiencies, competing mandates among ministries, and weak monitoring systems as barriers, affirmed also by Muhdiarta (2025). Thus, without strong institutional coordination, policy goals struggle to move beyond rhetoric. Table 2 summarizes the key challenges identified in the implementation of sustainability education in Namibia and Ghana.

In comparative literature, similar challenges have been observed in other sub-Saharan counties. For instance, studies in Kenya and Tanzania also report teacher under-preparedness and curriculum overload as major barriers (Majoni, 2017; Muthoka & Mbirithi, 2025; Rodriguez-Segura & Mbiti, 2022). This reinforces the view that curriculum innovation must be accompanied by institutional strengthening, pedagogical transformation, and cross-sector policy alignment.

Table 2 Comparative barriers in national education policies and curricula

Challenge	Namibia	Ghana
Teacher capacity	Limited training and professional development opportunities	Insufficient training and outdated resources
Curriculum integration	Fragmented and superficial incorporation of sustainability topics	Inconsistent embedding across subjects
Resource constraints	Shortage of materials and financial support	Lack of teaching aids and digital resources
Overcrowded curriculum	Limited instructional time for additional topics	Pressure to cover extensive subject matter
Stakeholder engagement	Minimal involvement of educators in curriculum development	Limited collaboration with communities and indigenous knowledge systems
Cultural and societal attitudes	Environmental issues not prioritized in some communities	Misconceptions about the relevance of sustainability education
Policy and institutional challenges	Ambiguous policies and bureaucratic inefficiencies	Lack of cohesive policies and institutional coordination

Source Authors' own elaboration (2025)

Interpretation reveals that the effective implementation of sustainability education in Namibia and Ghana faces multifaceted challenges that span teacher capacity, curriculum design, resource availability, stakeholder engagement, cultural perceptions, and policy frameworks. Addressing these challenges requires a holistic approach that encompasses comprehensive teacher training, coherent curriculum integration, adequate resource allocation, inclusive stakeholder collaboration, cultural sensitivity, and robust policy development. These findings directly contribute to the objective of critically examining barriers to sustainability education and provide actionable insights for reform.

4.3 Policy Innovation and Strategic Opportunities

To combat the persistent barriers in green curriculum implementation, this section explores comparative strategies and institutional frameworks that support the integration of sustainability in education systems across Namibia and Ghana. The key research question guiding this inquiry is: *What policy innovations and opportunities exist for strengthening green curriculum development in Namibia and Ghana?* Empirical findings reveal a growing alignment between national education reforms and global sustainability agendas, especially through curriculum reviews and policy alignment with SDG 4 and SDG 13. However, challenges persist in effective policy implementation, weak inter-ministerial coordination, limited teacher training on climate content, and insufficient educational resources. Addressing these gaps requires holistic education sector reforms, investment in teacher capacity building, and the mainstreaming of environmental literacy through participatory and context-specific approaches.

In Namibia, the adoption of the National EE/ESD Policy in 2019 marked a significant milestone. This policy, the first of its kind in SADC region, provides a comprehensive framework for integrating environmental education across all levels of formal and informal education. It emphasizes the development of environmentally literate citizens who can contribute to SDGs. The policy's implementation is overseen by an ESD Task Force comprising representatives from government, private sector, and NGOs, ensuring a multi-stakeholder approach to sustainability education.

> Interview responses from education stakeholders in Namibia revealed several strategic innovations. For instance, a policy advisor at the Ministry of Education explained, "*The integration of green content in primary and secondary school curricula is increasingly informed by data from climate risk assessments and regional environmental monitoring systems.*" These data-driven approaches allow for curriculum development that is responsive to national environmental priorities.

Namibia's Environmental Education Centres (EECs) are another innovation. They serve as learning hubs for experiential sustainability education, providing teacher training, community workshops, and student outreach programs. A school principal in the Erongo region noted, "*Our school's partnership with the EEC has transformed the way students understand water conservation and biodiversity. These programmes bring the curriculum to life.*" Ghana has pursued similar reforms under its National Climate Change Policy and Education Strategic Plan (2018–2030). The 2019 introduction of green entrepreneurship modules at the senior high school level has created pathways for students to link environmental knowledge with livelihood opportunities. According to a curriculum officer at NaCCA, "*We aim to not only teach climate issues but to inspire students to create solutions. Entrepreneurship in green sectors is a policy focus.*"

The use of digital platforms has also emerged as a promising opportunity. In both countries, digital toolkits, e-learning modules, and radio-based sustainability education are being piloted to overcome geographical and resource constraints, particularly during the covid-19 pandemic era. In Ghana's Upper East Region, for instance, a teacher remarked, "*We use mobile apps and local FM radio to deliver green content. Students respond well when learning is connected to their communities and daily lives.*" Policy innovation has also been driven by donor and civil society partnerships. Ghana's Green School Initiative (GSI), supported by GGGI, promotes eco-schools through recycling clubs, green spaces, and student-led environmental campaigns. Similarly, Namibia's partnership with UNESCO has advanced teacher training through the Teaching and Learning for a Sustainable Future programme. Thus, a comparative summary is provided in Table 3.

These findings highlight that while both countries are committed to sustainability education, Namibia's policy innovation is anchored in systems-based and data-driven

Table 3 Comparative policy innovations and opportunities for green curriculum development

Innovation/opportunity	Namibia	Ghana
Key policy driver	EE/ESD Policy (2019)	Education Strategic Plan (2018–2030), Green Modules (2019)
Stakeholder mechanism	ESD Task Force; inter-ministerial collaboration	Curriculum reform teams; donor-partner platforms
Experiential learning	Environmental education centers (EECs)	Green school initiative (GSI)
Integration strategy	Climate data-informed curriculum design	Green entrepreneurship modules in high school
Technological enablers	E-learning resources; teacher toolkits	Mobile apps; FM radio content
International partnerships	UNESCO, SADC, WWF	GGGI, UNDP, USAID

Source Authors' own research (2025)

curriculum strategies, whereas Ghana is leveraging youth entrepreneurship and localized digital delivery platforms. These insights align with Leicht et al. (2018) and Clark (2022), who argue that ESD must be context-sensitive and future-facing.

What makes this study unique is its comparative perspective that moves beyond descriptive analysis to interrogate the strategic and operational dimensions of sustainability education. The empirical data demonstrate that innovation is not merely a policy intention but a multi-scalar process shaped by ecological conditions, institutional capacities, and pedagogical experimentation. The integration of community-centered and entrepreneurship-driven approaches further situates this research at the intersection of educational transformation and sustainable development.

In conclusion, these findings reinforce the potential of adaptive policy frameworks, cross-sector partnerships, and technology-enhanced learning in advancing green curriculum development. To build on these opportunities, both Namibia and Ghana must institutionalize policy coherence, prioritize teacher training, and scale successful innovations. The next subsection synthesizes lessons from this comparative analysis to inform regional and international discourse on sustainability education.

4.4 Drivers of Change for Sustainability Education

The mainstreaming of sustainability education (SE) in national education systems is not merely a product of policy design but a reflection of institutional capacities, leadership dynamics, funding mechanisms, and multilevel stakeholder coordination. This section presents the empirical findings addressing research objective 4 by exploring the institutional drivers and enabling factors that support the implementation of sustainability education in Namibia and Ghana. It addresses the question:

What institutional enablers and multilevel drivers are instrumental in mainstreaming sustainability education in Namibia and Ghana? Drawing on key informant interviews, curriculum analysis, and policy review, the findings reflect an emerging trend of decentralization, increased inter-ministerial collaboration, and capacity development within educational institutions. These outcomes were interpreted through the theme of "*Multilevel Institutional Mobilization for Sustainability Mainstreaming*."

In Namibia, a major institutional driver is the strong alignment between the Ministry of Education, Arts and Culture and the Ministry of Environment, Forestry and Tourism. Interview participants from both ministries highlighted the coordinated development of the EE/ESD Policy and the institutionalization of an ESD Task Force as critical steps. A senior policy official noted, "*We are no longer working in silos. Our collaboration ensures that sustainability themes are infused from early childhood education to teacher training colleges.*" This synergy is strengthened by the decentralization of implementation efforts through Regional Education Offices, which ensure that localized ecological and social concerns inform green curriculum delivery.

Namibia's institutional support is also bolstered by strong integration with regional platforms such as the Southern African Development Community (SADC) ESD programme (Mensah, 2025). These regional collaborations enable the country to access training, funding, and policy tools that reinforce local implementation. Several educators from rural regions such as Oshikoto and Kavango West noted that teacher training workshops on sustainability were made possible through SADC-UNESCO linkages. One principal remarked, "*Our teachers returned from regional training with materials that are now being used to teach climate change and biodiversity at the local level. This has improved student engagement significantly.*"

In Ghana, decentralization is equally pivotal. The Ghana Education Service (GES) and the National Council for Curriculum and Assessment (NaCCA) have increasingly empowered district-level education directorates to implement sustainability themes in ways that reflect local contexts. According to a curriculum planner at NaCCA, "*We give the districts the core guidelines, but the real success comes from how well local directorates adapt these ideas. Some districts are even innovating their own assessment tools for sustainability learning.*"

A key enabler in Ghana is the engagement of faith-based organizations and community groups in promoting environmental awareness in schools. This grassroots support structure supplements formal education with community-led activities such as school gardening, plastic recycling, and tree planting initiatives. An environmental officer from the EPA observed, "*Many rural schools are more responsive to sustainability themes when they are introduced through respected community leaders or religious groups. It brings legitimacy and relevance.*" This community embeddedness, often overlooked in policy frameworks, emerged as a strong factor in sustaining environmental learning at the grassroots level.

Capacity development within teacher training institutions also serves as a critical driver. In Namibia, the University of Namibia (UNAM) and NIED have begun integrating modules on sustainability education into pre-service training. In Ghana, several colleges of education now include climate literacy as part of professional development. A lecturer at Accra College of Education explained, "*We are seeing a growing appetite among young teachers to bring environmental issues into their classrooms. What they need now is mentoring and the right tools.*"

Another cross-cutting enabler is funding and technical assistance from international development agencies. In Namibia, organizations like the Hanns Seidel Foundation and GIZ support ESD implementation through funding for environmental clubs, resource materials, and experiential learning. Ghana's curriculum reform process has benefited from technical support from the World Bank and UNICEF, which have helped build institutional infrastructure and provide digital learning platforms. Interviewees consistently cited these partnerships as essential in bridging national policy ambitions and local delivery mechanisms.

Furthermore, leadership within schools emerged as a vital institutional factor. Head teachers who prioritized sustainability often mobilized internal resources to support green education initiatives. In one Namibian secondary school, the principal reallocated funds from sports to initiate a permaculture garden, linking food security with environmental education. In Ghana, a headteacher in the Volta region championed a solar energy project that served both as a power source and a teaching tool for science and sustainability. Thus, Table 4 presents a comparative summary of the institutional enablers supporting sustainability education in Namibia and Ghana.

These findings reinforce the importance of institutional readiness, multilevel alignment, and community engagement in driving sustainability education. What makes this study unique is its identification of the "*hidden curriculum of leadership and community mobilization*" that enables sustainability to move beyond policy intent and into lived educational practices. Unlike other studies that focus exclusively

Table 4 Comparative institutional enablers for mainstreaming sustainability education

Institutional enabler	Namibia	Ghana
Inter-ministerial coordination	Strong collaboration between education and environment ministries	Education-environment synergies through GES and NaCCA
Decentralized implementation	Regional Education Offices facilitate local contextualization	District education offices customize implementation and assessments
Higher education and teacher training	UNAM and NIED integrating SE into pre-service training	Teacher training colleges introducing climate literacy modules
Community engagement	Environmental clubs, community workshops	Faith-based organizations, local environmental campaigns
International partnerships	UNESCO, GIZ, Hanns Seidel foundation	World Bank, UNICEF, USAID support
School leadership and internal innovation	Principals initiating in-house green projects	Headteachers championing solar energy, recycling, gardening

Source Authors' own elaboration (2025)

on curriculum content or teacher training, this research provides a comprehensive analysis of how institutional ecosystems shape educational transformation.

Comparatively, studies from Kenya, Malawi, and Botswana (Kioupi & Voulvoulis, 2019, 2022; Lotz-Sisitka, 2016; Ngcoza, 2019) confirm that decentralized implementation and community-driven education significantly enhance the uptake of sustainability education. Namibia and Ghana offer instructive models of how such decentralization, when coupled with institutional capacity and leadership, can result in tangible outcomes. These results also affirm the growing importance of whole-institution approaches in mainstreaming sustainability.

In conclusion, mainstreaming sustainability education requires not only robust policy frameworks but also enabling institutional ecosystems. The results underscore that institutional drivers—when aligned across national, regional, and local levels—can serve as powerful levers for embedding sustainability within formal education. Strengthening teacher education, investing in school leadership, and leveraging international partnerships are critical for long-term impact. The final subsection synthesizes these comparative insights into a roadmap for scaling up sustainability education across sub-Saharan Africa.

4.5 Lessons and Policy Recommendations

The final subsection of empirical findings aims to critically reflect on the comparative evidence gathered from Namibia and Ghana and translate those insights into meaningful lessons that can inform regional, continental, and global strategies for green curriculum development. This objective draws together cross-cutting themes from the previous sections and situates them within a broader discourse on Education for Sustainable Development (ESD). It addresses the question: *What lessons can be learned from Namibia and Ghana's experiences to guide curriculum reform for sustainability education globally?*

The results show that while both countries have demonstrated commitment to sustainability education, their paths reflect differing strategies, constraints, and enabling factors. Several core lessons emerge. First, contextually grounded curriculum reform is essential. Namibia's experience underscores the value of embedding ESD within national development visions (e.g., Vision 2030 and the Harambee Prosperity Plan), thus ensuring coherence between education and broader sustainability goals. In Ghana, integrating green entrepreneurship in high school curricula illustrates how ESD can intersect with economic empowerment, aligning with the country's youth employment agenda.

Second, the importance of decentralized and participatory governance in curriculum design is evident. In both contexts, interviewees emphasized the need for educators to be actively involved in policy formulation. A Ghanaian teacher stated, "*We feel sidelined in these curriculum changes. If teachers were more engaged, implementation would improve.*" Namibia's ESD Task Force, in contrast, exemplifies how multi-stakeholder participation can strengthen ownership and accountability. This aligns with findings by Clark (2022) and Abdallah (2025), who advocate for inclusive governance structures in ESD reform.

Third, experiential and localized learning is pivotal. Programmes such as Namibia's Environmental Education Centres and Ghana's Green School Initiative demonstrate the power of real-life, community-linked activities to reinforce sustainability concepts. This echoes UNESCO (2020) and Dutra (2025), which emphasizes the value of action-oriented pedagogy in ESD. A Namibian student from the Otjozondjupa region reflected, "*We learned about water recycling by doing it ourselves in school. That lesson stuck more than reading a textbook.*"

Fourth, capacity development for educators must be institutionalized. In both countries, limited training in sustainability content remains a persistent issue. Policymakers and education stakeholders stressed that short-term workshops are inadequate. A senior official from Ghana's Ministry of Education commented, "*If sustainability is a priority, then our colleges of education must make it a core subject—not an optional workshop.*" This aligns with the ESD Roadmap's call for systemic capacity building (Abdallah, 2025; UNESCO, 2020).

Fifth, effective monitoring and evaluation (M&E) systems are missing in both countries. Despite ambitious policy statements, neither Namibia nor Ghana has fully developed metrics to assess the impact of ESD on learning outcomes or behavioral change. This absence hinders long-term planning and policy refinement. Introducing M&E indicators—such as student attitudes toward climate issues, integration of green practices in schools, or teacher competency in ESD—would support adaptive policy learning. A comparative synthesis of the key lessons is summarized in Table 5.

In conclusion, Namibia and Ghana's experiences provide valuable insights for countries seeking to embed sustainability into their education systems. Their innovations—though uneven—demonstrate that sustainable curriculum reform requires

Table 5 Summary of lessons for advancing sustainability education policy globally

Key lesson	Namibia	Ghana	Global relevance
Policy coherence	Aligned with vision 2030 and SDGs	Integrated into youth entrepreneurship and climate policy	Align ESD with national SDGs agendas
Stakeholder engagement	ESD task force enables participation	Teacher voices underrepresented	Foster inclusive curriculum governance
Experiential learning	EECs promote local action projects	Green School Initiative links schools and communities	Prioritize place-based pedagogy
Teacher development	Still underdeveloped; ESD workshops ongoing	Sporadic training; not institutionalized	Integrate ESD into teacher education curricula
M&E systems	Lacking standardized metrics	Weak monitoring capacity	Develop ESD impact indicators

Source Authors' own research (2025)

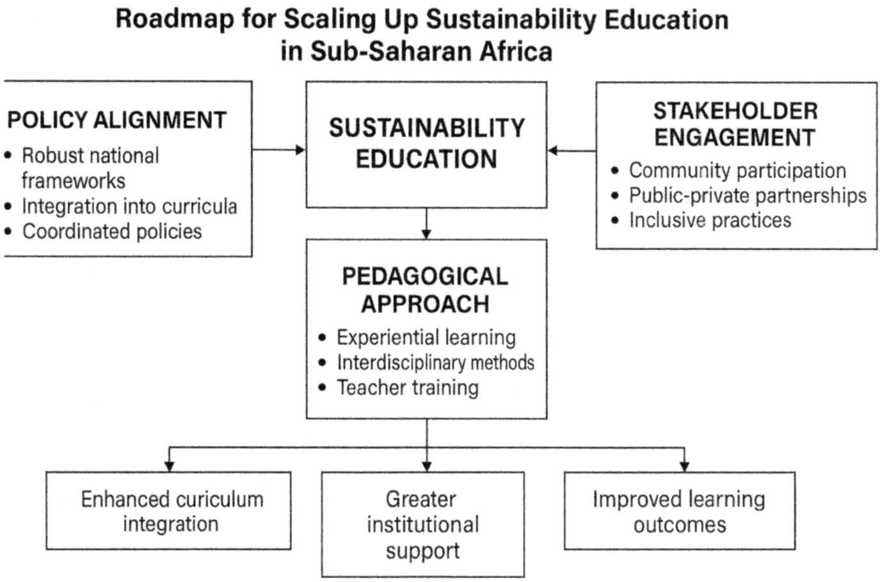

Fig. 3 Roadmap for scaling up sustainability education in sub-Saharan Africa. *Source* Authors' own research (2025)

more than policy ambition. It demands alignment with local contexts, sustained stakeholder participation, teacher empowerment, and robust evaluation. These lessons offer practical guidance for future ESD efforts worldwide, especially in the Global South, where resource constraints and competing development priorities often hinder educational reform. Thus, the study sum-up with a roadmap for scaling up sustainability education across sub-Saharan Africa in Fig. 3.

5 Conclusion

This chapter has examined the integration of sustainability education within national curricula in Namibia and Ghana, focusing on policy innovations, institutional challenges, and strategic opportunities for green curriculum development. Situated within the global imperative of achieving the SDGs, particularly SDG 4 (quality education) and SDG 13 (climate action), the study adopted a qualitative comparative case study design grounded in document analysis and semi-structured interviews. This methodological approach enabled an in-depth exploration of how two sub-Saharan African countries are embedding sustainability into education frameworks, and the degree to which policy intent translates into practice.

Empirical findings from Namibia and Ghana revealed a shared commitment to sustainability education, reflected in national policy frameworks such as Namibia's National Environmental Education and ESD Policy (2019) and Ghana's National

Curriculum Framework (2018). However, both countries face persistent challenges, including limited teacher capacity, fragmented curriculum integration, inadequate resources, and weak institutional coordination. Despite these constraints, the study identified innovative practices, including the use of environmental education centres, digital learning platforms, and green entrepreneurship modules, which offer promising models for scaling sustainability education in resource-constrained settings.

The implications of this study are both context-specific and globally relevant. It contributes to the emerging scholarship on education for sustainable development (ESD) by offering a comparative African perspective that is often underrepresented in global discourse. The findings underscore that meaningful green curriculum reform requires not just policy design, but coordinated action across sectors, grounded in pedagogical transformation and community engagement. By highlighting pathways for integrating local knowledge systems, digital tools, youth entrepreneurship, and STEM education (Science, Technology, Engineering, and Mathematics) into curricula, the study adds innovative dimensions to global sustainability education models.

This research also offers practical implications for policymakers, educators, and development partners. Strengthening teacher education, investing in contextualized learning materials, and fostering multisectoral collaboration are pivotal for mainstreaming ESD across African education systems (Bourn, 2021). Moreover, the study provides a framework for other low- and middle-income countries (LMICs) facing similar challenges, making its findings transferable to a broader global audience interested in inclusive and transformative sustainability education.

However, the study has several limitations. It primarily focused on two countries, which, although diverse in context, may not fully represent the complexity of ESD implementation across Africa. The qualitative nature of the study, while rich in depth, may limit generalizability. Furthermore, time and resource constraints restricted the inclusion of more grassroots voices, such as students and community-based educators, whose perspectives are vital in evaluating the impact of curriculum reforms. Despite these constraints, the findings offer a valuable foundation for future research that could expand to include longitudinal assessments, quantitative evaluations of learning outcomes, and the role of non-formal education in sustainability transitions.

Looking ahead, this study opens avenues for future scholarship to examine how regional policy harmonization, South-South collaboration, and interdisciplinary teacher training models can accelerate sustainability education. As climate, social, and economic pressures intensify globally, equipping learners with sustainability competencies becomes a universal imperative. The comparative insights from Namibia and Ghana underscore the urgent need for locally informed yet globally connected education reforms, and lay a conceptual and empirical roadmap for achieving this goal.

References

Abdallah, A. K. (2025). Local to global: Place-based learning in sustainability education. In *Legal frameworks and educational strategies for sustainable development* (pp. 233–260). IGI Global. https://doi.org/10.4018/979-8-3693-2987-0.ch013

Adom, D., Asare-Aboagye, I., Thompson, J., & Kquofi, S. (2024). Integration of indigenous knowledge systems in African higher education in Ghana: A post-CESA (2016–2025) review. In *What next for Africa's higher education?: What should follow the continental education strategy for Africa (CESA 2016–2025)* (p. 161).

African Union. (2015). Continental education strategy for Africa, CESA 2016–2025. Addis Ababa: African Union. https://ecosocc.au.int/sites/default/files/files/2021-09/continental-strategy-education-africa-english.pdf

Agbedahin, A. V. (2019). Sustainable development, education for sustainable development, and the 2030 agenda for sustainable development: Emergence, efficacy, eminence, and future. *Sustainable Development, 27*(4), 669–680. https://doi.org/10.1002/sd.1931

Agyemang, R. O. (2020). Embedding sustainable development in teacher education in the central region of Ghana. Ph.D. thesis, University of Iceland. https://www.academia.edu/download/91927434/Richard_20Opoku_20Agyemang_20Maters_20Thesis.pdf

Ahovi, E. (2025). Assessing the integration of sustainable development goal 13 (climate action) in Ghana's standards-based curriculum. *Journal of Education and Practice, 16*(3), 24–34. https://doi.org/10.7176/JEP/16-3-03

Anlimachie, M. A. (2022). Sankofa education: Bridging the rural-urban gap in Ghana's education outcomes through culturally responsive pedagogy to decolonise schooling and meet community needs. Ph.D. thesis, University of South Australia. https://hdl.handle.net/11541.2/30288

Anyolo, E. O. (2015). Implementing education for sustainable development in Namibian schools. In *Dissertations in education, humanities and theology* (pp. 1–147). University of Eastern Finland. https://erepo.uef.fi/bitstreams/b2d10c50-3087-49fd-92d2-021f48091b1d/download

Atuguba, J. A. (2019). Environmental education in colleges of education in Ghana: Experiences and perceptions of environmental educators. Ph.D. thesis, University of Birmingham. http://etheses.bham.ac.uk/id/eprint/9462

Bourn, D. (2021). Pedagogy of hope: Global learning and the future of education. *International Journal of Development Education & Global Learning, 13*(2), 65–78. https://doi.org/10.14324/IJDEGL.13.2.01

Burdett, N., & O'Donnell, S. (2016). Lost in translation? The challenges of educational policy borrowing. *Educational Research, 58*(2), 113–120. https://doi.org/10.1080/00131881.2016.1168678

Buzzelli, C. A. (2015). The capabilities approach: Rethinking agency, freedom, and capital in early education. *Contemporary Issues in Early Childhood, 16*(3), 203–213. https://doi.org/10.1177/1463949115600030

Chasokela, D. (2024). Fostering global citizenship through ESD in higher education: A study of international student mobility and sustainability awareness. *Journal of Education for Sustainable Development Studies, 1*(2), 117–133. https://doi.org/10.70232/jesds.v1i2.7

Clark, K. R. (2022). Education for sustainable development, curriculum reform and implications for teacher education in a small island developing state. *Discourse and Communication for Sustainable Education, 13*(1), 145–153. https://doi.org/10.2478/dcse-2022-0011

Darkwa, S. (2016). Environmental education for sustainable development: The case of University of Cape Coast-Ghana. *British Journal of Education, Society & Behavioural Science, 16*(3), 1–9. https://doi.org/10.9734/BJESBS/2016/25158

Duggan, M. S., Smith, T. F., & Thomsen, D. C. (2015). Organizational approaches to the facilitation of education for sustainability: An interpretive case study. *Sustainability, 7*(6), 7011–7030. https://doi.org/10.3390/su7067011

Dutra, A. (2025). Critical pedagogy for a sustainable future: Investigating paulo freire's influence on education for sustainable development. Retrieved May 20, 2025, from http://uu.diva-portal.org/smash/record.jsf?pid=diva2:1932306

Furst-Bowe, J. (2011). Systems thinking: Critical to quality improvement in higher education. *Quality Approaches in Higher Education, 2*(2), 2–5. https://obecurriculumsessions.wordpress.com/wp-content/uploads/2014/01/day-one_systems-thinking1.pdf

Griffiths, T. G. (2021). Education to transform the world: Limits and possibilities in and against the SDGs and ESD. *International Studies in Sociology of Education, 30*(1–2), 73–92. https://doi.org/10.1080/09620214.2020.1854829

Gyimah, E. K., & Ayinselya, R. A. (2022). Teacher professional learning, a tool for managing teachers to improve pupils' learning: The views of teachers in Ghana. *Practice, 4*(1), 12–32. https://doi.org/10.1080/25783858.2022.2065923

Hailombe, O. (2011). Education equity and quality in Namibia: A case study of mobile schools in the Kunene region. Ph.D. thesis, University of Pretoria. http://hdl.handle.net/2263/24256

Hanson, M. (2001). Institutional theory and educational change. *Educational Administration Quarterly, 37*(5), 637–661. https://doi.org/10.1177/00131610121969451

Henry, J. (2020). Rethinking schooling for development in wageless times. *African Geographical Review, 39*(3), 240–251. https://doi.org/10.1080/19376812.2019.1696214

Ishtiaq, M. (2019). Book Review Creswell, JW (2014). Research design: Qualitative, quantitative and mixed methods approaches. Sage. *English Language Teaching, 12*(5), 40. https://www.academia.edu/download/72214301/Ishtiaq_2019.pdf

Jin, H., Mikeska, J. N., Hokayem, H., & Mavronikolas, E. (2019). Toward coherence in curriculum, instruction, and assessment: A review of learning progression literature. *Science Education, 103*(5), 1206–1234. https://doi.org/10.1002/sce.21525

Kanyimba, A. T., & Coetzer, I. A. (2011). The integration of sustainability education in Namibian colleges of education. *Africa Education Review, 8*(1), 133–147. https://doi.org/10.1080/18146627.2011.586157

Kioupi, V., & Voulvoulis, N. (2019). Education for sustainable development: A systemic framework for connecting the SDGs to educational outcomes. *Sustainability, 11*(21), 6104. https://doi.org/10.3390/su11216104

Kioupi, V., & Voulvoulis, N. (2022). Education for sustainable development as the catalyst for local transitions toward the sustainable development goals. *Frontiers in Sustainability, 3*, Article 889904. https://doi.org/10.3389/frsus.2022.889904

Klawinski, T. P. (2022). Achieving environmental literacy: An evidenced-based model for public schools. Ph.D. thesis, Wilmington University (Delaware).

Kopnina, H. (2014). Revisiting education for sustainable development (ESD): Examining anthropocentric bias through the transition of environmental education to ESD. *Sustainable Development, 22*(2), 73–83. https://doi.org/10.1002/sd.529

Kwarteng, E. (2024). Early-grade teachers' pedagogical knowledge in the implementation of the standard-based curriculum in Atwima-Kwawoma district. Ph.D. thesis, University of Education. http://41.74.91.244:8080/handle/123456789/4137

Leicht, A., & Byun, W. J. (2021). UNESCO's framework ESD for 2030. *IBE on Curriculum, Learning, & Assessment,* 5(1)*89.*

Leicht, A., Combes, B., Byun, W. J., & Agbedahin, A. V. (2018). From agenda 21 to target 4.7: The development of education for sustainable development. *Issues and Trends in Education for Sustainable Development, 25.*

Lotz-Sisitka, H. (2016). Reviewing strategies in/for ESD policy engagement: Agency reclaimed. *The Journal of Environmental Education, 47*(2), 91–103. https://doi.org/10.1080/00958964.2015.1113915

Lotz-Sisitka, H., & Lupele, J. (2016). ESD, learning and quality education in Africa: Learning today for tomorrow. In *Schooling for sustainable development in Africa* (pp. 3–24). Springer International Publishing. https://doi.org/10.1007/978-3-319-45989-9_1

Maguire, M., & Delahunt, B. (2017). Doing a thematic analysis: A practical, step-by-step guide for learning and teaching scholars. *All Ireland Journal of Higher Education, 9*(3). 3351–33514. https://doi.org/10.62707/aishej.v9i3.335

Majoni, C. (2017). Curriculum overload and its impact on teacher effectiveness in primary schools. *European Journal of Education Studies, 3*(3), 155–162. https://doi.org/10.5281/ZENODO.290597

Malua, A., & Olvitt, L. (2024). The emergence of education for sustainable development in Namibia's senior primary English language curriculum: A morphogenetic case study. *Journal of Education for Sustainable Development, 18*(1), 56–76. https://doi.org/10.1177/09734082241282891

McLeod, S. (2024). Peer debriefing in qualitative research. *Simply Psychology,* 1–8. https://doi.org/10.13140/RG.2.2.33017.17761

Mensah, P. (2025). Strengthening maritime security for ocean sustainability within Namibia's green economy transition. *Academia Environmental Sciences and Sustainability, 2*(3). https://doi.org/10.20935/AcadEnvSci7792

Menzli, L. J., Smirani, L. K., Boulahia, J. A., & Hadjouni, M. (2022). Investigation of open educational resources adoption in higher education using Rogers' diffusion of innovation theory. *Heliyon, 8*(7). https://doi.org/10.1016/j.heliyon.2022.e09885

Mezirow, J. (2018). Transformative learning theory. In *Contemporary theories of learning* (pp. 114–128). Routledge.

Ministry of Education, Arts and Culture. (2016). *The national curriculum for basic education.* National Institute for Educational Development [NIED], pp. 1–53. https://www.nied.edu.na/assets/documents/05Policies/NationalCurriculumGuide/National_Curriculum_Basic_Education_2016.pdf

Ministry of Environment, Forestry and Tourism. (2019). *National environmental education and education for sustainable development policy* [National EE/ESD]. Namibia: MEFT, pp. 1–21. https://nadeet.org/sites/default/files/National%20EE%20ESD%20Policy.pdf

Ministry of Environment, Science, Technology and Innovation. (2016). *National climate change and green economy learning strategy.* Ghana: MESTI, pp. 1–5. https://www.igreengrowthsolutions.com/content/media/store/posts/postfile/file/National_Climate_Change_and_Green_Economy_Learning_Strategy.pdf

Ministry of Education. (2018). *National pre-tertiary education curriculum framework.* Ghana National Council for Curriculum and Assessment [NACCA], pp. 1–94. https://nacca.gov.gh/wp-content/uploads/2019/04/National-Pre-tertiary-Education-Curriculum-Framework-final.pdf

Mohammad, A. R., & Adam, R. K. (2022). Challenges to environmental education in senior high schools in Kumasi, Ashanti Region, Ghana. *Journal of Environmental and Geographical Studies, 1*(1), 27–39. https://doi.org/10.58425/jegs.v1i1.63

Mokski, E., Leal Filho, W., Sehnem, S., & Andrade Guerra, J. B. S. O. D. (2023). Education for sustainable development in higher education institutions: An approach for effective interdisciplinarity. *International Journal of Sustainability in Higher Education, 24*(1), 96–117. https://doi.org/10.1108/IJSHE-07-2021-0306

Morrow, R. A. (2008). Paulo Freire, indigenous knowledge and Eurocentric critiques of development: Three perspectives. In *Social justice education for teachers: Paulo Freire and the possible dream* (pp. 81–100).

Muhdiarta, U. (2025). Investigating the challenges of bureaucratic reform in developing countries: A case study approach. *Ilomata International Journal of Social Science, 6*(2), 689–702. https://doi.org/10.61194/ijss.v6i2.1687

Munasi, K. R. (2024). Evaluating the integration of education for sustainable development in the life sciences curriculum: A case study of grade 11 teachers in Vhembe east district. *Sustainability, 16*(24), 10918. https://doi.org/10.3390/su162410918

Muthoka, J. M. & Mbirithi, D. M. (2025). Teachers' preparedness and its effect on the implementation of competency-based curriculum in public primary schools in Taita Taveta County, Kenya.

International Academic Journal of Social Sciences and Education (IAJSSE), 2(4), 69–76. https://iajournals.org/articles/iajsse_v2_i4_69_76.pdf
Ngcoza, K. M. (2019). Education for sustainable development at the problem-posing nexus of re-appropriated heritage practices and the science curriculum. *Southern African Journal of Environmental Education, 35*. https://doi.org/10.4314/sajee.v35i1.9
Oe, H., Yamaoka, Y., & Ochiai, H. (2022). A qualitative assessment of community learning initiatives for environmental awareness and behaviour change: Applying UNESCO education for sustainable development (ESD) framework. *International Journal of Environmental Research & Public Health, 19*(6), 3528. https://doi.org/10.3390/ijerph19063528
Pacadaljen, L. M. (2024). Hurdling obstructions on instructional management of science teachers in schools with challenging laboratory resources. *Environment and Social Psychology, 9*(10), 1–12. https://doi.org/10.59429/esp.v9i10.3027
Parra, G., Hansmann, R., Hadjichambis, A. C., Goldman, D., Paraskeva-Hadjichambi, D., Sund, P., et al. (2020). Education for environmental citizenship and education for sustainability. In *Conceptualizing Environmental Citizenship for 21st Century Education* (pp. 149–160). Springer International Publishing. https://doi.org/10.1007/978-3-030-20249-1_10
Phillips, D., & Ochs, K. (2004). Researching policy borrowing: Some methodological challenges in comparative education. *British Educational Research Journal, 30*(6), 773–784. https://doi.org/10.1080/0141192042000279495
Phuthi, B. (2021). From international plans to national practice: How are issues of inequality addressed in education for sustainable development in teacher training? A case study research on how issues of inequality are addressed through education for sustainable development in teacher training in Botswana, Namibia, and Zimbabwe. Master's thesis, OsloMet-Storbyuniversitetet. https://hdl.handle.net/11250/2773690
Pinar, W. F. (2013). The reconceptualization of curriculum studies. In *Curriculum studies reader E2* (pp. 153–161). Routledge. https://files.eric.ed.gov/fulltext/ED141929.pdf
Porter, T., & Córdoba, J. (2009). Three views of systems theories and their implications for sustainability education. *Journal of Management Education, 33*(3), 323–347. https://doi.org/10.1177/1052562908323192
Rajapakse, N. (2016). Amartya Sen's capability approach and education: Enhancing social justice. *Revue LISA/LISA e-Journal. Littératures, Histoire des Idées, Images, Sociétés du Monde Anglophone–Literature, History of Ideas, Images and Societies of the English-speaking World, 14*(1). https://doi.org/10.4000/lisa.8913
Ramrathan, L., Le Grange, L., & Shawa, L. B. (2017). Ethics in educational research. In *Education studies for initial teacher education* (pp. 432–443). https://www.researchgate.net/publication/312069857_Ethics_in_educational_research
Rodriguez-Segura, D., & Mbiti, I. (2022). Back to the basics: Curriculum reform and student learning in Tanzania. In *RISE Working Paper 22/099* (pp. 1–43).
Salmen, A. (2018). A cross-case analysis of ESD learning and development in UK business schools. Ph.D. thesis, University of Southampton. http://eprints.soton.ac.uk/id/eprint/433257
Setordzi, S. (2023). Towards implementing the standards-based curriculum exploring teachers' experiences with developing the core competencies of pupils. Ph.D. thesis, University of Education. http://41.74.91.244:8080/handle/123456789/3494
Shephard, D. D., Ellersiek, A., Meuer, J., Rupietta, C., Mayne, R., & Cairney, P. (2021). Kingdon's multiple streams approach in new political contexts: Consolidation, configuration, and new findings. *Governance, 34*(2), 523–543. https://doi.org/10.1111/gove.12521
Shulla, K., Filho, W. L., Lardjane, S., Sommer, J. H., & Borgemeister, C. (2020). Sustainable development education in the context of the 2030 Agenda for sustainable development. *International Journal of Sustainable Development & World Ecology, 27*(5), 458–468. https://doi.org/10.1080/13504509.2020.1721378
Stone, D. (2004). Transfer agents and global networks in the 'transnationalization' of policy. *Journal of European Public Policy, 11*(3), 545–566. https://doi.org/10.1080/13501760410001694291

Suleiman, M. M. (2014). *Teachers' experiences of teaching science with limited laboratory resources*. GRIN Verlag.
Tubaundule, G. M. (2014). Evaluative research of the implemented secondary school curriculum in Namibia. Ph.D. thesis. University of the Free State. https://scholar.ufs.ac.za/bitstream/handle/11660/808/TubaunduleG.pdf?sequence=1
Uljens, M., & Ylimaki, R. M. (2017). Non-affirmative theory of education as a foundation for curriculum studies, didaktik and educational leadership. In *Bridging educational leadership, curriculum theory and didaktik: Non-affirmative theory of education* (pp. 3–145). Springer International Publishing. https://doi.org/10.1007/978-3-319-58650-2_1
UNESCO. (2020). Education for sustainable development: Learning to act for people and planet. Retrieved May 20, 2025 from https://www.unesco.org/en/sustainable-development/education
Warren, A., Archambault, L., & Foley, R. W. (2014). Sustainability education framework for teachers: Developing sustainability literacy through futures, values, systems, and strategic thinking. *Journal of Sustainability Education, 6*(4), 23–28.
Wicke, B., Moll, H. C., & Rebel, K. T. (2024). Integrative sustainability education: Emerging concepts and approaches. *Journal of Integrative Environmental Sciences, 21*(1), 2395226. https://doi.org/10.1080/1943815X.2024.2395226
Witoszek, N. (2018). Teaching sustainability in Norway, China and Ghana: Challenges to the UN programme. *Environmental Education Research, 24*(6), 831–844. https://doi.org/10.1080/13504622.2017.1307944
Young, M. D., & Diem, S. (2018). Doing critical policy analysis in education research: An emerging paradigm. In *Complementary research methods for educational leadership & policy studies* (pp. 79–98). Springer International Publishing. https://doi.org/10.1007/978-3-319-93539-3_5
Zavala, M. (2016). Decolonial methodologies in education. In *Encyclopedia of educational philosophy & theory* (pp. 1–6). Springer. https://doi.org/10.1007/978-981-287-532-7_498-1
Zickafoose, A., Ilesanmi, O., Diaz-Manrique, M., Adeyemi, A. E., Walumbe, B., Strong, R., et al. (2024). Barriers and challenges affecting quality education (Sustainable Development Goal# 4) in sub-Saharan Africa by 2030. *Sustainability, 16*(7), 2657. https://doi.org/10.3390/su16072657

Philip Mensah is a Research Assistant under the Fostering Research and Intra-African Knowledge Transfer through Mobility and Education (FRAME) project at the Namibia University of Science and Technology (NUST). He holds a Master's degree in Spatial Science at NUST and Bachelor's honors degree in Geography and Rural Development from the Kwame Nkrumah University of Science and Technology (KNUST), Ghana. His research focuses on policy development planning, sustainable regional resilience, spatial data infrastructure and integrating digital solutions into development planning. With experience in development policy, disaster management and curriculum sustainability, Philip has contributed to research addressing sustainable development challenges. Passionate about policy-driven development, he aims to bridge the gap between spatial science and sustainable development goals through innovative, evidence-based strategies.

Eric Yankson is an Associate Professor of Spatial Planning at NUST with extensive experience in spatial analysis, sustainable development and urban governance. He is also an Advanced Researcher with the Global Center of Spatial Methods for Urban Sustainability at the Technical University of Berlin. He holds a Ph.D. in Urban and Public Affairs from the University of Louisville and has contributed significantly to policy research addressing urban resilience and environmental management. Prof Yankson has authored multiple scholarly publications and supervised graduate-level research projects. His work emphasizes evidence-based policy frameworks and innovative urban solutions for emerging economies. He is actively involved in international collaborations on spatial planning and sustainable urban transitions.

Derrick Frimpong Buabeng holds a Master's degree in Spatial Science (Land Administration) at NUST, with a thesis exploring the comparative roles of traditional authorities in customary land governance in Namibia and Ghana. He earned a First-Class BSc in Real Estate from the KNUST, focusing on maintenance regimes in municipal assemblies. Derrick serves as a Graduate Research Assistant, contributing to book chapters and journal articles on sustainable development and governance. A FRAME scholarship recipient, he is also a published researcher, with works addressing land reclamation and blue economy sustainability. He has presented at conferences on coastal land reclamation and customary land governance. His professional experience includes roles as a teaching assistant and research mentor. Derrick's research interests encompass land administration, governance and real estate finance, with technical expertise in AutoCAD, SPSS, and MAXQDA.

Integrating Education for Sustainable Development in Teacher Training Curricula: A Comparative Analysis of Finland and Ghana

Hansen Owusu Tawiah, Matthew Biniyam Kursah, and Diana Yinime

Abstract This chapter compares the integration of Education for Sustainable Development (ESD) in the teacher education curricula of Finland and Ghana, with a focus on curriculum development for sustainability. The chapter aims to unearth the drivers, barriers, and practical indications of ESD within curricular design and pedagogical strategies in both countries. By sticking to established frameworks in sustainability in education, a directed qualitative content analysis approach was employed to analyze policy documents. Additionally, interviews were conducted with teacher educators from the two countries. The findings show that while both countries recognize the importance of sustainability in education, their approaches vary in some key aspects. Finland relies on a decentralized model that gives universities and educators a high degree of autonomy, enhancing innovation, but causing inconsistent implementation of ESD competencies. Ghana, on the other hand, has in place established national guidelines that correspond to global sustainability goals. However, effective implementation of ESD is limited due to inadequate resources and institutional support. In both systems, the essential role of the educator as a determinant remains evident. Educators are inspired by their passion and initiative to integrate sustainability competencies and components in their teaching when established frameworks and assessment protocols are not in place. By contributing to the global discussion on sustainability and teacher education, the chapter bridges the gap between pedagogy and policy. In addition, the chapter provides valuable insights to improve curriculum development in different educational contexts. The study concludes by advocating

H. O. Tawiah (✉)
Faculty of Education and Welfare Studies, Åbo Akademi University, Vaasa, Finland
e-mail: hansen.tawiah@abo.fi

M. B. Kursah
Department of Geography Education, University of Education Winneba (UEW), Winneba, Ghana
e-mail: mkursah@uew.edu.gh

D. Yinime
Department of Geography Education and College for Distance and e-Learning (CODeL), University of Education Winneba (UEW), Winneba, Ghana
e-mail: yinimediana@gmail.com

W. Leal Filho et al., *Curriculum Development for Sustainable Development*, World Sustainability Series, https://doi.org/10.1007/978-3-032-22146-9_10

for the deliberate integration of effective mechanisms that ensure that sustainability-driven elements are captured in teacher training curricula to establish sustainability, a significant and enduring part of teacher education globally.

Keywords Curriculum development · Education for sustainable development · Finland · Ghana · Sustainability education · And teacher training

1 Introduction

The urgent need for Education for Sustainable Development (ESD) is crucial in a dynamic world experiencing various phases of transition. In an era confronted with complex global challenges such as climate change, biodiversity loss, social inequalities, and environmental degradation, education systems are being called upon to intervene by playing a transformative role in building a sustainable future. Key to this task is the strategic incorporation of ESD into the design of curriculum, especially in teacher education, as teachers impact individual learners and society at large (Jekabsone & Ratniece, 2024; Rieckmann, 2019; Rieckmann et al., 2022; Salīte et al., 2024). ESD constitutes a fundamental component of the United Nations (UN) Sustainable Development Goal (SDGs) Target 4.7, aimed at ensuring that all learners obtain the knowledge and skills requisite for advancing sustainable development through education (Waltner et al., 2018). This chapter explores how Finland and Ghana have integrated sustainability into their curriculum, providing a useful analysis of the requirements for creating teacher education initiatives that really equip future teachers for the challenging requirements of sustainable development.

ESD holds promise in making the world an ideal place to live for present and future generations by equipping individuals with the knowledge, skills, values, and attitudes needed to address environmental, economic, and social issues holistically (Mckeown, 2014; UNESCO, 2017). ESD is not just about environmental awareness; it goes beyond that. It involves encouraging learners to understand systems, challenge existing systemic protocols, and ultimately strive for sustainable futures. ESD, therefore, is relevant to providing practical solutions to current and future world challenges (Ssossé et al., 2021).

To highlight the indispensable value of education in offering solutions for these problems, the UN's SDGs give Target 4.7 top priority on the need for ESD (Agbedahin, 2019). Teachers and principals significantly shape the mentality of students, impacting their values and attitudes (Bilbokaitė et al., 2024, 2025; Rieckmann, 2019; Rieckmann et al., 2022; Salīte et al., 2024). As a result, integrating ESD into teacher education curricula goes beyond a requirement to a necessity. If ESD should move from the policy level into the classroom and finally into life through attitude formation, then teachers must be well-versed in issues of sustainability to include such ideas in their day-to-day classroom activities. Despite this, research indicates that the effective implementation of ESD in teacher education programs remains inconsistent, as many programs, even those that highlight sustainability in theory, usually fail to reflect it in curriculum, pedagogy, and structure (Giridharan, 2020; Kioupi & Voulvoulis, 2019).

This chapter looks at how Finland and Ghana, two very different countries, have incorporated ESD into their teacher training programs. Finland is renowned for its high-performing and equity-driven educational model, which is based on a decentralized, trust-based approach to teacher education. To better meet the goals of both national and international development, Ghana, a nation in West Africa, has been making major changes to teacher education. This comparative study seeks to find out how ESD is integrated into teacher training curricula in Finland and Ghana, and the similarities and differences in ESD integration approaches between the two countries.

By stressing not only the policies and ideas in place but also the lived reality, institutional dynamics, and pedagogical practices that mold how future teachers are ready to interact with sustainability, the chapter seeks to add to the worldwide debate on sustainability in education.

2 Conceptual and Historical Foundations of ESD

ESD is rooted in the larger framework of sustainable development as expressed in the 1987 Brundtland Report, which defined sustainable development as meeting the needs of the present without compromising the ability of future generations to meet their own needs (WCED, 1987). From then on, ESD has changed from focusing on environmental education to a complete approach addressing ecological integrity, social justice, and economic viability.

UNESCO's Decade of Education for Sustainable Development (2005–2014) (UNESCO, 2017) catalyzed efforts worldwide, pushing education systems to incorporate sustainability themes across curricula and to rethink pedagogical models (Buckler & Creech, 2014). The follow-up Global Action Program (GAP) and the current ESD for 2030 Roadmap have further stressed the integration of sustainability into education policy, teacher education, and institutional culture (UNESCO, 2020).

ESD is built on key pedagogical principles: learner-centered approaches, interdisciplinary learning, action competence, and local–global linkages (Sterling & Orr, 2001; Tilbury, 2011). It promotes critical thinking, systemic understanding, and the ability to imagine and realize alternative futures (Wals, 2011). This implies that ESD advocates for integrated learning experiences where education is shifted from the traditional transmissive approaches to a transformative one.

2.1 Theoretical Backbone of ESD Integration

The transformative learning theory, developed by Mezirow (1997), provides a theoretical underpinning for this ESD. According to the theory, meaningful learning results from a close examination of presumptions and ideas, altering perspective. Applied to policies that propose a paradigm shift in teacher training, transformative learning fits very well with the objectives of ESD. This is because it seeks to transcend simple content delivery to generate reflective practitioners able to question accepted paradigms and advance social change.

Contemporary and future challenges like climate change and inequality must be addressed through the classroom by teachers, making sustainable practices an attitude in individuals, not merely knowledge. If teacher education is to respond meaningfully to the climate change crisis and global inequality, Wolff et al. (2022) contend that it must support a significant worldview change. Teachers themselves must thus participate in transformative learning, welcoming uncertainty, plurality, and value-based communication.

Transformative learning advocates for action-oriented individuals who contribute pragmatically to cause change. Teacher education, which serves as a link between national educational policies and their implementation at the individual level, serves as a useful context to assess the integration of ESD in a country's educational system. Following this, transformative learning provides an effective assessment lens to check whether teacher education promotes an in-depth engagement with sustainability or just superficial content inclusion.

2.2 Teacher Competencies and Professional Identity in ESD

The integration of ESD in teacher education fundamentally entails the development of educators' professional identities. The UNESCO (2020) roadmap clearly states that ESD should not be viewed as an isolated subject but rather as a comprehensive perspective integrated into educational practices, necessitating a re-evaluation of the educator's role in modern society (UNESCO, 2020).

UNESCO (2017) highlights some essential competencies as critical for ESD. These skills include systems thinking, anticipatory thinking, normative reasoning, teamwork, critical thinking, self-awareness, and integrated problem-solving. Empirical studies by Rieckmann (2012) and Millican (2022) have helped to underline the relevance of these skills in teacher training education.

Rieckmann (2012) and Millican (2022) detail core competencies essential for educators involved in ESD, including critical thinking, collaboration, action-oriented competence, systems thinking, and value-based reasoning. These competencies require individual engagement that goes beyond the theoretical pedestal to include both ethical and cognitive dimensions. Laurie et al. (2016) noted with much emphasis that curricular reforms stand the risk of remaining at the superficial level without purposefully imparting and fostering a sustainability mindset among educators. This assertion glories in the fact that, without educating educators on sustainability through ESD, they cannot satisfy the key tenets of ESD, limiting their role as agents of transforming the transformers.

Key requirements for the development of these competencies include a curriculum rooted and anchored in students' local settings, as well as multidisciplinary and value-driven ones. According to research, ESD runs the danger of being reduced to tokenistic inclusion instead of a transforming power without deliberate curriculum redesign and pedagogical innovation (Hofman-Bergholm, 2019; Laurie et al., 2016).

Globally, the effective integration of ESD in any curricula calls for the reshaping of institutional cultures and preparation practices to those that will end up developing educators who are empowered, reflective, and sustainably literate. Therefore, effectively integrating ESD into the teacher education curriculum for both Finland and Ghana requires massive changes in both institutional practices and curricula design.

2.3 *Opportunities and Barriers in Teacher Education for ESD*

Studies have attributed the advancement of ESD goals to the crucial role of teachers and, for that matter, teacher education (Almazroa et al., 2022; Bourn et al., 2017; Brandt et al., 2019; Fischer et al., 2022; Imara & Altinay, 2021; Kopnina & Meijers, 2014; Rieckmann, 2018). For instance, Bourn et al. (2017) noted that the role of teachers goes beyond conveying or transmitting knowledge but also serves as role models for ethical practices and sustainability. Nevertheless, integrating ESD into teacher education encounters notable obstacles. Pegalajar-Palomino et al. (2021) observed that, although student teachers generally have positive views on sustainability, their training programs often do not provide the depth or coherence required to foster practical ESD competencies effectively.

Wolff et al. (2022) identified inconsistencies in the Finnish educational setting regarding policy objectives and their execution. Notable in the Finnish educational context is their endorsement of global sustainability principles. However, ESD often remains sidelined as the responsibility to implement it lies in the hands of individual institutions and teachers. This decentralized implementation, promoted by their decentralized educational system, gives rise to inconsistent and non-uniform practices across various institutions due to limited institutional accountability (Hofman-Bergholm, 2019).

In Ghana, official publications such as the National Teachers' Standards (NTC, 2017) and the National Pre-Tertiary Education Curriculum Framework (MoE, 2018) explicitly delineate competencies associated with ESD, including critical thinking, environmental awareness, and active citizenship (MoE, 2018; NTC, 2017). Nketsia et al. (2020) and Agyemang (2020) identified significant discrepancies between policy objectives and educational practices attributed to curricular overload, insufficient training, and a deficient conceptual understanding of ESD among teacher educators. Agyemang (2020) indicates that educators generally include sustainability intermittently through extracurricular initiatives rather than integrating it constantly into classroom settings.

2.4 Integration of ESD Across Countries

Recent empirical studies show how ESD has been practically approached across various systems, especially when it comes to teacher education. Despite Finland's progressive educational system, ESD has not been routinely included in teacher training programs. Wolff et al. (2022) claim that national policies emphasize global responsibility but lack specific ESD requirements, allowing the implementation to be at the discretion of the faculty. In the same vein, Hofman-Bergholm (2019) noted that, although knowledge of sustainability exists, it is sometimes seen as an optional or peripheral issue rather than a cross-cutting topic. Teacher educators are, therefore, confronted with implementation challenges owing to institutional support and conceptual clarity.

Ghana's educational system is more centralized. Policy documents, including the National Teachers' Standards (NTC, 2017) and the National Pre-Tertiary Education Curriculum Framework (MoE, 2018), specifically address sustainability and advanced values, including environmental stewardship and critical thinking. However, Agyemang (2020), after conducting a study where he interviewed teachers in three Ghanaian colleges of education, concluded that actual classroom practices show little integration of ESD. Major challenges mentioned included overloaded curricula, inadequate training, and lack of institutional incentives. Nketsia et al. (2020) confirmed this by pointing out a rift between teacher educators' knowledge of ESD concepts and policy intentions.

A similar thread also shows in Balinggan (2023), who conducted a comparative analysis of the integration of ESD in the Philippines and Finland. Balinggan's study revealed that the integration of ESD is highly dependent on the culture and leadership of the institution as decisive agents. The policy environment in Ghana is conducive; however, community involvement and localized leadership are mostly missing. In the Finnish context, academic autonomy allows educators to be creative but does not provide coordinated supervision.

Laurie et al. (2016) synthesized studies from 18 countries and found that ESD integration largely improves academic quality, especially when connected to reflective and participatory pedagogies and integrated across subjects. They highlighted the need for professional development, institutional dedication, and assessment techniques that support sustainable learning outcomes.

Although real teaching strategies were often not in line with ESD competencies, attitudes toward sustainability were found to be present when Rieckmann et al. (2022) surveyed university teachers in Latin America. These results imply that, across the globe, there is a trend of awareness that exceeds pragmatic integration. Those in Finland and Ghana are no exception to this global trend.

Weighing all the ideas from these findings, they imply that the meaningful integration of ESD into the curricula of teacher training programs requires more than mere curriculum content. Additionally, it calls for institutional willingness to support teacher readiness and efficacy, and pedagogical innovation. These have been established by research to be the way to proceed and be successful, whether in Ghana, Finland, or any part of the world. The path forward is simply summarized as grassroots involvement coupled with institutional change.

2.5 The Educational System in the Study Countries

The study was conducted in Finland and Ghana (Fig. 1). Finland is a country in Northern Europe that shares borders with Russia to the east, Sweden to the northwest, Norway to the north, the Gulf of Bothnia to the west, and the Gulf of Finland to the south. It lies between longitudes 19.1°E and 31.6°E and latitudes 59.8°N and 70.1°N. It covers an area of about 338,400 km^2. About 5.64 million people are living in the country (Statistics Finland, 2025). In Finland, education is free at all levels, including universities, with an emphasis on the idea of equal and free access. At the basic education level, which spans grades 1 through 9 (ages 7–16), the emphasis is on learning support, student cooperation, and overall well-being. Following basic education, pupils can select between the general upper secondary (academic course, which culminates with matriculation exams) and the vocational education (skills-based and career-oriented). There are two main categories in higher education. That is, universities that are focused on theory and research, as well as institutions of applied sciences that focus on practical and career-oriented skills. Highly qualified teachers are valued in Finland's educational system. Teaching is a renowned and independent job since teachers are highly esteemed and need to possess a master's degree. With short school days, little homework, and creative flexibility, the focus of education is on balanced learning. Individual learning support is given priority, and schools modify their curricula to accommodate a range of student requirements. It is inclusive and student-centered.

Ghana is situated north of the equator in West Africa. It is located between longitudes 3.1°W and 1.2°E and latitudes 4.5°N and 11.0°N. It is bordered to the south by the Gulf of Guinea, to the north by Burkina Faso, to the east by Togo, and the west by Côte d'Ivoire. It covers an area of about 238,500 km^2. There are about 30.8 million people in the country (GSS, 2021). Primary, Junior High School (JHS), Senior High School (SHS) or Technical and Vocational Education and Training (TVET), and University or College comprise the 6-3-3-4 year structure of Ghana's educational system. Basic education consists of primary and JHS. Under the Free Compulsory Universal Basic Education (FCUBE) policy, education is free and mandatory at this point, and it ends with the Basic Education Certificate Examination (BECE). Despite

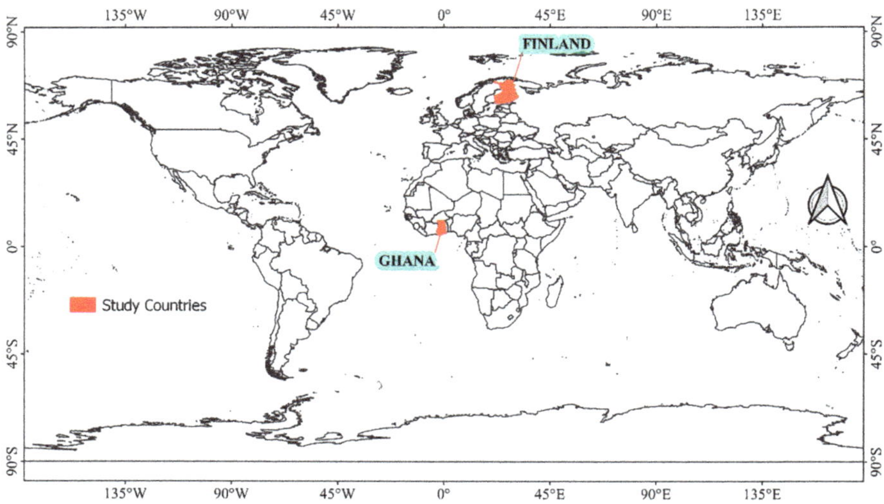

Fig. 1 World map showing Finland (Europe) and Ghana (Africa). *Source* Researchers' construct (2025)

the fact that basic education is compulsory, there are still school-age children nationwide who do not attend school, revealing weak policy implementation. Under the 2017 Free Senior High School policy, secondary education, which consists of three-year SHS and TVET, is free, and the level ends with the West African Senior School Certificate Examination (WASSCE) for SHS and the Commission for Technical and Vocational Education and Training (CTVET) for TVET students. The tertiary education level, which includes universities, technical universities (formerly polytechnics), and colleges of education, is where those who pass the WASSCE/CTVET and fulfil the admission requirements proceed. The Ghana Tertiary Education Commission (GTEC) oversees the tertiary education, which offers various bachelor's degrees, diplomas, and postgraduate programs. Some SHS graduates choose to enroll in nursing colleges. High student–teacher ratios in many schools, infrastructural and teacher shortages, particularly in rural areas, and regional disparity in access and quality are some of Ghana's educational problems. The expansion of Technical and Vocational Education and Training (TVET), competency-based and learning-focused curriculum, and a concentration on STEM education and digital literacy are the main features of the most recent educational reforms in the country.

3 Methodology

Using a qualitative comparative case study approach, this chapter explores how ESD is conceptualized, included, and used in teacher education curricula in Finland and Ghana. Leveraging on the comparative approach, the study seeks to capture the specific contextual reality of both countries as well as the overall trends impacting ESD in teacher training education. This approach was employed because it allows comprehensive and context-sensitive research of ESD integration in the two different national environments of Finland and Ghana. This method aids in gathering rich qualitative data from policy documents as well as practitioner experiences, enabling a sophisticated comparison of how teacher training programs approach global sustainability agendas such as ESD integration.

3.1 Data Collection

Data was collected through two main sources: national policy documents on teacher education and sustainable development from Finland and Ghana, and semi-structured interviews with four teacher educators, two from each country. Documents having relevance to ESD and their influence on teacher education were chosen from both countries. These documents included the Finnish Ministry of Education policy guidelines, reports on curriculum development, national education strategies, as well as Ghana's National Pre-Tertiary Education Curriculum Framework (MoE, 2018) and National Teachers' Standards (NTC, 2017).

Four in-depth interviews were conducted with teacher educators who were purposively sampled based on their professional engagement with sustainability-related teaching or curriculum development. This was done to capture the lived experiences and professional insights from these educators. The interviews lasted between 30 and 50 min, and were done face-to-face and online. Each interview was audio-recorded following permission and transcribed.

3.2 Analytical Approach

A directed qualitative content analysis approach was employed in the analysis of the documents, guided by established frameworks in sustainability education, such as UNESCO (2017) and the Transformative Learning Theory (Mezirow, 1997) principles. These frameworks provided include: (i) policy framing and vision, (ii) curriculum content and integration, (iii) pedagogical strategies and innovation, (iv) teacher competence development, and (v) institutional support and assessment. Coding began deductively using these categories but was later supplemented by inductive codes that emerged from the data, particularly from the interview transcripts.

The study did not solely depend on a single source's information. Thus, data triangulation was used to enhance validity as interviews were repeatedly compared with the official documentation to discern the connections or disconnections between the ideas. The statements of an interviewee during an interview frequently prompted us to re-evaluate a section of the curriculum framework or standards from a new perspective.

The study did not anticipate receiving everything simultaneously. Several iterations of reviewing the information were necessary, and occasionally, the categories that had been deemed robust needed adjustment when the data indicated otherwise. This method of working, involving consistently revisiting, verifying, and analyzing the policy documents while attentively considering the educators' perspectives, enabled the researchers to remain anchored. To make cross-case comparison easier, thematic matrices were created, matching each theme to the two main research questions of the study.

3.3 *Ethical Considerations*

Participants' identities were anonymized, and all data were securely stored. Ethical clearance was obtained from the authors' affiliated institutions where required. Informed consent was sought before interviews, and participants retained the right to withdraw at any point.

4 Results and Discussions

Having discussed the theoretical and analytical insights, this section now proceeds to present the success stories or empirical insights drawn from the document's analysis and interviews, purposely to address the research questions raised in Sect. 1.

4.1 *Integration of ESD into Teacher Training Curricula in Finland*

The document analysis (Table 1) of Finland's teacher education policy and reforms reveals a familiar pattern in education policy discourse: strong vision statements at the national level, but inconsistent and fragmented implementation at the institutional level. The lived experiences of the teacher educators interviewed for this study mirrored and elaborated on these trends.

The Ministry of Education's 2016 policy document (MEC, 2016) outlines a progressive vision centered on ethical awareness, teacher autonomy, and global responsibility. While the term "sustainability" is not frequently foregrounded, its underlying themes, equity, inclusivity, and lifelong learning align well with ESD principles. Yet, as Participant A shared,

Table 1 Document analysis table (Finland and Ghana)

Theme	Finland	Ghana
Policy vision and ESD commitments	The 2016 Ministry of Education and Culture (MEC, 2016) policy document lays out a strategic vision for teacher education reform that promotes ethical awareness, global responsibility, equity, and lifelong learning. Although the term 'sustainability' is not consistently emphasized, the overarching values reflect ESD-aligned priorities. The document proposes comprehensive professional development and envisions teachers as reflective practitioners prepared for diverse and evolving challenges (MEC, 2016, pp. 5, 10)	The National Pre-Tertiary Education Curriculum Framework (MoE, 2018) explicitly promotes sustainable development as one of its cross-cutting themes. It identifies climate change, environmental responsibility, and sustainable practices as vital components of national education goals. The framework ties ESD to national development and the SDGs, stating that Ghana's curriculum must reflect these global imperatives (MoE, 2018, pp. iii, 10, and 22)
Core teacher competencies for ESD	The Wolff et al. (2022) article points out that Finnish teacher education institutions enjoy a high level of curricular freedom. However, without binding policy mandates, sustainability themes tend to be incorporated inconsistently. Instead of being a core part of teacher education, ESD is frequently viewed as an elective or auxiliary subject. In order to achieve revolutionary education practices that incorporate ESD principles, the authors propose that educational change must go beyond inclusiveness (Wolff et al., 2022, pp. 4–5)	The National Teachers' Standards (NTC, 2017) position teachers as responsible citizens and change agents. They are expected to exhibit competencies related to environmental stewardship, inclusivity, and critical thinking. Teachers must promote sustainable values and integrate them into classroom practice, especially through inquiry-based and participatory methods (NTC, 2017, pp. 13–15)
Curriculum and pedagogical guidelines	The Wolff et al. (2022) article points out that Finnish teacher education institutions enjoy a high level of curricular freedom. However, without binding policy mandates, sustainability themes tend to be incorporated inconsistently. ESD is often treated as an elective or peripheral element rather than a central component of teacher education. The authors suggest that pedagogical reform must go beyond inclusion and aim at transformative education practices that embody ESD principles (Wolff et al., 2022, pp. 4–5)	ESD principles are woven into the general curriculum structure in Ghana. The framework (MoE, 2018) highlights sustainability themes such as responsible consumption, democratic participation, and ethical decision-making. It outlines competencies like problem-solving, personal development, and global citizenship as essential outcomes for learners (MoE, 2018, pp. 5, 26–29)

(continued)

Table 1 (continued)

Theme	Finland	Ghana
Pedagogical and instructional strategies	The strong autonomy granted to Finnish universities results in diverse interpretations and applications of ESD within teacher education (Wolff et al., 2022). While some faculties integrate ESD principles into courses and projects, others show minimal engagement. The lack of national oversight or unified evaluation tools leads to fragmented implementation. This disparity reflects both the strengths and weaknesses of the decentralized Finnish model (Wolff et al., 2022, pp. 5–7)	Teachers are encouraged to adopt learner-centered, reflective, and experiential teaching strategies. The framework and standards advocate for differentiated instruction, inclusive pedagogies, and the use of ICT and group work approaches that align with ESD's transformative goals. These strategies are intended to promote deep understanding and active citizenship (MoE, 2018, pp. 30–34; NTC, 2017, pp. 16–18)
Challenges in ESD integration	The integration of ESD into Finnish teacher education is hindered by vague policy directives and limited accountability mechanisms. According to Wolff et al. (2022), although national strategies acknowledge the importance of sustainability, there is a noticeable gap between aspiration and execution. ESD is not systematically embedded into course content, learning outcomes, or assessments, which weakens its institutional impact (Wolff et al., 2022, pp. 6–7)	Agyemang (2020) found that, while Ghana's curriculum and standards articulate strong sustainability visions, teacher education institutions face implementation barriers. These include insufficient faculty training, limited resources, curriculum overload, and lack of leadership in advancing sustainability goals. Many lecturers rely on their creativity to embed ESD topics (Agyemang, 2020, pp. 34–39)
Assessment and monitoring of ESD, and alignment with global frameworks (UNESCO SDGs)	While Finland's policy framework generally aligns with global commitments such as the UN SDGs and UNESCO's ESD goals, practical translation remains uneven. Lavonen et al. (2020) acknowledge that although teacher education institutions are encouraged to integrate themes like global responsibility, the degree to which UNESCO's ESD competencies are embedded varies greatly across institutions. Others have highlighted this inconsistency (Lavonen et al., 2020, pp. 9; Wolff et al., 2022, pp. 7)	Although ESD themes appear in learning outcomes, there is limited evidence of systematic assessment tools or indicators used to evaluate sustainable learning in teacher education. Agyemang (2020) and the national standards both mention formative assessment but do not provide detailed frameworks for assessing ESD competencies (Agyemang, 2020, pp. 42–44; NTC, 2017, pp. 22–23)

(continued)

Table 1 (continued)

Theme	Finland	Ghana
Recommendations for enhancing ESD	Both the Ministry of Education and Culture (MEC, 2016) document and Lavonen et al. (2020) stress the need for teacher education reform to support interdisciplinary collaboration, reflective pedagogy, and stronger ties between sustainability and teacher identity. They advocate for more coherent national strategies, institutional incentives, and ongoing professional development focused on ESD, suggesting these are necessary to move from policy rhetoric to practical transformation (Lavonen et al., 2020, pp. 10; MEC, 2016, pp. 12)	The analyzed documents emphasized continuous professional development, integration of ESD into teacher education curricula, and institutional leadership as key to successful implementation. Agyemang (2020) calls for structured collaboration across departments and more targeted national support to operationalize sustainability education in practice (Agyemang, 2020, pp. 45–48; MoE, 2018, pp. 36–38)

> We sort of let that [vision] guide our direction … but without much top-down accountability.

This reflects the tension between policy ideals and actual teaching practice.

The national development program documented by Lavonen et al. (2020) adds another layer of insight. It highlights efforts to embed competencies such as collaboration, global citizenship, and sustainability within teacher education reform. However, the uptake of these competencies depends heavily on institutional will. Participant A noted that, while courses like "Multicultural Education" and "Culturally Responsive Pedagogy" deeply engage social sustainability themes:

> We do not highlight education for sustainability as a singular criterion for evaluation... but it's there.

This quotation sums up the often latent and scattered nature of ESD integration. Wolff et al. (2022) offered a critical perspective on this matter by noting that the greater autonomy possessed by Finnish universities to shape their curricula can sometimes be the reason for not placing sustainability as part of their priority list. Participant B confirmed this by observing that

> It varies a lot across subject areas and programs. For instance, sustainability is more visible in natural science programs but nearly absent elsewhere. Some of the modules that used to exist have been cut.

This points to a troubling trend of deprioritizing ESD at the structural level. These interview insights also give voice to the lived contradictions of autonomy. On one hand, teacher educators can design and teach courses aligned with their interests. On the other hand, there is a risk that sustainability never becomes institutionalized. As Participant A put it,

> How we integrate those [policy values] into everyday teaching relies on teachers themselves.

While there are funding incentives and thematic research grants focused on sustainability, there is little in terms of direct curricular enforcement.

When it comes to pedagogy, both participants offered rich descriptions of ESD-relevant strategies. Experiential learning, intercultural dialogue, and inclusive pedagogy were recurring themes. Participant B emphasized "project-based and phenomenon-based teaching" and described taking students "sailing or hiking" to promote experiential and place-based sustainability learning.

Participant A advocated for the inclusion of crosscutting perspectives in classroom discussions. These perspectives, according to Participant A, should revolve around equity, stereotyping, and global power relations as they will end up encouraging students to be active change makers who challenge dominant narratives. Even when ESD themes are captured in teaching, Participant A noted that those topics are not measured in any formal way and that the topics are not clearly outlined as objectives. She noted,

> We do not assess it as a separate goal.

Instead, students encounter sustainability through broader topics like inclusion or equality. This makes it difficult to know how well future teachers understand or can apply these ideas.

There are other layers of challenge too. Participant A spoke openly about the added difficulty of working as a non-native speaker in Finnish universities, describing how

> Very often, I feel like I have to fight for myself.

This reveals how institutional culture, not just curriculum, affects how sustainability is taught. Committing to a demanding task like the inclusion of aspects of ESD into teaching requires educators to be supported and given a sense of belongingness. When these are not in place, it becomes difficult for them to effectively commit to the task.

Notwithstanding these challenges, the educators interviewed spoke of significant efforts they are making to make sustainability visible to their students. Participant A discussed working with colleagues from many departments, including science and technology, to investigate how artificial intelligence (AI) and digital tools might support sustainability education. Participant B offered a different strategy, taking students outside for practical lessons in the countryside. These illustrations demonstrate how ESD might adopt many strong and imaginative forms when given space to flourish.

All taken together, analyzing policy documents and the voices of educators in Finnish teacher training institutions exposes not only shortcomings but also aspirations, challenges, and commitments. The policy vision is strong. Institutional autonomy is real. However, without dedicated space in the curriculum, consistent assessment strategies, and systemic support for diverse educators, sustainability risks becoming a well-meaning afterthought.

To move forward, Finland must bridge the distance between policy and practice. ESD should no longer be something that "depends on the teacher" but a shared institutional priority. What is needed is not just curriculum reform but structural commitment from leadership, programs, and the state itself.

4.2 Integration of ESD into Teacher Training Curricula in Ghana

The policy documents (Table 1) that have been examined, specifically the National Teachers' Standards (NTC, 2017), Ghana's National Pre-Tertiary Education Curriculum Framework (MoE, 2018), and Agyemang (2020) empirical research, demonstrate a consistent and intentional attempt to incorporate Education for Sustainable Development (ESD) into teacher training and broad curriculum reform. However, the experiences of the educators interviewed in this study show that the road between vision and practice is not always smooth or straightforward.

It is evident at the policy level that Ghana is inclined and dedicated to ESD and matters related to sustainability. Sustainability is not represented as an isolated issue but rather as a guiding concept that cuts across several disciplines and competencies. The curriculum framework directly captures topics related to climate awareness, environmental responsibility, and ethical citizenship. The Teachers' Standards (NTC, 2017) reinforce these topics as values that reflect professionalism, including inclusivity, critical thinking, and responsiveness to environmental and social issues, which are clearly outlined in it. Yet, as Participant C explained,

> And not that I know (regarding formal institutional strategy), it is recently that we wrote to the registrar's office to have a policy on pollution. And we know that when we have that policy, then sustainability would definitely be included in it. So, we are about to work on that.

This aligns with a reality highlighted in Agyemang's (2020) findings regarding the ambition of national policy, which he noted is not always matched by local structures or administrative follow-through. This indicates that the execution of these policies is problematic since institutions are aware of the expectations, but implementation is often left to chance or personal initiative. Participant D was even more reflective:

> We do not see sustainability as something separate… it should be part and parcel of us.

He described how, in his department, ESD is not housed in a standalone course but threaded through multiple courses or disciplines from environment and society to regional planning and transport studies. Participant D added that:

> Sustainability must be in us, not just known but lived.

This strong personal conviction highlighted a crucial insight: teachers' attitudes and values may carry more weight in the classroom than formal structures.

Both educators pointed to examples of sustainability work in action. Participant C described collaborative projects with other departments and outside institutions, including tree planting drives and pollution research funded by a consultancy. Participant D shared stories of student-led innovation:

> One of my students started turning plantain and banana peels into plates and bags. He got sponsorship and showcased his work in Dubai before dropping out to work for an Airline.

While exceptional, this anecdote reveals the impact a well-positioned course or motivated educator can have on student agency.

Despite these pockets of innovation, challenges remain, some persistent, others structural. Participant D admitted that:

> The numbers are too high in the class. You give them work; they present it, but you cannot always go deep.

He also criticized the national tendency toward theoretical rather than practical assessment:

> Our students may know ten things and practice only one, two, or sometimes three.

This concern was mirrored by Participant C, who noted that the assessment of sustainability was indirect, often based on a general understanding, with limited tools to track real-world applications.

> Understanding in writing is different from understanding in practice, and the understanding in practice, we are not the ones to test that [since their high numbers only allow us to test their understanding in writing].

Another theme that emerged strongly was the importance of locally rooted sustainability education. Participant D explained how traditional practices like sacred forests or taboos against polluting rivers were based on environmental logic.

> It was not about dwarfs chasing you, he said with a smile. It was science; it was rational.

He emphasized that Ghana's traditional ecological knowledge should be honored in curriculum design, not treated as folklore.

When asked what kind of support would enhance their ability to teach sustainability, both educators gave answers that were more philosophical than technical. Participant C mentioned funding for projects and smaller class sizes. Participant D, however, approached it from another angle:

> Support must come from us. If we cannot afford to write our own policy, we cannot implement it if someone else sponsors the writing of the policy. If we do not have funds to write policy documents, then we certainly will not have funds to implement them if it is externally sponsored.

He emphasized ownership, agency, and internal commitment as prerequisites for meaningful reform. He warned against waiting for external donors to define or fund ESD strategies, saying:

> If we do not perceive that we are the problem or a contributor to it, we cannot solve it.

The conversation ended with a shared belief in the power of modeling. Participant D stressed that educators need to live what they teach:

> If students see their lecturer ride a bicycle to work, they will remember that more than Power Point slides or presentations on climate change.

In summary, the Ghanaian case shows a strong policy foundation, a wealth of local knowledge, and educator passion. What is missing in the equation necessary for ESD to take the right shape in the country's teacher training institutions is not awareness, but instead structural elements, including coherence and capacity. As noted by Agyemang (2020) and both participants, institutional leadership, better assessment tools, and interdepartmental collaboration are essential. At the top of the priority list, educators must be supported to embody ESD as they establish themselves as role models, not only as teachers of it. The next generation will be impacted in lasting ways when sustainability becomes a part of the daily lives of educators. Simply said, when educators practice what they teach, learners will practice what they learn.

4.3 *Comparative Analysis of ESD Integration in Finland and Ghana*

The integration of ESD into teacher education takes on markedly different forms in Finland and Ghana since the integration in each country is shaped by institutional culture, policy orientation, and the day-to-day realities of teacher educators. While the two countries show some level of commitment to sustainability, their approach to that commitment regarding its implementation and experience is not the same. Figure 2 summarizes the thematic comparative analysis between the two countries.

4.3.1 Policy Framing: Shared Intent and Different Clarity

The national education frameworks of the countries show that both countries have a clear commitment to sustainability. In Finland, even though the term sustainability is not explicitly used, the policy documents revealed values related to ESD, such

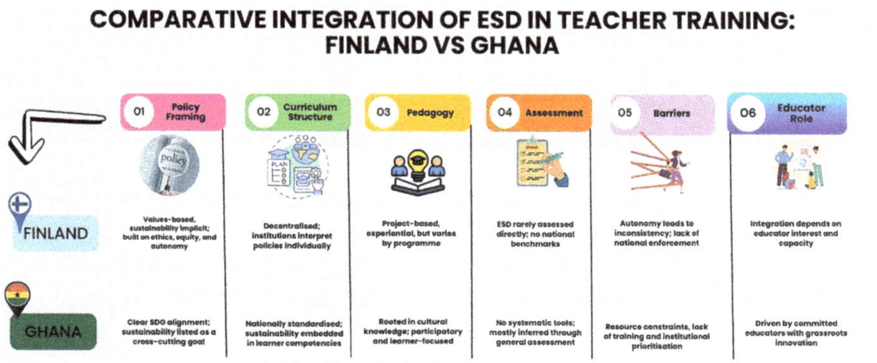

Fig. 2 Comparative integration of ESD in Finland and Ghana. *Source* Researchers' construct (2025)

as ethical responsibility, equity, and global awareness. Additionally, reflective and globally competent educators who relate to ESD competencies have been highlighted and encouraged in the Ministry of Education and Culture (MEC, 2016) document and subsequent development strategies. On the other hand, there is a direct mention of ESD in Ghana. The National Pre-Tertiary Education Curriculum Framework (MoE, 2018) refers to sustainable development as a cross-cutting theme and aligns clearly with the Sustainable Development Goals.

In essence, Ghana's policy language is more explicit, whereas Finland's is embedded within a broader discourse of educational equity and professional autonomy. However, in both cases, this commitment is aspirational. As Participant A from Finland put it,

> We sort of let that [vision] guide our direction.

But there is no clear mandate. Similarly, Participant C in Ghana noted,

> There is no real formal strategy at the institutional level yet.

In both systems, much of the heavy lifting is left to individual educators.

4.3.2 Curriculum and Competency Integration

Certain learner competencies that align with UNESCO's ESD objectives, including problem-solving, personal development, and global citizenship, have been outlined in Ghana's curriculum. Also, in alignment with ESD competencies, the Teachers' Standards (NTC, 2017) highlight the advancement of inclusivity, critical thinking, and environmental awareness as values educators should embrace. However, these intentions are captured across different courses rather than in a single content area.

In contrast, Finland promotes curriculum autonomy. Although national development plans, such as Lavonen et al. (2020), show sustainability competencies, the choice to incorporate ESD mostly rests with the faculties or even individual educators. This flexibility allows innovation, but it also creates gaps. As Participant B explained,

> It varies a lot.

The variation depends on the program. Sustainability might be strongly embedded in one subject/course/program and absent in another.

4.3.3 Pedagogical Practice and Delivery

In both countries, teachers are encouraged to use learner-centered, inclusive, and inquiry-based methods and approaches that naturally support ESD. In Ghana, ICT integration, group work, and differentiated instruction are promoted by the curriculum. Participant D spoke quite passionately about how local cultural knowledge and hands-on activities had been mixed to promote sustainability by noting that:

> It was not about dwarfs chasing you in the evil forest… It was a science of conservation embedded in myths; it was rational.

Conversely, the approach used in Finland stresses project-based and phenomenon-based learning. Participant B said that one kind of experiential education for ESD is getting pupils to sail or hike. Participant A spoke on the inclusion of critical points of view in classroom debates. Still, it is seen in both nations that these pedagogical approaches lack consistent structural support. Often, instead of institutional frameworks, teachers depend on improvisation or personal dedication.

4.3.4 Assessment and Monitoring

Neither nation has strong tools for assessing teacher education's sustainability competencies. Results related to ESD are not often expressed in course objectives or evaluation criteria in Finland. Participant A noted,

> We do not assess it as a separate goal.

This problem is also evident in Ghana. Participant C said that assessment is indirect and depends more on general knowledge than on methodical tools to gauge the development of sustainability. Agyemang (2020) supports this claim, noting that despite strong policy signals, there is little follow-through on the tracking or assessment of sustainability outcomes in teacher training in Ghana.

4.3.5 Barriers and Institutional Support

The challenges both countries face are strikingly similar, though they stem from different systemic roots. In Ghana, barriers are practical: lack of training, funding, and administrative prioritization. Participant D spoke of large class sizes and theoretical overload:

> Our students may know ten things and practice only one, two, or none because the practical aspect is generally anemic.

In Finland, the issue is decentralization. While autonomy encourages creativity, it could also cause ESD to be marginalized. Participant B noted that the influence of the educator often determines how sustainability is addressed. Regarding leadership, both systems have challenges. Finnish Participant A highlighted the difficulties minority teachers have with language and cultural gatekeeping. Participant D (Ghana) stressed the need for internal ownership:

> If we cannot write our own policy, we cannot implement someone else's.

4.3.6 Educator Agency

What emerges strongly in both cases is the role of the educator. Whether in Finland or Ghana, teachers and lecturers drive ESD integration, not because they are required to, but because they believe in it. Participant C from Ghana recounted embarking on local sustainability projects without any formal mandate. Participant A from Finland reported doing a similar thing by incorporating critical viewpoints into lessons due to how important they were, and not because any formal policy directs their inclusion. These trends highlight the fact that, while formal policy establishes the framework, effective integration is dependent on educators' dedication and initiative. The challenge here is that if the integration of sustainability is left at the mercy of goodwill, then there will not be any consistency in its implementation.

4.4 Comparative Perspectives on ESD Implementation

When comparing ESD practices between the two countries, both differences and similarities exist. Regarding factors that influence how ESD is adopted, Balinggan (2023) pointed to the local context, which includes government policies, funding, and teacher autonomy. With this as a measuring yardstick, it becomes evident that the decentralized educational system of Finland supports innovation and creativity but is deficient in uniformity owing to the absence of established national monitoring guidelines. On the other hand, Ghana's centralized educational system guarantees homogeneity, yet its flexibility to local educational settings may be limited.

Including sustainability in higher education calls for robust quality control systems (Holm et al., 2011; Javed & Alenezi, 2023). Although Finland's present quality control systems help at the policy level, they provide difficulties in properly applying these guidelines in the classroom. On the contrary, in the case of Ghana, the development of a quality assurance framework is confronted with the challenge of reliably monitoring outcomes related to ESD.

Transformative pedagogies play a crucial role in effective ESD implementation. According to Rieckmann et al. (2022), the development of critical sustainability competencies, including anticipatory thinking, normative judgment, and systems analysis, depends on creative teaching strategies, including project-based learning, interdisciplinary collaboration, and community engagement. Though these methods are being increasingly combined, Finland has shown a more methodical approach to research than Ghana.

5 Limitations and Future Prospects

This chapter offers insight into the integration of ESD into teacher education curricula in Finland and Ghana. However, it possesses some limitations that must be acknowledged. First, although the reflections offered by the four interview participants were detailed and multifaceted, a broader sample across multiple institutions may offer greater representativeness and allow for more generalized conclusions. Additionally, the study mainly focused on policy documents at the national level and did not include institutional-level course materials, which may detail how ESD is taught and assessed in practice. Further analysis involving these can be a future research focus.

Secondly, adopting the qualitative case study approach in the study was ideal to capture complexities in context. However, not including student voices places a limitation on obtaining a holistic perspective of how pre-service teachers experience sustainability education. Future studies should consider including students' voices to obtain a comprehensive understanding of how curriculum design and pedagogical strategies influence learning in practice.

Notwithstanding these limitations, the chapter offers interesting directions for future study and application. Follow-up research has a great possibility to track how ESD training affects newly qualified teachers' application of sustainability ideas into their classroom teaching and learning process. Comparative research across additional countries, or within multilingual education systems, could also enrich the global dialogue on curriculum development for sustainable development.

6 Conclusion

This chapter is not just intended to understand the appearance of ESD in teacher education curricula in Finland and Ghana through a comparative analysis; it transcends that scope by including the role the national policies, local institutional cultures, and the agency of educators play in shaping the inclusion of ESD. The chapter primarily aimed to explore how curriculum development for sustainable development is achieved or hindered in the two educational landscapes (Finland and Ghana).

The autonomy of institutions and teachers, which influences the implementation of the curriculum at all educational levels, is what sets apart the decentralized Finnish educational system. The autonomy granted to educators allows them to be innovative. However, such innovation sometimes leads to the inconsistent integration of sustainability-related topics and competencies. Additionally, some departments have adopted sustainability as an integral part of their practice, while others treat it with less relevance as they regard it as optional. Without national coordination or mandatory assessment, the commitment to ESD varies greatly across institutions.

Limited resources, coupled with inadequate training and overpopulated classrooms, leave a larger part of the work in the hands of committed educators in Ghana.

This, sometimes, pushes instructors to go beyond the threshold of their job description, leveraging local knowledge, community engagement, and personal convictions to make sustainability a part of their classroom practice.

What both countries have in common is that ESD depends heavily on the teacher. Policies can create the conditions, but it is the everyday decisions of educators, their creativity, dedication, and sense of purpose that make ESD real for learners.

The chapter's comparative lens made it evident that, in the development of curriculum for ESD, it is not enough to mainly concentrate on policy directives or the goodwill of educators. Instead, systematic structures that incorporate sustainability competencies into objectives, pedagogical strategies, and assessment instruments are a necessity. This implies that deliberate embedding of ESD into teacher education at the structural level needs to be considered by institutions for curriculum development to effectively contribute to sustainable development. ESD integration cannot be left to chance. At every educational level, it must be approached with all seriousness and intentionality.

Ultimately, for Finland and Ghana, and all educational systems, the integration of ESD into teacher education must be considered as a foundational block of curriculum development. This chapter adds to the global discussion by giving practical insights into how teacher education can be restructured into a transformative force capable of addressing the challenges of our time.

References

Agbedahin, A. V. (2019). Sustainable Development, Education for Sustainable Development, and the 2030 Agenda for Sustainable Development: Emergence, efficacy, eminence, and future. *Sustainable Development, 27*(4), 669–680. https://doi.org/10.1002/sd.1931

Agyemang, R. O. (2020). *Embedding sustainable development in teacher education in the Central Region of Ghana [Thesis]*. Faculty of Education and Diversity, University of Iceland.

Almazroa, H., Alotaibi, W., & Alrwaythi, E. (2022). Sustainable development goals and future-oriented teacher education programs. *IEEE Transactions on Engineering Management, 71*, 13517–13530. https://doi.org/10.1109/TEM.2022.3165686

Balinggan, L. F. C. (2023). *A long way to go: A comparative study on the implementation of education for sustainable development in secondary education in finland and the philippines.* Faculty for Education and Welfare Studies Åbo Akademi University. https://www.doria.fi/handle/10024/186953

Bilbokaitė, R., Bilbokaitė-Skiauterienė, I., Fjodorova, I., Kravale-Pauliņa, M., Oļehnoviča, E., Szplit, A., & Zbróg, Z. (2024). Global competence and sustainable development in teacher training: Preparing for a changing world. *Journal of Teacher Education for Sustainability, 26*(1), 196–217. https://doi.org/10.2478/jtes-2024-0012

Bilbokaitė, R., Bilbokaitė-Skiauterienė, I., Fjodorova, I., Kravale-Pauliņa, M., Oļehnoviča, E., Szplit, A., & Zbróg, Z. (2025). Global competences for principals: Enabling change for sustainable education. *Journal of Teacher Education for Sustainability, 26*(2), 141–161. https://doi.org/10.2478/jtes-2024-0020

Bourn, D., Hunt, F., & Bamber, P. (2017). *A review of education for sustainable development and global citizenship education in teacher education* (Global Education Monitoring Report No. ED/GEMR/MRT/2017/P1/8). United Nations Educational, Scientific and Cultural Organization (UNESCO).

Brandt, J.-O., Bürgener, L., Barth, M., & Redman, A. (2019). Becoming a competent teacher in education for sustainable development: Learning outcomes and processes in teacher education. *International Journal of Sustainability in Higher Education, 20*(4), 630–653. https://doi.org/10.1108/IJSHE-10-2018-0183

Buckler, C., & Creech, H. (2014). *Shaping the future we want: UN decade of education for sustainable development*. United Nations Educational, Scientific and Cultural Organization. https://unesdoc.unesco.org/ark:/48223/pf0000230171

del Pegalajar-Palomino, M. C., Burgos-García, A., & Martinez-Valdivia, E. (2021). What does education for sustainable development offer in initial teacher training? A systematic review. *Journal of Teacher Education for Sustainability, 23*(1), 99–114. https://doi.org/10.2478/jtes-2021-0008

Fischer, D., King, J., Rieckmann, M., Barth, M., Büssing, A., Hemmer, I., & Lindau-Bank, D. (2022). Teacher education for sustainable development: A review of an emerging research field. *Journal of Teacher Education, 73*(5), 509–524. https://doi.org/10.1177/00224871221105784

Giridharan, B. (2020). Learning for a sustainable future: Developing key competencies. In E. Sengupta, P. Blessinger, & T. S. Yamin (Eds.), *Introduction to sustainable development leadership and strategies in higher education* (world; Vol. 22, pp. 135–144). Emerald Publishing Limited. https://doi.org/10.1108/S2055-364120200000022013

GSS. (2021, September 22). *2021 population and housing census (PHC)*. Ghana Statistical Service (GSS). https://census2021.statsghana.gov.gh/

Hofman-Bergholm, M. (2019). Changes in thoughts and actions as requirements for a sustainable future: A review of recent research on the finnish educational system and sustainable development. *Journal of Teacher Education for Sustainability, 20*(2), 19–30. https://doi.org/10.2478/jtes-2018-0012

Holm, T., Sammalisto, K., & Vuorisalo, T. (2011, September 19). *Education for sustainable development and management systems in higher education in Finland and China: A comparative study*. 2nd Sino-Finland Forum on Higher Education.

Imara, K., & Altinay, F. (2021). Integrating education for sustainable development competencies in teacher education. *Sustainability, 13*(22), 22. https://doi.org/10.3390/su132212555

Javed, Y., & Alenezi, M. (2023). A case study on sustainable quality assurance in higher education. *Sustainability, 15*(10), 10. https://doi.org/10.3390/su15108136

Jekabsone, I., & Ratniece, L. (2024). Integrating sustainability into curriculum of legal education in Latvia: An insight into the main issues. *Journal of Teacher Education for Sustainability, 25*(2), 60–75. https://doi.org/10.2478/jtes-2023-0016

Kioupi, V., & Voulvoulis, N. (2019). Education for sustainable development: A systemic framework for connecting the SDGs to educational outcomes. *Sustainability, 11*(21), 21. https://doi.org/10.3390/su11216104

Kopnina, H., & Meijers, F. (2014). Education for sustainable development (ESD): Exploring theoretical and practical challenges. *International Journal of Sustainability in Higher Education, 15*(2), 188–207. https://doi.org/10.1108/IJSHE-07-2012-0059

Laurie, R., Nonoyama-Tarumi, Y., Mckeown, R., & Hopkins, C. (2016). Contributions of education for sustainable development (ESD) to quality education: A synthesis of research. *Journal of Education for Sustainable Development, 10*(2), 226–242. https://doi.org/10.1177/0973408216661442

Lavonen, J., Mahlamäki-Kultanen, S., Vahtivuori-Hänninen, S., & Mikkola, A. (2020). A collaborative design for a finnish teacher education development programme. *Journal of Teacher Education and Educators, 9*(2), 241–262.

Mckeown, R. (2014). The leading edge of teacher education and ESD. *Journal of Education for Sustainable Development, 8*(2), 127–131. https://doi.org/10.1177/0973408214548366

MEC. (2016). *Guidelines of developing teacher education: Ideas and suggestions of the teacher education forum*. Ministry of Education and Culture (MEC). https://julkaisut.valtioneuvosto.fi/handle/10024/75553

Mezirow, J. (1997). Transformative learning: Theory to practice. *New Directions for Adult and Continuing Education, 1997*(74), 5–12. https://doi.org/10.1002/ace.7401

Millican, R. (2022). A rounder sense of purpose: Competences for educators in search of transformation. In P. Vare, N. Lausselet, & M. Rieckmann (Eds.), *Competences in education for sustainable development: Critical perspectives* (pp. 35–43). Springer International Publishing. https://doi.org/10.1007/978-3-030-91055-6_5

MoE. (2018, May). *National pre-tertiary education curriculum framework*. National Council for Curriculum and Assessment (NaCCA), Ministry of Education.

Nketsia, W., Opoku, M. P., Saloviita, T., & Tracey, D. (2020). Teacher educators' and teacher trainees' perspective on teacher training for sustainable development. *Journal of Teacher Education for Sustainability, 22*(1), 49–65. https://doi.org/10.2478/jtes-2020-0005

NTC. (2017, September). *National teachers' standards for Ghana: Guidelines*. National Teaching Council (NTC).

Rieckmann, M. (2012). Future-oriented higher education: Which key competencies should be fostered through university teaching and learning? *Futures, 44*(2), 127–135. https://doi.org/10.1016/j.futures.2011.09.005

Rieckmann, M. (2018). Learning to transform the world: Key competencies in education for sustainable development. In A. Leicht, J. Heiss, & W. J. Byun (Eds.), *Issues and trends in education for sustainable development* (pp. 39–59). UNESCO Publishing. https://unesdoc.unesco.org/ark:/48223/pf0000261445

Rieckmann, M. (2019). Education for sustainable development in teacher education. An international perspective. In S. Lahiri (Ed.), *Environmental education* (pp. 33–48). BioGreen Publication.

Rieckmann, M., Pérez, M. L., Vanegas, A. Á., Aguirre, P., & Corres, A. (2022). Education for the sustainable development goals: A Latin-American survey on university teachers' competences. In *Education in a changing world: The impact of global realities on the prospects and experiences of educational research* (pp. 1–15).

Salīte, I., Fjodorova, I., & Ivanova, O. (2024). How do teachers promote the viability of education? *Journal of Teacher Education for Sustainability, 25*(2), 1–4. https://doi.org/10.2478/jtes-2023-0012

Ssossé, Q., Wagner, J., & Hopper, C. (2021). Assessing the impact of ESD: Methods, challenges, results. *Sustainability, 13*(5), 5. https://doi.org/10.3390/su13052854

Statistics Finland. (2025, April). *Preliminary population statistics*. Statistics Finland. https://stat.fi/en/statistics/vamuu

Sterling, S., & Orr, D. (2001). *Sustainable education: Re-visioning learning and change* (Vol. 6). Green Books for the Schumacher Society.

Tilbury, D. (2011). *Education for sustainable development: An expert review of processes and learning*. UNESCO Publishing. https://sdgs.un.org/publications/education-sustainable-development-expert-review-processes-and-learning-17696

UNESCO. (2017). *Education for sustainable development goals: Learning objectives*. UNESCO Publishing. https://doi.org/10.54675/CGBA9153

UNESCO. (2020). *Education for sustainable development: A roadmap*. United Nations Educational, Scientific and Cultural Organization (UNESCO). https://doi.org/10.54675/YFRE1448

Wals, A. E. J. (2011). Learning our way to sustainability. *Journal of Education for Sustainable Development, 5*(2), 177–186. https://doi.org/10.1177/097340821100500208

Waltner, E.-M., Rieß, W., & Brock, A. (2018). Development of an ESD indicator for teacher training and the national monitoring for ESD implementation in Germany. *Sustainability, 10*(7), 7. https://doi.org/10.3390/su10072508

WCED. (1987). *Our common future*. World Commission on Environment and Development (WCED).

Wolff, L.-A., Laherto, A., Cheah, S., Vivitsou, M., & Autio, M. (2022). Transformation toward sustainability in Finnish teacher education policy: Promises and shortcomings. *Frontiers in Education, 7*, 1–19. https://doi.org/10.3389/feduc.2022.856237

Hansen Owusu Tawiah is an experienced geography educator from Ghana with over eleven years of teaching at both the senior high school and undergraduate levels. He is the author of *Understanding Physical Geography*, a supplementary textbook approved by the National Council for Curriculum and Assessment (NaCCA) for Senior High Schools in Ghana. Hansen holds a bachelor's degree in Geography and Resource Development with Political Science from the University of Ghana, Legon, and a Post-Graduate Diploma in Education from the University of Education, Winneba (UEW). He has completed his Master of Philosophy in Geography Education at UEW. He holds a Master of Arts degree in Education (Teaching and Learning) from Åbo Akademi University, Finland, and is currently a doctoral researcher in Educational Sciences at Åbo Akademi University. Hansen is an active member of the Ghana Geographers Association, having served as the Vice President for the Greater Accra Region (2021–2022) and as the National Vice-Secretary (2022–2024). His professional and academic work has consistently reflected a commitment to sustainability, and he has been a strong advocate for tree-planting initiatives in schools and within his community.

Matthew Biniyam Kursah is a Senior Lecturer at the Department of Geography Education, University of Education Winneba (UEW) with over thirteen (13) years of teaching and research at the tertiary level and over twenty-five (25) peer-reviewed publications. He had his Bachelor's Degree in Geography and Resource Development at the University of Ghana, Legon, and obtained two master's degrees in Spatial Planning and Environmental Engineering and Sustainable Infrastructure (EESI) at Kungliga Tekniska Högskolan (KTH), Stockholm, Sweden. He has a Post-Graduate Diploma in Teaching and Learning in Higher Education (PGDTLHE) from the University of Education Winneba (UEW), Ghana, and a Ph.D. in Remote Sensing and Geographic Information Systems (GIS) at the School of Resources and Environment, University of Electronic Science and Technology of China (UESTC), Chengdu, China. Matthew's research focuses on Remote Sensing and GIS modeling, spatial analysis, environmental management, sustainability, and education. Matthew also integrates cutting-edge geospatial technologies with community-driven research to address pressing environmental, sustainability, and developmental challenges. Matthew is one of the writers of the new Geography Curriculum for Secondary Education in Ghana, which incorporates sustainability in all facets of the learning process, for the National Council for Curriculum and Assessment (NaCCA), Ghana. Matthew is a member of the Ghana Geographical Association.

Diana Yinime is a passionate and forward-thinking graduate in Geography Education from the University of Education Winneba (UEW), Ghana. With a good foundation in Remote Sensing, GIS, spatial analysis, environmental management, and education, Diana integrates cutting-edge geospatial technologies with community-driven research to address pressing environmental, developmental, and educational challenges. As a Teaching and Research Assistant at the College for Distance and e-Learning (CODeL) and Department of Geography Education, she supports academic instruction, student mentoring, Project Works and Dissertation processing for assessment, and research in areas ranging from land use and landcover mapping to environmental sustainability and education. Her commitment to environmental advocacy is evident in her active participation in organizations such as the Clean Ghana, Ocean Club, and the Green African Youth Organisation. Diana has completed a specialized course on Climate Resilient Health Systems from the Global Consortium for Climate and Health Education (GCCHE) and the Association of Schools of Public Health in the European Region (ASPHER). Her field experience spans classroom teaching, field data analysis using QGIS and SPSS, and youth-focused sustainability initiatives in education and the community.

Co-Developing European Curricula and Resources for Sustainability (Teacher) Education: The Case of Teacher Academy Project-Teaching Sustainability

Rachel Bowden and Elena Revyakina

Abstract This chapter considers European curricula and resources for Education for Sustainable Development (ESD), through a case study of the co-development of Learning and Teaching Packages (LTPs) as part of the Erasmus+ Teacher Academy Project–Teaching Sustainability (TAP-TS). In total, seven open-access Learning and Teaching Packages (LTPs) were co-designed by educators from across European countries, and piloted and revised through a series of 11 professional learning events in 2023 and 2024. The LTPs address transdisciplinary themes in relation to sustainability, are mapped to the European Commission's sustainability competence framework 'GreenComp', linked to ECTS and published online as Open Education Resources (OERs).

Keywords Education for sustainable development · Teacher education · Erasmus+ teacher academy

In this chapter, the context and process of curriculum and materials development is described, along with key tensions experienced. In addition, drawing on a secondary analysis of qualitative evaluation data and insider reflections, the chapter explores how involvement in curriculum and resource development supported teacher learning, agency, and interdisciplinary collaboration for ESD. Findings indicate that

R. Bowden (✉)
Centre of Teacher Education and Education Research, TU Dresden, University of Technology, Dresden, Germany
e-mail: rachel.bowden@tu-dresden.de

E. Revyakina
Research Associate, Pedagogical University Vienna, Vienna, Austria
e-mail: Elena.Revyakina@phwien.ac.at

W. Leal Filho et al., *Curriculum Development for Sustainable Development*, World Sustainability Series, https://doi.org/10.1007/978-3-032-22146-9_11

ESD curriculum co-development, as part of transdisciplinary communities of practice, can be a powerful driver for professional transformation and educational innovation. The chapter concludes with recommendations for designing sustainable, adaptable, and contextually grounded ESD curricula and resources within and beyond international teacher education projects.

1 Introduction

Exponential growth in socio-economic activity, industries and technologies over the past 150 years has pushed many of the ecological systems on which our lives depend beyond planetary boundaries (Raworth, 2017; Steffen et al., 2015). As a result, severe and unpredictable consequences, such as extreme weather, wildfires, floods and droughts, are increasingly experienced, especially by people who are least responsible (IPCC, 2021; Sultana, 2022). The term sustainable development (SD) represents international recognition of the urgent need to reorient human activity for socio-ecological justice globally and across diverse domains, as detailed in the 17 Sustainable Development Goals (SDGs) and 169 related targets (UN, 2015). However, beyond this level of consensus, SD presents a complex web of uncertainties, tensions and contradictions (Henkel et al., 2023). As a result, "social learning processes" (Rieckmann, 2020; Wals, 2007), including reflection, dialogue and action within and across communities, are inherent to SD and necessary for positive change towards uncertain and emergent futures.

Formal education has a vital role as part of SD (Sterling, 2001; UNESCO, 2020), including preparing learners with competences for participation in social learning processes. A number of competence frameworks for SD have been developed (e.g., Bianchi et al., 2022; Wiek et al., 2011; UNESCO, 2017), which indicate the transdisciplinary scope of education for sustainable development (ESD). SD competences typically span values and attitudes, critical, systemic and futures thinking, and individual and collective action in relation to socio-ecological phenomena and so are relevant within, between and beyond disciplinary, institutional and national boundaries (Kondratjuk et al., 2023; Renn, 2019). Transdisciplinarity can be a driver for educational transformation (Mittelstraß, 2003), but is challenge for formal education systems which largely operate within disciplinary boundaries. This is seen in the tendency of educators to consider ESD relevant only to certain subjects.

Certainly, education for sustainable development (ESD), demands more than the implementation of competence frameworks. This is because, in order to fulfill its potential to catalyse change towards more equitable and sustainable futures, education itself must be transformed (Orr, 2004; UNESCO, 2021a). Formal education systems are embedded in current socio-economic systems and tend to reproduce rather than challenge the values and assumptions on which they are based. Teacher education is an important driver of educational transformation for ESD (Fischer et al., 2022), but, in order to catalyze transformative learning, must go beyond introducing new content and techniques to fostering mindsets, assumptions, and expectations to be more "inclusive, discriminating, open, reflective and emotionally able to change" (Mezirow, 2009, p. 93). Transformative teacher education for ESD should also address educational practices, processes, and structures (Kondratjuk et al., 2023) and engage teachers as part of professional communities (Villegas & Lucas, 2002a, 2002b).

These insights have important implications for ESD curriculum design and teacher education, which may be seen as integrated through curriculum as practice (Green, 2022). Rather than training educators in the implementation of predetermined ESD curricula, co-developing curricula through reflective, participatory, and values-led processes may catalyse teacher professional learning and educational transformation for ESD.

Erasmus+ teacher Academies

In alignment with this argument, the European Commission (EC) has funded a series of Erasmus+ Teacher Academies (ETA) under Erasmus+ K2 Action. The academies address educational priorities aligned with the Commission's vision for establishing the European Education Area (EEA) by 2025 (Sorensen & Graf, 2024), including sustainability as a key priority. With 27 projects initiated by January 2025 and further 17 Academies currently in the start-up phase, a total of 44 Academies have been co-financed to date. In the first and second Calls launched 2022, 10 of 27 ETAs focused primarily on sustainability. Other academy themes such as inclusion, citizenship and democracy, digital learning, and STEAM are also connected to ESD.

The primary objectives of the ETAs include: improving teacher education policies and practices by creating networks and communities of practice; developing and testing various models of mobility (virtual, physical, and blended) in initial teacher education and continuous professional development; fostering sustainable collaboration between teacher education providers; and enhancing the European dimension and internationalisation of teacher education through innovative and practical collaboration. A small number of ETA projects are discipline-centred but most are interdisciplinary and transversal in their interests (Galvin et al., 2024).

Teacher Academy Project 'Teaching Sustainability'

The Erasmus+ Teacher Academy Project 'Teaching Sustainability' (TAP-TS), which ran from 2022 until 2025, aimed to strengthen pre-service and in-service primary and secondary school teachers' competences for teaching sustainability, with a focus on environmental sustainability, social inclusion, digitalisation, entrepreneurship and pedagogical approaches. Theoretically, the project reflects an understanding of teachers as reflective practitioners and the potential of communities of practice for teacher professional learning (Trayner, 2020; Wenger-Trayner & Wenger-TAP-TS, 2022). Teacher professional learning is understood to be related to professional autonomy which is both an individual and collaborative endeavor, which supports transformative practice and relates to teachers as a professional group (Clark & Hollingsworth, 2002).

TAP-TS is transdisciplinary (Kondratjuk et al., 2023) in a number of ways. The consortium comprises 11 teacher education organizations, across 7 European countries (Austria, Belgium, Cyprus, Germany, Ireland, Portugal, and Sweden), including schools, universities, a government agency, an educational enterprise, and a civil society organization. Partners bring a diverse range of expertise across the natural sciences, technology, business, socio-linguistics, and digital and media education, as do project participants, which include teachers of a range of curriculum subjects. Over the duration of the project, project partners and participants collaborated to produce seven open access 'learning and teaching packages' (LTPs) for use in teacher education and in schools, exploring digitality, entrepreneurship, STEAM, multilingual education, decoloniality, critical media literacy, and whole-institution approaches in relation to sustainability and education for sustainable futures. The LTPs are linked to the European Commission's GreenComp framework of sustainability competences (Bianchi et al., 2022).

This chapter considers European ESD curricula and resource development through a case study of the co-development of Learning and Teaching Packages (LTPs) as part of the Erasmus+ Teacher Academy "Teacher Academy Project-Teaching sustainability" (TAP-TS). The chapter begins with an account of the process of curriculum and materials development as part of TAP-TS. This is followed by a description of the methodology and methods used in this case study, and then findings from a secondary analysis of project evaluation in relation to the impact on teacher professional learning are presented. The findings are then discussed in relation to the literature, followed by a summary of the tensions experienced in developing international ESD learning and teaching curricula and resources. The chapter closes with tentative recommendations for others engaged in this field.

2 TAP-TS Learning and Teaching Packages in the Context of Curriculum Development for ESD

Curriculum development for ESD does not only involves designing instructional materials, but also reimagining what education is for and whose voices shape it. In this sense, the Learning and Teaching Packages (LTPs) created under TAP-TS are not simply teaching tools and materials, they represent a curriculum as practice (Green, 2022): a dynamic, participatory process grounded in critical reflection, dialogue, and action. The design process, co-led by international and transdisciplinary teacher educators, teachers, student teachers, and civil society organizations, reflected a collective effort to challenge dominant knowledge hierarchies and to make room for locally relevant, inclusive, transdisciplinary and critically engaged curriculum content. By centering GreenComp, contextual relevance, and plural knowledge systems, TAP-TS moves beyond curriculum-as-delivery toward curriculum as co-constructed and emergent. We will discuss how these challenges were approached by co-design of LTPs as part of TAP-TS.

The LTPs, which are one of the key outputs of TAP-TS, are learning and teaching programmes for ESD that include open and flexible learning and teaching materials, and guidance on use. Importantly, these programmes were co-developed and validated by the TAP-TS consortium and participants. LTP development followed an iterative design process encompassing several key stages: collaborative development of common ground, key ideas, concepts, design of first drafts of modules; piloting and testing of those; making changes based on the feedback and evaluation; and validation of Fine Versions of LTPs as Open Education Resources (OERs). School leaders, teachers and student teachers as participants of TAP-TS Learning Events were seen as co-designers of LTPs and key experts of their practice. In practice, this occurred through participation in one or more of the 11 TAP-TS teacher professional learning courses (4 fully online, 4 hybrid and 3 face-to-face).

The Content and Structure of TAP-TS LTPs

LTPs pursue two key goals. Firstly, they are planned as formal modules that can be taken and freely adopted/adapted by teacher education institutions across the EU. Secondly, they include materials and resources that teachers can implement directly in their classrooms with the necessary adaptations to reflect context and learning objectives. The core purpose of the LTPs is to motivate, build confidence, and provide teacher educators, teachers, and student teachers with high-quality materials, ideas, and other practical support on educating for a more sustainable Europe. Indeed, central to these LTPs is a model of professional learning that follows participatory and activity-oriented design contributing to collaborative knowledge building (Pischetola et al., 2023), and that is 'deeply reflective' and 'values-led' (Goodwin et al., 2023).

ESD could be understood as a politically conceived educational programme that brings together many social challenges (Wilhelm & Kalcsics, 2023). Therefore, the LTPs attempt to address many of those, by encompassing seven transdisciplinary themes in relation to sustainability: digitality; entrepreneurship; STE(A)M and environmental education; culturally and linguistically sensitive education; (digital and analogue) serious games for teaching sustainability; critical media education; and green citizenship in/for Europe through whole school approach. The content and structure of the LTPs was guided by an in-depth understanding of the goals of teacher education for ESD: (1) to consider the complexity of our world/context; (2) to envision different solutions/situations for the future of the world/context; and (3) to plan and carry out action to bring about the envisioned futures.

The following questions were used to guide the development of LTP curricula and resources:

1. What are the most pressing issues/themes related to education for sustainability for teachers, children, parents in Europe and worldwide?
2. Across these concerns, which concepts and values cross these themes and form the essence of sustainability?
3. How can these concepts and values acquire meaning in teachers' and learners' everyday lives?

4. How can we connect ESD and teacher education processes, including content knowledge and pedagogical content knowledge?

In addressing these questions, the principal challenge was to ensure coherence, within and across LTPs, whilst also enabling flexibility so that LTPs could be used by educators with diverse experiences, backgrounds, and levels of prior knowledge.

Three strategies were applied to promote coherence across LTPs. The first strategy to ensure a consistent structure for each LTP unit, which contains start-up, development, consolidation, and follow-up activities, and LTP as a whole. The second was the development of a TAP-TS Roadmap, as a process description for ESD resource development. This was developed with partners in the first year of the project, and subsequently published as an OER, for educators developing ESD resources. Core elements of the roadmap are summarised below:

1. Purposeful, Future-Oriented Learning: The core aim is to enable both teachers and learners to think and act sustainably, aligning with Sustainable Development Goals (SDGs) and the GreenComp Framework; and emphasize reflection for action—learning is not just about content, but also about equipping students to envision and create sustainable futures.
2. Active, Participatory Pedagogy that emphasizes collaborative projects, future workshops, research & analysis, participative discussions
3. Assessment as Reflective Practice that focuses on assessments that foster critical reflection, problem-framing, and ethical decision-making; use more innovative ways of teacher professional development recognition.
4. Inclusivity & Learner-Centered Design to ensure that materials are learner-centered, inclusive and interdisciplinary, designed to reinforce collaboration and empower diverse learners.
5. Open Sharing and Sustainability. The aim is to share all teaching materials under Creative Commons licenses to promote open access and sustainable educational practices; and encourage a culture of resource-sharing among educators through Communities of Practice.
6. Build Networks and Collaborate by working with regional and national partners; integrating various perspectives from the world of work; and building on existing expertise and resources.

The rationale behind these principles lies in approaches that promote critical thinking, systems thinking, and empower teachers and learners to become change agents.

The third strategy to ensure coherence was the use of the GreenComp framework (Bianchi et al., 2022), as a means to anchor all 7 LTPs to sustainability competences in the four key areas of embodying sustainability values, embracing complexity, envisioning sustainable futures, and acting for sustainability. In the LTPs, all activities and units are connected to one or more competences and thus provide examples of competences in practice.

To allow for flexibility, LTPs were designed as stand-alone units (from three to five per each LTP) to be used independently or sequentially as part of single lessons, a series of lessons or whole courses. In this sense, they follow a flexible modular structure that supports both adaptation and coherence. While each unit can be implemented independently, they are conceptually connected This structure enables teacher educators to tailor LTPs across diverse initial and continuous teacher education programmes inviting educators to reflect on local and cultural contexts in relation to global and European issues and contexts.

Another core element that was integrated into TAP-TS LTPs is a Model for Reflecting in/on/for Instructional Activity (Revyakina et al., 2024). Our rationale to include reflexivity as a key element was guided by Saari et al. (2024)'s argument that critical discussions and reflective co-design of pedagogical experiences are essential for a more transformative sustainability education (Saari et al., 2024). The necessity becomes even more urgent in the context of international mobilities for teachers that was an important part of TAP-TS learning journey, where further steps were required to transform this reflection-for-action into reflection-in-action and reflection-on-action (Schon, 1983).

TAP-TS professional learning courses and mobilities

The LTPs were co-developed with educators from across Europe through 11 TAP-TS professional learning courses, including 4 online workshops (one week duration), 4 active learning events (3 weeks online and three days face-to-face duration) and three summer schools (5-day duration). As such they represented virtual and physical mobilities, and the international dimension of the project introduced a further level of transdisciplinarity as a potential driver for teacher professional learning (Mittelstraß, 2003). As Parmigiani et al. (2022) and Flores (2023) argue, international experiences provide participants with a wide array of opportunities to reflect on their learning experiences, including issues of identity development, language, teaching competencies and self-reflection attitudes.

What needs to be stressed in conclusion to this section is that TAP-TS was not primarily conceived of as a curriculum or materials development project but as an international teacher education project which puts ESD (Rieckmann, 2020) and teacher professional learning (Clark & Hollingsworth, 2002) principles, as well as Europeanisation of Teacher Education Curriculum (Symeonidis, 2021) at its core. That said, the process of curricula and resource co-design proved a central focus of and catalyst for teacher professional learning. In the next section, the methodology and methods of this study are detailed, followed by an evidence-based consideration of how the LTPs supported teacher professional learning and practice for ESD.

3 Methodology and Methods

Case study is an approach to social research, which allows for investigation of phenomena in real world-contexts (Yin, 2009). For this study, the case study design allows us to consider the LTPs as part of the teacher professional development courses in which they were developed, and TAP-TS as a whole. The case is understood as unique, but also of relevance to other similar cases, such as other international teacher education projects developing resources for ESD. Findings cannot be directly transferred, but may be critically interpreted for other contexts. A central concern of the case study, is how the LTPs supported teacher professional learning for ESD, and their ESD practice. To address this question, secondary analysis (Vartanian, 2011) of qualitative data from the three annual TAP-TS project evaluation reports (KRE, 2023, 2024, 2025) was conducted. For the secondary analysis, the three annual evaluation reports were reviewed in relation to the research questions, and relevant data and analysis were extracted. Sub-categories were formed inductively and interactively by each author independently. The authors met three times during the process to review and adapt the emergent categories.

The TAP-TS evaluation was led by a small team of academics and education experts who were part of the project from the beginning. The evaluation was conceived as a formative or developmental evaluation, following Patton (2008). The aim of the TAP-TS evaluation was as follows:

> (…to) ensure the quality of project activities and outputs, in line with desired impact and outcomes, contribute to the ongoing learning and adaptation of project activities toward greatest impact, within and beyond the duration of the project. (K&R Education (KRE), 2024, p. 6)

One focus of the evaluation was to inform the development of LTP materials, by observing their use during teacher professional learning events and gathering feedback from participants. Quantitative and qualitative data collection methods were employed, including participant observation, document review, questionnaires, and focus groups. Regular feedback was provided to the project management team following each activity and through annual written reports (KRE, 2023, 2024, 2025).

This chapter draws on secondary analysis of qualitative evaluation data, enriched by the insider perspectives of the authors. While this offers depth and insight into the processes of curriculum co-design, it also includes limitations. The data were not collected to address the research questions of this particular study, and so relevant observations and responses may have been missed. Moreover, the authors of this paper are project insiders, which risks the biased selection and interpretation of data. The comments of participants who opted for voluntary focus groups and interviews form a significant part of the evaluation reports, especially in year three of the project when the teacher professional learning events were already completed. This may result in bias as participants who experienced positive results from the project were more likely to agree to be interviewed. In addition, accounts of using the LTPs rely on teachers' self-reports and are as a result, somewhat subjective.

4 Findings

In this section findings are presented in relation to the question of how the LTPs influenced teacher professional learning for ESD, and their subsequent practice in schools. The first aspect is how the LTPs fostered teacher professional learning for ESD. A clear finding here is that professional learning occurred for most or even all respondents, who reported their professional learning to be substantial, and even transformative (KRE, 2023, 2024, 2025). As the evaluators concluded in their final report:

> All interview participants unanimously state that regardless of whether they participated in events in year 1, 2, or 3, they experienced personal learning and professional development. What is particularly emphasized is that even teachers with many years of professional experience feel that TAP-TS has provided them with a deeper understanding of teaching and the learning process. (…) One participant said: "*This project has made my lessons different. I use this all the time. I have learned more through this project than I have in the past ten years.*" (2025, p. 17)

The evaluators emphasised the role of LTPs in participants' engagement in courses and professional learning. For example, in the year two evaluation report, the evaluators compared sessions in professional learning events linked to LTPs and those which were not:

> All the three learning events have been great experiences for the participants. The basis for this is undoubtedly the LTPs. In all three events participants rate events based on LTPs as more valuable for personal development and having better usefulness in their own practice. (KRE, 2024, p. 22)

The data also indicate which aspects of LTPs and their use in courses were most significant. A first factor is the extent to which the activities connected to participants' personal values, beliefs and experiences in relation to sustainability, which positively influenced participants' individual motivation, and sense of community within the course. This is exemplified through an observation:

> The exercise stimulates all participants to engage and be curious. There is a constant buzz and lively discussions. It is a great way to get to know each other but also a great way to capture everyone's input and then connect each person's values. This activity is both inclusive and engaging for the participants as they are involved, active, and collaborate with each other (... it is) a professional way to both tap into personal values and then transition to the topic of sustainability. (KRE, 2023, p. 41)

A related factor was the emphasis on inclusion, which increased participation and engagement and served as an experiential model for inclusive pedagogy to inform teachers' future practice:

> The material developed in the LTPs to be tried out by participants works well to promote inclusion and engagement. This was evident through our observations of the sessions, where engagement was visible through joy, enthusiasm, and activity. The course leaders guided the participants through the session, demonstrating inclusion in the pedagogical setup. (KRE, 2024, p. 17)

A further factor was the emphasis on interdisciplinary and international perspectives on ESD, in combination with opportunities to consider subject specific and local connections. The evaluators noted that this process was "linked to the development of new teaching skills" (KRE, 2024, p. 19), presumably as participants identified alternative approaches for teaching familiar content, which they were then able to use in their own schools. In fact, across all events, all participants were able to identify connections between their national curricula and at least some of the LTP resources. Not surprisingly, connections differed between national contexts, and this indicated the extent to which ESD is already an established part of teacher education (or not) (KRE, 2023). Opportunities for teachers to consider relevance for their contexts was positively linked to learning, as indicated in this citation:

> Learning is most evident when participants reflected on content linked to their own classrooms, and they perceived that the activities conducted during the workshop would become beneficial for them in the future. (KRE, 2024, p. 18)

Participants adopted two complementary roles in relation to the LTPs during workshops, on one hand, experiencing activities as learners, and on the other, reflecting on the relevance of activities for ESD in their own contexts and providing feedback on possible improvements (KRE, 2023). In the role of learners, participants experienced activities first-hand and this was both a way to engage them personally with sustainability themes and equip them to use/adapt activities for their students. The value of this approach was confirmed by the evaluation team:

> The importance of participants actively engaging in activities rather than just discussing them is significant. Active participation leads to personal experiences that can be referenced and reflected upon, rather than reflections based on preconceived notions. Participants experience and reflect on these activities, leading to deeper learning and understanding of how to use the activities/LTPs with their students. The pedagogical setup of the sessions, focusing on active learning, generates involvement, engagement, and more lasting learning. (KRE, 2024, p. 19)

This active approach increased the chance that teachers will use activities with their own students. For instance, one teacher reported: "*feeling more comfortable teaching about the topic because we got to try out the activities and contents ourselves*". (KRE, 2023, p. 31) At the same time, participants particularly valued their role as co-developers of materials, which led to different forms of engagement and recognised their professional expertise. As the evaluators noted: "When participants can reflect and test the materials, they feel included, and their engagement increases" (KRE, 2024, p. 18).

The evaluation data also provide detail on the learning which occurred. Data indicate that participants' engagement with the LTP materials resulted in new and often broader and more complex understandings of sustainability and ESD, and the identification of connections to a wide range of school subjects. For example, in a focus group conducted in the third year of the project one teacher commented: "*I was a bit skeptical about how I would be able to implement this in my classroom, but I was positively surprised to see that it could be applied to my subject and in connection with the curriculum. It has broadened my perspective on sustainability and teaching*." (KRE, 2025, p. 18) During a focus group in the first year of the project, another teacher reflected: "*It was important because we deepened the conviction that sustainability is multifaceted*." (KRE, 2023, 41). Another participant also reported a fundamental shift in understanding of ESD, as relevant to all subjects and also their personal life:

> We might have it within us, but I believe it can be applied in all subjects. This is something that needs to exist everywhere. Sustainability is everywhere. I try to find different places where it appears so I can include it in my teaching. I'm pretty sure of that now, but I didn't believe it at first. It's about the thought process, built from the subject. This should be usable in every subject. It's a mindset and a lifestyle. (KRE, 2025, p. 18)

The connection between increased understanding of sustainability and ESD and the willingness and confidence to act, suggested above, is echoed through all three annual reports. For example, in a post-course focus group in the first year of the project, a teacher stated: "*By better understanding what education for sustainability is, we feel we have developed skills and become more confident that it is possible and urgent to address education for sustainability at school*." (KRE, 2023, p. 14) Another teacher stated that the course: "(…) *reinforced our awareness of how important this topic is, which is a big motivating factor for working with students and showing them the importance too*" (KRE, 2023, p. 31–2). A participant on a different course stated that: "*The definitions for the different aspects of sustainability were an eye-opener for me. I will definitely use them in my work with the students*." (KRE, 2023, p. 14).

A further focus of the evaluation was to identify how teachers used the LTPs in their own contexts, during and/or following their participation in TAP-TS training courses, and the impact on their learners. A clear finding is that, across post-course focus groups and interviews participants report having used at least parts of the TAP-TS materials in their schools (KRE, 2023, 2024, 2025). Moreover, in follow-up interviews and focus groups with TAP-TS course participants indicate that the use of LTP materials at school has led to new learning and teaching practices, including interdisciplinary teaching and learning, and teaching and learning in out of school places. As one teacher said: "*This is also part of the development — that as a teacher, you start collaborating with others. I have worked together with geography, biology, and technology teachers. I've modified material and then shared it with teachers in other subjects.*" Another teacher stated: "*This time, several subjects are involved, and we are planning to have lessons outdoors.*" A further teacher reported that: "*I don't just teach inside the classroom anymore, but also outside — and you can learn outside the classroom*." (KRE, 2025, p. 17–8).

It is evident in the data that participants used parts of LTPs, which they considered most relevant for their subjects. This indicates a level of reflective engagement with the resources, in order to identify where, when and how to apply them. Interestingly, the evaluators note that most participants used materials from several different LTPs and this indicates that the transdisciplinary nature of the LTPs provided diverse possibilities for use across curriculum subjects (KRE, 2025). A further important finding, which is consistent across the three reports is that participants primarily used the LTPs which they had had the opportunity to test during one or more of the

TAP-TS events they attended. Regarding this finding, the evaluators note that "This highlights the importance for teachers to personally test and experience a practical, educational setup before applying it with their own pupils/students." (KRE, 2025, p. 25) Taken together these two findings reinforce the importance of participants' doing activities and reflecting on their application in local contexts afterwards for their subsequent application.

Another factor which supported teachers' use of LTPs was the consistent structure of each unit across all seven LTPs. The evaluators note that: "Participants also emphasized the importance of the recognizability in the structure of the LTPs—once you've tested one, it's easier to try a new one because of the familiar format" (2025, p. 25). Thus, having a consistent structure across the seven very different LTPs was a means to enable participants to access and try out activities which they had not experienced in courses. Two further features of LTPs which positively influenced teachers' use of them, was the extent to which they are seen as "practical and valuable in their classrooms" and flexible and adaptable. As one participant said: *"It's good to have ready-made material that can be adapted for the class or group of students you have in front of you." another commented: "The material is very teacher-friendly — it's excellent. You can take it as far as you want."* (2025, p. 25). This also resulted in teachers sharing the resources with other teachers at school (KRE, 2025).

Finally, we consider the impact of teachers' use of LTPs on their students. Teachers who have used LTP resources emphasized that LTP activities fostered new forms of engagement and positive attitudes to learning. For example, a teacher who has tried several activities with students said: *"The students became more involved, because these units allow them to be active — it's truly hands-on learning. It also made the student groups work actively and relate to what is happening in the world."* Other teachers noted that students enjoyed the activities, which were a change from when "*students just sit and watch the board while the teacher speaks*" (KRE, 2025, p. 19–20). Other participants noted that LTP material "*encourages students to be active in their own learning* (…and) *engage students by allowing them to be active and relate to real-world contexts*". (KRE, 2025, p. 25).

5 Discussion

The experience of LTP development as part of TAP-TS confirms many established findings in relation to teacher professional learning, such as the importance of reflection on, in and through action (Schon, 1983). In the context of ESD, two further dimensions seem to have been of particular relevance. The first of these is the personal dimension, which was emphasised in LTPs and reflected in the TAP-TS courses, so that participants reflected on their personal values, beliefs and actions in relation to sustainability. This seemed to increase motivation through personal commitment and learning, and an increased sense of belonging in an ESD community of practice. The second is the international dimension, as represented by the GreenComp framework of sustainability competences (Bianchi et al., 2022). GreenComp provided a conceptual framework for ESD and promoted coherence between LTPs, and which participants were asked to consider critically in relation to their beliefs, practices and contexts. In both cases, these dimensions introduced additional reflective journeys for teachers, as they considered them in relation to their professional beliefs and practices, their curricula and classroom contexts. Indeed, participants responded well to the emphasis on teachers as reflective practitioners, and this gave them agency to identify and adapt resources for use in their own schools and classrooms.

It is notable that participants responded well to the diversity of different perspectives about SD and ESD, which typically engaged rather than overwhelmed them. Supporting factors, such as practical and experiential focus of LTPs, the consistent design of LTPs, and anchoring the LTPs to GreenComp were helpful. Perspectives such as racism and colonialism in relation to sustainability are often challenging for teachers to engage with (UNESCO, 2021b). Addressing these themes as part of LTPs and in TAP-TS courses was surprising for some participants (Bowden, 2025), but participants responded positively and were motivated to act and learn more (KRE, 2023, 2024). It may be that the inclusive community setting of courses, outside of their habitual work settings, enabled participants to engage in uncertainties and find their own positions in relation to such challenging ESD.

Transdisciplinarity is important for ESD, both in order to be able to address the inherently transdisciplinary phenomena of sustainability (Mittelstraß, 2003; Renn, 2021) and in order to catalyse the transformation of educators' mindsets and beliefs, and the organisation of learning and teaching (Kondratjuk et al., 2023; Mezirow, 2009). The LTP resources cover a range of transdisciplinary topics in relation to ESD, and were explored with educators as part of transdisciplinary and international courses. Teachers have used them in schools in ways which have fostered transdisciplinary learning and teaching arrangements, including cross-curricula projects and learning in out of school places. It can be argued that there lies an important and potentially transformative shift while moving from a subject-specific TE towards coherence across disciplines, through increasing *multi/inter/transdisciplinarity* (e.g. Roiha & Heinonen, 2024). This more holistic model of teacher preparation might encourage the co-creation of knowledge across multiple disciplines to address complex, real-world challenges that cannot be solved within the confines of a single academic field. Within transdisciplinary teacher education, teachers and students are expected to *collaboratively engage* with diverse perspectives, enabling both to become active participants in shaping solutions for the global challenges of the twenty-first century.

What tensions have we experienced?

SD comprises numerous contradictions and tensions (Henkel et al., 2023). The development of the TAP-TS LTPs raised some particular tensions, which were both challenging and generative. The first tension to note is between partners within the consortium, which (as described in the introduction) was interdisciplinary and intersectoral. As such there was great potential for learning from the different expertise which partners brought, and an ongoing need for dialogue around basic understandings and approaches, which took place during workshops, and the co-authoring of a project charter in teacher professional learning and ESD. The evaluation team noted tensions in particular areas:

- "Theoretical anchoring of the material and teaching process – where some put much emphasis on the theoretical grounding while others place more effort on practical validity.
- University-based versus Consultancy-based partner organizations. Partly linked to the previous area, but this also spills into other areas.
- Teaching philosophy – base for how the teaching is done. Although there is great overlap, this still can cause friction when co-developing material and courses.
- Subject areas – from more "soft" social sciences to more "hard" natural sciences. While the social sciences often have a critical perspective, natural sciences are more normative. Clearly both perspectives are needed."

  (KRE, 2025, p. 7–8)

TAP-TS was inherently transdisciplinary, as the need to co-create the LTPs and to design and run teacher professional learning events provided continual impetus and opportunity for dialogue and cooperation between partners. In relation to the LTPs, a key tension was between creating resources which reflected international and interdisciplinary understandings of ESD, and the need for participants to identify and adapt activities which were relevant to their curricula, subjects and schools. This tension was negotiated and even generative by engaging teachers as reflective practitioners in the critical consideration and co-development of resources. A further tension was the push to develop finished materials, and the continual adaptation of materials through feedback and as experience. The solution here was to publish 'fine' versions, which remain open to ongoing adaptation and development. A final tension of relevance to ESD is the pedagogical tension between engaging learners with the urgencies and complexities of SD, and making SD accessible and not overwhelming and so that learners feel able to act. In this regard, from TAP-TS we have learned the importance of approaching ESD within inclusive community learning settings, where participants feel able to explore and enquire.

Finally, the TAP-TS Learning and Teaching Packages contribute to the intersection of sustainability science and subject-specific didactics (Wilhelm & Kalcsics, 2023) in three key ways. First, through the reception of empirical and theoretical knowledge, the LTPs expose educators to up-to-date sustainability research and subject-specific educational frameworks, contextualized within diverse disciplines. Second, in terms of transformation, these materials enable the translation of complex sustainability concepts into context-responsive teaching strategies, adaptable across subject boundaries and cultural settings. Third, the collaborative process of LTPs development supports the production of new didactic knowledge, as the co-design, testing, and reflective adaptation of LTPs by educators generates practical insights and theoretical refinements at the nexus of sustainability education and disciplinary pedagogy. This dynamic interplay advances both sustainability science and curriculum theory, positioning teachers not only as implementers but as co-producers of knowledge in sustainability education.

6 Conclusions

This chapter considered curricula and resources for Education for Sustainable Development (ESD), through a case study of the co-development of Learning and Teaching Packages (LTPs) as part of the Erasmus+ Teacher Academy Project–Teaching Sustainability (TAP-TS). The project, which was funded under the European Commission's Erasmus+ Teacher Academies action, provided an enabling framework for international and interdisciplinary cooperation in teacher education, while also presenting the challenge of designing curriculum resources usable across diverse education systems, institutions and subjects.

Through a secondary analysis of qualitative evaluation data and insider reflections, the chapter explored how involvement in curriculum and resource co-development supported teacher learning, agency, and interdisciplinary collaboration for ESD. Findings suggest that when educators participate as co-developers within transdisciplinary communities, curriculum co-development becomes a powerful driver for professional transformation and educational innovation. Indeed, the development of the LTPs can be considered a social learning process (Rieckmann, 2020; Wals, 2007), which involved iterative cycles of reflection, dialogue and action between the consortium and with our community of participants. It was never perfect, but led to learning and change for those involved.

We hope that the experience documented in this chapter may inform the work of others in this field. To this end, we close the chapter by making the following tentative recommendations for creating ESD curricula and resources as part of international teacher education projects:

- Develop inclusive, international and interdisciplinary communities of practice through a foundation of mutual respect, including respect for differences, and opportunities for dialogue and cooperation;
- Engage educators as reflective practitioners in materials development processes, as part of inclusive learning communities;
- Engage educators with interdisciplinarity, intersectoral and international perspectives in relation to their national and local contexts;
- Support of cross-sectoral, cross-border and cross-disciplinary collaboration contributing to rich, values-based, action-oriented curriculum resources.
- Encourage policies that support teachers, school leaders, and teacher educators in co-developing sustainability materials. This increases ownership, contextual relevance, and engagement with ESD;
- Provide opportunities for action, reflection and interaction; and
- Co-develop learning and teaching resources which are accessible, flexible and adaptable.

The experience from the TAP-TS project reported here indicates that ESD curriculum and resource co-design can catalyse professional learning and wider systemic change. This has important implications for teacher professional development and curriculum design, in particular the benefits of linking these processes and engaging with educators as reflective practitioners within international and interdisciplinary communities of practice.

References

Bianchi, G., Pisiotis, U., & Cabrera Giraldez, M. (2022). GreenComp The European sustainability competence framework. In Y. Punie & M. Bacigalupo (Eds.), *EUR 30955 EN*, Publications Office of the European Union. ISBN 978-92-76-46485-3. https://doi.org/10.2760/13286, JRC128040.

Bowden, R. (2025). Linguistically and culturally responsive pedagogy for sustainable futures: Learning from a European Teacher Education Project. *Education Sciences, 15*(6), 647. https://doi.org/10.3390/educsci15060647

Clark, D., & Hollingsworth, H. (2002). Elaborating a model of teacher professional growth. *Teaching and Teacher Education, 18*, 947–967. https://doi.org/10.1016/S0742-051X(02)00053-7

Fischer, D., King, J., Rieckmann, M., Barth, M., Büssing, A., Hemmer, I., & Lindau-Bank, D. (2022). Teacher education for sustainable development: A review of an emerging research field. *Journal of Teacher Education, 73*(5), 509–524.

Galvin, C., Madalinska-Michalak, J., & Revyakina, E. (2024). The European Union Erasmus+ Teacher Academies action. In V. Symeonidis (Ed.), *Enhancing the value of teacher education research: Implications for policy and practice* (pp. 170–197). Brill.

Goodwin, A. L., Madalińska-Michalak, J., & Flores, M. A. (2023). Rethinking teacher education in/for challenging times: Reconciling enduring tensions, imagining new possibilities. *European Journal of Teacher Education, 46*(5), 840–855.

Green, B. (2022). Understanding curriculum as practice, or on the practice turn(s) in curriculum inquiry. *Curric Perspect, 42*, 77–83. https://doi.org/10.1007/s41297-022-00160-0

Flores, M. A. (2023). Teaching and learning in teacher education: Agency, reflection and identity. *European Journal of Teacher Education, 46*(4), 557–560. https://doi.org/10.1080/02619768.2023.2266256

Henkel, A., Berg, S., Bergmann, M., Gruber, H., Karafyllis, N. C., Mder, D., Müller, A.-K., Siebenhüner, B., Speck, K., & Zorn, D.-P. (Eds.). (2023). *Dilemmata der nachhaltigkeit.* (Dilemmas of sustainability) Nomos Verlagsgesellschaft mbH & Co. KG.

IPCC. (2021). *Climate change 2021: The physical science basis. The working group I contribution to the sixth assessment report.* Retrieved January 20, 2024, from https://www.ipcc.ch/report/sixth-assessment-report-cycle/

Kondratjuk, M., Epp, A., Gabriel, S., Gasterstädt, J., Hinrichsen, M., Koevel, A., Rüger, S., Terstegen, S., Wanka, A., & Zschach, M. (Eds.). (2023). *Transdisziplinarität in der Bildungsforschung. Perspektiven und Herausforderungen theoretischer, method(olog)ischer und empirischer.* (Transdisciplinarity in education research. Theoretical, methodological and empirical perspectives and challenges) Springer VS. Wiesbaden.

KRE (2023). *Teacher academy project teaching sustainability: Evaluation of year one activities* [unpublished].

KRE (2024). *Teacher academy project teaching sustainability: Evaluation of year two activities* [unpublished].

KRE (2025). *Teacher academy project teaching sustainability: Evaluation of year three activities* [unpublished].

Mezirow, J. (2009). Transformative learning theory. In J. Mezirow, & E. W. Taylor (Eds.), *Transformative learning in practice: Insights from community workplace, and higher education* (pp. 18–32). Jossey-Bass.

Mittelstraß. (2003). *Transdisziplinarität—Wissenschaftliche zukunft und institutionelle wirklichkeit.* (transdisciplinarity, scientific future and institutional reality) UVK Universitätsverlag Konstanz GmbH.

Orr, D. W. (2004). *Earth in mind: On education, environment, and the human prospect*. Island Press.
Parmigiani, D., Jones, S.-L., Kunnari, I., & Nicchia, E. (2022). Global competence and teacher education programmes. A European perspective. *Cogent Education, 9*, 2022996. https://doi.org/10.1080/2331186x.2021.2022996
Pischetola, M., Møller, J. K., & Malmborg, L. (2023). Enhancing teacher collaboration in higher education: The potential of activity-oriented design for professional development. *Education and Information Technologies, 28*, 7571–7600. https://doi.org/10.1007/s10639-022-11490-x
Patton, Q. M. (2008). *Developmental evaluation. Applying complexity concepts to enhance innovation and use.* The Guildford Press.
Raworth, K. (2017). *Doughnut economics: Seven ways to think like a 21st century economist.* Chelsea Green Publishing.
Renn, O. (2021). Transdisciplinarity: Synthesis towards a modular approach. *Futures, 130*, Article 102744. https://doi.org/10.1016/j.futures.2021.102744
Revyakina, E., Branco, N., Cavadas, B., Colaço, S., & Galvin, C. (2024). How do you solve a problem like MaRIA? Designing opportunities for modelling and reflecting on instructional activity within EU TAP-TS learning & teaching packages and learning events. *Kwartalnik Pedagogiczny, 4* (274).
Rieckmann, M. (2020). Bildung für nachhaltige Entwicklung—Von Projekten zum Whole-Institution Approach. (Education for sustainable Development from projects to whole institution approach) in Kapelari, Suzanne (Hg.), *Vierte "Tagung der Fachdidaktik" 2019: "Interdisziplinäre fachdidaktische Diskurse zur Bildung für nachhaltige Entwicklung"* (fourth conference of subject didactics: interdisciplinary discourses for education for sustainable development) Innsbruck University Press. https://doi.org/10.15203/99106-019-210.15203/99106-019-2-03
Roiha, A., & Heinonen, P. (2024). Co-constructing multidisciplinary coherence in subject teacher education: Students' reflections in group discussions. In G. Doetjes, V. Domovic, M. Mikkilä-Erdmann, & K.Zaki (Eds.), *Coherence in European teacher education.* Springer VS, Wiesbaden. https://doi.org/10.1007/978-3-658-43721-3_7
Saari, M. H., Poulton-Busler, R., & Vladimirova, A. (2024). Does sustainability really start with teachers? Reflections on integrating environmental education in pre-service teacher education in Namibia and Finland. *The Journal of Environmental Education, 1–15.* https://doi.org/10.1080/00958964.2024.2375210
Schon, D. A. (1983). *The reflective practitioner: How professionals think in action.* Basic Books.
Sorensen, T. B., & Graf, L. (2024). A European experiment in governing teacher education and training: The case of the Erasmus+ teacher academies. *Education Inquiry, 1–28.* https://doi.org/10.1080/20004508.2024.2390713
Steffen, W., Richardson, K., Rockström, J., Cornell, S. E., Fetzer, I., Bennett, E. M., Biggs, R., Carpenter, S. R., de Vries, W., de Wit, C. A., Folke, C., Gerten, D., Heinke, J., Mace, G. M., Persson, L. M., Ramanathan, V., Reyers, B., & Sörlin, S. (2015). Planetary boundaries. Guiding human development on a changing planet. *Science, 347*, 1259855.
Sterling, S. (2001). *Sustainable education: Re-visioning learning and change* (Schumacher Briefing, 6, Band 6). Green Books.
Sultana, F. (2022). Critical climate justice. *The Geographical Journal, 188*, 118–124.
Symeonidis, V. (2021). *Europeanisation in teacher education a comparative case study of teacher education policies and practices.* Routledge.
TAP-TS. (2022). *Teacher academy project-teaching sustainability project charter* [Unpublished]. TAP-TS.
UN. (2015). *UN general assembly, transforming our world: The 2030 agenda for sustainable development*, A/RES/70/1. Retrieved January 9, 2025, from https://documents.un.org/doc/undoc/gen/n15/291/89/pdf/n1529189.pdf
UNESCO. (2017). *Education for sustainable development goals: Learning objectives.* UNESCO Publishing.
UNESCO. (2020). *Education for sustainable development: A roadmap* (p. 73). UNESCO Publishing.

UNESCO (2021a). *Reimagining our futures together: A new social contract for education. International commission on the futures of education.* Retrieved January 9, 2025, from https://unevoc.unesco.org/pub/futures_of_education_report_eng.pdf
UNESCO. (2021b). *Teachers have their say motivation, skills and opportunities to teach education for sustainable development and global citizenship.* https://en.unesco.org/inclusivepolicylab/system/files/teams/discussion/comments/2022/4/ESD%20teachers%20say.pdf
Vartanian, T. P. (2011). *Secondary data analysis.* Oxford University Press.
Villegas, A. M., & Lucas, T. (2002a). Preparing culturally responsive teachers: Rethinking the curriculum. *Journal of Teacher Education, 53*(1), 20–32. https://doi.org/10.1177/0022487102053001003
Villegas, A. M., & Lucas, T. (2002b). The culturally responsive teacher. *Educational Leadership, 64*(6), 28–33.
Wals, A. E. J. (Ed.). (2007). *Social learning towards a sustainable world: Principles, perspectives and praxis.* Wageningen Academic Publishers.
Wenger-Trayner, E., & Wenger-Trayner, B. (2020). *Learning to make a difference: Value creation in social learning spaces.* Cambridge University Press.
Wiek, A., Withycombe, L., & Redman, C. L. (2011). Key competencies in sustainability: A reference framework for academic program development. *Sustainability Science, 6*, 203–218. https://doi.org/10.1007/s11625-011-0132-6
Wilhelm, M., & Kalcsics, K. (2023) Diskussion einer Didaktik der Nachhaltigkeitswissenschaft. Auf der Suche nach einer Professionskompetenz zu BNE (Discussion around the didactic of sustainability science). *Journal für LehrerInnenbildung, 23*(3), 16–25. https://doi.org/10.25656/01:27970. 10.35468/jlb-03-2023-01
Yin, R. K. (2009). *Case study research: Design and methods* (4th ed.). Sage Publications.

Rachel Bowden is a research associate at the centre of teacher education and education research, at TU Dresden University of Technology Germany. From 2022 until 2025, she coordinated the Erasmus+ Teacher academy- Teaching sustainability, and a state-level ESD in teacher education initiative. Her research interests are linguistically and culturally sensitive education, education for sustainability and teacher professional development.

Elena Revyakina PhD in Education, is a researcher at the University College of Teacher Education, Vienna (Austria). Her research interests focus on teacher education policy, teacher professional development, digital technology in education, and sustainability. In 2024, she co-authored a Chapter *Education for Sustainability via, despite and in Digital Media: Ways of Teaching Sustainability in Digitality* in Belts Media Education, and Chapter *The European Union Erasmus+ Teacher Academies Action Complementing and Supplementing European Teacher Education and Teacher Education Research?* in Key Issues in Teacher Education: Policy, Research and Practice with J. Madalinska-Michalak and C. Galvin.

Investigating Teacher Turnover in Namibia's Education System: Implications for Sustainable Curriculum Development in the Kandjimi-Murangi Circuit

Karen Mwaka Lishomwa

Abstract Teacher turnover remains a critical challenge in Namibia's education sector, particularly in the Kandjimi-Murangi Circuit of the Kavango West Region. High attrition rates among teaching staff have significant implications for the sustainability and continuity of curriculum development, which is foundational to achieving long-term educational goals. This study investigates the factors contributing to teacher turnover and explores its impact on sustainable curriculum implementation and development, in alignment with national and global educational frameworks. Using a mixed-methods approach, the research gathers data from teachers, school administrators and education policymakers to identify key drivers of turnover, including inadequate incentives, poor working conditions and limited career growth opportunities. The study further analyses the implications of teacher attrition on student performance, curriculum continuity, and the systemic resilience of education systems. Findings indicate that addressing teacher turnover requires a holistic strategy that integrates policy reforms, teacher support programs and improved working conditions. The study proposes sustainable retention strategies, including professional development initiatives, better resource allocation, and targeted policy interventions to enhance teacher satisfaction and curriculum delivery stability. By aligning teacher retention efforts with sustainable curriculum development, this research contributes to ongoing discourse on educational sustainability and human capital retention in Namibia. The findings offer valuable insights for policymakers, educators, and stakeholders seeking to enhance teacher retention and improve long-term educational outcomes.

Keywords Teacher turnover · Sustainable curriculum development · Namibia education · Teacher retention · Policy reforms · Educational sustainability

---

K. M. Lishomwa (✉)
Faculty of Strategic Management and Business Administration, International University of Management, Windhoek, Namibia
e-mail: lishomwakaren@gmail.com

W. Leal Filho et al., *Curriculum Development for Sustainable Development*, World Sustainability Series, https://doi.org/10.1007/978-3-032-22146-9_12

1 Background to the Study

Quality education plays a pivotal role in promoting economic development, democracy, and social cohesion across nations (Camilleri & Camilleri, 2016). Central to the delivery of such quality education is the availability and retention of a well-functioning, motivated, and stable teaching workforce. Globally, the education sector has been grappling with persistent teacher turnover, often driven by high stress levels, inadequate resources, poor working conditions, and a lack of job satisfaction (Okeke & Mtyuda, 2017; Owusu, 2021). These factors not only compromise educational delivery but also threaten the overall goals of sustainable curriculum development, particularly in developing contexts such as Namibia.

In the broader global context, empirical evidence reveals that many teachers experience considerable occupational stress, leading to burnout and ultimately leaving the profession (Ryan et al., 2017; Richards et al., 2018; Räsänen et al., 2020; Owusu, 2021). In the Asia continent, for instance, Hong (2012) noted that between 25 and 50% of new teachers exit the profession within their first three years, primarily due to their inability to cope with the psychological demands of the job. This challenge is echoed across various education systems globally and appears particularly acute in sub-Saharan Africa, where limited resources, poor remuneration, and weak institutional support compound the difficulties experienced by educators (Zickafoose et al., 2024).

In Namibia, the teacher attrition rate has reached worrying levels, with serious implications for the functioning of the education system, especially in rural circuits such as Kandjimi-Murangi in the Kavango West Region. According to the Namibian National Teachers Union (NANTU), an estimated 2,000 teachers resign annually, while local training institutions can only produce approximately 1,200 new teachers each year (Engelbrecht, 2020; Janik, 2015; Janik & Marques, 2015). This imbalance creates a significant shortfall, especially in hard-to-staff schools, and disrupts the provision of continuous, quality education. NANTU attributes the high attrition rates to excessive workload, unattractive conditions of service, insufficient incentives, and a lack of institutional support—factors that severely affect teacher morale and retention.

The consequences of teacher turnover are particularly dire in rural and underserved areas. Hawene (2018) and Chirimbana et al. (2023) warns that resignation rates among Namibian teachers, particularly in remote locations, are so high that they threaten to paralyze rural schooling systems. In circuits like Kandjimi-Murangi, these conditions manifest in overcrowded classrooms, overstretched remaining staff, and a lack of consistency in curriculum delivery and pedagogical planning. This scenario undermines the national commitment to equitable and inclusive education as enshrined in the Education Act No. 16 of 2001, which mandates access to quality formal education for all children up to the age of sixteen (Chirimbana et al., 2023).

The teacher turnover crisis in Namibia cannot be viewed in isolation from the country's post-independence education reforms. Since gaining independence in

1990, the Namibian government has embarked on an ambitious journey to deconstruct the legacy of the Bantu Education system inherited from the apartheid regime. Central to this transformation has been the promotion of inclusivity, equity, and learner-centered pedagogy (Chirimbana et al., 2023). However, the reforms, while noble in intention, have imposed new demands on teachers who must now adapt to evolving curricular expectations, large class sizes, a lack of teaching aids, and ongoing administrative changes. As noted by Janik (2015) and Engelbrecht (2020), these systemic shifts have significantly increased the stress levels among Namibian teachers, many of whom feel overwhelmed, unsupported, and disillusioned by the realities of the restructured system.

Teacher turnover presents profound challenges to sustainable curriculum development. A sustainable curriculum, by definition, should be responsive, inclusive, and rooted in long-term educational goals aligned with broader sustainable development principles (Kioupi & Voulvoulis, 2019). It requires a consistent and well-trained teaching force to implement and assess its objectives effectively. Frequent teacher departures disrupt curriculum continuity, hinder the development of school-based innovations, and erode institutional memory. Moreover, high attrition rates undermine mentorship opportunities for novice teachers and hinder long-term school improvement strategies.

The Job Demands-Resources (JD-R) model offers a useful analytical lens through which to understand teacher turnover (Collie, 2023; Granziera et al., 2021; Rastegr & Zarei, 2023). This model posits that high job demands, when not adequately counterbalanced by sufficient job resources, result in burnout and disengagement. In the Namibian context, the JD-R model becomes particularly relevant given the national budgetary constraints that prevent the Ministry of Education from offering competitive remuneration, modern teaching infrastructure, and performance-based incentives. Teachers are expected to deliver high-quality outcomes in under-resourced environments, often without meaningful recognition or professional growth opportunities.

In rural circuits like Kandjimi-Murangi, where infrastructural and economic disadvantages are more acute, these challenges are magnified. Teachers must contend with a lack of basic teaching materials, poor housing conditions, inadequate transport, and minimal access to continuous professional development. These hardships not only diminish job satisfaction but also erode the sense of purpose and belonging that is essential for retaining committed educators. The significance of intrinsic motivational factors—such as the perceived meaningfulness of work—has been highlighted by scholars like Janik (2015) and Autin et al. (2022) as key protective psychological resources in such challenging contexts.

From an organizational behavior perspective, turnover has been the subject of extensive scholarly investigation (Kraft et al., 2016; Zhou & Ma, 2022), with numerous studies attempting to pinpoint its causes and implications. Adriano and Callaghan (2020) define turnover as the movement of individuals across the membership boundary of a social system—a definition that encompasses both exits and new hires. High turnover, particularly when driven by dissatisfaction and burnout, has been associated with decreased organizational performance, lowered morale, and a weakened ability to attract high-quality replacements. In the Ministry of Education, Arts and Culture, teacher turnover threatens not only the delivery of quality education but also the broader institutional mandate to serve all Namibian learners equitably.

Despite isolated interventions—such as rural hardship allowances and ad hoc incentives—intended to stem the tide of attrition, turnover rates in Namibia remain stubbornly high (Hawene, 2018). This raises critical questions about the underlying systemic and contextual factors driving this phenomenon, and about the effectiveness of current retention strategies in fostering curriculum continuity and reform. The lack of a clear, evidence-based policy framework on teacher retention, particularly one tailored to the needs of rural circuits like Kandjimi-Murangi, hampers national efforts to stabilize and improve the teaching profession.

Further complicating this issue is the limited availability of empirical data at the circuit level. While national-level statistics paint a broad picture of the attrition crisis, they often obscure local nuances that may influence teacher decisions to leave or remain in the profession. The Kandjimi-Murangi Circuit, located in one of Namibia's most underserved regions, represents a critical case for understanding how rurality, socio-economic challenges, institutional culture, and professional identity intersect to influence teacher turnover. Insights drawn from this circuit could inform broader national strategies aimed at developing and sustaining effective curricula by ensuring that schools retain capable, motivated, and professionally supported educators.

The importance of addressing teacher turnover in the context of sustainable curriculum development cannot be overstated. Education systems that fail to retain experienced teachers risk undermining the very reforms they aim to institutionalize. In the Namibian case, where inclusive education, learner-centered pedagogy, and curriculum resilience are key policy priorities, the high attrition rate poses a serious threat to achieving long-term goals (Minnie, 2015). Students are denied continuity in learning; schools struggle with fragmented implementation of curricular innovations; and the teaching profession suffers from reputational damage and a loss of morale.

This study is thus motivated by a critical need to investigate teacher turnover in the Kandjimi-Murangi Circuit in order to understand its underlying causes and far-reaching implications for curriculum sustainability. The research will also explore practical, context-specific strategies for improving teacher retention, drawing from both empirical data and stakeholder perspectives. In doing so, it aims to contribute to the development of a resilient, future-oriented education system in Namibia—one that values and invests in its teachers as fundamental agents of curriculum transformation and sustainability.

Ultimately, addressing teacher turnover in Namibia is not just a matter of workforce management; it is an educational, developmental, and social imperative. As Namibia seeks to meet its obligations under the United Nations Sustainable Development Goals (particularly Goal 4 on quality education), ensuring the stability and sustainability of its teaching force is essential. Without such stability, curriculum reforms will falter, learning outcomes will remain uneven, and the promise of education as a tool for national transformation will remain unfulfilled.

2 Emerging Concepts and Theories

This chapter explores key concepts, theories, and scholarly perspectives related to teacher turnover, with a focus on its implications for sustainable curriculum development. The literature reviewed is organized around five core themes: the concept of turnover, theoretical frameworks explaining turnover, the impacts of teacher turnover on education quality, major causes of teacher turnover, and implications for curriculum sustainability. This review provides a foundation for understanding the broader dynamics of teacher attrition and retention within Namibia's education system.

2.1 The Concept of Teacher Turnover

Teacher turnover refers to the exit or movement of educators from their current teaching roles, either by leaving the profession entirely (attrition) or by transferring to other schools or education sectors (migration). Sorensen and Ladd (2020) define teacher turnover as a "generic term for all changes in teacher status from one year to the next." Turnover may be voluntary, initiated by the teacher for reasons such as dissatisfaction, better opportunities, or retirement; or involuntary, initiated by the employer due to factors like poor performance or restructuring (Mary, 2014; Melesse, 2021). While some level of turnover is inevitable and may have positive implications—such as filtering out underperforming staff—excessive turnover (especially rates exceeding 10%) is largely viewed as detrimental. It can disrupt school functioning, undermine staff cohesion, and ultimately affect student outcomes (Zajac, 2016). The conceptual clarity around voluntary turnover, particularly attrition and migration, forms the basis for analyzing teacher dynamics within school systems.

2.2 Theoretical Perspectives on Teacher Turnover

A robust understanding of teacher turnover requires a multidimensional theoretical lens. Several theories have been developed to explain why individuals leave the teaching profession, these include:

Human Capital Theory: The human capital theory offers a foundational perspective on occupational choice. According to Hamdan (2013) and Marginson (2019), individuals make rational decisions to enter or exit professions based on perceived returns on investment—both monetary and non-monetary. In the context of teaching, educators may assess salary, professional growth, workload, and job satisfaction against other available options. If alternative careers offer better incentives or recognition, teachers may choose to leave. Hassen (2016) emphasize that such systematic evaluations continue throughout one's career, influencing retention.

Social Learning Theory: Yunus et al.'s (2024) social learning theory emphasizes that career decisions, including turnover, are shaped by the interaction of personal attributes, prior experiences, and environmental contexts. Key factors include:

- *Genetic endowment* (e.g., intelligence, personality traits)
- *Environmental conditions* (e.g., socio-cultural and economic circumstances)
- *Learning experiences* (e.g., professional development, exposure to new roles)
- *Task approach skills* (e.g., coping strategies, values, work ethic).

These factors interact to influence an individual's satisfaction and engagement, thereby impacting decisions to remain in or exit the profession.

Integrated Turnover Model: de Almeida Durães (2024) turnover model identifies three groups of predictive factors:

- *Demographic factors* (e.g., age, gender, tenure, education)
- *Uncontrollable factors* (e.g. job market alternatives, economic instability)
- *Controllable factors* (e.g., remuneration, working conditions, administrative support).

This model helps in distinguishing between factors school leaders can address and those that are external to the school system.

Organizational Conditions Framework: Ingersoll (2001) and Kraft et al. (2016) advances a theory linking teacher turnover to school organizational structures. He argues that institutional factors—such as administrative support, disciplinary climate, salary structure, and teacher autonomy—are primary determinants of turnover. Improving school climate and decision-making inclusion may significantly enhance retention.

2.3 Impact of Teacher Turnover on Education Quality

High teacher turnover undermines educational quality in several ways:

- *Instructional disruption*—Frequent staff changes disrupt curriculum continuity, forcing learners to adapt to varying teaching styles and potentially lowering academic performance (Manca & Delfino, 2021).
- *Resource drain*—Recruitment, training, and socialization of new staff incur significant costs, reducing resources for other academic activities (Watlington et al., 2016).
- *Staff demoralization*—Teachers who remain often carry increased workloads and face diminished morale. This can result in absenteeism, reduced effort, or further resignations.
- *Curriculum misalignment*—New or reassigned teachers may be compelled to teach subjects outside their specialization, reducing instructional quality (Ingersoll, 2020).

- *Diminished student support*—Consistent teacher turnover disrupts teacher-student relationships, impeding personalized instruction and pastoral care.

Since teachers are the key agents in implementing curriculum reforms, their stability is critical for the effective delivery of quality and sustainable education.

2.4 Major Causes of Teacher Turnover

A multitude of interrelated factors contribute to teacher attrition. Literature identifies several common causes:

- *Inadequate Salary and Incentives*: Low remuneration remains the most cited reason for teacher dissatisfaction. Studies by Al-Suraihi et al. (2021) emphasize that poor salary, lack of benefits, and disparities with other professions contribute significantly to turnover, particularly in developing contexts. Teachers often perceive their financial compensation as insufficient compared to their workload and qualifications.
- *Poor Administrative and Leadership Support*: Weak leadership, lack of recognition, and limited teacher participation in decision-making are noted as demotivating factors (Ingersoll, 2020). Where teachers feel unsupported or excluded from key institutional decisions, commitment to the school declines.
- *Inadequate Professional Development and Career Progression*: Limited opportunities for advancement, promotion, and training stifle motivation. Teachers who perceive stagnation in their career are more likely to seek alternative employment or leave the profession entirely.
- *Challenging Working Conditions*: Overcrowded classrooms, lack of teaching materials, student indiscipline, and infrastructure deficits contribute to teacher dissatisfaction (Al-Suraihi et al., 2021). Such conditions reduce job satisfaction and increase stress levels, especially in rural or under-resourced areas.
- *Societal Perception and Professional Status*: The perceived low prestige of the teaching profession affects recruitment and retention. Where society undervalues teachers, it becomes difficult to motivate individuals to enter or remain in the profession, especially when other sectors offer more attractive compensation and status.

2.5 Implications for Sustainable Curriculum Development

Teacher turnover has direct consequences for the sustainability of curriculum reforms and innovations. Sustainable curriculum development requires a stable, committed, and well-supported teaching workforce. Persistent turnover:

- Undermines policy implementation and curriculum consistency
- Disrupts institutional memory and long-term pedagogical planning

- Increases the reliance on temporary or underqualified personnel
- Limits reflective practice and curricular innovation.

Hence, addressing teacher turnover is not only a matter of human resource management but a strategic imperative for sustaining educational quality, equity, and innovation—particularly within circuits such as Kandjimi-Murangi, where rural and resource-limited conditions intensify the challenges.

The literature reveals that teacher turnover is a multidimensional issue influenced by individual, organizational, and systemic factors. Its impacts extend beyond workforce numbers, deeply affecting curriculum delivery and long-term educational outcomes. A theoretical understanding of these dynamics offers valuable insights into policy design and institutional reforms aimed at improving teacher retention and sustaining curriculum development efforts in Namibia's education system.

3 Research Locale and Methodology

This section constitutes the formal methods chapter of the study, outlining the detailed research design and methodological framework employed to explore the phenomenon of teacher turnover in Namibia's education system, with specific focus on the Kandjimi-Murangi Circuit. It presents the research locale, design, population, sampling procedures, data collection techniques, research instruments, data analysis procedures, validity and reliability considerations, ethical standards, and study limitations. The methods are grounded in a convergent parallel mixed-methods approach, integrating predominantly qualitative than quantitative data to ensure a robust and comprehensive understanding of the issues under investigation. This methodological structure aligns with the study's objectives and offers a transparent and replicable process for academic and policy application.

3.1 Research Locale

This study was conducted in the Kandjimi-Murangi Circuit, located in the Kavango West Region of Namibia. The region, situated in the northeastern part of the country, is largely rural and characterized by vast geographic expanses, poor road infrastructure, and limited access to resources and services. The Kandjimi-Murangi Circuit comprises a number of public primary and secondary schools that serve dispersed and often marginalized communities. The circuit was purposively selected for this research due to its ongoing challenges with teacher retention, curriculum delivery, and educational sustainability. According to preliminary reports from the Regional Directorate of Education, schools in the circuit have consistently recorded high levels of teacher turnover, particularly among early-career and specialist subject teachers, over the past five years (Nambundunga, 2016). This turnover has posed serious implications for continuity in teaching, learner outcomes, and the effective implementation of the national curriculum.

The socio-economic context of the Kandjimi-Murangi Circuit is significant in understanding the complexity of the teacher turnover issue. Many teachers are deployed from urban centers to these rural areas, only to face challenges such as inadequate accommodation, insufficient teaching and learning materials, geographic isolation, and limited opportunities for professional advancement. These conditions, coupled with the general lack of incentive structures or retention policies tailored to rural environments, contribute to the high attrition rates observed in the circuit (Nambundunga, 2016). Furthermore, the region's educational institutions often struggle to fill vacancies promptly, resulting in long-standing gaps in staffing, especially for core subjects. These realities make the Kandjimi-Murangi Circuit an ideal site for exploring the drivers and impacts of teacher turnover in relation to sustainable curriculum development. The localized data obtained from this circuit also holds wider relevance for rural education planning and teacher management across Namibia and other similarly structured education systems.

3.2 *Analytical Approach*

The study adopted a convergent parallel mixed-methods research design to comprehensively investigate the factors influencing teacher turnover and its impact on sustainable curriculum development in the Kandjimi-Murangi Circuit. A mixed-methods approach was deemed appropriate due to the multifaceted nature of teacher turnover, which involves both measurable variables (e.g., working conditions, job satisfaction) and contextual, subjective experiences (e.g., teacher motivation, school climate, and leadership support). In this design, quantitative and qualitative data were collected concurrently, analyzed separately, and then merged for interpretation to enhance the richness and validity of the findings.

The target population for this study included all teachers and school administrators operating within public schools in the circuit. From this population, a total sample of 47 participants was selected, comprising 40 teachers, 4 principal heads, and 3 education policymakers. A multi-stage sampling technique was employed to ensure representative and purposeful inclusion of participants. First, schools with a documented history of high teacher turnover were purposively selected based on records from the Regional Directorate of Education. These included both primary and secondary schools to reflect the different operational challenges across levels of education. Second, stratified sampling was used to ensure balance in gender, teaching experience, and subject areas among the teacher participants. Lastly, the principal heads and policymakers were included using purposive sampling due to their leadership roles and institutional knowledge of turnover trends, curriculum delivery, and school performance.

Data collection was conducted using a triangulation of instruments to capture both statistical trends and lived experiences. Quantitative data was gathered through structured questionnaires administered to the 40 teacher participants. The questionnaire items were organized around key variables such as job satisfaction, availability of teaching resources, workload, administrative support, professional development opportunities, and personal factors contributing to the decision to remain in or leave the profession. The questionnaire included both closed-ended items and Likert scale ratings to allow for numerical analysis of perceptions and correlations between variables. A pilot test was conducted with ten teachers from a neighboring circuit to assess the reliability and clarity of the questionnaire. Feedback from the pilot test was used to revise and refine the instrument. The reliability coefficient of the questionnaire, determined using Cronbach's Alpha, was 0.82, indicating a high level of internal consistency.

Qualitative data were collected through semi-structured interviews with the four principal heads, three education policymakers, and two focus group discussions (FGDs) with groups of teachers. The interviews were designed to probe deeper into school-specific experiences of turnover, administrative interventions to mitigate attrition, and the perceived effects on curriculum continuity. Each interview lasted between 30 and 45 min and was conducted in a private setting to ensure confidentiality. The FGDs, consisting of 6–8 teachers in each session (one group from primary and one from secondary schools), allowed for interactive discussions and collective reflections on shared experiences. The focus groups explored themes such as professional isolation, workload stress, peer support, and long-term career intentions. All qualitative sessions were audio-recorded with consent, transcribed verbatim, and analyzed using thematic content analysis. NVivo software was used to assist in the coding of recurring themes and categorization of data according to emerging patterns. Themes from interviews and FGDs were later compared with quantitative results to ensure coherence and completeness of interpretation.

In addition to primary data, secondary sources such as staffing reports, teacher resignation letters, performance evaluation documents, and relevant education policy frameworks were reviewed. These documents helped to contextualize the responses and provided institutional evidence of the challenges discussed by participants. The integration of both qualitative and quantitative data was carried out during the interpretation stage of the analysis. This convergence of data sources strengthened the study's credibility through triangulation and allowed for a more nuanced understanding of the causes and consequences of teacher turnover.

Data analysis for the quantitative portion involved descriptive and inferential statistics using SPSS (Statistical Package for the Social Sciences) version 25. Frequencies, percentages, and mean scores were calculated to describe patterns in the data, while Pearson correlation and regression analyses were used to examine relationships between teacher working conditions, motivation, and their intention to leave or remain. Qualitative data analysis, as mentioned earlier, was driven by thematic analysis. The major themes that emerged included poor infrastructure, limited career advancement opportunities, lack of professional recognition, community isolation, and psychological burnout. These themes were consistently linked to negative impacts on curriculum delivery, including teaching inconsistencies, learning gaps, and diminished school morale.

To ensure the validity and trustworthiness of the findings, several strategies were employed. In terms of content validity, the research instruments were reviewed by two experts in educational leadership and curriculum studies. Their feedback helped refine the tools to ensure alignment with the study's objectives. Triangulation of data collection methods—questionnaires, interviews, FGDs, and document analysis—ensured a robust cross-verification of findings. Member checking was also employed during qualitative data collection, where participants were asked to confirm or clarify interpretations of their responses, thereby increasing credibility. Furthermore, the study ensured reliability through standardized administration of tools and adherence to ethical data handling procedures. Thus, data analyses followed a more qualitative approach than quantitative approach where necessary in supporting data validity.

Ethical considerations were paramount throughout the study. Approval to conduct the research was obtained from the Research Ethics Committee of the International University of Management (IUM), and permission was granted by the Kavango West Regional Directorate of Education. Participation was voluntary, and informed consent was obtained from all respondents prior to data collection. Participants were assured of the confidentiality and anonymity of their responses, and no identifying information was recorded in any public presentation of findings. Respondents also retained the right to withdraw from the study at any time without any consequences. All data collected were securely stored and used solely for academic purposes.

Despite its robust design, the study acknowledges certain limitations. Firstly, while the findings offer valuable insights into rural teacher turnover, they may not be fully generalizable to urban or semi-urban educational contexts in Namibia. Secondly, self-reporting biases may have influenced some questionnaire responses, especially where teachers felt vulnerable discussing institutional shortcomings. Lastly, the relatively small number of principal heads included in the study limited the diversity of administrative perspectives. However, the triangulation of data sources and methods helped to minimize these limitations and enhance the reliability of the conclusions drawn.

In summary, the methodology employed in this study allowed for an in-depth exploration of the dynamics of teacher turnover and its impact on curriculum delivery in rural Namibia. The mixed-methods design facilitated the collection of both broad-based and nuanced data, while the rigorous sampling, ethical safeguards, and analytical procedures ensured a high standard of research integrity. The findings derived from this methodological approach offer evidence-based recommendations to inform policy, practice, and future research on educational sustainability.

4 Results and Discussion

Unmasking the Realities of Teacher Turnover

Having discussed the theoretical and analytical approach employed in this study, this chapter now proceeds to unearth the empirical realities surrounding teacher turnover in the Kandjimi-Murangi Circuit. Drawing on both quantitative trends and qualitative narratives, the section presents and interprets the key findings, highlighting the interplay between teacher attrition, systemic challenges, and curriculum sustainability. Through this lens, the chapter sheds light on the lived experiences of educators and school leaders, offering critical insights into the root causes, patterns, and consequences of teacher turnover in rural Namibia.

4.1 Teacher Attrition in Rural Namibia

To unveil the key factors contributing to teacher turnover in the Kandjimi-Murangi Circuit of the Kavango West Region, the study's findings reveal a set of interrelated drivers, grounded in both statistical data and stakeholder narratives. These include *inadequate incentive structures*—survey results revealed that 73% of respondents identified low remuneration as a primary driver of dissatisfaction. Teachers noted that their salaries did not reflect the workload, cost of living, or professional qualifications. The absence of performance-based incentives and stagnant salary progression created demoralization. Qualitative data reinforced this finding. One teacher remarked: "*It is disheartening when after years of service, your salary barely increases. There is no motivation to stay.*" Furthermore, rural teachers expressed frustration over the lack of rural hardship allowances or housing support, despite challenging living and working conditions. The theme *poor working and living conditions* emerged from the survey. The majority (68%) of survey participants reported dissatisfaction with the physical conditions of their schools. Classrooms were overcrowded, infrastructure was dilapidated, and teaching materials were insufficient. Schools lacked essential amenities such as running water, electricity, and internet access. An education officer explained: "*The infrastructure backlog is significant. Many schools have not seen major improvements in over a decade. This impacts morale.*"

In interviews, several teachers mentioned the psychological toll of teaching in isolated, under-resourced environments with minimal recreational or health services. Such conditions not only hinder professional satisfaction but also disrupt family life, especially for teachers posted far from their hometowns. Another factor includes *limited career development and growth opportunities*: A substantial 65% of respondents believed that limited opportunities for advancement contributed to their intent to leave the profession or transfer. Many cited a lack of in-service training, mentoring programmes, or transparent promotion criteria. A principal shared: "*Young teachers feel stuck. They want to grow, but the system is rigid. Unless a vacancy opens up, there is no way to move up.*" This stagnation fosters frustration and incentivizes exit from the profession or movement to better-resourced urban schools or different sectors altogether.

Administrative burdens and lack of support—teachers highlighted excessive administrative tasks, such as paperwork and compliance reporting, as detracting from their core instructional roles. 52% of surveyed teachers believed that bureaucratic overload was a source of burnout. Qualitative responses emphasized insufficient support from management. Several participants described school leadership as reactive rather than proactive in addressing grievances, mentoring, or promoting a supportive work culture. One teacher noted: "*We are not just educators—we are data clerks, social workers, and crisis managers. Without support, it is exhausting.*" *Geographical isolation and family separation* is another impeding factor. Geographical isolation emerged as a strong non-financial determinant of turnover. Teachers in the circuit often come from distant regions and face prolonged family separation. This was particularly challenging for female teachers with children. 56% of respondents stated that being far from their family and lack of opportunities to reunite contributed to their dissatisfaction. While decentralization has improved some regional staffing dynamics, teacher deployment policies still fail to accommodate family needs.

A teacher shared: "*I have not seen my children for months. If I get the chance to transfer closer to them, I will take it—even if it means resigning.*" Additionally, *community engagement and social integration* present another key factor contributing to teacher turnover. Teachers in remote communities often face social exclusion or lack of community support, especially if they are not indigenous to the area. 42% of respondents felt socially disconnected in their assigned communities, leading to feelings of isolation and alienation. This lack of integration into the local culture and insufficient recognition by community leaders undermined their sense of belonging and professional fulfilment. Lastly, *policy and governance gaps* reveal a key attribute where participants across all groups highlighted systemic policy challenges. While national education policies recognize teacher welfare, implementation was seen as fragmented and inconsistent. Delays in processing transfers, poor policy communication, and inconsistent enforcement of retention strategies weakened trust in the system. A regional official acknowledged: "*We have good policies on paper, but implementation is a major gap. There is no accountability when commitments are not met.*"

The findings reveal a multifaceted and interlinked set of factors contributing to teacher turnover in the Kandjimi-Murangi Circuit. The interplay between structural issues (e.g., poor infrastructure and low pay), systemic gaps (e.g., weak policy implementation), and personal challenges (e.g., family separation and limited growth) illustrates that teacher attrition is not caused by a single issue, but rather a convergence of multiple stressors. These challenges undermine not only the morale and well-being of teachers but also the broader goal of sustainable curriculum delivery. High turnover leads to classroom instability, loss of institutional memory, and inconsistent delivery of learning outcomes. Moreover, it affects curriculum planning cycles, increases the burden on remaining teachers, and complicates professional development pipelines.

To address the above findings, the study recommends:

- *Comprehensive Incentive Packages*: Including rural hardship allowances, performance bonuses, and housing support.
- *Improved Infrastructure*: Prioritizing school renovations and access to essential services.
- *Career Development Pathways*: Establishing mentorship schemes and clearer promotion criteria.
- *Administrative Reforms*: Reducing non-teaching workloads and strengthening school leadership support.
- *Family-Oriented Transfer Policies*: Considering teacher family dynamics in deployment decisions.
- *Community Integration Programs*: Promoting cultural orientation and local teacher recruitment.
- *Policy Accountability Mechanisms*: Ensuring timely and transparent implementation of teacher welfare policies.

These solutions, tailored to local contexts, are necessary for not only reducing teacher turnover but also ensuring curriculum continuity, student learning outcomes, institutional memory, and sustainability in education delivery in Namibia's rural circuits.

4.2 Disrupted Continuity and Compromised Outcomes

To combat teacher turnover on sustainable curriculum delivery and student performance, the results reveal a multifaceted and deeply entrenched challenge that impacts every level of the education delivery system in the Kandjimi-Murangi Circuit of the Kavango West Region. These include *teacher turnover trends and curriculum disruption*—the quantitative analysis revealed that over 60% of teachers surveyed has considered leaving their posts within the past two years, with 35% having already changed schools at least once during the same period. The primary reasons cited were infrastructure (78%), lack of housing and transport allowances (65%), and workload stress (52%). Qualitative data emphasized that frequent teacher exits lead to *curriculum fragmentation.* As one principal stated: "*Each time a teacher leaves midyear, the class loses direction. The replacement is usually delayed, and even when they arrive, they must start from scratch.*" This disrupts the "sequencing of content," "undermines lesson continuity," and contributes to "gaps in syllabus coverage," especially in core subjects like Mathematics, Natural Science, and English.

Decline in student performance—qualitative evidence showed a negative correlation ($r = -0.62$) between turnover frequency and student performance metrics, especially in Grade 7 national assessments. Schools with higher turnover rates recorded average pass rates of 45%, compared to 68% in schools with more stable teaching staff. Interviewees highlighted how sudden teacher exits lead to:

- Overcrowding, as students are distributed across remaining teachers.
- Reduced instructional time, due to time spent on onboarding new staff.

- Loss of student–teacher rapport, essential for engagement and learning motivation.

A regional education officer explained: "*Students are not just losing teachers—they are losing mentors. Trust takes time to build, and with frequent turnover, we are constantly resetting the clock.*"

Long-term implications for educational development suggest another ripple effect of teacher turnover on curriculum delivery and student achievement. The data suggests that turnover contributes to systemic instability. Educational planners struggle to implement long-term reforms due to persistent human resource vacuums. Curriculum innovations and pilot programmes often fail due to lack of consistent personnel. One Ministry official noted: "*We launched a digital curriculum platform last year, but the lack of information communication technology (ICT) trained teachers—exacerbated by turnover—meant the project stalled in most rural schools.*" This underscores the vicious cycle where turnover weakens implementation, which in turn diminishes trust in reform processes.

Gendered impact and vulnerable learner groups—results further indicated that female students are disproportionately affected, especially in subjects like Mathematics and Science where male teachers are more prevalent. When male teachers leave and female replacements are unavailable or delayed, girls' participation declines due to lack of role models or encouragement. Similarly, learners with disabilities face greater risks, as replacement teachers often lack specialized training, leading to exclusion or regression. *Coping mechanisms and adaptive responses* represent another observation from participants. Despite these challenges, some schools have developed coping mechanisms to mitigate turnover effects:

- Peer-led study groups to fill learning gaps.
- Shared lesson planning among staff to ensure curriculum alignment.
- Community volunteers stepping in for non-specialized roles.

However, such adaptations are stop-gap measures and not sustainable without institutional support.

Some possible *stakeholder recommendations* from the findings illustrate a clear call emerges for holistic retention strategies aligned with sustainable curriculum goals. Key suggestions include:

- Early-warning systems to detect and address teacher dissatisfaction.
- Incentive structures (e.g., hardship allowances, accommodation support).
- Continuous professional development, especially for rural teachers.
- Decentralized hiring practices to reduce replacement lag.

This section confirms that teacher turnover is not merely a staffing issue, but a core threat to educational sustainability. Its ripple effects on curriculum delivery, learner outcomes, and systemic development necessitate urgent, integrated policy responses. Strengthening teacher retention in rural Namibia is critical for sustaining gains in curriculum reform and long-term human capital development.

4.3 Stakeholder Perceptions of Teacher Retention Policies

To combat teacher turnover, this section addresses the research question: what are the views of key stakeholders regarding current teacher retention policies and support mechanisms in the Kandjimi-Murangi Circuit? Understanding these perceptions is crucial for developing effective strategies to enhance teacher retention and ensure sustainable curriculum delivery in the region. These include:

Inadequate incentives and remuneration—A significant majority of teachers (78%) expressed dissatisfaction with their current remuneration packages, citing them as insufficient to meet living expenses. Interviews with school administrators revealed that the lack of financial incentives contributes to low morale and increased turnover rates. Policymakers acknowledged budget constraints but emphasized ongoing efforts to review and adjust salary structures to retain qualified teachers.

Poor working conditions—Teachers reported substandard working conditions, including overcrowded classrooms, insufficient teaching materials, and inadequate infrastructure. Approximately 65% of respondents indicated that these conditions negatively impact their teaching effectiveness and job satisfaction. School administrators highlighted challenges in securing funding for facility improvements, while policymakers pointed to initiatives aimed at infrastructure development in rural schools.

Limited career growth opportunities—The absence of clear career progression pathways was identified as a deterrent to long-term commitment among teachers. Over 70% of teachers felt that opportunities for professional advancement were limited or unclear. School administrators noted the lack of structured professional development programs, and policymakers discussed plans to implement mentorship and training schemes to address this gap.

Lack of support mechanisms—Teachers emphasized the need for robust support systems, including mentorship programs and regular feedback mechanisms. Many reported feelings of isolation and a lack of guidance, particularly for novice teachers. School administrators acknowledged the importance of support structures but cited resource limitations as a barrier to implementation. Policymakers recognized the issue and mentioned the development of induction programs to support new teachers.

Stakeholder engagement and policy implementation—There was a consensus among stakeholders on the importance of collaborative policy development and implementation. Teachers and administrators advocated for greater involvement in decision-making processes to ensure policies are contextually relevant and effectively address on-the-ground challenges. Policymakers expressed a willingness to incorporate feedback from educators to refine retention strategies.

The study reveals that while there are ongoing efforts to address teacher retention challenges in the Kandjimi-Murangi Circuit, significant gaps remain in incentive structures, working conditions, career development opportunities, and support mechanisms. Enhanced stakeholder engagement and targeted policy interventions are essential to create a conducive environment for teacher retention and sustainable curriculum delivery.

4.4 *Sustainable Model for Retention and Curriculum Stability*

To combat teacher retention, this section proposes a sustainable model grounded in theoretical framework (Sect. "Theoretical Perspectives on Teacher Turnover"), all contextualized within the empirical realities of the Kandjimi-Murangi Circuit in Namibia. The data reveal that teacher turnover in the circuit is primarily driven by the lack of *consistent professional development*, *weak mentorship structures*, *challenging work conditions*, and *limited institutional support*. The majority of teachers reported dissatisfaction with the lack of in-service training and professional growth opportunities, a finding consistent with *Human Capital Theory*, which posits that investments in staff development are vital for performance and retention. Teachers felt professionally stagnant, with one respondent noting that *they had only received one training opportunity in over three years*. This stagnation breeds frustration and reduces motivation, particularly for teachers deployed to remote areas with limited access to instructional resources.

Further compounding this issue is the absence of structured mentorship programs. Teachers—especially early-career educators—expressed a need for guidance and role models within the system. In line with *Social Learning Theory*, which emphasizes the role of observational learning, the absence of mentorship undermines the transmission of effective teaching practices and institutional culture. Many teachers indicated that they often felt isolated and unsupported in their professional roles, which increased their desire to leave the profession or transfer to better-resourced areas. The empirical results also reveal considerable dissatisfaction with the organizational climate in schools. Teachers pointed to ineffective leadership, lack of recognition, and top-down decision-making processes as significant factors that demoralize staff. The *Organizational Conditions Framework* helps contextualize these concerns, illustrating that poor administrative support, limited autonomy, and weak communication structures contribute to high attrition rates. When leadership fails to create an inclusive, supportive, and participatory school culture, teachers are more likely to disengage and eventually exit the system.

Additionally, the study found a strong link between material working conditions and turnover. Teachers cited poor infrastructure, inadequate housing, insufficient teaching materials, and excessive workloads as major sources of burnout. These findings align with the *Integrated Turnover Model*, which explains that job dissatisfaction interacts with perceived external opportunities and personal constraints to predict turnover intentions. The data show a high correlation between substandard infrastructure and teachers' intentions to leave, suggesting that improving physical and professional conditions is crucial to retention. Importantly, these challenges have a direct and measurable impact on curriculum stability and student outcomes. Schools with high turnover rates reported declining student performance, disrupted curriculum delivery, and a lack of continuity in instructional planning. For example, student achievement dropped significantly in classrooms where teacher changes occurred multiple times in a single year. These gaps are particularly evident in critical subject areas like mathematics and science, where teacher shortages are more pronounced.

In response to these findings, the study proposes an integrative model comprising four mutually reinforcing components: (1) capacity development through regular professional training, (2) structured mentorship and peer support networks, (3) leadership and organizational reform to foster supportive school cultures, and (4) incentive and workload management strategies tailored to rural contexts. This model is both theoretically informed and empirically validated (see Fig. 1). It suggests that sustainable teacher retention depends on a holistic strategy that addresses not only individual motivation but also the broader organizational and systemic context in which teachers operate. Adopting this model has significant implications for educational policy and practice in Namibia. Short-term measures could include rural hardship allowances, teacher housing schemes, and regular professional development workshops. Medium-term strategies should focus on institutionalizing mentorship and leadership development programs. Over the long term, the model should be embedded into Namibia's Education Strategic Plan and monitored through teacher satisfaction audits and curriculum performance assessments.

In sum, the findings illustrate that teacher turnover is not merely a personnel issue but a structural and systemic challenge that affects the quality and continuity of education. By adopting a model that integrates *human capital development, social learning, organizational reform,* and *strategic retention mechanisms*, Namibia can build a more resilient and sustainable education system. These efforts will not only stabilize the teaching workforce but also enhance curriculum implementation and student learning outcomes across the country.

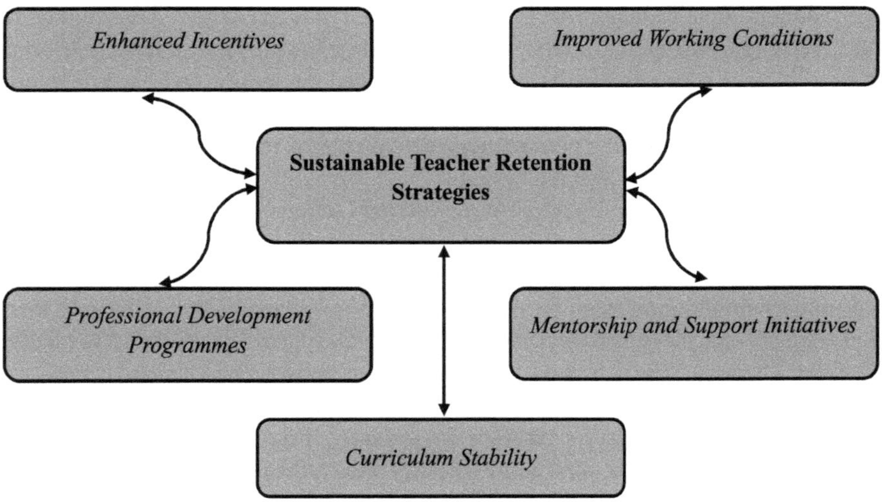

Fig. 1 Sustainable teacher retention and curriculum stability model. *Source* Author's construct (2025)

5 Implications and Conclusion

The findings of this study underscore critical implications for educational leadership, curriculum design, and policy sustainability within Namibia's education system, particularly in underserved and geographically challenging settings such as the Kandjimi-Murangi Circuit. Drawing from a convergent parallel mixed-methods research design, the study elucidated stakeholder perceptions—teachers, school administrators, and policymakers—on existing teacher retention strategies and their intersection with curriculum stability. The data revealed that teacher turnover is not merely an administrative inconvenience but a systemic disruption with cascading effects on student performance, curriculum delivery, and institutional coherence.

Through empirical analysis, the study identified key drivers of attrition: insufficient professional development support, lack of mentoring programs, poor working conditions in rural postings, and limited involvement of teachers in school-level decision-making. These findings, grounded in stakeholder narratives, echo the propositions of the Human Capital Theory and the Integrated Turnover Model, which suggest that investment in teacher capacity and satisfaction significantly reduces workforce mobility and reinforces curriculum continuity. The implications are multifaceted. First, teacher attrition must be treated as a policy-level challenge that threatens the realization of Sustainable Development Goal 4 (SDG 4) on inclusive and quality education. Second, improving teacher retention calls for a holistic intervention strategy—one that combines material support, professional growth pathways, community integration, and participatory leadership.

In conclusion, the study proposes a strategic retention framework that includes continuous professional development embedded in national plans, mentorship for early-career educators, participatory school governance, and infrastructural investment in remote schools. These recommendations are not only responsive to the lived realities documented in the Kandjimi-Murangi Circuit but also offer scalable solutions for national policy adaptation. Future research should examine longitudinal effects of implemented strategies, regional customizations of national retention policies, and how curriculum adaptation frameworks can proactively respond to shifting teacher dynamics. If appropriately contextualized and implemented, these approaches can transform Namibia's education sector into a more resilient and sustainably staffed system.

References

Adriano, J., & Callaghan, C. W. (2020). Work-life balance, job satisfaction and retention: Turnover intentions of professionals in part-time study. *South African Journal of Economic and Management Sciences, 23*(1), 1–12. https://hdl.handle.net/10520/ejc-ecoman-v23-n1-a27

Al-Suraihi, W. A., Samikon, S. A., Al-Suraihi, A. H. A., & Ibrahim, I. (2021). Employee turnover: Causes, importance and retention strategies. *European Journal of Business and Management Research, 6*(3), 1–10. https://doi.org/10.24018/ejbmr.2021.6.3.893

Autin, K. L., Herdt, M. E., Garcia, R. G., & Ezema, G. N. (2022). Basic psychological need satisfaction, autonomous motivation, and meaningful work: A self-determination theory perspective. *Journal of Career Assessment, 30*(1), 78–93. https://doi.org/10.1177/10690727211018647

Camilleri, M. A., & Camilleri, A. (2016). Education and social cohesion for economic growth. *International Journal of Leadership in Education, 19*(5), 617–631. https://doi.org/10.1080/13603124.2014.995721

Chirimbana, M., Kashweka, E., & Amunime, S. (2023). Assessing the effectiveness of the remoteness and hardship allowance on teacher retention in Oshana Region in Namibia rural schools: Teachers views. *Open Journal of Social Sciences, 11*(8), 383–397. https://doi.org/10.4236/jss.2023.118027

Collie, R. J. (2023). Teacher well-being and turnover intentions: Investigating the roles of job resources and job demands. *British Journal of Educational Psychology, 93*(3), 712–726. https://doi.org/10.1111/bjep.12587

de Almeida Durães, N. R. (2024). Employee turnover predictive model: A model to predict who is more likely to leave a company. Master's thesis. Universidade NOVA de Lisboa. https://run.unl.pt/bitstream/10362/175319/1/TGI3782.pdf

Engelbrecht, M. (2020). An examination of psychological well-being and burnout of primary and secondary school educators in the Omaheke region of Namibia. Ph.D thesis. University of Namibia. https://repository.unam.edu.na/bitstream/handle/11070/2918/

Granziera, H., Collie, R., & Martin, A. (2021). Understanding teacher wellbeing through job demands-resources theory. *Cultivating Teacher Resilience, 24*(3), 229–244. https://library.oapen.org/bitstream/handle/20.500.12657/41720/2021_Book_CultivatingTeacherResilience.pdf#page=233

Hamdan, H. E. B. (2013). The influence of rewards (monetary and non-monetary) toward intention to leave in financial institutions. Ph.D thesis. Universiti Utara Malaysia. https://etd.uum.edu.my/5043/1/s810816.pdf

Hassan, R. (2016). The role of human capital management in enhancing engagement and retention among top talent in organization: A critical literature review. *Journal of Emerging Economies and Islamic Research, 4*(4), 1–14. https://journal.uitm.edu.my/ojs/index.php/JEEIR/

Hawene, L. K. (2018). Effects of remoteness and hardship allowance in retaining teachers in rural schools in the Oshana Region in Namibia. Ph.D thesis. University of Namibia. https://repository.unam.edu.na/bitstream/handle/11070/2440/hawene2018.pdf?sequence=1

Hong, J. Y. (2012). Why do some beginning teachers leave the school, and others stay? Understanding teacher resilience through psychological lenses. *Teachers and Teaching, 18*(4), 417–440. https://doi.org/10.1080/13540602.2012.696044

Ingersoll, R. M. (2016). Teacher turnover and teacher shortages: An organizational analysis. *American Educational Research Journal, 38*(3), 499–534. https://doi.org/10.3102/00028312038003499

Ingersoll, R. M. (2020). Misdiagnosing the teacher quality problem. In *The state of education policy research* (pp. 291–306). Routledge. https://files.eric.ed.gov/fulltext/ED498327.pdf

Janik, M. (2015). Meaningful work and secondary school teachers' intention to leave. *South African Journal of Education, 35*(2), 1008–1008. https://doi.org/10.15700/saje.v35n2a1008

Janik, M., & Marques, L. (2015). The impact of work relationships and certain job factors on the work engagement and other positive work outcomes of primary school educators in Namibia. *Namibia Journal of Managerial Sciences, 1*(1), 119–144. https://journals.ium.edu.na/index.php/njms/article/view/53

Kioupi, V., & Voulvoulis, N. (2019). Education for sustainable development: A systemic framework for connecting the SDGs to educational outcomes. *Sustainability, 11*(21), 6104. https://doi.org/10.3390/su11216104

Kraft, M. A., Marinell, W. H., & Shen-Wei Yee, D. (2016). School organizational contexts, teacher turnover, and student achievement: Evidence from panel data. *American Educational Research Journal, 53*(5), 1411–1449. https://doi.org/10.3390/su142013314

Manca, S., & Delfino, M. (2021). Adapting educational practices in emergency remote education: Continuity and change from a student perspective. *British Journal of Educational Technology, 52*(4), 1394–1413. https://doi.org/10.1111/bjet.13098

Marginson, S. (2019). Limitations of human capital theory. *Studies in Higher Education, 44*(2), 287–301. https://doi.org/10.1080/03075079.2017.1359823

Mary, G. (2014). Individual, organizational factors and teacher turnover intentions among primary schools in Mbarara district. Ph.D thesis. Makerere University. http://hdl.handle.net/10570/3255

Melesse, F. (2021). The assessment of causes of teachers turnover at harvest academy as percieved by existing staff. Ph.D thesis. St. Mary's University. http://hdl.handle.net/123456789/5947

Minnie, A. (2015). Perceptions of Namibian primary school teachers' and stakeholders' in the Khomas education region, regarding curriculum requirements for sustaining a successful physical education program. Master's thesis. Itä-Suomen yliopisto. https://erepo.uef.fi/bitstream/handle/123456789/15352/urn_nbn_fi_uef-20150986.pdf

Nambundunga, M. (2016). Factors that influence teacher attrition in secondary schools in the Omusati education region, Namibia. Ph.D thesis. University of Namibia. http://hdl.handle.net/11070/1996

Okeke, C. I., & Mtyuda, P. N. (2017). Teacher job dissatisfaction: Implications for teacher sustainability and social transformation. *Journal of Teacher Education for Sustainability, 19*(1), 54. https://doi.org/10.1515/jtes-2017-0004

Owusu, D. (2021). Occupational stress, burnout, job dissatisfaction and turnover intentions among senior high school teachers in cape coast metropolis, Ghana. Ph.D thesis. University of Cape Coast. http://hdl.handle.net/123456789/6290

Richards, K. A. R., Hemphill, M. A., & Templin, T. J. (2018). Personal and contextual factors related to teachers' experience with stress and burnout. *Teachers and Teaching, 24*(7), 768–787. https://doi.org/10.1080/13540602.2018.1476337

Räsänen, K., Pietarinen, J., Pyhältö, K., Soini, T., & Väisänen, P. (2020). Why leave the teaching profession? A longitudinal approach to the prevalence and persistence of teacher turnover intentions. *Social Psychology of Education, 23*, 837–859. https://doi.org/10.1007/s11218-020-09567-x

Rastegr, B., & Zarei, A. A. (2023). Assessment literacy components predicting EFL teachers' job demand-resources: A focus on burnout and engagement. *International Journal of Language Testing, 13*(1), 44–66. https://files.eric.ed.gov/fulltext/EJ1385327.pdf

Ryan, S. V., von der Embse, N. P., Pendergast, L. L., Saeki, E., Segool, N., & Schwing, S. (2017). Leaving the teaching profession: The role of teacher stress and educational accountability policies on turnover intent. *Teaching and Teacher Education, 66*, 1–11. https://doi.org/10.1016/j.tate.2017.03.016

Sorensen, L. C., & Ladd, H. F. (2020). The hidden costs of teacher turnover. *AERA Open, 6*(1), 2332858420905812. https://doi.org/10.1177/2332858420905812

Watlington, E., Shockley, R., Guglielmino, P., & Felsher, R. (2016). The high cost of leaving: An analysis of the cost of teacher turnover. *Journal of Education Finance*, 22–37. http://www.jstor.org/stable/40704404

Yunus, N. M., Zainudin, Z. N., Mohamad, Y., Yusop, W. N., Engku, E. M., & Kamarudin, M. A. (2024). Understanding career decision-making: Influencing factors and application of Krumboltz's social learning theory. *International Journal of Academic Research in Business and Social Sciences, 14*(7), 36–51. https://doi.org/10.6007/IJARBSS/v14-i7/21562

Zajac, E. C. (2016). Teacher turnover and school reform: How teacher turnover affects urban secondary school improvement. Ph.D thesis. Boston University. https://open.bu.edu/bitstream/handle/2144/17111/Zajac_bu_0017E_12165.pdf?se

Zhou, H., & Ma, J. (2022). Organizational justice and teachers' turnover intention in primary and secondary schools: The importance of sustainable salary management. *Sustainability, 14*(20), 13314. https://doi.org/10.3390/su142013314

Zickafoose, A., Ilesanmi, O., Diaz-Manrique, M., Adeyemi, A. E., Walumbe, B., Strong, R., et al. (2024). Barriers and challenges affecting quality education (Sustainable Development Goal# 4) in sub-Saharan Africa by 2030. *Sustainability, 16*(7), 2657. https://doi.org/10.3390/su16072657

Karen Mwaka Lishomwa is an advocate for Business Technology Management Education at the International University of Management (IUM). She holds a Master's degree in Business Administration majoring in Management from IUM and a Bachelor honor degree in Business Information System from the same University. Her research expert focuses on education technology, business management and strategic skills.

The Pond as an Educational Context for Early Childhood Science and Sustainability Learning

Candela Rodríguez Conde, Nuria Castiñeira Rodríguez, Mercedes Varela Losada, and Maria A. Lorenzo Rial

Abstract Schools are key settings for promoting a shift towards environmental sustainability by equipping citizens with the knowledge, skills, and resources necessary to participate actively in decision-making processes. This chapter aims to contribute to curriculum development for sustainable development by presenting a practical example illustrating how integrating sustainability principles into early childhood science education can support sustainability education from a young age. Accordingly, a teaching experience focusing on the impact of pollution on the environment was designed and implemented in an Early Childhood Education classroom. The approach is grounded in a socioconstructivist paradigm and adopts a holistic, competence-based orientation aligned with the PISA frameworks (OECD, 2022) and the Framework for K–12 Science Education (NRC in A framework for K-12 science education: Practices, crosscutting concepts, and Core ideas, The National Academies Press, 2012). Methodologically, this was a pre-experimental case study with a qualitative approach, in which evidence of the children's evolving ideas was gathered through audio-visual recordings and document analysis. The findings highlight children's capacity to reflect on and discuss socio-environmental issues, as well as to establish causal relationships between human actions and environmental degradation or change. In this regard, the study underscores the potential of fostering scientific and environmental literacy from an early age, a stage that remains relatively underexplored in current scientific literature, while providing practical insights for curriculum developers and educators seeking to integrate sustainability principles effectively in early education.

C. R. Conde · N. C. Rodríguez · M. V. Losada · M. A. Lorenzo Rial (✉)
Area of Experimental Sciences Education, University of Vigo, Pontevedra, Spain
e-mail: marialorenzo@uvigo.es

C. R. Conde
e-mail: can.rodriguez@uvigo.es

N. C. Rodríguez
e-mail: nuria.castineira@uvigo.es

M. V. Losada
e-mail: mercedesvarela@uvigo.es

W. Leal Filho et al., *Curriculum Development for Sustainable Development*, World Sustainability Series, https://doi.org/10.1007/978-3-032-22146-9_13

Keywords Early childhood education · Science education · Education for sustainability · Competency approach

1 Introduction

Education is a key driver in enabling societies to transform and adapt in response to an unprecedented global environmental crisis. Processes such as climate change, pollution, and biodiversity loss are undermining the resilience of the Earth System (Richardson et al., 2023; Steffen et al., 2015). This crisis is marked by the impact of human activities on the environment, with negative consequences for both nature and people, and it affects all regions of the world simultaneously (IPCC, 2023; Richardson et al., 2023). In this context, there is an urgent need to engage children, even from preschool, in educational experiences that promote education for sustainability.

This chapter focuses on the integration of scientific and sustainability education from early childhood, aiming to foster more sustainable, just, and supportive ways of living, thinking, and relating from the earliest stages. This perspective seeks to lay the groundwork for the development of educational experiences that promote children's scientific and environmental literacy, as well as their ability to understand phenomena and engage critically with their environment.

1.1 Challenges of the Twenty-First Century Environmental Crisis

Various reports (IPCC, 2023; UNEP, 2019; UNEP, 2022) warn about the unsustainability of current global development trends and their future trajectories. We are facing an alarming situation that demands urgent action to address an unprecedented environmental crisis (UNESCO, 2021). Halfway to the 2030 Agenda deadline, the outlook is troubling: progress towards achieving the Sustainable Development Goals (SDGs) appears too slow and insufficient. Achievements for more than 50% of the targets are inconsistent and inadequate, while 30% are either stalled or have regressed to levels below the 2015 baseline (UN, 2023).

On the other hand, the Stockholm Resilience Centre, a group of scientists led by Rockström and Steffen, introduced the concept of Planetary Boundaries (Rockström et al., 2009; Steffen et al., 2015). The aim was to identify and quantify critical thresholds that must not be crossed if humanity is to remain within a "safe operating space" that ensures the stability of the Earth system. The intention was not only to describe the problems but to establish a practical sustainability framework that could guide informed decision-making by policymakers, scientists, businesses, and civil society based on ecological limits. In this regard, the proposal responds to the need to move from environmental diagnosis to coordinated global planning and action, with the goal of maintaining safe conditions for life. Additionally, this framework provides a systemic perspective on the global environmental crisis, warning of the risk of crossing irreversible tipping points in Earth's degradation. The most recent update (Richardson et al., 2023) reveals that six out of the nine planetary boundaries have already been transgressed and highlights the urgent need to transform our patterns of production, consumption, and global governance in pursuit of sustainability. Under current trends, environmental degradation will likely continue at a high rate, making it unlikely that the SDGs will be achieved (UNEP, 2019).

1.2 Addressing the Challenges of the Environmental Crisis through Education

In this context, the rapid decline in the planet's habitability demands a transformation of education (Davis & Elliott, 2024). Schools must therefore prepare children to confront the environmental crisis by fostering active and transformative engagement aimed at driving social, political, economic, and cultural change, in response to the urgent need to redesign education for sustainability (Assadourian, 2017).

To achieve this, the training of educators is both essential and urgent (Muro & Pérez, 2021; Sjöblom et al., 2023; Stouthart et al., 2023). It is necessary to rethink education with new priorities aimed at fostering the skills deemed essential for building a sustainable society capable of adapting to a changing planet (Assadourian, 2017). From this perspective, the complexity of the processes involved in global environmental change requires the development of competences such as critical and systems thinking, skills that enable analysis and understanding of the interrelationships among these processes, as well as an interdisciplinary and transformative approach that considers the social, economic, political, and ethical dimensions of environmental problems (Bianchi et al., 2022; Lorenzo-Rial et al., 2024; Wiek et al., 2011).

This approach aims to promote a transformative education that creates learning experiences designed to foster the competencies needed to tackle environmental challenges (Chapman & O'Gorman, 2022). In this regard, early childhood education plays a key role in engaging young children in caring for the environment, raising awareness of the impacts of human actions on nature, and contributing to the formation of an environmentally literate, responsible, and committed citizenry (Ardoin & Bowers, 2020; Muro & Pérez, 2021).

In recent years, publications on education for sustainability in early childhood education have increased. However, some reviews (Bascopé et al., 2019; Güler et al., 2021; Somerville & Williams, 2015) note that research on sustainability education in early childhood is largely confined to journals specialising in early childhood education, rather than appearing in journals within the field of environmental education. They also note that intervention-based studies are scarce, and that most practices are primarily adult-centred. This leads to the conclusion that children's participation has not yet been considered sufficiently important. In this respect, Davis and Elliott (2024) highlight the importance of implementing actions that recognise young children as thinking beings and active agents, capable of initiating changes in their own environments towards more sustainable practices and ideas.

1.3 Science Education for Sustainability in Early Childhood

In the context of global environmental change, science education plays a vital role in shaping a dynamic educational model that maintains continuous dialogue with both the natural and social environments (Jiménez et al., 2015; Karelovic & Kong, 2022). Scientific literacy must therefore address environmental issues from a multifactorial perspective and adopt a socio-scientific approach (Kumar et al., 2024).

From this standpoint, science education and education for sustainability share pedagogical foundations, as well as a global and cross-curricular approach that promotes holistic experiences in relation to the environment. Within this framework, children are encouraged to make decisions, develop ethical reasoning, and construct knowledge in an active and conscious manner (Martínez-Silva et al., 2021). Sustainability thus becomes the arena in which science education meets the need for transformation, serving as an enabling tool for action through the development of systems thinking, critical thinking, and social skills that equip learners to navigate a complex and changing world. This approach is grounded in the ethical perspective that underpins education for sustainability and promotes the development of scientific competences (Jiménez et al., 2015; Karelovic & Kong, 2022).

Science education for sustainability should offer holistic and integrated experiences from a unifying scientific perspective (Widiyawati, 2020). By capitalising on children's natural curiosity and interests (Martínez-Silva et al., 2021), complex and critical science education positions young children as active and critical agents for sustainability. It offers them opportunities to observe science in their own lives and in those of other living beings and creates spaces to share their knowledge. These experiences foster sustained and transformative interest and engagement in sustainability (Trott & Weinberg, 2020). To this end, science education for sustainability aims to support the development of competences for problem recognition and resolution through experiences related to sustainability. These experiences should originate from socio-scientific problems, offering opportunities to propose ideas, discuss, take action, and move towards a goal, based on people's relationships with science in diverse contexts (Karelovic & Kong, 2022).

Early science education contributes to the development of the scientific skills necessary for understanding and evaluating phenomena such as climate change or pollution (Bascopé et al., 2019). Young children learn through exploration and discovery of their immediate environment, making it an ideal context for addressing relevant socio-environmental issues from an educational perspective. In this regard, schools are key spaces for promoting change towards environmental sustainability (Pramling & Kaga, 2008), allowing children to acquire the knowledge, skills, and resources needed to actively participate in decision-making processes. Education at all levels should be participatory, critical, and transformative, which necessarily entails recognising children's right to express their views on sustainability-related issues (Güler et al., 2021; RodrIgues-Silva & Alsina, 2023; Somerville & Williams, 2015).

2 Objective of the Study

The objective of this study is to evaluate the didactic potential of an educational intervention that integrates science education for sustainability into early childhood education through an interdisciplinary approach. Specifically, it aims to: (i) analyse the evolution of preschool children's prior ideas and knowledge about the characteristics and habitat of living beings (pond frog) and the impact of pollution on ecosystem disruption, and (ii) examine the development of scientific competencies (NRC, 2012; OECD, 2023) such as question formulation, interpretation and explanation of phenomena, and decision-making, considered key learning outcomes in curricula focused on science and sustainability. The didactic design employs the pond as an interdisciplinary educational resource to connect natural sciences with education for sustainability, aiming to foster more sustainable, just, and supportive ways of living, thinking, and relating from the earliest stages of education, and from a systemic and critical perspective.

3 Methodology

From a methodological standpoint, this study is a case study understood as a descriptive and holistic analysis that considers the particularity and complexity of the concrete reality of the object of study (Priya, 2021). It adopts a qualitative approach and is based on a pre-experimental design, which aims to provide evidence regarding the impact of a specific intervention on a given population prior to conducting a larger-scale study (Campbell & Stanley, 2015). To provide an overview of this methodological framework, Fig. 1 presents a schematic representation of the main characteristics and sequential phases of the research process.

Within this structure, data collection focused on capturing rich and varied evidence of the children's learning process. Data were collected through audio recordings of children's utterances and documentary records, such as the teacher's observation diary and children's drawings. This enabled the collection of evidence regarding how students' initial ideas evolved before and after the educational intervention.

Given the developmental stage of the participants (pre-operational phase of cognitive development), the drawings, together with their utterances, illustrate the process of meaning-making (Papandreou, 2014), allowing their pedagogical potential to be harnessed in science lessons with preliterate students (Areljung et al., 2021). In this

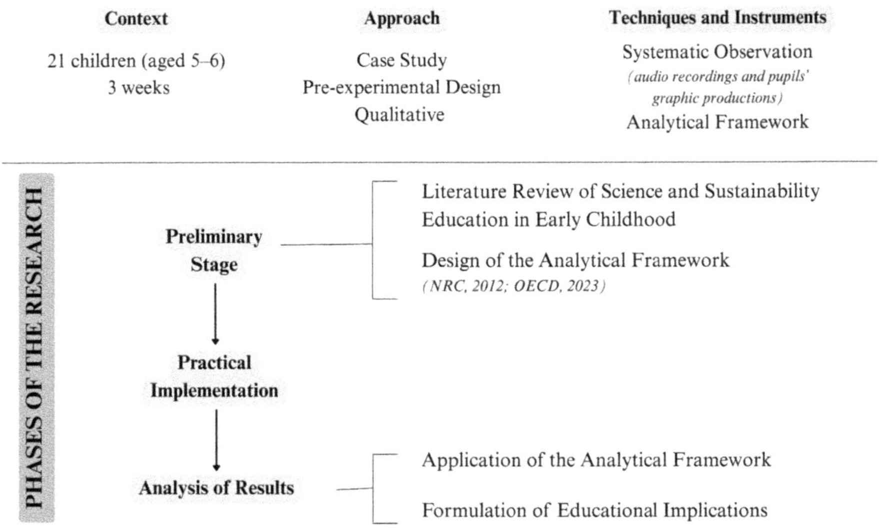

Fig. 1 Characteristics and phases of the research

regard, Hong et al. (2021) note that drawings "are a vehicle for children to represent science concepts such as predicting the outcomes of cause-and-effect events. Children can illustrate their understanding of quantity or how things work through the features they choose to include in their drawings" (p. 16). The children's utterances, originally expressed in Spanish, have been translated into English with contextual sensitivity in order to preserve their literal meaning and communicative intent.

The analysis of the results is thus undertaken from a comprehensive understanding of science and environmental education, integrating science-environment competences drawn from the PISA framework (OECD,2023) with scientific content outlined in the *Framework for K–12 Science Education* (NRC, 2012), specifically in the domain of Life Science. An analytical framework was therefore developed with a series of specific criteria, as show as Table 1, which guided the analysis of the results, taking into account the design of the intervention and the characteristics of the learners. The need to select and adapt different analytical frameworks is justified by the absence of a single model that integrates this educational approach to environmental science in early childhood (Larimore, 2020).

3.1 Description an Educational Proposal for Early Childhood Education for Sustainability

The design of the didactic proposal is grounded in the socioconstructivist paradigm, conceptualising learning as a process of meaning-making linked to interaction with others and with the environment (Mishra, 2023; Zajda, 2021). Furthermore, the

Table 1 Analytical framework for the results

PISA (OECD, 2023)	Framework for K-12 (NRC, 2012)	Specific analysis criteria
SC. Explain scientific phenomena: recall and apply appropriate scientific knowledge	LS1.A: structure and function	Recognise morphological traits of a living organism and alterations in them Establish relationships between the structures of the living organism and biological processes
SC. Explain scientific phenomena: recall and apply appropriate scientific knowledge	LS2.A: interdependent relationships in ecosystems	Identify beneficial and harmful elements in an ecosystem Establish relationships between the environment and the living beings that inhabit it, as well as with their biological processes
CCA. Explain the impact of human interactions with Earth systems: explain how social, cultural, or economic factors contribute to these impacts	LS4.D: biodiversity and humans	Explain the impact of human actions on the environment Recognise the consequences of anthropogenic actions on the lives of living beings
CCA. Make informed decisions to act based on evidence and systemic thinking to regenerate and sustain the environment	LS4.D: biodiversity and humans	Identify problems in their lifestyle and consumption habits Formulate personal and social solutions to reduce environmental degradation

sequence adopts a holistic, competency-based approach aimed at developing knowledge and skills within an authentic context, tailored to the needs of the learners and the demands of a globalised world (Catacutan et al., 2023). In this regard, contextualised learning situations are implemented, connected to real-life experiences, relevant to problem-solving, and linked to the development of key competencies (Ayu Dewi et al., 2021).

A socio-scientific issues (SSI) approach is employed (Kumar et al., 2024), which aligns with the PISA framework (OECD, 2023) concerning competencies in science and environmental science. These are seen as interrelated as show in Fig. 1, with an emphasis on the role of scientific knowledge in fostering a critically aware and responsible citizenry (Badeo et al., 2024). The proposal also draws upon the Framework for K–12 Science Education (NRC, 2012) as a design reference, incorporating scientific practices (e.g. Constructing Explanations), cross-cutting concepts (e.g. Cause and Effect: *Mechanism and Explanation*), and disciplinary core ideas (*Life Science*). Accordingly, a framework is developed that integrates these elements and is articulated through a set of learning questions, which are to be contextualised according to the characteristics of the target group (see Fig. 2).

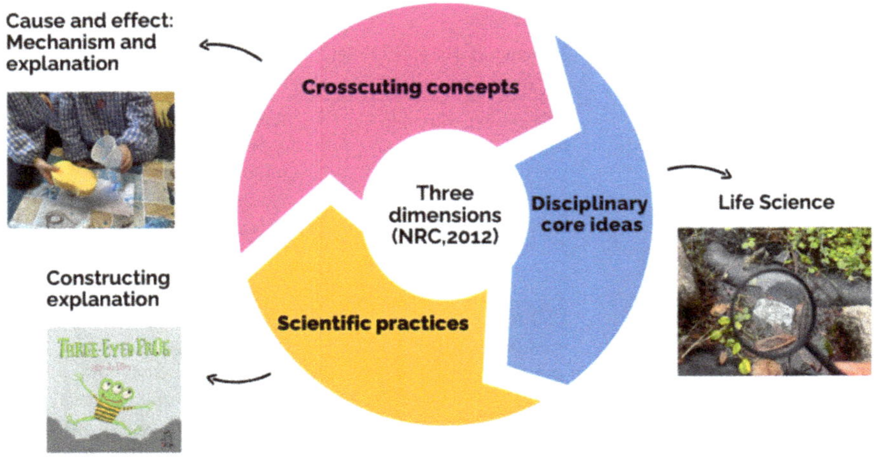

Fig. 2 Framework for the analysis of the results

With respect to the pedagogical strategies structuring the sequence, it is worth noting that activities are organised into phases associated with guiding questions, contextualised through environmental observation (Pahome, 2023; Rosenbaum, 2024) and the reading of environmental narratives (Laliena & Tabernero, 2023).

3.2 *Characteristics of the Learning Sequence*

The learning sequence is contextualised and centred around the observation of an ecosystem altered by human activity (the pond), using a relevant and situated problem addressed through guided inquiry and question-based exploration, within an *active school science* framework that engages critically with real-world contexts (Strat et al.,2023). The sequence unfolds through contextual observation and inquiry, structured into a series of phases that guide learners from discovering and understanding a living organism and its habitat to examining the impact of human actions upon them.

Various teaching strategies and resources are employed to support the development of students' scientific and environmental literacy, contributing to the attainment of the intended learning objectives (see Fig. 3).

Furthermore, these objectives are aligned with the development of key competencies and draw on the PISA frameworks (OECD, 2023) and the *Framework for K–12 Science Education* (NRC, 2012). In this sense, a didactic transposition of content and competencies is undertaken, adapted to the cognitive and developmental characteristics of five- and six-year-old children, using the pond as a concrete and meaningful learning context, introduced through the reading of an environmental narrative, as show in Table 2.

Fig. 3 Resources for fostering scientific and environmental literacy

In this case, the resource *Three-Eyed Frog* (see Fig. 4) by author and illustrator Olga de Dios (2017) was chosen as an example to visualise and contextualise the sequence. This resource was selected due to the relevance and connection of its graphic and textual content with the learning objectives of the didactic sequence. Environmental stories present significant potential, as their formative and socializing capacity can be used as a means for reflection and questioning of social practices that contribute to environmental degradation, thus serving as transformative instruments to educate for commitment to the environment (Laliena & Tabernero, 2023). In this way, environmental literature helps generate children's participation in knowledge and action regarding environmental problems, fostering sustainable attitudes and creating activists who are committed with a social intention (Harju & Rouse, 2018).

It should be noted that at this stage of development, children are in the pre-operational phase of cognitive development, which is characterised by an egocentric view of the world, symbolic play, and the emergence of basic linguistic skills. Accordingly, the activities, learning spaces, and resources were designed and adapted with these characteristics in mind, aiming to provide experiences that were meaningful and accessible to the children through experimentation and direct observation.

4 Results and Discussion

Through the analysis of the results of the proposal, it is observed that the students are able to recognise morphological traits "frogs have two eyes", identifying alterations in them "frogs don't have three eyes" and progressing, throughout the experience, in establishing relationships between the structures of the living being and biological processes "I know *how the frog breathes, I noticed it had some holes to breathe*" or functions "*this way it sees better* [the frog having three eyes]", identifying purposes *"through the holes in the skin* [the frog] *sucks water"* (see Fig. 5). Although it is also notable that through the establishment of these relationships, the students' anthropocentric view is observed in how they formulate their explanations *"it doesn't*

Table 2 Learning sequence

Phases of the sequence	Proposed activities	Learning objectives (Scientific-environmental)	Competencies (NRC, 2012; OECD, 2023)
Exploration of prior knowledge and ideas	Drawing and explanation about the habitat and morphology of the frog Creation of a mental map with images	Explore prior ideas about the morphology and habitat (the pond) of a living being (the frog) Identify characteristics	Identify characteristics
Creation of a contextualised interest context	Observation of a polluted pond Analysis of the cover of the book Three-Eyed Frog Reading of the book and group discussion Classification of pond elements	Identify alterations in the morphology and habitat of the frogIdentify beneficial and harmful elements in an ecosystem (the pond)	Observe Identify alterations Make predictions
Introduction of new information and perspectives	Searching for information about necessary habitat conditions for frogs Researching harmful elements in the pond	Recognise traits and alterations of a living being and its habita Identify the influence of human actions on the biosphere Recognise types of pollution and their sources	Discover cause–effect relationships
Search for solutions and critical reflection	Debate on human roles in ecosystem degradation Creation of a scale to assess polluting actions	Reflect on behaviours, actions, and their environmental impact	Reflect on and assess their actions
Dissemination and evaluation of the learning process	Preparation and delivery of a decalogue of "green actions Assembly to reflect on the process Final drawing and explanation of a frog and its habitat	Propose solutions to reduce anthropogenic impacts on ecosystems Evaluate the learning process and perform self-assessment	Propose solutions Reflect on the learning process

hear because it doesn't have ears [the frog]" since many times the concept of life in creatures is constructed using anthropocentric animistic cognitive schemas (Yorek et al., 2009).

As the experience develops, the children are able to establish relationships between living beings and their habitat "*it's the water where* [the frog] *lives*", "*this* [a fence] *is so the* [frog] *doesn't escape*" and "[the frog] *will want to escape to go somewhere*

Fig. 4 Excerpts from the environmental story Three-Eyed Frog

Fig. 5 Identify alterations in frog morphology related to permeability

else". Thus, these relationships evolve towards understanding cause-and-effect relationships between the elements of the environment and the processes of the living beings that inhabit it *"the insects left because the water is dirty"* advancing towards the discrimination of permanent and variable characteristics and attributes, both natural and derived from human action in an ecosystem *"the frog can't live there* [in an inflatable pool] *because what happens if* [people] *take away the pool"*. In this sense, during the development of the proposal, the students show the ability to identify beneficial and harmful elements within an ecosystem (pond) *"plastic harms the frog because it pollutes"*, *"it needs clean, warm, fresh, and chlorine-free water"* as well as in the types of pollution and waste that influence environmental degradation and their origin *"the smoke comes from the factory and the cars"*, *"the oil comes from the bottles that people throw away"*.

These results reveal an evolution in children's understanding of the impact of human actions on the environment (pond) and their consequences for the living beings that inhabit it (frog). Their knowledge evolves, recognizing beneficial and harmful elements in an ecosystem and advancing in understanding the relationships established between these elements and living beings. Other studies agree (Iliopoulou, 2018; Keifert & Reed, 2019), having obtained similar results, that school science offers children the opportunity to engage in experiences that allow them to tackle problems that are meaningful and relevant to their daily lives, starting from their prior ideas and knowledge and taking these into account in the learning process.

Moreover, this evolution of prior ideas and the incorporation of new learning is evident in the analysis of the graphic representations made at the beginning and at the end of the experience. Drawing thus constitutes a valuable tool for identifying and analysing the scientific knowledge acquired, especially at this stage, when children's communication skills are still limited and can be enhanced through graphic representations that facilitate the transmission of such knowledge (Hong et al., 2021; Muro & Pérez, 2021).

Similarly, after the development of the proposal, the students establish new relationships between the different elements that make up the environment, while introducing new ones or reformulating those represented in the initial drawings of the pond (Table 3). In this regard, the contribution and potential of analyzing these representations as a tool to evaluate the progress of prior ideas throughout the intervention is highlighted (Muro & Pérez, 2021). In this sense, it is noteworthy that reading the literary work served as a guide to promote the utterances and explanations made by the children.

On the other hand, the students are progressing in their understanding of the impact of human activities on the environment. This is a factor that was not present in their initial predictions "[the frog has three eyes] *because it is a toad"*. Thus, as the experience develops, they take it into account in their explanations "[the frog] *has many eyes because of the smoke"*. In this way, they manage to establish causal relationships between the habitat of a living being (the frog) and human actions *"throwing plastic is dangerous and animals can die"*, analyse the impact and consequences of their own actions on the physical environment *"if we write only on one side* [of the paper] *we pollute the frog"*, *"if people go to a pond we can accidentally step on a frog"*. Thus, they are able to identify problems by relating them to the impact of human actions on the environment *"if they throw oil in the water it gets dirty and the frog can no longer live there"*, *"chlorine and oil enter through the frog's skin holes and it dies"* and *"the car pollutes the frog with smoke"*. Likewise, their ability to identify problems is linked to the capacity to propose simple

Table 3 Evolution of representations and utterances during the proposal

Initial representations and explanations of the pond created by the children		

"There is the sea so it can bathe [...] the road to go see the frog"	*"It's so the frog can sit down when it's tired [the water lilies]"*	*"This [the fence] is so the frog doesn't escape"*
Children's final representations and explanations of the pond		

"[the frog] is hunting a bug with its tongue"	*"The frog is sad because it can't live here anymore because of the pollution"*	*"The frog hid its eggs here because the water is polluted"*

solutions *"we can put glasses* [on the frog] *so its eyes don't sting*", which become more complex *"bikes don't emit bad smoke because they don't have a little tube"*, *"the bus is not as bad as t*he car" and personal *"we can play outside so we don't buy so many toys"*. This shows that the students are capable of proposing actions linked to the development of sustainable habits *"with things you don't use you can make other things to not pollute the pond*", *"we have to always bring the same bottle to not have more trash*". Finally, some children were able to reflect on their learning process and their new knowledge "I thought it was fog, but it's factory smoke". Although boys and girls show a critical and reflective understanding of the natural environment based on observation and experimentation, it is important to acknowledge that their conceptions may vary individually and evolve over time (Iliopoulou, 2016, 2018). Furthermore, their responses about how to address environmental problems reflect both direct and concrete actions as well as more proactive and preventive proposals, demonstrating a diverse and developing approach to environmental protection. Thus, the children's ability to understand, explain, and question their relationship with the environment can be observed, as well as an increase in their commitment to conserving and protecting the environment through active participation, reflected in the children's interest in sharing their learning to transform their families' actions (Paulsen et al., 2021).

5 Conclusions and Educational Implications of Early Childhood Education for Sustainability

Scientific education from early childhood enables a more meaningful integration of science in the classroom, supporting children's curiosity and their right to know, explore, and make sense of the world around them, contributing to the development of skills to understand phenomena (Bascopé et al., 2019; Larimore, 2020). Children are exposed to and are part of the problems of today's world; therefore, school science should bring them closer to understanding these realities (Davis & Elliott, (2024). Thus, in the context of the major environmental crisis affecting the entire planet (IPCC, 2023; Richardson et al., 2023), the need for scientific education for sustainability in early childhood becomes evident, transforming children into agents of change through the development of critical, conscious, and active commitment (Assadourian, 2017).

This study implemented a learning sequence in preschool education designed to integrate scientific and environmental education. The proposal was developed in alignment with Spanish curriculum frameworks, similar to those of other countries (Laliena & Tabernero, 2023; Negrín & Marrero, 2021), and adapted from international reference documents, such as the PISA frameworks (OECD, 2023) and the Framework for K–12 Science Education (NRC, 2012), to suit the preschool context.

The results show the educational potential of integrating environmental issues into early science education (Larimore, 2020; Sanginova, 2024). Participation in such experiences fosters scientific literacy on climate change and strengthens environmental awareness, appreciation of conservation, and understanding of human impact (Ardoin & Bowers, 2020; Kumar et al., 2024; Trott & Weinberg, 2020).

The implementation of the proposal is justified by the need to provide opportunities for scientific and environmental literacy, encouraging informed actions by children, and offering possibilities to disseminate their learning to translate their knowledge into action for sustainability (Trott & Weinberg, 2020). Educational proposals such as the one presented here exemplify effective approaches for the development and implementation of practices aligned with the most recent national curriculum frameworks, as is the case of Spain (Laliena & Tabernero, 2023; Negrín & Marrero, 2021), which emphasize the integration of sustainability principles and socio-environmental issues into education from the earliest stages.

Regarding the educational implications of the research, this experience is expected to serve as a reference for early childhood educators, both in training and in practice. Furthermore, the proposal offers a useful example for designing and analysing educational experiences that integrate scientific and environmental education at early ages, using the PISA frameworks (OECD, 2023) and the Framework for K–12 Science Education (NRC, 2012) as guiding references. Since these frameworks cannot be used directly as reference frameworks for preschool children, the specification of proposals such as these can serve as a benchmark to guide practice as an "adapted framework" (Greenfield et al., 2017; Larimore, 2020). Concerning limitations, it is important to interpret the results in light of their specific and contextual nature, given that the data were collected from a particular sample of participants. Consequently, the findings cannot be generalized to the broader population. Individual differences among children at these ages lead to a wide range of achievement levels with respect to the proposed analysis criteria. Consequently, the results reflect and respect the individuality and diversity present among the children. Further research could adopt a longitudinal and experimental design to explore the long-term effects and wider applicability of this approach, assessing its potential to influence sustainable attitudes and behaviours over time.

Acknowledgements This research was funded by the Ministry of Science, Innovation, and Universities (Spain) through the CLIMABILITY project (Teacher training in the development of key competences in sustainability for Climate Action and Decarbonisation in the framework of Global Environmental Change, Grant PID2023-147800OB-I00 funded by MICIU/AEI/10.13039/501100011033 and by FEDER/EU).

References

Ardoin, N. M., & Bowers, A. W. (2020). Early childhood environmental education: A systematic review of the research literature. *Educational Research Review, 31*, 100353. https://doi.org/10.1016/j.edurev.2020.100353

Areljung, S., Due, K., Ottander, C., Skoog, M., & Sundberg, B. (2021). Why and how teachers make use of drawing activities in early childhood science education. *International Journal of Science Education, 43*(13), 2127–2147. https://doi.org/10.1080/09500693.2021.1953186

Assadourian, E. (2017). Earthed: Rethinking education on a changing planet. In *EarthEd. State of the world*. Island Press. https://doi.org/10.5822/978-1-61091-843-5_1

Ayu Dewi, C. C., Erna, M., Ikhfan Haris, M., & Kundera, I. N. (2021). The effect of contextual collaborative learning based ethnoscience to ıncrease student's scientific literacy ability: Research article. *Journal of Turkish Science Education, 18*(3), 525–541. https://doi.org/10.36681/tused.2021.88

Badeo, J. M. O., Duque, D. A. G., & Arnaldo, R. L. (2024). Teachers' implementation of socio-scientific issues-based approach in teaching science: A needs assessment. *Journal of Technology and Science Education, 14*(2), 363–375. https://doi.org/10.3926/jotse.1988

Bascopé, M., Perasso, P., & Reiss, K. (2019). Systematic review of education for sustainable development at an early stage: Cornerstones and pedagogical approaches for teacher professional development. *Sustainability, 11*(3), 719. https://doi.org/10.3390/su11030719

Bianchi, G., Pisiotis, U., & Cabrera, M. (2022). *GreenComp, El Marco europeo de competencias sobre sostenibilidad [GreenComp: The European sustainability competence framework]*. Publications Office of the European Union. https://data.europa.eu/doi/10.2760/094757

Campbell, D. T., & Stanley, J. C. (2015). *Experimental and quasi-experimental designs for research*. Ravenio books.

Catacutan, A., Kilag, O. K., Diano, J. F., Tiongzon, B., Malbas, M., & Abendan, C. F. (2023). Competence-based curriculum development in a globalized education landscape. *Excellencia: Internationa multi-disciplinary. Journal of Education, 1*(4), 270–282. http://multijournals.org/index.php/excellencia-imje/index

Chapman, S. N., & O'Gorman, L. (2022). Transforming learning environments in early childhood contexts through the arts: Responding to the United Nations sustainable development goals. *International Journal of Early Childhood, 54*(1), 33–50. https://doi.org/10.1007/s13158-022-00320-3v

Davis, J., & Elliott, S. (Eds.). (2024). *Young children and the environment: Early education for sustainability* (3rd ed.). Cambridge University Press.

de Dios, O. (2017). *Rana de Tres Ojos [three-eyed frog]*. Apila Ediciones. https://olgadedios.es/obra/rana-de-tres-ojos/

del Carmen, M.-S.,. M., Cruz-Sánchez, G. E., & Aparicio-Cid, R. (2021). La Educación Científica infantil con perspectiva ecológica y social Para la formación de ciudadanías participativas [early childhood science education with an ecological and social perspective for the development of participatory citizenship]. *Revista Iberoamericana de Educación, 87*(2), 35–50. https://doi.org/10.35362/rie8724608

Greenfield, D. B., Alexander, A., & Frechette, E. (2017). Unleashing the power of science in early childhood. *Zero to Three, 37*(5), 13–21. https://eric.ed.gov/?id=EJ1143042

Güler, T., Öztürk, N., İlhan, T., Aşkar, N., Banko, Ç., et al. (2021). Education for sustainability in early childhood education: A systematic review. *Environmental Education Research, 27*(6), 796–820. https://doi.org/10.1080/13504622.2021.1896680

Harju, M. L., & Rouse, D. (2018). "Keeping some wildness always alive": Posthumanism and the animality of children's literature and play. *Children's Literature in Education, 49*(4), 447–466. https://doi.org/10.1007/s10583-017-9329-3

Hong, S. B., Broderick, J. T., & McAuliffe, C. M. (2021). Drawing to learn: A classroom case study. *Early Childhood Education Journal, 49*(1), 15–25. https://doi.org/10.1007/s10643-020-01041-9

Iliopoulou, I. (2016). How young children think they can act for the environment: The case of forest and waste. *Education, 3-13, 46*(3), 249–263. https://doi.org/10.1080/03004279.2016.1236829

Iliopoulou, I. (2018). Children's thinking about environmental issues. *Educational Research, 60*(2), 241–254. https://doi.org/10.1080/00131881.2018.1453753

Jiménez, R., García, E., Azcárate, P., & Navarrete, A. (2015). Dimensión ética de la sostenibilidad curricular en el sistema de evaluación de las aulas universitarias. El Caso de la Enseñanza Aprendizaje de las ciencias [Ethical Dimension of Curricular Sustainability in the University Classroom Assessment System: The Case of Science Teaching and Learning]. *Revista Eureka Sobre Enseñanza y Divulgación de Las. Ciencias, 12*(3), 536–549. https://doi.org/10.25267/rev_eureka_ensen_divulg_cienc.2015.v12.i3.11

Karelovic, F., & Kong, F. (2022). Articulación entre Educación Ambiental y Educación Científica: Una mirada desde las competencias en sostenibilidad desarrolladas en la Formación Inicial Docente [Linking Environmental Education and Science Education: A Perspective from the Sustainability Competencies Developed in Initial Teacher Training]. *Pensamiento Educativo: Revista de Investigación Educacional Latinoamericana, 59*(1), 1–18. https://doi.org/10.7764/pel.59.1.2022.8

Keifert, D., & Reed, S. (2019). Inquiry as a members' phenomenon: Young children as competent inquirers. *Journal of the Learning Sciences, 28*(2), 240–278. https://doi.org/10.1080/10508406.2018.1528448

Kumar, V., Choudhary, S. K., & Singh, R. (2024). Environmental socio-scientific issues as contexts in developing scientific literacy in science education: A systematic literature review. *Social Sciences & Humanities Open, 9*, 100765. https://doi.org/10.1016/j.ssaho.2023.100765

Laliena, D., & Tabernero, R. (2023). Picturebooks and reader training in the 21st century. An ecocritical reading of canonical works of children's literature. *Frontiers. Education, 8*, 1304027. https://doi.org/10.3389/feduc.2023.1304027

Larimore, R. (2020). Preschool science education: A vision for the future. *Early Childhood Education Journal, 48*(6), 703–714. https://doi.org/10.1007/s10643-020-01033-9

Lorenzo-Rial, M. A., Varela-Losada, M., Pérez-Rodríguez, U., & Vega-Marcote, P. (2024). Developing systems thinking to address climate change. *International Journal of Sustainability in Higher Education, 26*(1), 83–100. https://doi.org/10.1108/IJSHE-12-2022-0404

Martínez-Silva, M. del C., Cruz-Sánchez, G. E., & Aparicio-Cid, R. (2021). La Educación Científica infantil con perspectiva ecológica y social para la formación de ciudadanías participativas. *Revista Iberoamericana de Educación, 87*(2), 35–50. https://doi.org/10.35362/rie8724608

Mishra, N. R. (2023). Constructivist approach to learning: An analysis of pedagogical models of social constructivist learning theory. *Journal of Research and Development, 6*(01), 22–29. https://doi.org/10.3126/jrdn.v6i01.55227

Muro, Á., & Pérez, J. M. (2021). La concienciación ambiental en el aula de Infantil Mediante El cine y los Cuentos [environmental awareness in the early childhood classroom through film and stories]. *Revista de Educación Ambiental y Sostenibilidad, 3*(1), 1–23. https://doi.org/10.25267/rev_educ_ambient_sostenibilidad.2021.v3.i1.1302

National Research Council (NRC). (2012). *A framework for K-12 science education: Practices, crosscutting concepts, and Core ideas*. The National Academies Press. https://doi.org/10.17226/13165

Negrín, M. Á., & Marrero, J. J. (2021). La nueva ley de Educación (LOMLOE) ante los Objetivos de Desarrollo Sostenible de la agenda 2030 y el reto de la COVID-19 [the new education law (LOMLOE) in the face of the sustainable development goals of the 2030 agenda and the challenge of COVID-19]. *Avances en Inspección Educativa. Revista de la Asociación de Inspectores de Educación de España, 34*, 140–182. https://doi.org/10.23824/ase.v0i35.709

OECD. (2023). *PISA 2025 science framework (draft)*. OECD. https://pisa-framework.oecd.org/science-2025/

Pahome, D. (2023). Observation sheet—An essential tool for facilitating learning. *Romanian Review of Geographical Education, 12*(1–2), 5–16. https://doi.org/10.23741/rrge20231

Papandreou, M. (2014). Communicating and thinking through drawing activity in early childhood. *Journal of Research in Childhood Education, 28*(1), 85–100. https://doi.org/10.1080/02568543.2013.851131

Paulsen, C. A., Carroll, E., Paulsen, O., & Andrews, J. R. (2021). Engaging children and families in active, environmental science learning through digital media. *International Journal of Early Childhood Environmental Education, 8*(2), 43–58. https://eric.ed.gov/?id=EJ1298025

Pramling, I. and Kaga, Y. (2008). *The contribution of early childhood education to a sustainable society*. UNESCO: París: France. https://unesdoc.unesco.org/ark:/48223/pf0000159355

Priya, A. (2021). Case study methodology of qualitative research: Key attributes and navigating the conundrums in its application. *Sociological Bulletin, 70*(1), 94–110. https://doi.org/10.1177/0038022920970318

Richardson, K., Steffen, W., Lucht, W., Bendtsen, J., Cornell, S. E., et al. (2023). Earth beyond six of nine planetary boundaries. *Science Advances, 9*(37), 2458. https://doi.org/10.1126/sciadv.adh2458

Rockström, J., Steffen, W., Noone, K., Persson, Å., Chapin, F. S., III, et al. (2009). Planetary boundaries: Exploring the safe operating space for humanity. *Ecology and Society, 14*(2) https://www.jstor.org/stable/26268316

Rodrigues-Silva, J., & Alsina, Á. (2023). STEM/STEAM in early childhood education for sustainability (ECEfS): A systematic review. *Sustainability, 15*(4), 3721. https://doi.org/10.3390/su15043721

Rosenbaum, L. F. (2024). Spotlighting spectatorship: Elevating observation-based learning in the design and evaluation of body-scale learning environments. *Educational Technology Research and Development, 72*, 2133–2157. https://doi.org/10.1007/s11423-024-10373-x

Sanginova, G. B. (2024). Improving organization in the first environmental education of preschool children. *Multidisciplinary Journal of Science and Technology, 4*(4), 440–447. https://www.mjstjournal.com/index.php/mjst/article/view/1318

Sjöblom, P., Wolff, L. A. & Sundman, J. (2023). Climate change as a socio-scientific issue in upper secondary education: Addressing wicked problems through crosscurricular approaches. En S. Harnow Klausen & N. Mård (Eds.), Developing a didactic framework across and beyond school subjects: Cross- and Transcurricular teaching (182–196). Routledge. https://doi.org/10.4324/9781003367260-17

Somerville, M., & Williams, C. (2015). Sustainability education in early childhood: An updated review of research in the field. *Contemporary Issues in Early Childhood, 16*(2), 102–117. https://doi.org/10.1177/1463949115585658

Stouthart, T., Bayram, D., & Van der Veen, J. (2023). Capturing pedagogical design capacity of stem teacher candidates: Education for sustainable development through socioscientific issues. *Sustainability, 15*(14), 11055. https://doi.org/10.3390/su151411055
Strat, T. T. S., Henriksen, E. K., & Jegstad, K. M. (2023). Inquiry-based science education in science teacher education: A systematic review. *Studies in Science Education, 60*(2), 191–249. https://doi.org/10.1080/03057267.2023.2207148
Steffen, W., Richardson, K., Rockström, J., Cornell, S. E., Fetzer, I., et al. (2015). Planetary boundaries: Guiding human development on a changing planet. *Science, 347*(6223), 1259855. https://doi.org/10.1126/science.1259855
Trott, C. D., & Weinberg, A. E. (2020). Science education for sustainability: Strengthening children's science engagement through climate change learning and action. *Sustainability, 12*(16), 6400. https://doi.org/10.3390/su12166400
UNEP. (2019). *Global environment outlook—Geo-6: Healthy planet, healthy people.* Cambridge University Press. https://doi.org/10.1017/9781108627146
UNEP. (2022). *Emissions gap report 2022: The closing window - climate crisis calls for rapid transformation of societies.* Nairobi. https://wedocs.unep.org/20.500.11822/40874
UNESCO. (2021). *Berlin declaration on education for sustainable development.*https://unesdoc.unesco.org/ark:/48223/pf0000381228
Widiyawati, Y. (2020). Global warming & climate change: Integration of socio-scientific issues to enhance scientific literacy. *Journal of Physics: Conference Series, 1511*(1), 1–11. https://doi.org/10.1088/1742-6596/1511/1/012071
Wiek, A., Withycombe, L., & Redman, C. L. (2011). Key competencies in sustainability: A reference framework for academic program development. *Sustainability Science, 6*(2), 203–218. https://doi.org/10.1007/s11625-011-0132-6
IPCC. (2023). Climate change 2023: Synthesis report. Contribution of working groups I, II and III to the sixth assessment report of the intergovernmental panel on climate change. In *Intergovernmental panel on climate change.* https://doi.org/10.59327/ipcc/ar6-9789291691647
Yorek, N., Sahin, M., & Haidin, H. (2009). Are Animals 'More Alive' than Plants? Animistic–Anthropocentric Construction of Life Concept. *Eurasia Journal of Mathematics, Science & Technology Education, 5*(4), 369–378. https://doi.org/10.12973/ejmste/75287
Zajda, J. (2021). Constructivist learning theory and creating effective learning environments. In: *Globalisation and education reforms. Globalisation, comparative education and policy research,* 25 35–50. Cham, Springer. https://doi.org/10.1007/978-3-030-71575-5_3

Candela Rodríguez-Conde is a predoctoral researcher at University of Vigo (Spain). Graduate in Early Childhood Education, with a specialization in physical education, and Master in Research and Innovation in Specific Didactics for Early Childhood and Primary Education at University of Vigo. Her lines of research include Science Education and Education for Sustainability.

Nuria Castiñeira-Rodríguez is a PhD Candidate in Equity and Innovation and researcher at University of Vigo (Spain). Graduate in Early Childhood Education and Master in Research and Innovation in Specific Didactics for Early Childhood and Primary Education. Her research includes Science Education, Education for Sustainability and Educational Technology.

Mercedes Varela-Losada is a PhD and professor in Experimental Science Didactics and Environmental Education at the Faculty of Educational Sciences, University of Vigo (Spain). She has conducted research in the field of sustainability for more than 15 years. Her research interests focus on Environmental Education for Sustainability and teacher training in Higher Education, particularly in relation to transformative learning, development of action competence, and climate education. In recent years, her work has also explored perceptions of the Global Environmental Change process and the development of key competences in sustainability.

María A. Lorenzo-Rial is a PhD and associate professor in Experimental Science Didactics at the Faculty of Educational Sciences, University of Vigo (Spain), of which she is the Coordinator of the Degree in Early Childhood Education. Her research interests focus on Science Education, Education for Sustainability, Educational Technology and Gender Perspective. In recent years, she has participated as a researcher in projects related to the educational approach to global environmental change, with a special focus on developing sustainability skills, and automating assessment processes in higher education.

Developing a Scalable Online Ecology Course with 4DEE

Nathan Ruhl and Sara J. Wright

Abstract Ecological science touches on all aspects of sustainable development and provides critical knowledge to address both socioeconomic and socioecological sustainability challenges. Online learning is a promising instructional format for addressing the global need for high quality ecology education. Ecology education is assisted by the Ecological Society of America's Four-Dimensional Ecology Education (4DEE) framework. Here we discuss the integration of the 4DEE framework into a scalable online ecology course that supports sustainability education. This online ecology course utilizes best-practices related to place-based and experiential education and focuses on authentic assessments. Online students are engaged in scientific inquiry, gain instructional flexibility (the course is semi-asynchronous with recorded lectures), and the course can be scaled to reach a larger audience. The course includes themes of ecological literacy, sustainability, and interdisciplinary thinking while emphasizing inquiry-based learning, accessibility, student engagement, and transferable skill development. This chapter demonstrates how ecology education can be effectively adapted for online environments and how 4DEE can be applied to online curriculum development that supports sustainable development goals.

Keywords 4DEE framework · Sustainability education · Ecological literacy · Online learning · Inquiry-based learning · Human-environment interactions · Sustainable development goals · Curriculum development · Interdisciplinary

1 Introduction

Preparing students to understand and respond to socioeconomic and socioecological sustainability problems is a growing priority in higher education (Husic, 2024). Addressing complex issues like climate change and loss of biodiversity requires not only technical knowledge but also systems thinking, ethical reasoning, ecological

N. Ruhl (✉) · S. J. Wright
Biological & Biomedical Sciences, Rowan University, NJ, USA
e-mail: Ruhl@Rowan.edu

S. J. Wright
e-mail: Wrights@Rowan.edu

W. Leal Filho et al., *Curriculum Development for Sustainable Development*, World Sustainability Series, https://doi.org/10.1007/978-3-032-22146-9_14

literacy, and communication skills (Wiek et al., 2011). To meet these educational challenges, sustainability instructors increasingly look to competency frameworks (Brundiers et al., 2021; Tomassi et al., 2024) to guide curriculum and foster the knowledge, skills, and values necessary for responsible future-oriented actions and decisions. Ecology, as a scientific discipline, plays a central role in education for sustainable development by providing a scientific foundation for understanding how living organisms (including humans) interact with each other and the abiotic environment. Ecology courses can cultivate ecological literacy, thereby equipping students with the ability to critically analyze complex natural systems and apply those insights to their own scientific pursuits and real-world decision-making (Klemow et al., 2019; Sorensen et al., 2024). Designing ecology courses to meet the current and future needs of students requires not only careful attention to course content, but also to instructional design, delivery format, and accessibility (Ruhl et al., 2022).

Online learning is an instructional format that has the potential to positively impact global education efforts (Levin, 2024), like sustainable development, and the use of this instructional modality is increasing (GMI, 2023). Online learning and traditional face-to-face instruction each offer distinct advantages and limitations. Traditional classrooms provide synchronous real-time interaction that facilitates immediate feedback, dynamic discussions, and hands-on experiential and collaborative activities that are especially valuable for teaching ecological science (Beltran et al., 2020). However, the traditional classroom can be limited by fixed schedules, geographic constraints, and resource availability such as space. Importantly, the traditional classroom reduces accessibility for certain types of learners, especially non-traditional students. Some aspects of the traditional classroom can be brought online through synchronization of learning activities and virtual student-student interaction (Blended or Hybrid Learning; Ashraf et al., 2021; Fitzgerald et al., 2022; Martins et al., 2025), but there is a tension between synchronization of the learning environment and learning accessibility in this style of course. Increased synchronization decreases accessibility for both distance learners, due to a reduced ability to participate, and non-traditional students, who might not be available when synchronous learning occurs.

The Ecological Society of America (ESA) recently developed a pedagogical framework for teaching ecology, called the Four-Dimensional Ecology Education (4DEE) framework (Fig. 1; ESA, 2018). 4DEE was originally intended for use in developing individual learning activities but has also been used to develop entire courses (Ruhl et al., 2022). The 4DEE framework supports not only disciplinary learning in ecology, but also the incorporation of ecology into non-majors courses (Rodgers et al., 2024), thereby broadly supporting the development of key competencies in sustainability education. The 4DEE framework integrates four dimensions of ecology:

1. **Core Ecological Concepts**: ideas critical to understanding ecology, such as population dynamics, nutrient cycling, energy flow, ecosystems, and species interactions.

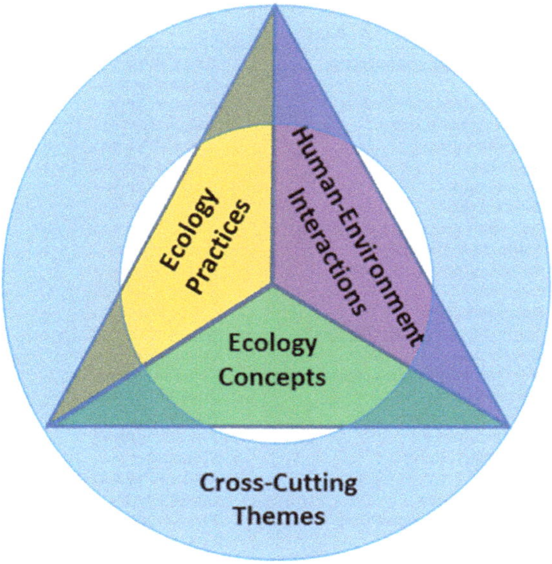

Fig. 1 Illustration of the ecological society of America's 4DEE Framework. Used with permission. The content of this publication has not been approved by the ecological society of America, its leadership or its committees and does not reflect the views of the community

2. **Ecology Practices**: how ecological science is conducted, including the scientific method, field and lab-based methods, quantitative reasoning, and science communication.
3. **Human-Environment Interactions**: focusing on the interdependence between humans, other species, and the environment to address socioecological and sustainability problems like climate change and biodiversity loss.
4. **Cross-Cutting Themes**: concepts that unite ecology with other biological and scientific disciplines, such as systems thinking, evolution, and spatio-temporal variation.

Sustainability education is crucial for meeting the United Nations Sustainable Development Goals (SDGs), and the 4DEE framework supports all 17 SDGs in some capacity, with some ecological subjects touching on many of the SDGs simultaneously (Table 1). In developing a scalable online ecology course using 4DEE, we sought to both introduce learners to the science of ecology and equip them with knowledge and skills necessary to engage constructively with a wide range of sustainability issues by connecting the course content to the lives of a global student population. This chapter outlines our approach to reformulating an in-person course (integrated with 4DEE and detailed in Ruhl et al., 2022) into an online course and links the 4DEE framework to sustainable development goals.

Table 1 Contrast of 4DEE dimensions against potential course topics and the 17 UN Sustainable Development Goals (SDGs)

Potential Ecology Course Topics	4DEE Category	Related UN Sustainable Development Goals
Biodiversity	Ecological Concepts	1, 3, 6, 10, 12, 13, 14, 15
Biomes, ecosystems and habitats	Ecological Concepts	3, 6, 10, 13, 14, 15
Nutrient and carbon cycling	Ecological Concepts	6, 12, 13, 14, 15, 17
Species interactions	Ecological Concepts	2, 3, 11, 14, 15
Disturbances	Ecological Concepts	9, 11, 12, 13
Urban systems	Ecological Concepts	6, 9, 11, 13
Disease ecology[a]	Ecological Concepts	3, 6, 11, 17
Organismal responses to climate change	Ecological Concepts	13, 14, 15
Trophic / feeding interactions	Ecological Concepts	2, 14, 15
Habitat fragmentation / connectivity	Ecological Concepts	10, 11, 15
Energy flow	Ecological Concepts	3, 7, 11
Primary productivity	Ecological Concepts	2, 7
Carrying capacity and population growth	Ecological Concepts	11, 12
Spatial mapping[a]	Ecological Practices	1, 3, 6, 7, 8, 10, 11, 13, 14, 15
Ecosystem, community, and population monitoring	Ecological Practices	1, 2, 6, 8, 11, 13, 14, 15, 17
Modeling & statistical inference	Ecological Practices	3, 7, 8, 9, 12, 13, 14, 15, 17
Inquiry-based education[a]	Ecological Practices	5, 10
Impact assessment	Ecological Practices	9, 16
Experimental field plots	Ecological Practices	2
Biotic indices[a]	Ecological Practices	6
Global observatories	Ecological Practices	17
Ecosystem services, management & conflict	Human-environment Interactions	1, 2, 3, 5, 8, 9, 10, 11, 12, 13, 14, 15, 16, 17
Sustainable yields	Human-environment Interactions	8, 11, 12, 14, 15, 17
Ecosystem degradation	Human-environment Interactions	1, 3, 14, 15, 17
Development and land-use change	Human-environment Interactions	6, 7, 8, 9, 11
Pollution	Human-environment Interactions	6, 12, 13, 14, 15
Zoonotic disease[a]	Human-environment Interactions	3, 6, 17
Restoration	Human-environment Interactions	8, 9
Agroecosystems[a]	Human-environment Interactions	2
Ecological literacy and information accessibility	Human-environment Interactions	4
Gender-based ecosystem interactions[a]	Human-environment Interactions	5
Ecological engineering	Human-environment Interactions	9
Systems	Cross-cutting themes	All SDGs
Space and Time	Cross-cutting themes	All SDGs
Interdisciplinarity[a]	Cross-cutting themes	All SDGs
Science policy[a]	Cross-cutting themes	All SDGs
Globalization[a]	Cross-cutting themes	All SDGs
Tipping points (non-linear feedbacks)[a]	Cross-cutting themes	All SDGs
Structure and function	Cross-cutting themes	3, 9, 11, 13, 14, 15
Evolution	Cross-cutting themes	2, 3, 13, 14, 15
Pathways and transformation of matter and energy	Cross-cutting themes	2, 6, 7, 12, 13
Open science[a]	Cross-cutting themes	4, 10, 16, 17

Note a: these ecology topics have applicability in a sustainability context but do not explicitly appear in 4DEE documentation

2 Methods

2.1 Online Course Development Context

The COVID-19 pandemic coincided with an effort to increase the number of online course offerings at Rowan University, in Glassboro, New Jersey, United States of America. As we redeveloped our in-person version of the Introduction to Ecology course using 4DEE (detailed in Ruhl et al., 2022), we simultaneously planned for the implementation of an online version of the new course, and used the same instructional alignment for both instructional modalities. Online courses at Rowan currently

(2025) receive minimal updates from semester to semester, and major changes happen every 3–5 years, if needed. The online version of the course was developed in spring/summer 2020 and taught for the first time in fall 2021, one year after the new in-person course first ran (Fig. 2). Introduction to Ecology is the last of a four-course sequence in introductory biology at Rowan University and students are able to take the course in any available format.

Certain online course features are encouraged and sometimes required at Rowan University. Online course materials are released to students two weeks at a time, with an overview of everything to be accomplished during each two-week period, making the course semi-asynchronous. All lectures must be pre-recorded and are viewed by students asynchronously. Lectures are recommended to be short, no more than 5–30 minutes in duration, but many short lectures can be assigned in a given week. Students have access to all lecture videos for the entire semester. Student-student interaction largely occurs through regular and required Discussion Questions (DQs), which are posted in a discussion board format. Students must create an original DQ post and also comment on the posts of their peers every 1–2 weeks, which demonstrates engagement with the course material, and also provides students with a chance for interaction with each other. Exams are time-limited and have a due date, but can be taken at any time after they are assigned, providing flexibility for the student. Exams are typically designed with large test banks of questions, with some questions required for all students and other questions selected and assigned to individual students at random. All exams are required to be open-book and open-notes. These features aim to reduce academic integrity violations and reports, though students are typically unmonitored during exams.

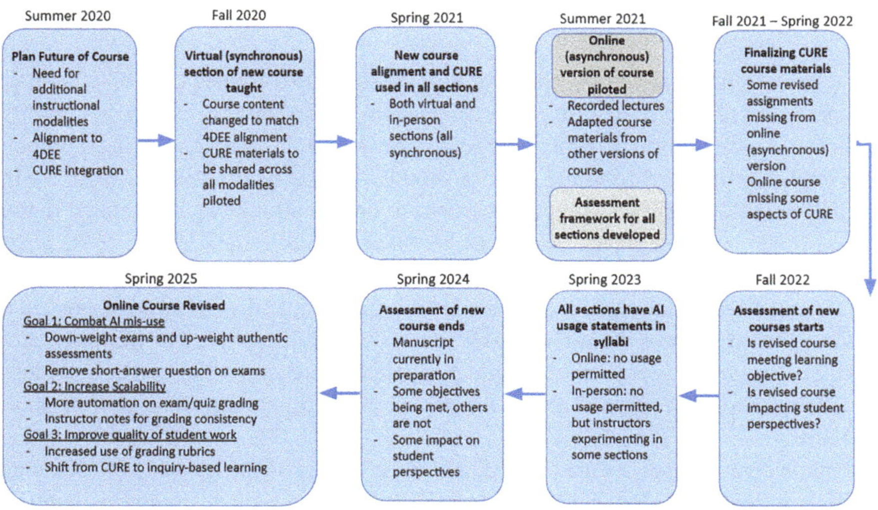

Fig. 2 Timeline of the collaborative redesign of the Rowan University Introduction to Ecology course using 4DEE and development of the online version of the course

2.2 Common Course Goals (in-Person and Online)

In addition to the course attributes imposed by our home institution (above), we also took care to ensure that course goals were maintained across instructional modalities. Most assignments and projects require students to integrate information across biological scales and we put strong emphasis on human-environment interactions. We also emphasized ecological practices and skills that are transferable to a variety of careers, especially data literacy and data limitations. We intentionally sought to showcase the field of ecology as accessible and beneficial for all, center students as scientists in training, and position students as contributors to our shared knowledge. Active and inquiry-based learning were utilized wherever it was practical.

The in-person and online versions of Introduction to Ecology both utilize a unifying biological example of urban evolution (Santangelo et al., 2022), structured as a semester-long project: cyanogenesis in white clover (*Trifolium repens*). A focus on cyanogenesis in white clover provides a current, real-world example that can be studied almost anywhere a student might be located because white clover has a global distribution. Students in both instructional formats consider the role of white clover around their home (a place-based emphasis) and complete field activities where they locate and observe white clover. Both course formats seek to give students the agency to "choose your own adventure" whenever possible, thereby allowing the course to stay fresh as students find and bring new information (i.e., people, news articles, scientific literature, data and results) to the course.

2.3 Production of Online Course Materials

In creating online course materials, we utilized the 4DEE framework and sought to balance accessibility, scalability, student engagement, and interdisciplinary thinking. Building course materials involved regularly revisiting the course alignment, which was explicitly mapped to 4DEE, and created by our curriculum team (Table S1 in Ruhl et al. 2022). Learning outcomes from our course alignment are restated in the course materials, as learning objectives, so that students understand what we want them to learn. These learning objectives are stated in the form of a bulleted list at the start of course materials. Individual lectures, assignments, DQs and exams each utilize all four dimensions of 4DEE whenever possible. Individual questions are regularly grounded in real-life ecological problems, many of which are directly tied to sustainability concepts and reflect the human-environment interactions dimension of 4DEE.

Online course materials are housed in Google Drive and disseminated to students through a Learning Management System (LMS; Canvas by Instructure). Lecture videos for the online course were created using voice recordings over Powerpoint slides and are housed in the LMS media gallery. Supplemental videos from YouTube were used in some cases, so a YouTube playlist was created and is shared with students to keep all supplemental videos in one location for easy reference. LMS discussion boards were used for DQ assignments. Ultimately, all work from online students, including exams, are submitted and graded through the LMS.

The freely-available browser-based Google Suite of applications, particularly Google Docs, is heavily utilized because it provides for accessibility, ease of sharing, collaboration, and the ability to monitor and provide feedback on student work. Bi-weekly overviews and instructions for assignments are shared as view-only Google Doc links in LMS modules that are released two weeks in advance of due dates. Students are able to make copies of view-only docs to fill in their responses for assignments. Those working documents can be shared with instructors. In some cases, online students are able to submit responses in the LMS directly, with a subset of questions being auto-graded for instant feedback. While some assignments require the full Microsoft Excel application, our students are able to install Microsoft applications for free on personal devices using their student accounts, which makes Excel an accessible analysis tool that additionally provides transferable skills for our diverse student population. These browser-based and software tools are also used in the in-person course.

Online assignments largely mirror assignments from the in-person course, emphasizing the use of peer-reviewed published literature and shared datasets, with questions asking students to apply knowledge, analyze complex systems, evaluate the impacts of perturbations to systems, and create hypotheses—tasks that elevate learning to higher levels of Bloom's Taxonomy. Active and inquiry-based learning are emphasized in both formats of the course but pose pedagogical challenges for the online format, as each student is effectively working independently, with minimal collaboration or ability to learn alongside peers. Nonetheless, questions are written to promote personal reflection and metacognition, encouraging students to make connections between ecology, sustainability, and prior learning in a variety of other courses.

To reinforce course concepts and help students help themselves, instructions for the semester-long writing assignment reference prior scaffolded assignments and encourage students to refer back for assistance. Grading rubrics were also used as a tool to help students understand the instructor's expectations and do well on the assignment (particularly DQ's). Turnitin is utilized for written work to assess originality and discourage plagiarism.

2.4 Update to Online Course for Scalability and Academic Integrity

The online course was first developed in 2021, before the release of generative AI (Artificial Intelligence). We updated the course during the 2024–2025 academic year. Our focus during this update was to address academic integrity concerns created by student reliance on generative AI and increase the scalability of the course. In our context, a scalable course is one that allows for larger class sizes, can be taught by new instructors, requires minimal course updates in terms of time and cost, maintains alignment to 4DEE and our in-person course, and maximizes both student outcomes and equivalency between instructional modalities (online versus in-person).

The first iteration of the course included many assignment prompts aimed at higher orders of learning, but there were also many quiz/exam questions where students needed to demonstrate knowledge by writing short-answers summarizing a particular concept or idea in ecology. These questions can now be readily answered with generative AI and so were removed from the course or reformulated into multiple-choice questions. The length (number of questions) and the time allotment for exams was standardized and calibrated to better match the exam experience of the in-person sections. Short-answer questions are no longer present in the online exams and about 20% of each exam is composed of multiple-choice questions that generative AI struggles to answer (i.e., questions focused on unexpected results or specific examples from the course). In our experience, Generative AI also struggles to create documents (like spreadsheets, papers, posters, or videos) following specific guidance/format/instructions (as of 2025), and so the updated course leans more heavily on authentic assessment of works that students have created. As a whole, the course was shifted higher on Bloom's taxonomy (Anderson and Krathwohl, 2001) in response to generative AI.

Removing the instructor's role in assessing summative short-answer questions in favor of automatically graded multiple-choice questions (particularly on exams) created time for the instructor to provide additional feedback in other areas of the course. This instructional time was allocated to replace a peer-review activity in which students reviewed the work of other students as part of the semester-long project. This was a difficult decision because the peer-review activity was an opportunity for peer-peer interaction, but this change was made because many students were outsourcing this experience to AI (defeating the peer-peer interaction we intended) and some students who would clearly benefit from corrective feedback were not getting the feedback they needed.

The semester-long project was also altered to shift from a CURE to inquiry-based learning. Students previously chose between a study of white clover urbanization (human-environment interaction theme using student-collected data) or a camera trap studying using data from Gorongosa National Park (Mozambique; open global science theme) for their semester-long project. Our hope with this assignment choice was to (1) demonstrate that science is a global activity that can be conducted anywhere, (2) science is accessible, (3) inspire students to follow their interests, and (4) allow online students to have a CURE experience equivalent to that of our in-person students. Most online students opted to do the camera-trap study rather than collect their own data on white clover, and unfortunately, very few students produced a high quality study using the camera-trap data due to the constraints of the online modality. Given this poor student outcome for the semester-long project in the online modality, we made the decision to utilize inquiry-based learning moving forward, and re-designed the semester-long project accordingly, keeping the focus on white clover cyanogenesis as the unifying theme.

Existing grading rubrics were adapted, new grading rubrics were instituted, and instructional documents were developed to guide new instructors teaching the online course. Instructional documents help to maintain consistency across online sections and between in-person and online versions of the course. Students and instructors both have access to the grading rubrics on all assignments where a rubric is used. The weight of different components of the course in a student's final grade was rescaled to incorporate these changes, with exams decreased in weight, and the semester-long project and other authentic assessments increased in weight. We believe that this issue was caused by a combination of limited data literacy among students and limited interaction with the instructor as the study was developed.

3 Results and Discussion

3.1 Overview and Major Units of the Online Course

The online course is composed of three units, combining the dimensions of 4DEE (Table 2) and highlighting transferable skills that are needed for both ecological and sustainability science:

Table 2 Overview of the online introduction to ecology course at Rowan University relative to the four dimensions of 4DEE

Course Unit	Course Week	Cross-cutting Themes	Ecological Concepts Foci	Ecological Practices (Transferable Skills) Foci	Human-Environment Interactions Foci
1	1-6	**These themes are present in all units:** Structure and Function, Pathways and Transformations of Matter and Energy, Systems, Evolution, Space and Time, Open Science	Environments, Cycles, and Disturbances	Synthesizing Scientific Literature and Experimental Design	White clover urban adaptation, climate change, sustainability, anthropocene
2	7-11		Energy, Trophic interactions, and Population growth	Modeling and Quantitative Thinking	White clover urban adaptation, COVID-19 disease ecology, anthropocene
3	12-15		Species interactions and Biodiversity over space and time	Communication and Presentation	White clover urban adaptation, sustainability, anthropocene

Unit 1: Synthesizing scientific literature and experimental design.

Unit 2: Modeling and quantitative thinking.

Unit 3: Communication and presentation.

In all units, lectures introduce ecological concepts, while assignments and related supplementary videos contextualize the concepts with real-world examples. In assignments, heavy emphasis is placed on developing transferable skills that are useful across disciplines, contexts, and scientific practices. All units and most assignments build to the semester-long project.

Assignments are scaffolded to allow students to practice and receive feedback as they develop their scientific skills. In Unit 1, students increase their ecological literacy and scientific skills as they distinguish and differentiate peer-reviewed versus non-peer-reviewed sources, locate and paraphrase primary and secondary sources using scientific research databases, and create both in-text and bibliographic citations. Students also use white clover scientific papers as examples to identify best practices for experimental design, taking into account scope and scale, and they propose testable hypotheses to address follow-up questions.

In Unit 2, students hone their analytical skills by critiquing and drawing conclusions from figures published in the primary literature. Students also develop spreadsheet, modeling, and statistical skills using Microsoft Excel and/or Google Sheets. Assignments in Unit 2 ask students to use published data to learn and practice basic statistical tests, write professional figure legends, and explore quantitative ecological models. To gain a deeper understanding of ecological and systems-level processes, students perturb components of ecological models. Unit 2 concludes by introducing students to the concepts of citizen science and open source science, emphasizing global scientific efforts and the strength of collaborative data collection.

Finally, Unit 3 requires students to put everything together and practice multiple types of scientific communication. They finalize their semester-long project, demonstrating their ability to propose hypotheses, make predictions supported by the scientific literature, and communicate ecological science in a variety of ways (a paper, poster, and video). Students are supported in these tasks by complementary assignments focused on open science and best practices for creating and presenting scientific posters.

3.2 Semester-Long Project

Students in the online course complete an inquiry-based scaffolded individual writing assignment. Online students write a research proposal for a study of white clover urbanization which is later reworked into an introduction for a scientific paper, test their own hypotheses on data generated by the in-person students, write a full scientific paper detailing their study, create a scientific poster summarizing their study, and present that poster in a recorded video presentation. As a scaffolded assignment, online students receive feedback on different pieces of their project over the course of the semester and iteratively improve their study. White clover ecology and urban adaptation are integrated throughout the online course in lectures and smaller assignments (Table 2). The instructor does not need to be an expert in plant biology to teach the course, but does need to have a strong background in ecology, experimental design, and hypothesis testing in order to guide students through their semester-long project.

The use of inquiry-based learning in the online course is a departure from the in-person course, where students are engaged in a CURE. In-person students identify and select plots of white clover plants, collect geo-tagged data in the field and laboratory, and test hypotheses on data they actually collect themselves. The in-person CURE emphasizes ecological practices in both the field and laboratory, transferable scientific and communication skills (including teamwork and collaboration), culminating in the creation of a scientific poster to share the result of their study. In-person students start their semester-long project as individuals but finish the project working in collaborative teams.

The switch to inquiry-based learning in the online course was made primarily to address poor student outcomes in the semester-long project but this change also supports improvements to course scalability and academic integrity. The instructor can swap out a single data file from semester to semester or section to section, thereby altering the outcome of the projects. This requires little time or effort from the instructor, allows the course to have timely updates, while also preventing students from recycling projects from prior semesters because results from past projects will not match the new data. Additionally, we plan for the white clover CURE that we developed for the in-person sections of the course to evolve over time, with new types of data collection enabling new questions to be asked, so the ability to swap out a single data file in the online sections ensures that the online student experience paces with the in-person sections.

Despite working from a dataset that we provide, online students still have agency to ask their own questions about white clover and students can choose to take their study to the next level by collecting additional complementary data, transitioning the activity to a CURE for those students, but only about 10% choose this path through the assignment. Online students are always encouraged to think outside the box, pursue their own ideas, and ask interesting new questions using complementary data that they find in the published literature or collect themselves.

3.3 Unique Features of the Online Course

In redesigning our Introduction to Ecology course to align with 4DEE, our curriculum team aimed to emphasize accessibility of ecological science and center students as active contributors to a vibrant scientific community. An obvious limitation of online learning is reduced student-student interaction, which can hinder efforts to build community and create a barrier to collaborative learning (see Sect. 3.4 for discussion on this point). Discussion Questions (DQs) were therefore designed with building community in mind and are unique to the online version of the course. All DQ assignments are initiated by a prompt from the instructor and allow students to “choose their own adventure” (follow their individual interests) while providing space for peer-to-peer interaction and learning.

Unit 1 DQs work to build community within the course and the global community of ecologists. Students make video introductions of themselves and, after reading an introductory white clover paper, share interesting tidbits about both themselves and white clover. “Profile an Ecologist” is a DQ where students identify a press release of interest, trace the news story to a specific scientific paper and ecologist, then spend time learning more about the person behind the science. Students share information about the ecologist’s research history, current level of training, and other details gleaned from websites and online profiles. Another Unit 1 DQ asks students to explore national and global research networks that are tackling large questions in ecology (e.g., NSF's National Ecological Observatory Network (NEON) and the ForestGEO Network for studying global forest dynamics).

Unit 2 DQs focus more on ecology practices, while also revisiting the broader community of ecologists. Two of the DQs focus on modeling - first an introduction to the wide variety of "models" used by scientists, and then a focus on specific models that enable scientists to address major ecological problems and solutions grounded in human-environment interactions: climate change, conservation, and disease spread. Students are introduced to various examples of citizen science and best practices for scientific posters and presentations in Unit 3 DQs.

We emphasized relational learning and sought to provide space for diverse global perspectives as we initially developed the online course and then worked to prepare the course for scaling to a wider audience. Promoting community and a sense of belonging were important considerations as we worked to include self-similar examples that help students see themselves as ecologists. A key example of this approach is the unit 1 DQ, "Profile an ecologist", which we see as an example of a flexible activity that promotes a sense of belonging. Promoting a sense of belonging has a positive effect on student outcomes (Fong et al., 2025) and retention (Bahnson et al., 2025). This activity generates comments from students that explicitly point to cultural connections or subject-matter interests that they find to be relatable. Self-similarity is an important consideration when selecting course resources like videos and textbooks (Richards et al., 2025), as well as designing "choose your own adventure" activities. We also emphasized the human-environment dimension in our redesign of the course, which we hope allows students to connect all four dimensions of ecology to their life by studying a plant that is likely to be growing in close proximity to their home. We encourage other instructors to similarly develop activities that help students see their own importance relative to the subject at hand.

3.4 Limitations of this Work

The 4DEE framework is very flexible, and all dimensions of ecology can be represented in an online course. That said, teaching some aspects of ecology is limited in the online modality, particularly the practices of an ecologist (or scientist more generally). Some online students (e.g., physically impaired, incarcerated) have limited capacity to go outdoors, obtain biological samples, or collect data from those samples. Most online students do not have dedicated laboratory spaces in their home where they can perform controlled experiments. Online students do not have access to specialized scientific equipment and cannot receive hands-on training in the use of that equipment in an online course.

Another key limitation in an online course is that the course material is usually asynchronous, making it very difficult to train students in teamwork and collaboration. It is one thing to say that it is good to work as a team and be collaborative and quite another thing to have students experience the challenges and benefits of working together through collaborative learning. Learning to work collaboratively in teams is a transferable skill sought by employers, required for success in many different fields, and is a point of major emphasis in the in-person version of our Introduction to Ecology course. We spend a great deal of time fostering community in the online version, but there are no activities where students work as a team or rely on one another to complete assignments, mainly because of the tension between accessibility and synchrony. Our online students do complete activities where peer-peer learning occurs.

Lastly, in our experience, it is more difficult to make updates or alter an online course after it has been made, relative to an in-person course. This limitation may not apply to all higher education institutions, but where development resources are limited, it can be very difficult to make on-the-fly changes to the course. Updates are also complicated in situations where an external or adjunct instructor is hired to teach the online course. These newly hired faculty might not be prepared or able to update the course content, have access to the instructors charged with curriculum development, appreciate the pedagogical intention for the course within a program of study, or understand how the course will be assessed. We have found the limitation on timely updates to be manageable, barring major societal shifts like the introduction of generative AI. We have also created course activities that allow for natural updates by giving students agency to bring current events or newly published studies to their work, keeping the course fresh for both students and instructors. We believe that the flexibility that we have built into the course increases inclusion and accessibility, but we did not explicitly use Universal Design for Learning (UDL; Super et al., 2021) principles, which is an area where the course could be improved.

Lastly, while the 4DEE framework is a very useful way to conceptualize the varying dimensions of ecological science, student outcomes need to be assessed, and that is something that we are currently working on with our online course. The 4DEE framework aims to enhance ecological literacy (Klemow et al., 2019; Rodgers et al., 2024), which would help close the science practice gap and support sustainability efforts. It is too soon to judge the success of 4DEE in meeting the goal of ecological literacy but this is an active area of inquiry in the ecological research community (e.g., Woods et al., 2025).

3.5 Extending the Scope of 4DEE for Sustainability Topics

Sustainability is included in the 4DEE framework as part of the human-environment interactions dimension, but sustainable development goals touch on all dimensions of ecology (Table 1). The topical lists governing the structure of 4DEE are a starting point, not an exclusive list. We therefore offer some ideas for additional topics, with particular interest to sustainability themes, that do not appear in official 4DEE documentation. Topics like interdisciplinarity (or transdisciplinarity), science policy, globalization, open science, and tipping points (non-linear feedbacks) might be particularly useful as cross-cutting themes in a sustainability context. Like other cross-cutting themes in 4DEE, these additional cross-cutting themes bridge disciplinary boundaries and serve as uniting concepts. Other topics that are not explicitly included in 4DEE, but would be appropriate to include and have utility in the context of sustainability education, include disease ecology (ecological concepts), agroecosystems (human-environment interactions), and spatial mapping (ecological practices).

4 Conclusion

This chapter presents a comprehensive case study of designing an online ecology course with the Four-Dimensional Ecology Education (4DEE) framework at Rowan University. The restructured online course includes themes of ecological literacy, sustainability, and interdisciplinary thinking and emphasizes inquiry-based learning, authentic assessments, accessibility, student engagement, and transferable skill development. The course successfully incorporates 4DEE's four dimensions of ecology despite inherent limitations of the online instructional modality.

This work demonstrates how ecology education can be effectively adapted for online environments and how 4DEE can be applied to curriculum development that supports SDGs. The 4DEE framework provides a useful tool for educators to integrate the dimensions of ecological science into their courses. A key attribute of a 4DEE activity or course is that all four dimensions are simultaneously (and we would argue, explicitly) explored, helping students to make connections and develop multidimensional systems thinking. Additionally, the four basic dimensions of 4DEE are easily translated to other disciplines (e.g., as disciplinary concepts, disciplinary practices, social dimensions of the subject, and cross-cutting themes), so the basic structure of 4DEE might be useful when applied to other disciplines.

References

Anderson, L.W. and Krathwohl, D.R., Eds. (2001) A taxonomy for learning, teaching and assessing: A revision of bloom's taxonomy of educational objectives. Complete Edition, Chap. 8–13, Longman, .

Ashraf, M. A., Yang, M. J., Zhang, Y. F., Denden, M., Tlili, A., Liu, J. Y., Huang, R. H., & Burgos, D. (2021). A systematic review of systematic reviews on blended learning: Trends, gaps, and future directions. *Psychology Research and Behavior Management, 14*, 1525–1541.

Bahnson, M., Godwin, A., Schunn, C., McChesney, E., & DeAngelo, L. (2025). Retention in engineering pathways: An ecological belonging intervention supports help-seeking and continued enrollment. *International Journal of STEM Education, 12*(1), 6.

Beltran, R. S., Marnocha, E., Race, A., Croll, D. A., Dayton, G. H., & Zavaleta, E. S. (2020). Field courses narrow demographic achievement gap in ecology and evolutionary biology. *Ecology and Evolution, 10*(12), 5184–5196.

Brundiers, K., Barth, M., Cebrián, G., Cohen, M., Diza, L., Doucette-Remington, S., Dripps, W., Habron, G., Harré, N., Jarchow, M., Losch, K., Michel, J., Mochizuki, Y., Rieckmann,

M., Parnell, R., Walker, P., & Zint, M. (2021). Key competencies in sustainability in higher education—Toward an agreed-upon reference framework. *Sustainability Science, 16*(1), 13–29.

Ecological Society of America. (2018). *4-dimensional ecology education framework*. Ecological Society of America. https://esa.org/4dee/

Fitzgerald, D. A., Scott, K. M., & Ryan, M. S. (2022). Blended and e-learning in pediatric education: Harnessing lessons learned from the COVID-19 pandemic. *European Journal of Pediatrics, 181*(2), 447–452.

Fong, C. J., Fathi, Z., Adelugba, S. F., Garćia, A. J., Garza, M., & Pinto, G. L. (2025). Belongingness moderates students mindsets' influence among first-generation, community college students of color. *Social Psychology of Education, 28*(1), 9.

Global Market Insights (GMI). 2023. E-Learning Market Size—By Technology, Provider, Application, Growth Forecast, 2023–2032. GMI215. Last accessed 6/2025. https://www.gminsights.com/industry-analysis/elearning-market-size

Husic, D. W. (2024). Reframing sustainability initiatives in higher education. *Sustainable Earth Reviews, 7*, 5.

Klemow, K., Berkowitz, A., Cid, C., & Middendorf, G. (2019). Improving ecological education through a four-dimensional framework. *Frontiers in Ecology and the Environment, 17*(2), 71.

Levin, R. C. (2024). Online learning & the transformation of global higher education. *Daedalus, 153*(2), 262–274.

Martins, F., Cezarino, L. O., Challco, G. C., Liboni, L., Dermeval, D., Bittencourt, I., Paiva, R., da Silva, A., Marques, L., Isotani, S., & Bittencourt, I. I. (2025). The role of hybrid learning in achieving the sustainable development goals. *Sustainable Development, 0*, 1–18.

Richards, J. H., Charton, K. T., McFarlane, S. L., Widell, A. F., Haddad, N. M., Wyer, M. B., & Damschen, E. I. (2025). Rethinking the undergraduate textbook as a tool to build a diverse community of ecologists. *Frontiers in Ecology and the Environment, 23*(3), e2819.

Rodgers, W. L., Scanga, S. E., St Juliana, J. R., Tietjen, E. S., Honea, J. M., Bryne, L. B., Leggett, Z. H., & Middendorf, G. (2024). Four-dimensional ecology education (4DEE) for everyone: Teaching ecology to non-majors. *Frontiers in Ecology and the Environment, 22*(6), e2749.

Ruhl, N., Crumrine, P., Oberle, J., Richmond, C., Thomas, S., & Wright, S. J. (2022). Harnessing the four-dimensional ecology education framework to redesign an introductory ecology course in a changing higher education landscape. *Ecosphere, 13*(2), e03857.

Santangelo, J.S., Ness, R.W., Cohan, B., Fitzpatrick, C.R., Innes, S.G., Koch, S., Miles, L.S., Munim, S., Peres-Neto, P.R. [...] & Johnson, M.T.J. (2022). Global urban environmental change drives adaptation in white clover. Science, 375(6586), 1275–1281.

Sorensen, A. E., Alred, A., Fontaine, J. J., & Dauer, J. M. (2024). A model-based instructional approach in a socio-ecological course-based undergraduate research experience (CURE). *Ecosphere, 15*(9), e70007.

Super, L., Hofmann, A., Leung, C., Ho, M. B., Harrower, E., Adreak, N., & Manesh, Z. R. (2021). Fostering equity, diversity, and inclusion in large, first-year classes: Using reflective practice questions to promote universal design for learning in ecology and evolution lessons. *Ecology and Evolution, 11*(8), 3464–3472.

Tomassi, A., Caforio, A., Romano, E., Lamponi, E., & Pollini, A. (2024). The development of a competence framework for environmental education complying with the European qualifications framework and the European green Deal. *Journal of Environmental Education, 55*(2), 153–179.

Wiek, A., Withcyombe, L., & Redman, L. (2011). Key competencies in sustainability: A reference framework for academic program development. *Sustainability Science, 6*(2), 203–218.

Woods, N. N., Leggett, Z. H., Spence, P. L., & Prevost, L. B. (2025). A 4DEE approach to evaluating ecological literacy in environmental science courses. *Journal of College Science Teaching, 54*(4), 345–352.

Nathan Ruhl (PhD) is an Associate Teaching Professor at Rowan University where he teaches a variety of ecology-oriented courses. Dr. Ruhl is an aquatic ecologist and eco-educator whose scholarship ranges across biological scales from behavioral studies of single species to whole-ecosystem monitoring. His teaching blends instructional techniques and modalities with a focus on improving ecology education and enhancing experiential learning with technology.

Sara J Wright (PhD) is an Associate Teaching Professor at Rowan University where she has a diverse teaching portfolio, including ecology, evolution, genetics, plant biology and data science courses. Dr. Wright emphasizes the important role of plants in society in her teaching and is particularly interested in developing pedagogies that support non-traditional and first-generation learners.

Serious Games as Pedagogical Practice(S) for Teaching Sustainability

Hélder S. Lopes and Virgínia Teles

Abstract Educating for sustainable development is one of the goals for achieving the Sustainable Development Goals (SDGs) by 2030, particularly in the implementation of its SDG 4—Quality Education, which aims to ensure inclusive, equitable quality education and promote lifelong learning opportunities for all. Acknowledging the importance of Education for Sustainable Development (ESD), there is an urgent need to transform the way we think and act. This implies a paradigm shift in education, especially by adopting innovative approaches that move away from traditional expository methods and foster a more active learning attitude among students. It also requires incorporating sustainability principles both in teaching practices and in institutional management. In an increasingly Volatile, Uncertain, Complex, and Ambiguous (VUCA) world, ESD represents an interdisciplinary field that seeks to empower individuals and societies to make informed and responsible decisions on environmental, social, and economic issues. The path toward sustainability, regardless of the many names it has assumed in recent decades, is the privileged battleground for those who do not give up on what is fair and necessary: that is, in essence, the challenge of education for sustainable development. This chapter explores the curricular potential of serious games as pedagogical tools for Education for Sustainable Development (ESD), based on an undergraduate course on Climate Change and Civil Protection at the University of Minho. The chapter presents a three-stage methodological approach (diagnosis, construction, and communication) that integrates climate vulnerability assessment, serious game design, and public engagement. During the 2023–2024 academic year, a pedagogical territorial intervention project was developed within a curricular unit on Climate Change and Civil Protection, part of two undergraduate programmes at the University of Minho. The activities took place in the municipalities of Felgueiras and V.N. de Famalicão, in the Northern Portugal,

H. S. Lopes (✉)
Lab2PT—Laboratory of Landscapes Heritage and Territory/IN2PAST—Associated Laboratory for Research and Innovation in Heritage, Arts, Sustainability and Territory, Department of Geography, Institute of Social Sciences, University of Minho, Campus de Azurém, Av. da Universidade, Guimarães, Portugal
e-mail: helderlopes@geografia.uminho.pt

V. Teles
CECS—Communication and Society Research Centre, Department of Geography, Institute of Social Sciences, University of Minho, Campus de Azurém, Av. da Universidade, Guimarães, Portugal
e-mail: vteles@ics.uminho.pt

W. Leal Filho et al., *Curriculum Development for Sustainable Development*, World Sustainability Series, https://doi.org/10.1007/978-3-032-22146-9_15

and integrated active learning methodologies based on problem- and project-based learning. The focus was on the impacts of Climate Change (CC), especially heatwaves. The methodology adopted followed a Diagnosis, Assessment, and Dissemination model, which included mapping territorial characteristics using environmental and socio-demographic indicators, assessing critical areas using SWOT analysis, and designing solutions adapted to the territories. Particularly noteworthy were the serious games, created by the students themselves and applied both in the classroom and with local and regional stakeholders, fostering information exchange and validation. These games were designed as tools to raise awareness and prepare for extreme events, facilitating learning about measures to adapt to and mitigate the effects of CC, in schools. Despite the successful implementation, it is acknowledged that future research is needed to evaluate the concrete impact of these practices on the sustainable behaviours of young people and local communities, as key agents of adaptation and change.

Keywords Education · Sustainability · Serious games · Pedagogical practices

1 Introduction

In an era marked by growing climate instability, deepening social inequalities, and heightened environmental risks, education has emerged as a critical lever for fostering sustainable transitions. The recognition of education as a central pillar in achieving the 2030 Agenda for Sustainable Development, particularly through SDG 4—Quality Education, underscores the urgency of rethinking how knowledge is produced, shared, and applied (UN, A/RES/70/1, 2015). Education for Sustainable Development (ESD) calls for transformative pedagogical practices that empower students to address complex socio-environmental challenges by cultivating critical thinking, civic engagement, and systems-based reasoning (Römhild, 2023). Among the active learning strategies increasingly embraced in this context, serious games have gained visibility as educational tools that integrate content knowledge with creative design and participatory engagement (Raybourn, 2014; Yamoul et al., 2024).

This chapter aims to contribute to curriculum development for sustainability by presenting a pedagogical intervention that integrates territorial analysis and game-based learning in higher education. The experience described is anchored in the TIP model (Territory–Intervention–Proposal) and aligns with SDG 4 (Quality Education) by promoting inclusive, contextualized, and transformative education practices.

This chapter presents a pedagogical intervention developed during the 2023–2024 academic year at the University of Minho, within the undergraduate course "Climate Change and Civil Protection". Designed around a three-stage methodology—diagnosis, construction, and communication—the intervention combined spatial analysis, vulnerability mapping, and the co-creation of serious games addressing heat-related risks in the municipalities of Felgueiras and Vila Nova de Famalicão, in northern Portugal. These two territories, characterized by distinct urban configurations and documented susceptibility to extreme heat events, provided the basis for an applied learning process rooted in territorial specificities and grounded in scientific and civic objectives. Students developed and tested game-based educational tools aimed at promoting awareness and preparedness for climate change impacts, particularly among younger audiences and local stakeholders. The serious games produced demonstrated originality, clarity, and potential for replication, while students reported gains in technical, communicative, and collaborative competencies.

Although the long-term effects on behavioral change remain to be assessed, this initiative illustrates how higher education can function as a catalyst for sustainability-oriented innovation. The chapter is structured as follows: it begins by situating the pedagogical approach within the broader framework of ESD and active learning, followed by a detailed account of the methodological process and local context (Sect. 3). It then presents the main results across diagnostic, pedagogical, and civic dimensions in Sect. 4, concluding with a reflection on the implications for replicability, partnerships, and future developments in sustainability education (Sect. 5).

2 Integration of Sustainability in Higher Education

The contemporary epoch is distinguished by a developmental paradigm that has given rise to substantial issues on a global scale, with profound and interdependent social and environmental ramifications. The phenomenon of these phenomena, which include hunger, extreme poverty, armed conflicts and the intensification of natural disasters, are caused by two main factors. Firstly, there is the degradation of the biosphere and the climate crisis that is being witnessed. Secondly, there is the worsening of social inequalities and the emergence of new forms of exclusion and social unrest, many of which are rooted in the dysfunctional dynamics of human interactions. These interactions include how we interact with each other and our relationship with our planet (UNRIC, 2016). In the opinion of Soromenho-Marques (2012, p. 103), the fact that the future of the world is uncertain in terms of knowledge does not exempt us from the ethical duty to safeguard the planet as a habitable home for future generations. This means that there is still a need for a transition to a sustainable development model. The sustainable development model was defined in 1987 by the Brundtland Report—'Our Common Future' (WCDE, 1987).

The concepts of sustainability and sustainable development have been defined and debated throughout a series of significant world congresses. Despite the presence of divergent and occasionally contradictory interpretations, these concepts presuppose that 'people, in the present and in the future, will achieve a satisfactory level of social and economic development, as well as human and cultural fulfilment, by using the planet's resources rationally, and by preserving biodiversity and natural habitats' (WCDE, 1987). It is evident that sustainable development is contingent upon the promotion of social justice, equity, eco-efficiency and social progress, in addition to the fundamental pillars of economy, society and environment. This is particularly pertinent within the context of the prevailing environmental and social crises, which exert a significant influence on contemporary societies. It is imperative to acknowledge the significance of territorial cohesion and regional solidarity at the national level (Soromenho Marques, 2012; Cachinho, 2012).

In terms of policies for sustainable development, a paradigm shift occurred with the ratification of the 2030 Agenda (*Transforming our world: the 2030 Agenda for Sustainable Development*), defined by the UN Summit on 25 September 2015, consisting of 17 Sustainable Development Goals (UN, A/RES/70/1, 2015). This agenda, the result of joint work between governments and citizens around the world, inspired by the progress made with the 8 Millennium Development Goals (UN, 2000), is bolder and promotes a global model that aims to end poverty, promote prosperity and well-being for all, protect the environment and combat climate change, and covers areas as diverse but interconnected as: equitable access to quality education and health services; the creation of decent jobs; energy and environmental sustainability; ocean conservation and management; the promotion of effective institutions and stable societies and the fight against inequality at all levels (UN, 2015).

Educating for sustainable development is one of the goals for realising the Sustainable Development Goals (SDGs) by 2030, particularly in terms of achieving SDG.4 - Quality Education, by guaranteeing access to inclusive, equitable quality education and promoting lifelong learning opportunities for all. In this sense, Education for Sustainable Development (ESD) will contribute to the formation of aware, critical and responsible citizens committed to a fairer and more sustainable society. It is therefore understood that a paradigm shift is needed in education (primary, secondary and university), especially in exploring innovative approaches from the definition of objectives and relevant content, to teaching practices that abandon expository methods and move towards a more active learning attitude on the part of the student, as well as including the principles of sustainability both in teaching practices and in the management of the institutions themselves. Trends show that technology, interdisciplinarity and active learning are determining factors in the implementation of ESD. Teaching is not the same as it was a decade ago; technological advances have profoundly changed the way people learn, access and produce knowledge. Education for the future will follow this path, with schools having to reinvent themselves and adopt creative and innovative teaching practices. These include problem-based and project-based approaches, which make learning more active, autonomous and meaningful. This stimulates critical thinking and fosters education for sustainability. In the future, education will follow this path. Schools will have to reinvent themselves by incorporating creative and innovative teaching practices, such as problem-based or project-based approaches. These approaches make learning more active, autonomous and meaningful. They stimulate students' critical thinking and foster education for sustainability.

This article aims to present and reflect on a pedagogical intervention based on serious games in the context of university teaching and it is interesting to see that serious games, in addition to their playful component, appear as active methodologies capable of solving real problems, contributing to the construction of knowledge, but also favouring the development of cognitive, affective and social skills, in other words, the use of serious games has contributed to greater involvement and transformation in students' attitudes and behaviours, committing them to sustainability.

3 Methods and Data

The pedagogical intervention carried out during the 2023–2024 academic year (namely in the second semester, between february and june of 2024), as part of the undergraduate course *Climate Change and Civil Protection* at the University of Minho was developed within a framework of educational transformation aligned with the goals of Education for Sustainable Development (ESD). This curricular unit, part of two undergraduate programs, was structured to combine scientific knowledge on climate change with active, hands-on pedagogical practices, which had materialized through the creation of serious games developed by the students themselves.

The intervention involved 48 undergraduate students, organized into 8 groups with maximum of 5 students. Weekly sessions were complemented by independent group work and regular formative feedback moments. A total of 30 contact hours were dedicated to project development, including workshops on data analysis using Geographic Information Systems (GIS) and Analytic Hierarchy Process (AHP), vulnerability mapping, and educational game design. Each group submitted an interim plan and a final descriptive report. Academic supervision was continuous, with emphasis on collaborative learning and peer review. Evaluation was rubric-based, assessing scientific accuracy, game functionality, creativity, and societal impact. The approach followed constructivist and participatory principles, respecting ethical guidelines for authorship and dissemination of student-produced content.

3.1 Case Study: Municipalities of Felgueiras and Vila Nova de Famalicão

The intervention took place in two municipalities in Northern Portugal: **Felgueiras** and **Vila Nova de Famalicão** (Fig. 1), with distinct urban and social characteristics, both affected by vulnerabilities to extreme heat. Felgueiras is marked by dispersed urbanization, significant industrial areas, and high pressure on natural resources, including river systems and green spaces. Vila Nova de Famalicão, in contrast, has a denser urban core, with compact residential neighborhoods that are increasingly exposed to impervious surfaces, low tree canopy cover, and high thermal stress in summer months.

The municipalities of Felgueiras and Vila Nova de Famalicão, located in northern Portugal, belong to the Tâmega e Sousa and Ave NUTS III sub-regions, respectively. According to the most recent population data from Statistics Portugal (INE, 2022), Felgueiras has 57,342 inhabitants, while Vila Nova de Famalicão has 132,801 residents.

Both municipalities are experiencing warming trends consistent with regional and national climate scenarios. In Felgueiras, observational data from the Luzim meteorological station confirms a steady increase in the number of very hot days

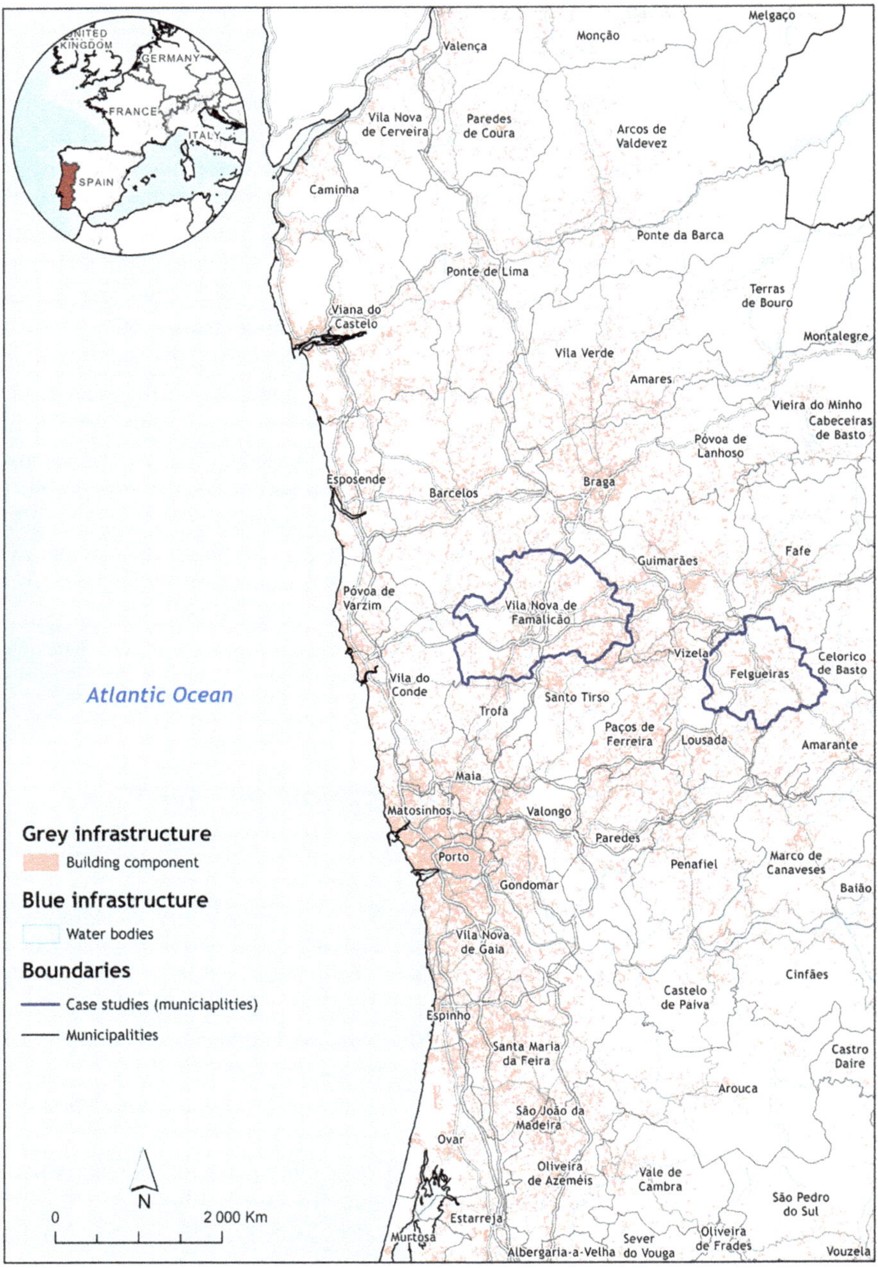

Fig. 1 Case study—municipalities of Felgueiras and Vila Nova de Famalicão. *Source* Authors' own elaboration

(Tmax >35 °C) and tropical nights (Tmin >20 °C), particularly during the June–October period. These trends are supported by projections indicating a rise in average and maximum summer temperatures by +1.1 – +1.6 °C. The Municipal Climate Adaptation Plan of Felgueiras (PMAC-Felgueiras; Município de Felgueiras, 2024) highlights these phenomena and classifies the municipality's susceptibility to heatwaves as moderate to high, especially within the broader Tâmega e Sousa sub-region (Município de Felgueiras, 2024).

Similarly, Vila Nova de Famalicão shows a significant intensification of summer heat, with an increase of 2.1 very hot days per decade and 2.3 tropical nights per decade. The number of frost-threshold days (Tmin <5 °C) is declining by approximately 7.9 days per decade. These thermal changes are acknowledged in the Municipal Climate Adaptation Plan of Vila Nova de Famalicão (PMAC-VNF; Município de Vila Nova de Famalicão, 2024), which underscores the municipality's exposure to urban heat stress and the need for tree canopy expansion, improved public space design, and energy efficiency measures (Município de Vila Nova de Famalicão, 2024). Table 1 shows the main climatic characteristics of the municipalities.

Table 1 Main climatic characteristics of the municipalities of Vila Nova de Famalicão and Felgueiras

Indicator	Municipalities	
	Vila Nova de Famalicão	Felgueiras
Minimum temperatures < 5 °C	Decrease of **7.9 days per decade**	Not specified
Maximum temperatures ≥ 30 °C	Increasing number of days per year	Monthly and annual increase observed (IPMA data for Luzim)
Maximum temperatures > 35 °C	Average of **5 days/year**; increasing by **2.1 days/decade**	Increasing number of very hot days (especially June–October)
Tropical nights (min. temp > 20 °C)	Increasing by **2.3 nights per decade**	Expected increase in summer (linked to higher heatwave risk)
Heatwaves	Not directly assessed	Becoming more frequent and intense; region classified as **moderate to high susceptibility**
Seasonal changes	Slight temperature rise in **spring and summer**; fewer cool days (<20 °C) in spring (–5 days/decade)	Strongest warming projected for **summer and autumn**; maximum temperatures to rise by **1.1–1.6 °C**
Regional data availability	Local temperature trends well documented	Based on Luzim station; limited official data on heatwaves, but temperature metrics confirm trends

Source Authors' own elaboration, based of PMAC Felgueiras and PMAC VNF (Município de Felgueiras, 2024; Município de Vila Nova de Famalicão, 2024)

These municipalities illustrate the broader climatic transition occurring in northern Portugal (from a temperate Atlantic to a warmer, Mediterranean-like regime), highlighting the urgent need for locally tailored adaptation policies that consider urban density, land-use patterns, and socio-environmental vulnerabilities.

In both municipalities, **heatwaves** were selected as the primary environmental risk under analysis, given their rising frequency and intensity and their critical impacts on public health, urban-rural comfort, and energy demand. The project aimed to diagnose, communicate, and contribute to the mitigation of this risk through a pedagogical strategy combining territorial analysis and the design of educational tools.

3.2 Framework in Three-Stage Model

The methodological framework adopted in this project aligns with the TIP model (Territory—Interventions - Proposal) proposed by Lopes et al. (2022) and Lopes (2024), which emphasizes the integration of place-based learning, social engagement, and action-oriented pedagogy to address climate risks through co-constructed knowledge and educational tools. The three-stage model—Overcoming Preconceptions, Reasoned Construction, and Verification Through Facts—was designed to support students' progressive development of autonomy, critical thinking, and civic engagement, grounded in the concrete challenges of their local territories. These dimensions implemented in three stages were synthetized in Table 2.

The first stage, Overcoming Preconceptions, involved defining the project's research object and goals, selecting priority intervention areas, and formulating detailed action plans. At this stage, students conducted an extensive literature review and consulted technical documents on climate change, urban vulnerability, and environmental education, including sources such as Portuguese Institute for Ocean and Sea (IPMA), Statistics of Portugal (INE), and local spatial and climate adaptation plans (PMAC).

The AHP method was introduced, enabling students to construct pairwise comparison matrices and critically weight vulnerability factors.

Fieldwork enabled direct observation and qualitative assessment. Students produced thematic maps and conducted SWOT analyses to identify territorial strengths, weaknesses, opportunities, and threats. This stage aimed to deconstruct intuitive or superficial understandings of local risks and stimulate analytical reasoning, as suggested by the "diagnostic dimension" of the TIP framework (Lopes et al., 2022).

The second stage, Reasoned Construction, focused on the creation of serious games by student groups. These games served as communicative and educational devices aimed at disseminating territorial diagnostics and raising awareness about climate adaptation. Following the TIP model's emphasis on collaborative and action-based learning, students defined target audiences (e.g., primary school pupils, families, youth), set pedagogical goals, designed rules and components, and developed low-cost, physical prototypes. In Felgueiras, game scenarios addressed the scarcity

Table 2 Three-stage framework developed with the students of University of Minho in UC climate change and civil protection

Section	Sub-section	Content
PART I ǀ Building through reason – diagnosis, characterisation and climate projections	A. Data collection, observation, priority setting, and surface temperature calculation	1. Selection and territorial description of the case-study municipality (V.N. Famalicão or Felgueiras)
		2. Analysis of summary indicators using official statistics, reports, and remote sensing data
		3. Use of GIS tools to identify Urban Heat Island (UHI) and model key problems
		4. Calculation of Land Surface Temperature (LST) for summer date (cloud cover $\leq$10%) using LANDSAT 8 or 9
	B. Determination of risk levels in the study area	1. Identification of variables for risk analysis – hazard, exposure, and vulnerability (susceptibility + adaptive capacity)
		2. Application of Analytic Hierarchy Process (AHP) methodology to support weighting of vulnerability variables
		3. Summary identification of risk assessment methodology for heat (optional use of flowchart)
		4. Assessment of extreme heat risk in the study area
PART II ǀ Verifying through reason – serious game: design, development and application in extreme heat risk study	A. Conceptualisation and development of the serious game based on Part I structure	1. Planning: Clear definition of game objectives, target audience, and specific needs
		2. Design: Game structure including selection of cards, boards, tokens, images, and rules
		3. Development: Game prototype creation and rule definition, followed by testing

(continued)

Table 2 (continued)

Section	Sub-section	Content
		4. Testing and adjustments: User testing and feedback collection for game improvement
		5. Production of the final game version
PART III I Overcoming (PRE)conceptions – communicating results and compiling final report	A. Writing a clear descriptive report demonstrating synthesis and reasoning	1. Formal organisation of content (introduction, sources, methods, results, choices, references)
		2. Structured introduction: a) thematic context; b) study object; c) application site; d) brief method; e) results overview; f) societal implications; g) structure outline
		3. Description of methods and data used in Phase A
		4. Description of Phase B methods, including game design and functionality
		5. Connection between Phase B game development and Phase A outcomes
		6. Chronological and spatial organisation of information and result exploration
		7. Writing and visual layout of game instruction manual (explanation of all components)
		8. Clear conclusions: limitations, relevance to degree, societal impact, and future directions

Source Authors' own elaboration.

of urban shading and the ecological recovery of degraded public spaces. In Vila Nova de Famalicão, the focus was on heat stress in dense neighbourhoods and insufficient climate-resilient infrastructure. Each prototype was pre-tested with family members and occasionally local government staff, providing critical feedback on clarity, usability, and educational potential, a process akin to the TIP model's social experimentation dimension (Lopes, 2024).

The third stage, Verification Through Facts, culminated in the production of a comprehensive descriptive report by each student group. These reports integrated the project's context, methodological trajectory, territorial diagnosis, and game development process. They also included user manuals with clear instructions, game components, and suggestions for pedagogical integration into school curricula or community workshops. This stage aligns with the "reflexive appropriation" pillar of the TIP model, reinforcing critical analysis and the consolidation of civic competencies through structured outputs (Lopes et al., 2022).

By applying this three-stage structure, the project successfully merged academic learning with situated social responsibility, enabling students to act as knowledge co-producers and communicators in the face of local climate risks.

A rubric-based formative assessment was used to evaluate each group's work, focusing on (1) scientific accuracy, (2) game functionality and pedagogical clarity, (3) creativity, and (4) societal relevance. Peer assessment and self-assessment were also integrated to promote reflective learning.

4 Educational Intervention with Serious Games—Main Results

The results of this educational intervention unfold across four interrelated dimensions: territorial diagnosis, pedagogical product development, transversal skill acquisition, and replicability of the methodology. This multi-dimensional outcome aligns with the pedagogical orientation proposed by the TIP model (Lopes et al., 2022; Lopes, 2024), which emphasizes the co-construction of knowledge in response to real-world risks, and is consistent with the principles of Education for Sustainable Development (ESD), particularly in fostering critical reflection, civic engagement, and agency among learners (Schmitz, 2022). In a world increasingly characterized by volatility, uncertainty, complexity and ambiguity (VUCA), such learning experiences are crucial to preparing students to navigate dynamic socio-environmental challenges (Ryazanova et al., 2020; Sum, 2022).

The territorial analysis carried out by students revealed important spatial patterns of heat-related vulnerability in the municipalities of Vila Nova de Famalicão and Felgueiras. These analyses were conducted using official demographic data from INE and BGRI, combined with satellite imagery from Landsat 8 and 9 and geoprocessing techniques implemented in ArcGIS, following the proposal of Lopes et al. (2023, 2025). Students applied an AHP approach to structure their assessments, incorporating hazard, exposure, sensitivity and adaptive capacity dimensions to produce detailed thematic maps. In Felgueiras, areas of critical concern included industrial peripheries adjacent to densely populated neighbourhoods, where the lack of tree canopy and high surface temperatures, along with social vulnerability indicators such as an ageing population and outdated housing, created significant cumulative risk. In Vila Nova de Famalicão, the spatial configuration of risk was different:

residential expansions from the late twentieth century showed high impermeability and low green infrastructure, resulting in urban microclimates with limited summer thermal comfort.

The process of variable selection within the AHP method reflected both convergence and diversity in the analytical perspectives of student groups. All groups selected Land Surface Temperature (LST) as the main hazard indicator, using a consistent Landsat-based workflow that included radiance conversion, brightness temperature, and emissivity correction. One Felgueiras group of students supplemented this with official IPMA temperature data. For exposure, population density was a near-universal choice, with two groups in Felgueiras incorporating refined overlays of residential land use. Sensitivity indicators, especially the proportion of residents aged over 65 or under 15, were common across projects, while metrics such as building age, surface impermeability, and verticality (e.g., average storey height) appeared more selectively. Adaptive capacity indicators were typically chosen for their clarity and relevance - educational attainment, and proximity to health services or shaded green spaces were recurrent. These mapping exercises, supported by critical reflection, allowed students to spatialize abstract climate vulnerabilities and thereby engage with the complexities of local risk geographies (Table 3).

These spatial analyses were then translated into pedagogical outputs through the development of serious games, a process grounded in the co-creative and action-oriented ethos of the TIP model. Each group designed a game directly informed by the vulnerabilities identified in their municipality, seeking to communicate adaptation needs in an accessible and engaging way (see the list in Table 4). Examples included *Acerta ou Queima [Hit or Burn]*, a knowledge-based quiz game linking correct answers to thermal safety, and *Calor Capital*, a resource management game where players simulate interventions to mitigate Urban Heat Islands (UHI). Other games, like *Jogo das Alterações Climáticas [Climate Change Game]*, integrated role-playing dynamics to encourage debate and decision-making among fictional actors (*e.g.*, citizens, NGOs, planners). In other serious games, students responded to urban density issues with formats such as *Jenga Climática [Climate Jenga]*, where each block represents a structural or environmental stressor, and *Cartas contra calor extremo [Cards Against Extreme Heat]*, a card-based game promoting collective neighbourhood resilience. Additional titles like *PlanetQuest* and *Planeta em Ação [Action Planet]* encouraged strategic thinking and teamwork in developing adaptation pathways.

Each game was accompanied by a printed manual that included pedagogical objectives, detailed instructions, component lists, and proposed usage scenarios. These materials were designed to be visually accessible and scientifically accurate, enabling use in formal classroom settings or informal community events. The process of game creation involved iterative design, peer feedback, and physical prototyping, allowing students to refine both mechanics and educational messaging. The act of transforming complex climate data into interactive narratives required a balance between scientific rigor and creative storytelling, reinforcing ESD's emphasis on interdisciplinary thinking and communication.

Table 3 Risk dimensions following by the students in the mapping exercise—Stage 1: building through reason

Risk dimension		Variable	Groups using It	Municipalities	Justification
Hazard		Land-surface temperature (°C)	8	Famalicão, Felgueiras	Used as the primary proxy for heat hazard due to spatial precision and availability from Landsat imagery; chosen by all groups
		Mean air temperature (IPMA)	1	Felgueiras	Included by one group to complement surface temperature with official meteorological data from Luzim station
Exposure		Population density (Instituto Nacional de Estatística, 2022)	7	Famalicão, Felgueiras	Used to represent the number of people potentially exposed to extreme heat; chosen for its high spatial resolution
		Residential and Industrial land use	2	Felgueiras	Used to better isolate residential exposure areas by overlaying CORINE and COS land use
		Working population	1	Felgueiras	Used to reflect exposure of workers in industrial zones during peak heat hours
Vulnerability	Susceptibility	Population aged ≥ 65	8	Famalicão, Felgueiras	Used as a core indicator of heat vulnerability based on age-related physiological sensitivity

(continued)

Table 3 (continued)

Risk dimension		Variable	Groups using It	Municipalities	Justification
	Susceptibility	Population aged 0–14	7	Famalicão, Felgueiras	Infants identified as highly sensitive to heat; used in combination with elderly data
	Susceptibility	Buildings pre-1919 / pre-1945	4	Famalicão, Felgueiras	Used to capture structural vulnerability of older buildings lacking insulation
	Susceptibility	Buildings 1971–1980	3	Famalicão	Referenced as a thermally inefficient concrete construction era by Famalicão teams
	Susceptibility	Impervious surface (250 m buffer)	2	Felgueiras	Used as a proxy for local surface heat retention, especially in dense urban zones
	Susceptibility	Average storey height	1	Famalicão	Proposed to estimate thermal load and ventilation inefficiency in tall blocks
	Adaptive capacity	% dwellings with piped water	4	Famalicão, Felgueiras	Chosen as a basic indicator of housing adequacy and ability to maintain hydration
	Adaptive capacity	% dwellings with mansonary	4	Famalicão, Felgueiras	Used to reflect minimum comfort and hygiene conditions relevant to heat exposure
	Adaptive capacity	% population with tertiary education	4	Famalicão, Felgueiras	Selected to represent social awareness and knowledge-based resilience

(continued)

Table 3 (continued)

Risk dimension		Variable	Groups using It	Municipalities	Justification
	Adaptive capacity	Distance to health facility (<1 km)	3	Famalicão, Felgueiras	Used to assess accessibility to emergency care during heat-related health episodes
	Adaptive capacity	Proximity to shaded green space (<300 m)	2	Felgueiras	Used to reflect potential access to cooling refuges in the urban fabric
	Adaptive capacity	Municipal social support index (CSS)	1	Famalicão	Included as a synthetic vulnerability score reflecting social dependency and income

Source Authors' own elaboration

In parallel with game development, students demonstrated meaningful acquisition of transversal competencies. They developed technical skills in GIS and data analysis, alongside soft skills such as collaborative planning, critical synthesis, and persuasive communication. These competencies are particularly relevant in the context of ESD and essential for navigating VUCA environments, where solutions require adaptive, systems-based and participatory approaches (Sutil-Martín & Otamendi, 2021). The experience pushed students beyond traditional academic tasks, placing them in the role of knowledge mediators capable of engaging non-specialist audiences.

This civic orientation culminated in a public dissemination event on June 4, 2024, at the University of Minho (Fig. 2). Entitled A Câmara Municipal na UMinho—Felgueiras, the session brought together municipal officials, academic staff and community members for a showcase of the developed games. Students presented their work in a public forum, reinforcing their sense of authorship and relevance. Municipal representatives, particularly from Felgueiras City Council, expressed interest in adapting the games for use in local schools and awareness campaigns, highlighting the potential for real-world impact. Community feedback further validated the games' clarity, visual design and originality.

Feedback from attendees was encouraging with this type of initiative during the presentation, analysis and exibhition (Fig. 3).

Municipal staff highlighted the potential applicability of the games in schools and public education programmes. Several expressed interest in further collaboration with the university to adapt the materials for wider dissemination. Parents and community

Table 4 Serious games developed with student groups during the course of climate change and civil protection

Game title	Municipality	Typology	Target audience	Implementation
Acerta ou Queima [Hit or Burn]	Felgueiras	Quiz board game	Primary and lower secondary students	Teams answer climate-related questions; incorrect answers trigger 'heat penalties', linking knowledge and vulnerability
Calor capital [Capital Heat]	Felgueiras	Simulation game with tokens and resource management	Students and urban planning stakeholders	Players simulate interventions in an urban heat island context. Focus on trees, water, and public participation
Jogo das Alterações Climáticas [Climate Change Game]	Felgueiras	Board game with role-play	Students aged 10–14	Players assume roles (citizens, planners, NGOs) and make decisions about urban adaptation. Encourages discussion and strategy
Jenga Climática – Calor Extremo [Climate Jenga – Extreme Heat]	V. N. Famalicão	Modified Jenga game	Teenagers and families	Each block removed includes climate facts or challenges. Instability represents urban vulnerability to heatwaves
Cartas contra calor extremo [Cards Against Extreme Heat]	V. N. Famalicão	Card game	Youth and community groups	Card-based game highlighting risk situations and adaptive strategies. Players build 'resilient neighborhoods' over rounds
Caminhada Climática [Geographical Walk]	Felgueiras	Path-based game	School groups	Participants 'walk' a circuit with stations presenting climate challenges and actions
PlanetQuest	V. N. Famalicão	Adventure strategy board game	Upper secondary and general public	Players travel through different neighborhoods, managing resources and planning climate solutions. Promotes systems thinking
Planeta em Ação [Action Planet]	V. N. Famalicão	Action-based board game	Children aged 8–12	Players perform sustainable actions to cool down cities. Rewards teamwork and eco-behaviors through challenges

Source Authors' own elaboration.

Fig. 2 Public dissemination in workshop entitled "A Câmara Municipal na UMinho... Felgueiras". *Source* Poster produced by authors for public dissemination of the workshop

Fig. 3 Records of the dissemination workshop. (A) Presentation of practical projects by the students; (B) Feedback on the serious games from local government entities; (C) Exhibition of the serious games. *Source* Photos taken by authors, in June fourth, 2024

members who participated in preliminary test rounds praised the clarity and visual quality of the materials, as well as the originality of the gameplay mechanics.

While no large-scale implementation has yet occurred, the games are fully documented and ready for adaptation. Their modular, low-cost structure allows customization for different age groups, educational levels, or even other climate-related risks. More importantly, the integrated methodology—linking spatial analysis, participatory design and civic education—demonstrates a replicable model for fostering climate literacy and local engagement. By merging diagnosis with action and communication, this intervention exemplified how climate education can move beyond knowledge transmission toward the development of informed, empowered and active citizens.

5 Conclusion

The pedagogical intervention implemented during the 2023–2024 academic year clearly demonstrated that higher education institutions can act as transformative agents in promoting sustainability through innovation in teaching practices. By integrating territorial diagnostics with game-based learning, the project offered students a rich, place-based educational experience, rooted in real-world climate challenges and oriented toward public engagement. This approach reflects the principles of Education for Sustainable Development (ESD), aligning learning objectives with critical competencies such as systems thinking, collaboration, and action-oriented reflection (Schmitz, 2022).

The structured three-stage methodology-diagnosis, construction, and communication -proved effective in mobilizing student agency. It enabled participants to move from abstract environmental knowledge to the co-creation of concrete, playable tools with societal relevance. Serious games emerged not only as vehicles of environmental literacy, but also as platforms for social dialogue, capable of engaging diverse audiences in understanding and addressing climate risks, particularly those associated with extreme heat in Felgueiras and Vila Nova de Famalicão.

The chapter also contributes to curriculum development for sustainability by offering a replicable educational model that integrates spatial analysis, participatory design, and civic engagement. This methodological triangulation enabled students to become active co-creators of knowledge and agents of change, while promoting curriculum transformation within ESD.

Although formal partnerships with schools and municipalities are still being consolidated, the project succeeded in producing robust educational resources that are ready for testing and dissemination. The active involvement of students in the co-design process contributed to a sense of ownership and civic responsibility, positioning them as co-authors of climate resilience narratives. In this way, the university fulfilled a critical role as a hub for innovation, public participation, and knowledge transfer, key dimensions in any effective sustainability strategy.

This project reaffirmed that sustainability education must transcend theoretical instruction. When students become designers of learning tools aimed at their own communities, they gain technical and communicative skills and embody the ethical dimension of ESD: transforming knowledge into collective action. Serious games, when embedded in territorial realities and pedagogical structures, can turn classrooms into laboratories of resilience and students into agents of sustainable transformation.

Limitations and future directions for sustainability education.

Despite its promising outcomes, the project faces several limitations that must be acknowledged. The most critical is the absence of systematic implementation in school settings, which prevents a full pedagogical assessment of the games' impact on learning outcomes. Moreover, while the games were publicly presented, structured evaluation from educators, municipal partners, and learners remains limited, which constrains the capacity to improve and scale the initiative based on evidence.

Future directions for sustainability education inspired by this experience can be structured into short, medium, and long-term actions:

- Short-term (next academic year).

1. Implement pilot sessions of the serious games in local primary and secondary schools, in coordination with municipal education departments.
2. Conduct teacher training workshops to support the integration of these tools into classroom practice.
3. Develop basic feedback instruments (surveys, observation rubrics) to assess student engagement and learning outcomes.

- Medium-term (next 3 years).

1. Establish formal school–university partnerships for curricular integration of serious games within ESD programs.
2. Create an open-access digital repository hosting manuals, templates, and pedagogical resources for replication.
3. Promote interdisciplinary research on the efficacy of serious games in ESD through academic-public collaborations.
4. Foster student-led initiatives for climate literacy in collaboration with civil society actors and local governance.

- Long-term (until 5 years).

1. Develop a Climate Education and Serious Games Hub based at the university, dedicated to research, innovation, and outreach.
2. Scale the model nationally and internationally by partnering with networks focused on ESD, resilience education, and digital pedagogy.
3. Advocate for the inclusion of game-based learning as a recognized strategy within national and regional climate education policies.

4. Evaluate long-term behavioral and civic impacts of game-based sustainability interventions across age groups and territories.

The broader goal is to consolidate a replicable and evolving educational model that positions learners not only as receivers of knowledge, but as contributors to sustainable futures. By advancing from isolated interventions to systematized strategies, the project opens a pathway toward institutionalizing climate literacy and participatory learning at scale.

References

H Cachinho. (2012). A EDS no Currículo Nacional [ESD in the National Curriculum]. In *Painel III—A EDS no Currículo e nas Práticas Escolares: Fundamentos e Investigação. Actas do Seminário "Educação Para o Desenvolvimento Sustentável, Audição sobre Educação Para o Risco"* [panel III—ESD in the curriculum and school practices: Foundations and research. Proceedings of the seminar "education for sustainable development, hearing on risk education"] (pp. 157–179). Lisboa: Lisboa 2011. https://www.cnedu.pt/content/antigo/files/pub/EducDesenvSustent/EducDesenvSustent.pdf. Accessed 4 Jul 2025.

Instituto Nacional de Estatística (2022). Censos 2021—Resultados definitivos [2021 Cenus—Final Results]. https://www.ine.pt/xurl/pub/65586079. Accessed 13 July 2025.

Lopes, H. S. (2024). Educational sustainability initiatives in higher education: An integrative approach to urban areas in northwestern Portugal. *Societal Impacts, 4*, 100093. https://doi.org/10.1016/j.socimp.2024.100093

Lopes, H. S., Remoaldo, P., Ribeiro, V., & Martín-Vide, J. (2022). The use of collaborative practices for climate change adaptation in the tourism sector until 2040—A case study in the Porto metropolitan area (Portugal). *Applied Sciences, 12*(12), 5835. https://doi.org/10.3390/app12125835

Lopes, H., Remoaldo, P., Ribeiro, V., & Martín-Vide, J. (2023). The impacts of climate change on human wellbeing in the municipality of Porto—An analysis based on remote sensing. In *Climate change and health hazards: Addressing hazards to human and environmental health from a changing climate* (pp. 135–172). Springer Nature Switzerland. https://doi.org/10.1007/978-3-031-26592-1_8

Lopes, H. S., Silva, P. F., Pinto, B. G., Cuevas, P. D., Ribeiro, V., & Remoaldo, P. (2025). Urban heat stress in the context of socioeconomic and environmental challenges: Heat risk analysis and online surveys in northwestern Portugal. *International Journal of Disaster Risk Reduction, 120*(10538), 105384. https://doi.org/10.1016/j.ijdrr.2025.105384

M. Schmitz (2022). Sustainable futures of business–future studies meets sustainable management education. In Shaping a sustainable future (pp. 87–108). Nomos Verlagsgesellschaft mbH & KG. https://www.nomos-elibrary.de/10.5771/9783748933090-87.pdf.

Município de Felgueiras (2024). *Plano Municipal de Ação Climática de Felgueiras (PMAC-Felgueiras)* [Felgueiras Municipal Climate Action Plan (PMAC-Felgueiras)]. Município de Felgueiras. https://cm-felgueiras.pt/viver/ambiente-e-acao-climatica/. Accessed 4 July 2025.

Município de Vila Nova de Famalicão (2024). Plano Municipal de Ação Climática de Vila Nova de Famalicão (PMAC-VNF) [Município de Vila Nova de Famalicão. Vila Nova de Famalicão Municipal Climate Action Plan (PMAC-VNF)]. https://www.famalicao.pt/acao-climatica. Accessed 4 Jul 2025.

Raybourn, E. M. (2014). A new paradigm for serious games: Transmedia learning for more effective training and education. *Journal of Computational Science, 5*(3), 471–481. https://doi.org/10.1016/j.jocs.2013.08.005

Ryazanova, N., Dikova, T., & Semak, A. (2020). The potential of the agile technology application in university education. In *7th international conference on education, language, art and intercultural communication (ICELAIC 2020)* (pp. 6–12). Atlantis Press. https://doi.org/10.2991/assehr.k.201215.292

Römhild, R. (2023). Learning languages of hope and advocacy-human rights perspectives in language education for sustainable development. *Human Rights Education Review, 6*(1), 9–29.

Soromenho-Marques, V. (2012). Alterações climáticas-a verdadeira crise do século XXI. *JANUS.NET, 3*(2), 165–168.

Sum, N. (2022). School leaders' perceptions of their roles during the pandemic: An Australian case study exploring volatility, uncertainty, complexity and ambiguity (VUCA leadership). *School Leadership & Management, 42*(2), 188–207. https://doi.org/10.1080/13632434.2022.2045268

Sutil-Martín, D. L., & Otamendi, F. J. (2021). Soft skills training program based on serious games. *Sustainability, 13*(15), 8582. https://doi.org/10.3390/su13158582

UN (2015). General Assembly—Resolution A/RES/70/1. https://docs.un.org/en/A/RES/70/1 (Accessed 14 Jul 2025).

UN (2000). *United Nations Millennium Declaration.* https://www.ohchr.org/en/instruments-mechanisms/instruments/united-nations-millennium-declaration. Accessed September 5, 2026.

UN (2015). *The Millennium Development Goals Report.* United Nations.

Yamoul, S., Ouchaouka, L., & Moussetad, M. (2024). Towards a wider adoption of serious games in higher education: Deepening the analysis of benefits and challenges. *International Journal of Information Science and Technology, 8*(4), 16–25. https://doi.org/10.57675/IMIST.PRSM/ijist-v8i4.284

Hélder Silva Lopes holds a degree in Geography and Planning [with a grade of 17/20 (ECTS A)], a master's degree in Geography—Spatial Planning and Management [with the maximum grade of 20/20 (ECTS A)] and a PhD in Geography and Regional Planning from the University of Minho. Also, a PhD degree in Geography, Spatial Planning and Environmental Change, from the University of Barcelona. Throughout his educational and professional career, he has received 10 scientific awards. He is an invited assistant professor at the Department of Geography at the University of Minho and ESEPF—Higher Education School of Paula Frassinetti. He is also a contributor to Meteored—tempo.pt., where he writes weekly about environment and climate change. He is an integrated researcher at the Lab2PT (Laboratory of Landscapes, Heritage, and Territory)/IN2PAST (Associated Laboratory for Research and Innovation in Heritage, Arts, Sustainability, and Territory) and member of the Compor Mundos Network - Humanities, Well-being, and Health in the twenty-first Century, of the Fernando Pessoa Foundation. His main research interests are Climate Change, Environmental Perception, Citizen Science, Nature-based Solutions and Geostatistics. He has published several articles in international journals indexed in WoS and SCOPUS, as well as book chapters in international publishing companies. He has carried out extracurricular training at national and international level to update his skills in some international university institutions in the fields of health, urban planning, and qualitative and quantitative methods for data analysis.

Virgínia Teles is Assistant Professor at the Department of Geography, Institute of Social Sciences, University of Minho. She is a full member of CECS (Communication and Society Research Centre). With a 30-year academic career, she has developed work in the field of Geomorphology, particularly Karst and Fluvial Geomorphology; Natural Hazards in Urban Environments, especially in mapping flood susceptibility and vulnerability; Risk Perception; Climate Change and Spatial Planning; as well as on Geomorphological Heritage and new territorial interpretations through virtual territories (geocaching). She was a member of the National Geography Commission and was a part of the board of the Portuguese Association of Geomorphologists (APGeom). She is currently the director of the first cycle of the degree in geography and planning at Department of Geography at the University of Minho.

Sustainable Development in Practice: Applying a Learner-Centred Approach to Sustainable Development e-Curricula

Michael Stanley-Jones and Claire Egehiza Obote

Abstract *Poverty-Environment Action for Sustainable Development Goals*, a joint initiative of the United Nations Development Programme (UNDP) and the United Nations Environment Programme (UNEP), sought to place environmental sustainability, inclusivity, and poverty reduction at the heart of the sustainable development agenda. Poverty-Environment Action incorporated two decades of experience in implementing measures to address the many challenges that arise from the interaction between poverty and environmental problems. This 'poverty-environment nexus' was identified and addressed in more than 30 Least Developed Countries. Having brought together this rich experience in their 2023 *Sustainable Development in Practice* handbook, UNDP and UNEP created an interactive, multimedia platform based on the handbook and other resources to provide continuous learning opportunities for poverty-environment practitioners and secure the legacy of Poverty-Environment Action. Key components of the platform are two e-learning courses, *Introduction to Sustainable Development in Practice* and *Advancing Sustainable Development in Practice*, that can be easily adapted into curriculum modules or used as teaching resources. Adopting the Analyze, Design, Develop, Implement, and Evaluate (ADDIE) model to distance learning, the e-learning courses stimulated numerous poverty-environment actions in communities that the original programmes could not reach through conventional capacity-building activities, while supporting practitioners in fostering integrated thinking and real-world application of the Sustainable Development Goals. The 2023 handbook and e-learning courses have been recommended as resource materials in the UNEP Poverty and Inequality Guidance. The chapter reviews the development of Poverty-Environment Action capacity development materials incorporated into two UN CC e-Learn courses. It explores how the limitations distance learning imposes on learner-centred education

M. Stanley-Jones (✉)
Governance, Security and Justice Program, United Nations University Institute for Water, Environment and Health, Richmond Hill, On, Canada
e-mail: michael.stanleyjones@unu.edu

C. Egehiza Obote
University West Sweden, Trollhätten, Sweden
e-mail: claireegehiza@gmail.com

W. Leal Filho et al., *Curriculum Development for Sustainable Development*, World Sustainability Series, https://doi.org/10.1007/978-3-032-22146-9_16

were addressed through a variety of engagements with practitioners and community members. Particular attention is given to the extensive outreach to stakeholders conducted during curriculum development. The participation of government officials and community stakeholders in the project helped shape the content of the e-learning courses. Analysis of the results of a survey of enrollees in the e-learning courses provides evidence of their effectiveness as a capacity-building tool. A fuller learner-centred approach to e-learning, which combines participatory learning with self-paced study, could further enrich the learning experience.

Keywords Capacity-building · COVID-19 pandemic · Distance learning · Learner-centred approach · Poverty-environment nexus · Poverty-environment mainstreaming · Poverty reduction · Sustainable development curricula

1 Introduction

For many low-income individuals, the environment is a crucial source of income and daily sustenance. Improved management of the environment and natural resources contributes directly to poverty reduction, more sustainable livelihoods, and inclusive growth. Economic development should not be viewed as being in competition with environmental sustainability, as the two need to work hand in hand to end endemic poverty. To fight poverty, promote security, and preserve the ecosystems that many communities depend on for their livelihoods, governments must place equitable economic growth and environmental sustainability at the centre of economic policies, planning systems, and institutions. Eradicating multidimensional poverty is an indispensable requirement for sustainable development.

Building on the legacy of the UNDP–UNEP Poverty-Environment Initiative (2005–2018), UNDP and UNEP jointly launched the Poverty-Environment Action for Sustainable Development Goals in September 2019, to place environmental sustainability, inclusivity, and poverty reduction at the heart of the sustainable development agenda. Together, Poverty-Environment Initiative and Poverty-Environment Action incorporated two decades of experience of implementing measures addressing the poverty-environment nexus gained from more than 30 Least Developed Countries. The poverty-environment nexus consists of issues of intractable poverty, inefficient, unsustainable, or inequitable natural resource management, and environmental degradation that require they be addressed in an integrated way—optimizing their trade-offs and synergies. The programmes involved hundreds of regional, national, and sub-national agencies.

This chapter explores the creation of an online capacity-building platform, culminating in 10 h of courseware focused on integrated approaches to the poverty-environment nexus. The "Sustainable Development in Practice" e-learning courses drew from decades of experience across four regions of the world and were developed through a unique partnership among international agencies committed to poverty-environment action. The method followed in this exploration includes a historical and descriptive narrative, tracing the development of the online platform and its tools, and a brief analysis of a survey of enrollees following their completion of the e-learning courses. It draws from a case study of the application of the Analyze, Design, Develop, Implement, and Evaluate (ADDIE) instructional design model to the constraints of a global e-learning environment. The chapter illustrates how sustainable development can be implemented in education through a learner-centred approach, drawing on widely-held United Nations pedagogical principles applied by Poverty-Environment Action. It also highlights how the programme sought to adapt educational content to users' needs, even in contexts with limited resources and institutional capacity.

2 Background

Poverty-Environment Nexus, the PEP, and Poverty-Environment Initiative

Launched in 2005, the joint UNDP and UNEP Poverty-Environment Initiative grew out of the Poverty-Environment Partnership (PEP), an informal network of bilateral and multilateral development agencies, United Nations organisations, and international non-governmental organizations. The PEP sought to integrate poverty reduction, environmental sustainability, and climate resilience in global, national, and local development agendas. The PEP operated as a 'network of networks' and played a catalytic role, focusing on agenda-setting and improving development policy and practice through: (i) joint analytic work; (ii) networking; and (iii) shared (South-South and South-North) learning. PEP members collaborated through working groups on priority and emerging poverty-environment and climate issues, such as environmental fiscal reform, to ensure poor groups would not be further marginalized and promote more inclusive, equitable, integrated and sustainable policy, planning, and implementation at all levels.

The PEP 'Reference Group' consisted of members who met virtually between the annual meetings. To guide and review the progress of joint activities and the PEP's overall strategy and direction. The PEP Coordinator was based at the International Institute for Environment and Development (IIED) in London, UK.

PEP functioned as a network focused on joint analytical work, networking, shared learning, and advocacy, and served as an informal counterpart to the Organisation for Economic Co-operation and Development (OECD) Development Assistance Committee, to whose attention it brought many needs and ideas. PEP meetings held annually or biannually from the early 2000s to 2017 provided an informal and open forum for networking and sharing lessons. Many PEP members were directly involved in capacity building and technology transfer. PEP members supported outreach activities and knowledge sharing through their circles.[1] PEP-facilitated

---

[1] At its peak, the PEP Reference Group included the Asian Development Bank, Austrian Development Agency, *Deutsche Gesellschaft für Internationale Zusammenarbeit* (GIZ), Development Alternatives, Durban City, Ecosystem Services for Poverty Alleviation Programme (ESPA), Finland

exchanges led to collaborations among like-minded organizations, amplifying their impact in developing countries through joint programming.

As part of a novel arrangement, UNDP and UNEP launched the Poverty-Environment Initiative on a pilot basis in sub-Saharan Africa in 2005. Subsequently, the full first phase of the Poverty-Environment Initiative was launched in 2008, followed by a second phase (2013–2017) covering approximately 30 countries across four regions: Africa, Asia and the Pacific, Europe and the Commonwealth of Independent States, and Latin America and the Caribbean. Coordination of the initiative was managed by the Poverty-Environment Facility, a joint UNDP–UNEP unit based in Nairobi.

Building a country's capacity to meet both environmental sustainability and poverty eradication commitments is a vital element in transforming current growth paths to achieve the Sustainable Development Goals (SDGs). This requires adopting lessons from integrated policy approaches used in joint United Nations programmes.

Poverty-Environment Action operated at country, regional, and global levels through a two-pronged strategy of deepening and broadening support to countries on poverty-environment and climate mainstreaming. This two-track approach aimed:

- to deepen in-country efforts to integrate environmental sustainability and climate objectives into development planning, budgeting, monitoring systems, and both public and private finance and investment, with a focus on poverty eradication;
- to broaden the dissemination and application of country-level experiences with integrated poverty-environment approaches through enhanced knowledge management, targeted technical assistance, South–South cooperation, and strategic engagement with global and regional actors supporting SDG implementation (UNEP, 2023).

The mainstreaming of poverty-environment objectives into policy, budgeting, programming, and investments by Poverty-Environment Action demonstrated how improved environmental sustainability can address the challenge of integration and contribute to poverty eradication.

Country-level projects were implemented in eight countries: Bangladesh, Lao PDR, Myanmar, Malawi, Mauritania, Mozambique, Nepal, and Rwanda. Technical assistance projects on Green Bonds, the Blue Economy, gender, and capacity building were conducted in Indonesia and Tanzania.[2]

Ministry for Foreign Affairs, Government of Bhutan, International Centre for Climate Change and Development (ICCCAD), International Institute for Environment and Development (IIED), Korea International Cooperation Agency (KOICA), Organisation for Economic Co-operation and Development (OECD), The World Bank, UK Department for International Development (DFID), United Nations Development Programme (UNDP), United Nations Environment Programme (UNEP), World Resources Institute (WRI) and WWF International.

[2] The UNDP–UNEP Poverty-Environment Action for Sustainable Development Goals project received financial support from the governments of Austria, Belgium, Norway, and Sweden, and the European Union.

Through partnerships with GIZ and UN Women, Poverty-Environment Action extended the reach of its technical assistance project to South Africa (Green Bonds) and sub-Saharan Africa (Gender).

Technical Assistance Interventions aimed to catalyse and broaden the use of poverty-environment mainstreaming tools through wider UNDP and UNEP programmes to contribute to Poverty-Environment Action's outcomes. Support was delivered through the development and application of poverty-environment mainstreaming tools, using environmental, social, and economic data to generate concrete evidence for government partners. Over 50 such tools have been developed and widely disseminated (Bass et al., 2023).

Poverty-Environment Action accelerated its support to governments to enhance shifting government budgeting processes, finance, and investment towards environmental sustainability and climate objectives for poverty eradication and improving the environmental sustainability of current investments.

Poverty-Environment Action's support contributed to the strong integration of poverty-environment and climate objectives in national development plans, with partial inclusion in sectoral and subnational plans. For example, Tanzania's Third Five-Year Development Plan (FYDP III) and Rwanda's annual sector and district development plans incorporated poverty-environment objectives. In Rwanda and Indonesia, increased annual and medium-term budget allocations—both at national and subnational levels—integrated environmental sustainability and climate priorities for poverty eradication. In South Africa, Poverty-Environment Action supported the development of a Technical Handbook on Issuing Municipal Sustainable Bonds, followed by capacity-building activities for municipalities exploring green bond investments (Table 1).

Countries have maintained a coordinated approach to addressing poverty-environment linkages by engaging in SDG implementation monitoring committees and environment and natural resources management sector working groups. For example, Mozambique worked to align its Nationally Determined Contributions (NDCs) with national development plans. In Nepal, Poverty-Environment Action supported SDG progress assessment through the development of the "Leave No One Behind" Framework and the SDG Localization Source Book. In Malawi, the project helped establish and coordinate the Agriculture Sector Working Group.

Virtualizing the 2015 Handbook

UNDP and UNEP launched the capacity-building handbook, *Mainstreaming Environment and Climate for Poverty Reduction and Sustainable Development*, at a meeting of Poverty-Environment Initiative Africa in Kigali in September 2015 (United Nations Development Programme–United Nations Environment Programme Poverty-Environment Initiative, 2015). The meeting included a 2 h training session on climate finance. Notably, the training did not draw directly from the handbook but was based on research conducted independently by the IIED. The experience raised questions about the applicability of the new handbook for capacity building and training activities.

Table 1 Selected examples of integrated poverty-environment and climate action

Selected examples of poverty-environment and climate actions
• In Africa, national development plans with integrated poverty-environment goals and targets were produced in all poverty-environment initiative (PEI)/poverty-environment action countries except Kenya. Sector plans with mainstreamed poverty-environment issues and specific goals and targets were developed in Burkina Faso, Kenya, Malawi, Mauritania, Mozambique, Rwanda and Tanzania. District and other subnational plans with mainstreamed poverty-environment issues and specific goals and targets were completed in Burkina Faso, Malawi, Mali, Mauritania, Mozambique, Rwanda, Tanzania and Uganda • In support of gender mainstreaming, by 2018, 36 policies and monitoring frameworks across Africa had gender mainstreamed into them • Green villages demonstrating an integrated spatial planning and development approach were pioneered with PEI/poverty-environment action support in Rwanda. By 2021, every district in Rwanda had built one of more green villages • Bangladesh introduced a climate budget code with indicators for use in future budgets, so that it can track spending continuously across all government departments. Mozambique's introduction of budget codes for cross-cutting issues (like climate) allows decision-makers to verify allocated resources and which agencies specifically responded to environmental and climate change objectives • In Malawi, existing national planning directives and guidelines were modified to achieve poverty-environment outcomes, notably in its "Guidelines for Integrating Environmental Sustainability and Natural Resource Management in Policy-Making and Planning in Malawi" • Rwanda's 2018 national investment guidelines, developed with PEI/poverty-environment action technical assistance, require public projects to demonstrate that environmental and social safeguards are in place before a government institution can receive internal or external finance. An environmental and climate checklist is included as an annex to the budget call circular issued annually by the Ministry of Finance and Economic Planning. All sectors and districts must not only plan for environmentally sustainable and climate change-resilient actions but, critically, also budget for their implementation

Initially, the 2015 handbook was made available in English (hardback) and English, French, and Spanish (softback), reflecting the primary languages of countries engaged in the Poverty-Environment Initiative. To improve accessibility, the Poverty-Environment Facility developed a brochure and distributed digital copies on memory sticks in all three languages, replacing the hard copies.

Translations into Arabic and Russian were completed in 2016 and 2017, respectively.

Since June 2020, with the launch of the new Poverty-Environment Action website, all materials have been published and shared exclusively in digital format, further supporting accessibility and flexible learning.

In 2017, the Secretariat developed an electronic digest of the 2015 handbook, complete with chapter-by-chapter checklists and exercises to create a more accessible and interactive learning experience, laying the groundwork for the approach later adopted in the 2023 handbook and accompanying e-learning courses.

Impact of the COVID-19 Pandemic on Capacity Building, South-South Cooperation, and Information Exchange

Limitations on international meetings and South-South exchanges during the COVID-19 pandemic led to the suspension of most conventional capacity-building activities conducted by UNDP and UNEP Poverty-Environment Action throughout 2020 and 2021. For example, a planned South-South exchange on green bond financing between national and municipal officials from Indonesia and South Africa was cancelled due to COVID-19 restrictions. In-country travel restrictions further hindered capacity-building. The United Nations Office at Nairobi (UNON) and UNEP Headquarters remained closed to all but essential personnel from March 2020 to mid-2022. During this period, the core management team of Poverty-Environment Action met virtually via Zoom and MS Teams platforms and delivered capacity-building webinars as an alternative method for knowledge sharing and programme implementation.

Operating from UNEP Headquarters in Nairobi provided the advantage of being at the center of management decision-making for what had become a bi-regional project. However, it also created a disconnect from country programme activities and the communities they served. Corporate reporting relied on information gathered from national and regional offices, submitted on a quarterly and annual basis. This form of reporting inevitably filtered out human-interest elements and nuanced insights into the project's impact on local communities. These limitations were further compounded by the onset of the COVID-19 pandemic, which severely restricted travel to participating countries and effectively shut down most operations at UNON from March 2020 to September 2022. As a result, documenting and assessing the project's impact became even more challenging during this restrictive period.

3 Implementation and Knowledge Tools: Poverty-Environment Action Online Capacity-Building Platform

At the midpoint of the original four-year term of Poverty-Environment Action, the UNEP Evaluation Office conducted an independent assessment of the project. While the evaluation rated technical contributions highly, it highlighted shortcomings in strategic and financial management, as well as in knowledge management and the dissemination of lessons learned (UNDP–UNEP, 2021). The review found that knowledge management, communication, and visibility, as well as sustainability, could be enhanced through capacity building, South–South exchanges, and other forms of information sharing. Engagement within the institutions in general and of the poverty–environment nexus was judged to be weak vis-à-vis many broad areas of work at the global and at the corporate level.

Partnerships it advised should also be seen as a way to imbed or increase sustainability, to leverage greater influence and awareness of what the project is trying to achieve or is achieving, and to increase the visibility of achievements.

Due to COVID-19 related restrictions, more cost-efficient knowledge delivery, such as using more online and at a distance modality in certain settings, was further recommended. Knowledge management can also encourage South-South cooperation through the exchange of experts, and knowledge transfer (including knowledge management products), best practices and lessons learned generated throughout the project.

With the project slated to conclude at the end of 2022, the host organizations and their intergovernmental and national partners sought to address these shortcomings and capture lessons learned by enhancing dissemination of accumulated knowledge, strengthening partnerships, and supporting the continued application of the poverty-environment nexus approach beyond the back-to-back joint programmes.

At a Global Partnership meeting held virtually in May 2021, United Nations partner agencies including UNEP, UNDP, United Nations Office for Disaster Risk Reduction (UNDRR), UNITAR, and UN Women discussed how to extend the knowledge and skills developed through nearly two decades of experience in poverty-environment mainstreaming. At the meeting, UNEP presented a package of measures aimed at securing the legacy and expanding the impact of Poverty-Environment Action.

With support from their partners, UNDP and UNEP developed a set of tools to sustain poverty-environment action at the country level beyond 2022. These tools collectively formed an ambitious online capacity-building platform, designed to facilitate knowledge delivery and experience-sharing even in the absence of renewed short-term funding for staffing the platform.

The key components of the Poverty-Environment Action online capacity-building platform were:

1. An updated handbook covering key themes and methodologies applied by Poverty-Environment Initiative and Poverty-Environment Action, as well as related nexus programmes (notably the Partnership for Action on Green Economy—PAGE), to provide an evidence base for poverty-environment action;
2. A public peer review process supporting handbook development, launched at the Poverty-Environment Action Global Retreat held in Nairobi (online and in person) in September 2022;
3. Field case studies and video interviews showcasing the impacts of poverty-environment action at the country level;
4. Expert video interviews highlighting the unique contributions of the poverty-environment nexus approach in addressing environmental, climate, gender, and poverty-reduction challenges in national policy making;
5. Online, self-paced UN CC:e-Learn training modules derived from the case studies, handbook, video interviews, and related resources[3];

[3] The One UN Climate Change Learning Partnership (UN CC:e-Learn) is a joint initiative of more than 30 multilateral organizations helping countries to achieve climate change action both through

6. Peer review of the UN CC:e-Learn course via two interactive webinars and a survey instrument, allowing participants to provide feedback to the course design team;
7. A dedicated web environment hosting key knowledge products developed under the Poverty-Environment Initiative and Poverty-Environment Action on the Green Policy Platform.[4] This included access to UN CC:e-Learn courses, articles, guidance documents, and videos, curated with minimal oversight by a designated host institution.[5]

By linking the contents of Poverty-Environment Action on the Green Policy Forum to the UN CC:e-Learn website, Poverty-Environment Action enabled its knowledge to be discovered by a much wider audience and ensured continuity of access to the products it had produced.

Field Visits and Country Impact Reporting

Field visits to select countries were organized by UNITAR and UNEP in May, October, and November–December 2022. A programme management officer and an Audiovisual and multimedia creative producer were dispatched to conduct interviews with government officials, non-governmental leaders, and community members to assess what measures worked and the impacts of the programmes in three countries: Indonesia, Malawi and Rwanda.

The selection of countries for field visits was made by UNITAR and UNEP and was based on several considerations, including programme longevity, impact at national, and local level, and ongoing government commitment.

Rwanda was chosen as it represented a mature, full country programme in both Poverty-Environment Initiative and Poverty-Environment Action and featured several highly successful initiatives which included strong community participation.

Malawi exemplified multi-agency cooperation in Poverty-Environment Action between UNDP, UNEP, FAO and UN Women that supported 'Delivering as One UN'. It also featured the participation of women's agricultural cooperatives in the programme.

Indonesia had operated a full country programme under Poverty-Environment Initiative which transitioned in 2018 to a technical assistance programme under Poverty-Environment Action. By design, technical assistance programmes were of

general climate literacy and applied skills development. It provides 64 free e-learning courses on the UN CC:e-Learn platform. UN CC:e-Learn is supported by the Swiss Agency for Development and Cooperation (SDC).

[4] The Green Policy Forum is operated by managed by the Green Growth Knowledge Partnership (GGKP). The GGKP is led by the Global Green Growth Institute (GGGI), Organisation for Economic Co-operation and Development (OECD), United Nations Environment Programme (UNEP), United Nations Industrial Development Organization (UNIDO) and World Bank Group.

[5] The Government of Belgium, through Belgian Cooperation, was the principal funder of the Poverty-Environment Action online capacity-building platform. Additional contributions were provided by the Government of Austria (via the Austrian Development Agency), the Government of Sweden (through the Swedish International Development Cooperation Agency—Sida), and the European Union.

18 months duration and received limited programmatic funding. Thus, Indonesia provided a window into how the foundation laid by Poverty-Environment Initiative was extended through Poverty-Environment Action and national government commitments to continue addressing the poverty-environment nexus using resources primarily mobilized at the national level.

Significantly, the actual itinerary of each visit was determined by in-country UNDP staff and the national authorities. The selection of sites for the visits and of individuals for the interviews was made by those whose knowledge of the country programmes was greatest.

Products of the County Site Visits

Four Poverty-Environment Action (PEA) impact stories covering Lao PDR, Indonesia, Malawi, and Rwanda were drafted in 2023, drawing from the three field visits plus inputs from the UNDP country office in Lao PDR (Box 1).

The first mission to Rwanda resulted in an impact video that was previewed during the Poverty-Environment Action Global Retreat on 15 September 2022. Two additional country impact videos covering Indonesia and Malawi were later produced, along with an overall project impact video summarizing all 11 Poverty-Environment Action countries. The inclusion of interviews with community members in these videos provided personal stories of learning and the impact of those directly affected by the programme.

More than a dozen field interviews with Ministerial officials and local stakeholders were conducted by UNITAR and UNEP between May and December 2022. The impact stories were shared with UNEP and UNDP staff for review and revisions were made based on their feedback. A total of 39 bibliographic references were appended to the four country impact stories.

The county impact stories and videos were later posted to the Green Policy Forum (2023), a repository of research, case studies, toolkits, and learning products supported by the Green Growth Knowledge Partnership (GGKP). The partnership with GGKP provided the technical means for ensuring the sustainability of Poverty-Environment Initiative/Poverty-Environment Action's knowledge products, as well as a means for their dissemination, following the closure of the project.

Box 1. Poverty-Environment Action Country Impact Stories

Poverty-Environment Action strengthens women-led cooperatives in Malawi

During a visit to Malawi in October 2022, UNDP Malawi introduced our team to the main tools used to support Poverty-Environment Action, with a special focus on agricultural and gender reforms. Three seminal reports and policy briefs developed through the Poverty-Environment Action Malawi—the *Climate Smart Aquaculture Toolkit*, *Soil Loss Mitigation Action Plan and Strategy,* and the companion Policy Brief, *Route to Sustainable Land Management in Malawi: Soil Conservation and Restoration*—highlight opportunities for addressing aquaculture and soil and nutrient losses issues in Malawi. The *Cost–Benefit Analysis of the ADAPT PLAN* report and policy brief demonstrate the efficiency and impact of various climate adaptation interventions on community livelihoods, environment and natural resources. The findings of the report helped build the case for strengthening support for women-led cooperatives in Malawi.

Poverty-Environment Action aids Indonesian sustainable investments

In late November 2022, a team from UNEP and UNITAR set out in Coast Guard pontoon boats to observe up close installations of solar photovoltaic buoys and lighthouses which serve to safely guide navigation in Jakarta Bay. The solar installations are part of the effort to put the country on the path to decarbonisation and meet commitments set in Indonesia's *Enhanced Nationally Determined Contribution*, in fulfilment of the Sustainable Development Goals. The joint project UNDP–UNEP Poverty-Environment Initiative supported government efforts to develop and implement tools to support measures which qualify for Green Bond/Green Sukuk (Islamic Bond) investments. A Green Sukuk issuance in 2019 funded the Double Track Railway Project in North Java Line Double track railway, making transportation in and out of the Capital more sustainable.

Rwanda's green villages

A team from UNEP and UNITAR visited Rwanda in May 2022 to learn how Rwanda's efforts to achieve sustainable development would be sustained after 2022. The team examined how Rwanda's "Green Village" projects address poverty-related environmental problems such as soil erosion, inadequate access to water, deforestation and unsustainable land use and clean energy. The *National Climate and Environment Fund* (FONERWA) helped provide sustainable financing of poverty-environment linkages at the national and local level and turbocharged the Green Villages, which evolved to become "Integrated Development Project Model Villages". The interventions have enabled their communities to increase livelihoods, improve food security, protect natural resources, advance gender equality and enable children, particularly girls, to attend school.

Thirteen stakeholder video interviews were also uploaded to the UN CC:e-Learn YouTube gallery, each also accompanied by a short description.[6]

[6] Among the performance indicators for evaluating the results of the UNITAR capacity-building sub-project were requirements that:

- Outreach and learning materials were supported by national and international partners, and
- Poverty-Environment Action knowledge products were sustainably documented and showcased.

The series of country impact videos, the overall project video, and the articles were incorporated into the UN CC:e-Learn courses, to provide a multimedia and stakeholder perspective on the impact of Poverty-Environment Action.

Development of the 2023 Handbook: *Sustainable Development in Practice: Integrating Environment, Climate and Poverty Reduction*

The Poverty-Environment nexus experience is documented in the *Sustainable Development in Practice: Integrating Environment, Climate and Poverty Reduction* handbook (Bass et al., 2023).[7] The handbook distills key lessons and methodologies from these programmes into a practitioner-oriented resource.

Poverty-Environment Initiative had developed a flexible programmatic model for poverty-environment mainstreaming, which incorporates gender and a rights-based approach, and is comprised of three components:

(i) finding the entry points and making the case, e.g. by conducting preliminary poverty, social and environmental assessments, and developing country-specific evidence;
(ii) mainstreaming in national planning and budgeting processes, e.g. through expenditure reviews and preparation of budget guidance notes;
(iii) mainstreaming into sector and subnational planning and budgeting, monitoring and private investment through budgets and financing options (economic instruments, expenditure reviews).

Building upon this foundation, Poverty-Environment Action sought to increase the impact of poverty-environment mainstreaming by expanding the components of planning and budgeting to cover the finance and stakeholder inclusion needed for successful implementation.

Key lessons learned since the 2009 and 2015 handbooks included the need to strengthen financing mechanisms after planning and budgeting stages. Another major lesson was the importance of fostering inclusive decision-making that engages the poor.

In support of the first point, the following national partner organizations participated in producing the outreach materials: Directorate General of Sea Navigation, Ministry of Transportation, Indonesia; Fiscal Policy Agency, Ministry of Finance, Indonesia; Sululu Horticultural Cooperative, Malawi; and Rwanda Environment Management Authority (REMA). The following international partners participated in the production of the outreach materials: International Institute for Environment and Development (IIED), UNDP Bangkok Regional Hub, UNDP Indonesia, UNDP Lao PDR, UNDP Malawi, UNDP Rwanda, UNEP Africa Office, UNEP Asia and the Pacific Office, and Women's Economic Empowerment Portfolio, UN Women, Malawi. All visual materials are branded with the UNDP–UNEP Poverty-Environment Action for Sustainable Development Goals logotype and the logos of its donors: Austrian Development Agency, Belgian Cooperation, Swedish International Development Cooperation Agency (Sida), and the European Union.

[7] The 2023 handbook was principally authored by Steve Bass, Senior Associate, IIED (UK), with additional contributions from co-authors David H. Smith and Michael Stanley-Jones, both long-time UNEP staff affiliated with the Poverty-Environment Initiative and Poverty-Environment Action.

The 2023 *Sustainable Development in Practice* handbook presents the drivers of integration and institutional bridges that build trust and enable integration; and illustrates for readers how to achieve integration across the policy cycle.

The 2023 handbook served as the foundation for online, self-paced learning modules developed by UNITAR and DelphianLogic, with major contributions from Steve Bass and Michael Stanley-Jones. These were the UN CC:e-Learn courses, *Introduction to Sustainable Development in Practice/Advancing Sustainable Development in Practice* (Fig. 1).

E-Learning Courses: *Introduction to Sustainable Development in Practice/ Advancing Sustainable Development in Practice*

UNITAR, in collaboration with UNDP–UNEP Poverty-Environment Action, developed two specialised self-paced e-learning courses focused on applying an integrated approach to address poverty and environmental challenges in the development process. A self-paced e-learning course was deemed a good solution for the sharing of best practices and lessons learned, and ensuring the lasting legacy of the PEI/ Poverty-Environment Action cost-effectively in the absence of ongoing staff support for the programme.

These courses, *Introduction to Sustainable Development in Practice* and *Advancing Sustainable Development in Practice* (United Nations Institute for

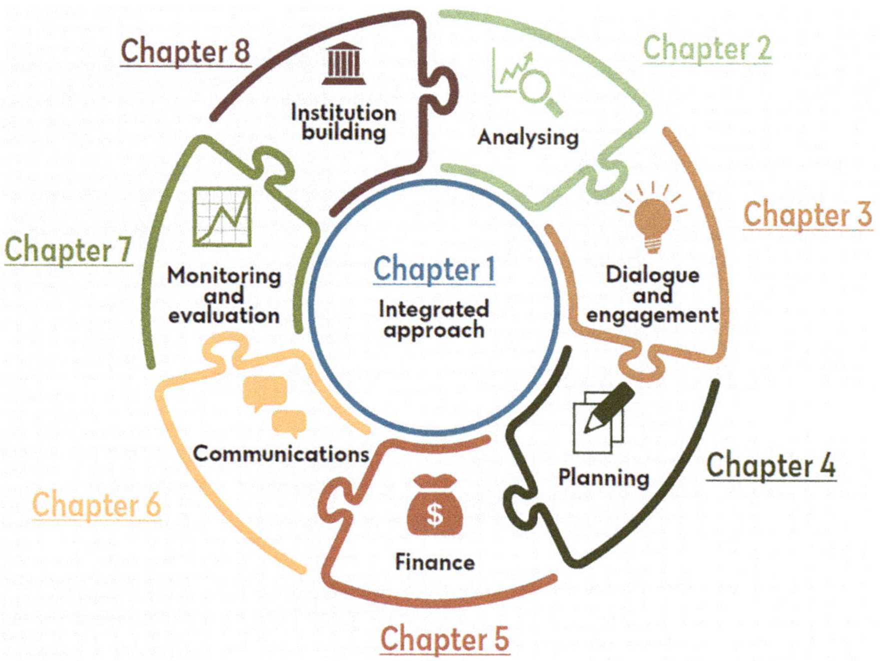

Fig. 1 Chapters of *Sustainable Development in Practice: Integrating Environment, Climate and Poverty Reduction. Source* Bass et al. (2023)

Training and Research, 2023a, b), are based on the UNDP–UNEP handbook *Sustainable Development in Practice: Integrating Environment, Climate and Poverty Reduction* (Bass et al., 2023) and are hosted on the UN CC:e-Learn platform[8]:

- *Introduction to Sustainable Development in Practice*—a 2 h interactive introductory course targeted to a general audience;
- *Advancing Sustainable Development in Practice*—an 8 h interactive course tailored for current practitioners, following 8 Chapters in the 2023 Handbook.

The introductory course teaches learners what an integrated approach to sustainable development is and why it is needed. It explains the rationale, drivers, challenges, and opportunities for this approach in addressing poverty-environment issues often referred to as the "poverty-environment nexus": the complex and mutually reinforcing relationship between poverty, the quantity, quality, and access to natural resources, and environmental degradation.

The advanced course builds upon the introductory course, teaching learners how to apply in practice an integrated approach to sustainable development throughout the various stages of the policy and project cycle, including analysis, dialogue, planning, financing, communicating, monitoring, and evaluation. The course contains tools, tips, and methods for applying sustainable development in practice (see Table 2).

The UN CC:e-Learn courses incorporated country-level impact stories, expert and stakeholder interviews, videos, and a comprehensive documentary to enrich the learner experience. Detailed syllabi outlined objectives, audiences, structure, methodology, and completion criteria.

A self-paced e-learning course was deemed a good solution for sharing of best practices and lessons learned and ensuring the lasting legacy of Poverty-Environment Initiative/Poverty-Environment Action cost-effectively in the absence of ongoing staff support for the programme.

Ultimately, by strengthening capacities for applying an integrated approach to sustainable development, the e-learning courses aim to accelerate the delivery of the 2030 Agenda and the Sustainable Development Goals.

4 Peer to Peer Exchanges on e-Learning Experience

Effective distance education relies on good teaching practices such as collaborative learning, active engagement, fostering social presence, and a supportive instructor role. These elements enhance both interaction and the overall learning experience (Spatioti et al., 2022).

To enrich the online learning experience of *Sustainable Development in Practice* and complement the e-learning courses, a series of peer learning and exchange events were organized by UNITAR in late 2023. These virtual sessions enabled learners from around the world to connect, share experiences, and engage in discussions with

[8] The UN Climate Change:e-Learn programme was funded by the Government of Switzerland.

Table 2 Introduction to sustainable development in practice and advancing sustainable development in practice learning objectives

Introduction to *sustainable development in practice* and *advancing sustainable development in practice* learning objectives
The introductory course aims to raise awareness and build capacities for adopting and implementing an integrated approach to tackling poverty and environment issues and achieving the sustainable development goals. The course will explain the rationale, drivers, challenges, and opportunities in integrating environment, climate, and poverty issues, while laying the foundations for a more advanced course focusing on policy analysis, integrated planning, designing interventions, financing, communications, and monitoring and evaluation of poverty and environment issues. After completing the introductory course, learners will be able to • Discuss the poverty and environment nexus and related concepts • Identify related international developments, global and regional trends, and explore their significance in a local context • Explain the rationale and drivers behind integrating policy planning • Identify challenges and opportunities for advancing an integrated approach • Give examples of enabling policy conditions for integrating poverty and environment issues Building upon the foundations of the introductory course, the advanced course looks at how to take an integrated approach to policy analysis, integrated planning, designing interventions, financing, communications, and monitoring and evaluation of poverty and environment issues at national and subnational levels. After completing the advanced course, learners will be able to • Examine and promote evidence-based approaches to poverty and environment issues • Set up effective sustainable development dialogue process with key stakeholders • Develop strategies for embedding poverty and environment issues into policy processes and plans • Demonstrate the integration of poverty and environment issues into public and private finance, including fiscal policy • Design an effective communication strategy supporting poverty and environment issues integration • Formulate approaches for effective monitoring and evaluation of poverty and environment actions • Discuss institutional change approaches to enabling integrated sustainable development

leading experts on the more nuanced aspects of an integrated approach to poverty and environmental issues.

The two learning events were organized as webinars, aligned with the Introductory and Advanced e-courses respectively.[9] Each webinar followed a moderated fireside chat format featuring renowned subject-matter experts with diverse areas of expertise. In addition to the moderated discussions, strong emphasis was placed on audience participation through Q&A sessions and interactive elements such as polls and quizzes.

The peer learning workshop participants confirmed their interest in more experience sharing events and training opportunities on how to achieve sustainable development through integrating environment, climate and poverty reduction in development

[9] First Peer Learning Event: Introduction to Sustainable Development in Practice, 18 September 2023 (1 h); Second Peer Learning Event: Advancing Sustainable Development in Practice: Preparing an Ideal Project Proposal, 14 November 2023 (1.5 h).

Table 3 Breakdown of enrolled learners

Introductory course: 11,953 enrolled learners 3,804 certificates of completion issued Completion rate: 32%
Advanced course: 5,800 enrolled learners 1,169 certificates of completion issued Completion rate: 20%
Gender Introductory course enrollees: Male—50.6% Female—49.0%
Advanced Course: Male—55.8% Female—43.7%

Source UNITAR (2024)

processes (UNITAR, 2024). The peer learning workshops demonstrated the importance and the interest of learners from around the world to connect to experts in real time and share experiences with each other.

Enrolment, certifications, types of learners and their geographic distribution

Since their launch on the UN CC:e-Learn platform in July 2023 and October 2023, the introductory and advanced e-learning courses on Sustainable Development in Practice have enrolled a combined 43,311 learners and issued 13,522 certificates as of April 2025.

The narrative report on the project contains key testimonials from a diverse range of beneficiaries from the courses (Table 3).

In both course groups the organizational affiliations of learners were similar, with 32.8% and 32.3% of learners indicating academic, and 22.7% and 22.4% private sector affiliations, respectively. 9% and 11% of enrollees were drawn from nongovernmental organizations, respectively.

National, state or local government affiliation was reported cumulatively by 2% of enrollees in the Introductory e-learning course; and by 3% of those taking the Advancing e-learning course.

1.8 and 2.5% were drawn from the United Nations system and nationally recruited United Nation staff, non-UN international organizations, or regional organizations.

50.6% of Introductory course enrollees were male; 49.0% female. For the Advanced Course, 55.8% were male; 43.7% female.

Introductory course learners were drawn from 186 countries; and 166 countries for the Advanced course. The highest number of Introduction e-learners were from India, China, Indonesia, UAE and Nigeria; India, China, Nigeria, Kenya and United Arab Emirates for the Advanced course.

Notably, none of the top 5 countries from which learners were drawn participated in the Poverty-Environment Action programme; and only one (Kenya) participated in Poverty-Environment Initiative, and then only briefly.

This demonstrates that the demand for knowledge and skills related to the poverty-environment nexus extends far beyond the 30+ countries that were engaged in either of the two programmes. The demand for this knowledge also extends beyond those groups typically thought of as being in the development assistance community.

Together, international and regional organizations and NGOs comprise a minority of learners with combined 10.8% and 13.5% of enrollees, respectively, or roughly half the rate of those enrollees drawn from the private sector.

Private sector demand for knowledge of the poverty-environment nexus was higher than anticipated and unforeseen by the project managers, who did not include the private sector among the target audiences in the Terms of Reference for the UNITAR e-learning course contract.

The relatively low rate of government organization participation may indicate that a key target audience was missing. The rationale for providing online, self-paced training grew out of experience with COVID-19 pandemic, which largely shut down face-to-face capacity-building activities, which traditionally had been core elements of the Poverty-Environment Initiative and Poverty-Environment Action programme delivery. By 2023, as the pandemic receded, government staff could resume the rounds of meetings, and aid agencies the in-person training which characterized much of their activities, thus lessening the demand for on-line training.

Ideally, capacity building and exchange of experience would occur both in-person and through online engagements. The e-learning courses were designed to be used as stand-alone courseware but also could be incorporated in workshops and other in-person training environments.

Testimonials

The UNITAR impact survey reveals how beneficiaries have been applying the knowledge and skills acquired through the e-courses and how the training has supported career advancements. The impact survey is level 3 of the Kirkpatrick Evaluation Model (demonstrates behavioural change).

The narrative report on the project contains key testimonials from a diverse range of beneficiaries from the courses. Testimonial evidence gathered by UNITAR suggests that the impact of the courses were widely positive and had influenced public policy and private sector approaches to sustainable development. The survey documented further accounts of how the courses had supported individual professional development.

A survey of 563 enrollees found 91% of them had applied what they had learned, 84% frequently or occasionally.[10]

97% reported they had acquired necessary skills and knowledge to take climate action. 89% reported they had changed the way they think about climate change and sustainable development.

[10] The response rate to the survey was 10%.

One respondent told how the course had supported his organization to develop a Disaster Risk Reduction (DDR) plan for its projects. A Somalian learner applied the knowledge he had gained to the review of national energy policy that his team was undertaking, and saw the team adopt sustainable development goals, principles and actions.

60% reported that they chose the course for professional reasons. Several respondents reported how they had received a financial benefit, such as a bonus, a promotion or otherwise professional advancement as a result of completing the course.

An Indian learner told of being admitted to a master's degree in sustainable development programme at the University of Surrey following completion. A learning from Kenya reported that the course had enabled her to join a private consultancy firm for her internship. A Zimbabwean learner found his organization recognized the certificate issued for completing the course and had been asked to represent his agency in climate action discussions as a result.

5 Applying the ADDIE Model to e-Learning

Key principles of education for sustainable development undergird the design and delivery of poverty-environment education curricula, including being action-oriented, justice-promoting, and incorporating accurate, comprehensive and relevant examples of sustainable development in practice. The normative framework of the United Nations guided the development of the programme and is reflected in its courseware (UNESCO, 2024, pp. 21–23). Within this framework, the UN CC:e-Learn courses followed a holistic approach with poverty-environment mainstreaming serving as the entry point to addressing sustainable development (Fig. 2).

UNEP's Terms of Reference for the UN CC:e-Learn courses envisioned a blended training approach, combining online and face-to-face capacity-building. However, given uncertainties around agency funding and long-term support for poverty-environment nexus training, the e-learning courses were designed to be comprehensive enough to function independently in the absence of future blended learning opportunities.

Both e-learning courses developed by UNITAR applied innovative, learner-centred methodologies that emphasized interaction, problem-solving, and active engagement. High-quality use of multimedia elements such as short films, audio clips, and graphics has been shown to enhance learner motivation, sustain interest, increase participation, and promote both interaction and flexibility (Spatioti et al., 2022).

These approaches employed in the e-learning courses were guided by the Analyze, Design, Develop, Implement, and Evaluate (ADDIE) instructional design model. The ADDIE model is particularly well-suited to distance education and proved practical in extending the knowledge presented in *Sustainable Development in Practice* to learners beyond the scope of Poverty-Environment Action programme. ADDIE provides learning that responds to the audience's needs. Learning objectives,

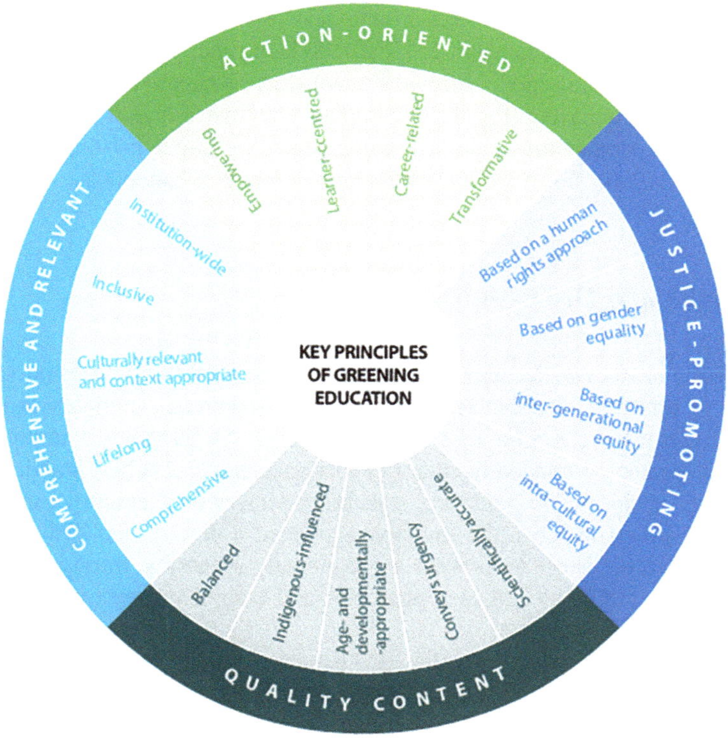

Fig. 2 Key principles of greening education. *Source* UNESCO (2024)

methodologies, and evaluations are aligned to ensure achievement of the learning goals, leading to behavioural change, and ultimately impact (Kirkpatrick 4 levels of evaluation).[11]

Effective teaching practices under this model include the use of multimedia presentations, timely feedback, diverse interactive activities, and a blended learning strategy that combines individualized and collaborative approaches. The pedagogical approach was not fully employed in onsite training, however, due to the wrapping up of the project. A more thorough evaluation of its potential as a training tool for sustainable development cannot, therefore, be made.

A synchronous format, where participants communicate and engage with each other simultaneously, is often preferred in distance learning. This format was used to a limited extent in the development of the 2023 handbook and evaluation of the e-Learn courseware. The value of a self-paced course, by contrast, is that it is

[11] The fourth level of Kirkpatrick Training Evaluation Model measuring results ensures that the training is achieving its intended goals and contributes to the organization's success. The ADDIE training model is similar to the Kirkpatrick Model in that it uses a structured process to evaluate training programs and can be employed in tandem with the ADDIE Model. University of San Diego (n.d.), USA. The Kirkpatrick Training Evaluation Model.

asynchronous, allowing people from around the world in different time zones to take the courses at their convenience over long periods of time.

The ADDIE training model offered Poverty-Environment Action a streamlined and structured framework for developing an effective e-learning product. This online training programme could be complemented by in-person training sessions, presentations, or the *Sustainable Development in Practice* handbook.

6 Results and Discussion

Synthesizing nearly two decades of work across more than 30 countries into a single package of practical, scalable learning tools was a uniquely complex endeavour. Poverty-Environment Action rose to the challenge by adopting a multi-layered knowledge strategy grounded in country experience and global best practices.

Drawing on normative guidance, stakeholder interviews, country case stories, videos, and expert insights, the initiative produced a diverse and dynamic set of learning resources. These tools were made globally accessible through the GGKP Green Policy and UN CC:e-Learn platforms and complemented by targeted peer learning exchanges.

At the heart of this effort were the 2023 handbook and the interactive UN CC e-Learn courses, developed using the ADDIE instructional design model. The process of doing the e-learning courses honed the authors understanding of its audiences and key messages.

Key components of the learning package, including the 2023 handbook and *Sustainable Development in Practice* courses, were developed in partnership with stakeholders. Extesnive interviews with government and community stakeholders were conducted on-site and the resulting materials incorporated into the curricula. Poverty-environment nexus practitioners were invited to review draft core chapters of the 2023 handbook and the e-learning courses drawn from the handbook were tested and shared through webinars and virtual exchanges with peer reviewers and stakeholders. A survey of enrollees provided further feedback from the community, offering insights into the application of the e-learning courses. These efforts not only helped improve the quality of the tools but also encouraged their uptake across academia, government, international agencies, civil society, and the private sector.

In April 2025, DelphianLogic and UNITAR were awarded the Learning Excellence Awards 2025 *Environment and Sustainability* award for the *Sustainable Development in Practice* e-learning courses at Imperial College, London (Sector Projects Plus Limited, 2025).

Future Applications

UNEP featured the 2023 handbook and e-learning courses as recommended resources in its forthcoming 2025 guidance on poverty and inequality for United Nations and government agency training. The guidance aims to strengthen the integration of the social and economic pillars of sustainable development into the work of UNEP. This will be done by supporting the implementation of the environmental pillar with a particular focus on the 2030 Agenda principle of 'Leaving No One Behind'.

The proposed work will entail poverty to be approached from a multi-dimensional perspective. This will include incorporating Poverty-Environment Action's people-oriented, social, and economic development approach into relevant strategic documents and project design and implementation, enabling UNEP to further internalize development to ensure that the environmental goals of UNEP projects and initiatives are at the same level of relevance and importance as 'Leaving No One Behind' objectives (UNEP, 2023).

In delivering the Poverty-Environment Action e-learning courses and related capacity building materials, the partnership between UNDP and UNEP, the principal poverty-environment content providers, UNITAR, a specialized training and research agency, and Green Growth Knowledge Partnership, a knowledge management initiative, provides a model of how United Nations capacity-building resources may be optimized through a clear division of labour, in which each agency contributes what it does best. Current processes and practices do not enable the United Nations system to achieve such a clear division of labour (UN, 2025). In May 2025, a work group chaired by the United Nations University Rector and UNITAR Executive Director brought together six United Nations entities to examine research and training functions within the United Nations system. It considered aligning training entities under UNITAR and research entities under UNU—with a central coordination mechanism overseeing both functions (UNU, 2025). The experience of the Poverty-Environment Virtual Capacity-Building Platform provides evidence of the value of functional capacity-building partnerships between specialized entities brings to the United Nations system.

Limitations and Opportunities

The approach taken by the United Nations developers of Sustainable Development in Practice was not without its limitations. Learner-centredness considers learners as autonomous and emphasizes their active construction of knowledge rather than being the recipients of a passive transfer of knowledge (UNESCO, 2024). The UN CC:e-Learn course design could only meet this objective to a partial degree, as the courses were constructed to be self-paced and taken without an instructor or peer group to facilitate fully participatory, experiential learning. The e-learning courses are nonetheless suitable for blended learning which allows for instruction to take fuller advantage of the learning-centred approach.

The e-learning courses were intended for all levels of education, both formal and non-formal, as well as for lifelong learning. The resulting educational experience has reached tens of thousands of learners worldwide, extending impact far beyond traditional United Nations capacity-building channels and audiences. Notably, the inclusion of field-based stakeholder interviews brought real-world depth and testimonial evidence to the learning experience capturing voices often missed by formal monitoring and evaluation processes.

The success of this integrated approach shows that impactful, inclusive, and scalable capacity development is possible even in complex development contexts. The legacy of Poverty-Environment Action lies not just in what was delivered, but in how it continues to inspire, inform, and influence sustainable development practice around the world.

Education for sustainable development aims to foster methods that can routinely help people address the range of environmental, economic, social, and governance issues in an integrated manner. It needs to embed robust and interdisciplinary ways of handling complex interacting problems and recognise and sometimes reconcile different stakeholder views and completing worldviews.

Poverty-Environment Initiative and Poverty-Environment Action have been pioneers in innovating such methods. The new handbook and e-learning courses provide a strong basis for growing the institutions and capacities sorely needed to face a sustainable future.

References

Bass, S., Smith, D. H., & Stanley-Jones, M. (2023). *Sustainable development in practice: Integrating environment, climate and poverty reduction.* United Nations Development Programme–United Nations Environment Programme Poverty-Environment Action.

Green Policy Forum. (2023). *Poverty-environment action sustainability page.* United Nations Development Programme–United Nations Environment Programme Poverty-Environment Action for Sustainable Development Goals.

Sector Projects Plus Limited. (2025). *Learning excellence awards 2025.* Stratford-upon-Avon.

Spatioti, A. G., Kazanidis, I., & Pange, J. (2022). A comparative study of the ADDIE instructional design model in distance education. *Information, 13*(9), 402.

UNESCO. (2024). *Greening curriculum guidance: Teaching and learning for climate action* (p. 23). https://unesdoc.unesco.org/ark:/48223/pf0000390022

United Nations. (2025). *Report of the mandate implementation review, UN80 Workstream 2.* White Cover Publication (subject to review). UN_MIR_2025_1120_Web.pdf

United Nations Development Programme–United Nations Environment Programme Poverty-Environment Initiative. (2015). *Mainstreaming environment and climate for poverty reduction and sustainable development, Nairobi.*

United Nations Development Programme–United Nations Environment Programme. (2021). *Poverty-environment action for sustainable development goals project mid-term review, Nairobi.*

United Nations Environment Programme. (2023). *Contribution of poverty-environment action to implementation of the United Nations system-wide plan of action for the United Nations decade for the eradication of poverty (2018–2027), Nairobi.*

United Nations Institute for Training and Research. (2023a). *Introduction to sustainable development in practice.* UN CC:e-Learn.

United Nations Institute for Training and Research. (2023b). *Advancing sustainable development in practice.* UN CC:e-Learn.

United Nations Institute for Training and Research. (2024). *Capacity development and outreach on Poverty and Environment Action (PEA) for Sustainable Development Goals: Final Narrative Report* (UNITAR Ref: C2022.TARGC064). UNEP.

United Nations University. (2025). Update on UN80. August 1, 2025 communication to staff by Dr. Tshilidzi Marwala, UNU Rector.

Michael Stanley-Jones is an Environmental Policy and Governance Fellow with the United Nations University Institute for Water, Environment and Health, based in Redmond Hill, Ontario, Canada. He served in the United Nations Economic Commission for Europe in Geneva, Switzerland, and United Nations Environment Programme in Nairobi from 2004 to 2022. He is co-author of *Sustainable Development in Practice: Integrating Environment, Climate and Poverty Reduction.*

Claire Egehiza Obote is a Graduate Student in Sustainable Development at University West in Trollhättan, Sweden. She served as an Intern with the United Nations Human Settlements Programme (UN-Habitat) in Nairobi in 2018 and supported the launch of the United Nations Alliance for Sustainable Fashion at the United Nations Environment Assembly in 2019 and the public review of *Sustainable Development in Practice* in 2022.

How to Make a Difference: Capacities and Processes for Teaching Community Engagement in Higher Education

Nora Timmerman, Leah Mundell, Alder Keleman Saxena, Lissy Goralnik, Lindsey Falkenburg, and Jennifer Lee Johnson

Abstract Scholars and practitioners have long recognized that sustainability is both a technical and a social issue. Identifying, measuring, and ameliorating environmental harms; developing and implementing seventh-generation relationships with environmental "resources;" and building political will to steward more sustainable solutions all require sustainability professionals to engage with broader communities. This challenges sustainability educators to develop curriculum that includes the capacities and processes needed to participate in, catalyze, and evaluate community engagement initiatives. This chapter describes three distinct approaches for teaching community engagement, drawn from established community sustainability programs at Northern Arizona University and Michigan State University. For each approach (undergraduate research program, undergraduate minor, and stand-alone course), we offer concrete instructional skills, curricular structures, and examples to help practitioners envision how they can develop curriculum for community engagement. Drawing from these experiences, we share five lessons learned: (1) dedicate time to building community partnerships; (2) create support systems to facilitate student-community relationships; (3) develop skills for communication,

N. Timmerman (✉)
Sustainable Communities, Northern Arizona University, Arizona, USA
e-mail: Nora.Timmerman@nau.edu

L. Mundell · A. K. Saxena
Anthropology and Sustainable Communities, Northern Arizona University, Arizona, USA
e-mail: Leah.Mundell@nau.edu

A. K. Saxena
e-mail: Alder.Keleman-Saxena@nau.edu

L. Goralnik · J. L. Johnson
Community Sustainability, Michigan State University, Michigan, USA
e-mail: goralnik@msu.edu

J. L. Johnson
e-mail: john8259@msu.edu

L. Falkenburg
City of Redmond, Washington, USA
e-mail: lef237@nau.edu

W. Leal Filho et al., *Curriculum Development for Sustainable Development*, World Sustainability Series, https://doi.org/10.1007/978-3-032-22146-9_17

conflict, and collaboration; (4) balance and (re)evaluate community, student, and faculty goals; and (5) develop reflexivity and mindfulness practices. Our goal is to provide models, best practices, and guiding questions for community engagement sustainability curriculum that makes a difference.

Keywords Higher education · Community engagement · Sustainability education · Curriculum and pedagogy · Community partnerships

1 Introduction

For some academics, education has always been a community-based project (i.e. ethnic and feminist studies (Hill Zuganelli, 2017; Sheridan & Jacobi, 2014; Williams et al., 2021)). However, it is largely since the early 2000s that the language and values of community engagement have been more broadly adopted at colleges and universities (Kelly & Given, 2024). In the US and elsewhere, universities and colleges are including community engagement as a pillar in their strategic planning, departments are developing language for their promotion and review documents to recognize and reward community engagement in faculty work, and academic degree programs are finding ways to integrate community engagement into their pedagogy (Carnegie Foundation, 2025; Janke et al., 2023). In short, community engagement is increasingly being recognized as a necessary component of postsecondary education, especially in sustainability fields. What does this mean for sustainability education curriculum?

Sustainability is not only a technical issue but is also, fundamentally, social. Identifying, measuring, and ameliorating environmental harms; developing and implementing seventh-generation relationships with environmental "resources;" and building political will to steward more sustainable solutions all require sustainability professionals to engage with broader communities (Engle & Halsell, 2017; Loew, 2014). This challenges sustainability educators to teach our students—and learn for ourselves—the capacities and processes needed to participate in, catalyze, and evaluate community engagement initiatives. Written from our perspective as educators in long-standing, degree-granting community sustainability programs in two U.S. public universities, this chapter describes our methods for teaching community engagement and distills several best practices for educators looking to develop community engaged curriculum and pedagogy in their sustainability programs.

First, what do we mean by community engagement? The Carnegie Foundation (2025), the leading organization responsible for classifying and evaluating community engagement in U.S. higher education, defines it as: "collaboration between institutions of higher education and their larger communities (local, regional/state, national, global) for the mutually beneficial exchange of knowledge and resources in a context of partnership and reciprocity." This Carnegie definition gives a helpful baseline and highlights the importance of reciprocity, collaboration, and exchange of knowledge and resources in community engagement.

Building upon this, our approach to community engagement is grounded in transformative and critical pedagogies and pushes against problematic assumptions of value-neutrality common in some community engagement literature (Pellow, 2012; Richards, 2011). We understand community engagement to be a political experience, not in the sense of electoral politics, but in terms of grappling with power and part of a larger movement toward more equitable power relations. Thus, our approach to community engagement requires power analyses: discussing, identifying, and deciding how to navigate the systems of power that manifest in our communities (Routledge & Derickson, 2015). Our transformative and critical approach also orients us toward collaboration versus service. Collaboration is empowering; it shows through action that student voice and agency matters. Finally, community engagement is relational; it asserts that everyone can and must learn the skills of how to work well together (Hill & MacDonald, 2022): negotiating conflict, facilitating meetings, understanding privilege, collaborating across differences, active listening, and becoming responsible and accountable. For us in sustainability education, community engagement is how we understand, practice, and co-create visions of sustainable living. Sustainability is a communal, community-based project, one that requires us to grapple with its meaning and work toward its emergence collaboratively, engaged together.

Developing curriculum and pedagogy for deep, transformative community engagement is challenging. Building and maintaining community partnerships conducive to this kind of work is time-intensive, requires negotiating different goals and understandings, and necessitates extensive skill development. Further, these challenges are iterative and need ongoing attention. Given these challenges, how can educators cultivate student-community partnerships that are mutually rewarding and make a difference?

This chapter answers this question by drawing on our experiences and insights from teaching community engagement. We begin with a section describing the approaches and methods we use to teach community engagement in our respective programs. This section details the pedagogical frameworks, instructional strategies, and programmatic structures of three university-level sustainability programs that prioritize community engagement. The following section distills the lessons we have learned through these programs and shares these results in the form of best curricular and pedagogical practices for community engagement in sustainability higher education programs. The chapter closes with reflections on the limitations of our work and areas for further thought, research, and practice. Our goal is to both highlight the practical, ethical, and pedagogical challenges of community engagement work, and to provide models and institutional structures that can work through those challenges to support quality community engagement for sustainability programs in higher education.

2 Approach and Methods: Community Engagement Pedagogies in Sustainability Programs

In this section, we explain how we do community engagement in our sustainability programs. Specifically, our approach and methods consist of pedagogical strategies, instructional tools, and programmatic structures, each of which come together to create unique opportunities for postsecondary students to tie their learning to concrete goals and practices within the broader community. Across the different contexts in which we work, the approaches and methods we use for community engagement have at their core a commitment to mutually beneficial exchanges of knowledge, time, and resources. As programs focused on sustainability, our community engagement methods also connect students to community leaders who make the lives of everyday people healthier, more resilient, more democratic, and more environmentally just.

To illustrate the variety of approaches one can take to community engagement in sustainability programs, we have described below three cases from two different state-based institutions, Northern Arizona University and Michigan State University. Within each case we describe the structural details of how the course/program works, as well as stories of the course/program in action to bring to life the curricular, pedagogical, and structural details that make them possible.

2.1 Example 1: An Interdisciplinary Undergraduate Community-Based Research Program

Box 1: CUPI Program Structure

Overall Program

- Students enroll in 3 credits of 400-level undergraduate research credit and can repeat as much as they like.
- CUPI is partnered with 8+ departments so that student participation in the program substitutes for degree requirements in their majors.
- A team of two coordinators runs the program—one faculty member who receives 10% of their workload for coordination and one graduate student (who is also a facilitator) who works 10 h a week.

Research Pods

- Each pod has: a community partner, 6–15 undergraduate students, 1–2 graduate facilitators, 1 faculty advisor.
- Pods meet for 1.5 h/week during class and additionally outside of class at community events, field trips, sub-working groups sessions, etc.
- Graduate facilitators are the main contact for undergraduate students, organizing weekly meetings, connecting them to research resources, and liaising with the community partner.

- Faculty advisors help the pod with research skill training, assignment and reading development, and advising graduate facilitators as needed.
- Students apply to the program and rank the pods they're most interested in; students stay with the same pod for the whole semester.

Community University Public Inquiry (CUPI) is an undergraduate research program housed within the Sustainable Communities Program at Northern Arizona University. CUPI brings community partners, students, and faculty together to address community needs through applied interdisciplinary research. Each academic year we enroll approximately 80 undergraduates, providing them with community-based, interdisciplinary, collaborative, and empowering research experiences. We recruit students from majors all over the university, and most are able to fulfill degree requirements through their participation in CUPI. Our goals for the program are big and visionary: make meaningful improvements in our local communities, build lasting collaborative relationships with community members, facilitate an experience of collective agency and empowerment amongst our students, and, overall, participate in the steady, hard work of social transformation.

CUPI is organized around "pods," groupings of undergraduate students, graduate students, a faculty member, and a community partner (see Box 1). Our relationships with community partners span multiple semesters and research projects often unfold over years. Behind the scenes we have a coordinating team that recruits students, manages enrollment and substitutions, fundraises and hires for the graduate student facilitator positions, trains graduate facilitators weekly, ensures community partnerships are mutually beneficial, and coordinates assessment and evaluation.

Pods engage in iterative determinations of their focus, starting the semester with 1–2 umbrella research questions developed by the community partner, faculty, coordinator, and sometimes graduate facilitator. These questions are specific enough to elicit information needed by the community partner, but broad enough for students to exert their agency as well, identifying and pursuing sub-questions related to students' disciplinary expertise. By the end of the semester, pods create reports, presentations, and research posters outlining their findings, which they present to both the academic community and their community partner.

Graduate students are core to the CUPI program. While they do have significant help from faculty—one faculty advisor per pod who comes to many of the weekly meetings to teach research and inquiry skills—the graduate students are in primary leadership positions for facilitating the complex, dynamic, and experiential community-based learning for their pod's undergrads. Moreover, graduate students also develop the leadership potential of the undergrads themselves so that they can co-lead the group. As the university system in the United States turns more toward career-readiness and austerity in its funding, graduate student leadership has also been a strategic tool to both support graduate students financially and provide them with professional leadership experience.

Having graduate students take such big leadership positions, and having the graduates support the undergraduates in also taking leadership, is part of our larger orientation toward democratic empowerment in the CUPI program. These put into practice an ethos of radical democracy—the notion that we all have power and that teaching about power is part of our purpose. In a community-engaged research program, that means decentralizing decision-making and teaching our students about how power operates in research practices (positionality, decolonial research, ethical conflicts, whose knowledge is considered valid, citation politics, how research is used in advocacy, etc.). Our goal is for students to understand how to use research/inquiry ethically and civically, as part of a larger project of making our world more just and sustainable.

It is worth commenting on the terms "inquiry" and "research." We use them somewhat interchangeably in CUPI, but they have notable differences. The US federal government's "common rule" defines research as "systematic investigation, including research development, testing, and evaluation, designed to develop or contribute to generalizable knowledge" (Steneck, 2007, p. 39). At US-based universities, research that involves collecting information from people (human subjects) must generally be cleared through an Institutional Review Board (IRB) before research can begin. Meanwhile, "inquiry" tends to refer to a process of fact-gathering, which may be systematic but may not necessarily adhere to the steps of design, testing, and evaluation, or be intended to produce generalizable knowledge. Both terms can also cover investigative processes which do *not* collect information from human subjects, such as literature reviews, content analyses of documents, map-making, etc. In short: all research is inquiry, but not all inquiry is research.

Some pods are better served by approaching the work as "research," others as "inquiry" (see Box 2). IRB coverage allows the pod to produce and disseminate information publicly and make certain claims about generalizability. For example, one pod working with a municipal government determined that generalizable knowledge that could be shared publicly in city council would be most useful for the policy change work they were doing. To put an IRB in place, the faculty advisor worked with the city government and graduate facilitators over four semesters to develop research questions and methods flexible enough for students to tailor them to their own interests; apply for the IRB (and get approval); and implement the methods systematically, including training students in how to document informed consent and manage data generated in the research. However, running a pod under an IRB also introduces a degree of inflexibility, requiring students to complete lengthy institutional training and requiring IRB approval (taking at least a month) for any substantive changes to research procedures. The focus on documented informed consent can also come across as over-formalized to community members, which takes the focus away from long-term relationship-building. Given these limitations, some pods approach their work as non-generalizable inquiry. For example, many pods do internal evaluative work (program evaluation), background research on a policy issue, and/or participatory action research that develops the organization.

Box 2: Questions to Ask for Considering Seeking Ethics Approval

- Does research need to be systematic and generalizable to achieve the community partner's goals?
- Does the community partner need to present the findings publicly or will they be used internally?
- Will formalized consent procedures inhibit relationship-building, and does this likelihood outweigh benefits of systematic data collection or public presentation?

Internal review boards (IRBs) may interpret these issues differently across institutions, and ethics approval for research in Europe (largely governed by the General Data Protection Rule, or GDPR) may also means EU-based institutions may be subject to different requirements or foci. Informally consulting with your ethics board is an important first step to determining answers to these questions.

CUPI draws inspiration from many powerful university-community research collaborations around the world that center reciprocity, relationships, interdisciplinarity, and collaboration. What makes the CUPI model unique is its scaffolded, near-peer mentoring, the years-long relationships with community partners, the large and diverse body of undergraduate students enrolled, and the combination of centralized and decentralized support. CUPI demonstrates how undergraduate research does not have to be individualized or filtered through faculty, but instead can be a group effort between faculty, graduate students, undergraduates, and community partners to discover knowledge that makes a meaningful sustainability impact in our community.

2.2 *Example 2: A Self-Contained Minor in Community Engagement*

The Community Engagement Minor (CEM) at Northern Arizona University teaches students the purpose, craft(s), and power of community engagement. Rather than a service-learning or volunteer-based approach, the CEM puts students in the position of co-directing their action through community organizing and/or advocacy work. This 18-credit undergraduate minor degree, housed within the Sustainable Communities Program at Northern Arizona University, was created in response to student demand for educational experiences that gave them the chance to engage directly with the issues that most concern them. Since 2017, 55 students have graduated with a Community Engagement Minor and many more have enrolled in CEM courses and participated in the community engagement experiences available through the program. Also during that time, dozens of community organizations mentored and benefited from the support of students.

The minor pairs 3-credit theory courses with 1-credit practice courses each semester (Box 3). Theory classes teach the history of organizing and social movements, analysis of structures of power, approaches to social change, and critical thinking strategies. Hands-on community engagement happens in the 1-credit fieldwork class as students take action in direct collaboration with community partners. Each semester, students are given several choices of partners to work with, amongst which they rank their favorites. We use their rankings to place students into "action teams," collaborative groups with whom they work the whole semester. Most action teams consist of ongoing collaborations with community partners, though some are new, and students can also initiate collaborations. Partners have included campus clubs such as No More Deaths and Fossil Free NAU, or student leadership programs such as Arizona Students' Association (which offers stipends and additional training), or even local campaigns, such as a recent political initiative to raise the minimum wage, which are short-term and focused on an immediate goal. Amongst many successful campaigns and projects, our students have helped increase student voting rates on campus, established a rapid response network to support immigrant families, and educated the community about the impacts of snowmaking and ski resort expansion for Indigenous residents.

Box 3: Community Engagement Minor Structure

The Community Engagement Minor combines the theory and practice of community engagement through the following 18 credits:

- Elective (3 credits) selected from >50 courses across campus
- SUS 230 Leadership for Social Change (3 credits)
  - *paired with* SUS 208 Fieldwork Experience (1 credit)
- SUS 320 Community Engagement from Local to Global (3 credits)
  - *paired with* SUS 208 Fieldwork Experience (1 credit)
- SUS 330 Community Engagement in American Democracy (3 credits)
  - *paired with* SUS 208 Fieldwork Experience (1 credit)
- SUS 450—Community Engagement Internship (3 credits, 10 h/week)) with a nonprofit, organizing group, government agency or business.

To set community partner expectations and build relationships, we host a pre-semester meeting with all of the community representatives. We ask past partners to share successes and challenges and have faculty describe the curriculum arc and desired outcomes. Together we brainstorm possible actions for the semester that fulfill community partner goals while steering clear of projects that simply provide volunteer capacity for the partner. If either the community partner or the student group is not fulfilling shared expectations as the semester progresses, faculty or graduate student mentors join action team meetings to help facilitate, check in regularly with everyone involved, and offer support for how to negotiate conflicts. End of semester feedback surveys from students and community partners also provide insight into future opportunities, adjustments for improvement, and indicators as to whether the partnership is coming to a close.

Learning outcomes for the CEM are grounded in a relationship-building approach to community engagement. That is, the CEM rests on the assumption that mentorship, organizing, engagement, critical thinking, in-depth understanding and listening, etc. are strongest in the context of trusting, genuine relationships. We therefore structure the minor to enhance relationship building wherever we can. For example, coordination of the minor and instruction of its courses are team efforts. We pair faculty with graduate students from the Sustainable Communities MA program to mentor one another and undergraduate student leaders. We do our best to encourage students from diverse political backgrounds to be part of the minor and to develop relationships across political differences. The curriculum we teach our students is based on a relational community organizing structure employed by civil rights leaders and long-standing grassroots organizations (Alinsky, 1971; Engler & Engler, 2016; Ganz, 2024; Payne, 1996). And lastly, because this work can feel overwhelming, we have been experimenting with integrating more contemplative practices into courses, similar to what our co-author Goralnik does in her classrooms (see Box 4).

Box 4: Contemplative Practices in the Classroom

Contemplative classroom practices can take many shapes. In undergraduate sustainability courses, Goralnik starts each class with a 3–5 min pause, which she introduces as a sustainability problem-solving skill, an opportunity for self-care, and a chance to put everything else down to be present with one another:

- She turns down the lights, invites students to close their eyes or glance toward the floor, and guides them through simple deep breathing.
- Each class she introduces a new contemplative practice—e.g. breathwork, mantra, sound healing, mindful attention—demonstrates it, and then sits down to practice with students until the timer chimes.
- On the screen she shows an image of the natural world or plays a short video to accompany the practice, e.g. a starling murmuration.
- The practice is optional; the only expectation is that students be still and quiet to allow others to enjoy the moment.

On-the-ground, one of the practices the CEM teaches for relational community organizing is the “relational meeting.” In the community organizing literature and practice, a relational meeting is a crucial tool used to assess aligned interests and the potential for collaboration. Relational meetings can be useful to understand needs, interests, and whether there is sufficient overlap to establish a collaboration/partnership. The relational meeting is defined by its focus on deep questions eliciting story, emotion, and potential action (Chambers, 2003; Ganz, 2024). At only thirty-minutes, the relational meeting asks questions like "what angers you about this situation?" "When did you first feel anger about this issue; what happened?" "What strategies is your organization using to address it?" In each CEM class, students are required to practice and use relational meetings with each other and community members to build connection and capacity for influential action in the community.

The CEM's theoretical grounding and established community partnerships have shared values of equity, liberation, uprooting oppression, and challenging dominant systems. While these values are not partisan, they have at times been perceived as ideologically aligned, making it challenging for some students to participate, despite concerted efforts to make the minor an inclusive space. Yet, one of our most powerful action teams was a group of students who did work together across difference. This group consisted of primarily white students who collaborated with a local Indigenous organizing group. The CEM students and faculty followed the lead of their community partner in prioritizing learning from Indigenous community members over rapid action. Ultimately, the group was able to elevate the work of Indigenous high school students toward changing the colonial name of a sacred peak near Flagstaff to the mountain's Hopi name, *O'omawki*. For our students, this was a transformative experience in unlearning their own cultural assumptions, intentionally playing a supporting role in work with Indigenous leaders, and exploring possibilities for solidarity across cultural differences.

In a world increasingly characterized by individualism, divisiveness, and fears of powerlessness, the CEM offers a model for teaching students the theory of democratic systems and historical examples of organizing that have made communities across the world more just and sustainable. An entire degree program devoted to helping students work together to enact change, advocate for what they believe in, and try out their ideas in the "real world" goes a long way towards combating apathy and disillusionment. The minor's suite of classes creates repeated opportunities for students to put theory into action, speak publicly on issues that affect them, create solidarity within communities, and ultimately become engaged citizens.

2.3 Example 3: A Community Engagement Course to Build Skills Without Leaving the Classroom

Community engagement is a vital component of skill development in sustainability education, yet teaching community engagement effectively can present significant logistical challenges, particularly for faculty new to a region. This example explores an innovative approach to fostering community engaged research and skill development within the confines of a classroom, as implemented in Johnson's undergraduate course *CSUS 301: Community Engagement for Sustainability* at Michigan State University (MSU). This course, a requirement for all students majoring in any one of the three undergraduate degrees housed within MSU's department of Community Sustainability, aims to equip students with both conceptual understanding and practical skills in community engagement for sustainability.

As a faculty member new to East Lansing, Johnson's connections to local community groups were limited. This prompted the design of a course that drew on her existing relationships from another city to deliver authentic community engagement experiences without relying on a new, in-person partnership. The solution combined short lectures, student-led current event presentations, active in-class discussions, multiple guest speakers from local community NGOs or regional governments, and significant in-class time dedicated to a community-engaged research project. Primary assessments included weekly question posing, short reading response papers, and draft and final versions of students' group projects. Central to the course's pedagogical approach were three texts: a theoretical exploration of contemporary crises and case studies of community-driven sustainability (Malin & Kallman, 2022), a practical guide to participatory community engagement for sustainability (Sarkissian et al., 2009), and an investigative memoir by community partner Sherry Roe (2022).

Profits over people: A story of tragedy and corporate malfeasance by Sherry Roe, proved to be the cornerstone of the community engagement project. Published in 2022, Roe's book details her family's and community's struggles with industrial contamination and related health impacts in Kokomo, Indiana, including the tragic cancer deaths of her mother and father. Her narrative deeply resonated with students, providing a powerful and accessible human-centered entry point into the complexities of environmental injustice in the US Midwest.

Kokomo, a city of approximately 60,000 residents located 50 miles north of Indianapolis, holds a rich industrial history, known as "the city of firsts" for its pioneering technological developments, particularly in the automobile industry. However, part of this legacy also includes closed factories like Continental Steel and PPG, which left behind significant land and groundwater contamination. Today, Kokomo remains home to major manufacturing facilities, including Haynes International, a top U.S. polluter, and plants for Stellantis, General Motors, BorgWarner, and Delphi. The impending launch of new multi-billion dollar electric vehicle battery plants by Stellantis and Samsung SDI has further amplified residents' concerns regarding legacy pollution, cultural shifts, and potential health impacts. This context provided a vivid, real-world backdrop for student research into the intersection of industry, environment, and community health. It also highlighted how, in the still-industrialized Midwest, factories and residential areas are often co-located, creating zones of intertwined living, working, and environmental concern.

To initiate the project, students first read Roe's book. Roe then visited the classroom virtually to discuss her book and give the class its charge: create an interactive and informative website that documents the historical and contemporary status of known and suspected contaminated sites in Kokomo. Together, Roe and Johnson developed and launched a community survey to identify sites of concern to residents. The sites suggested from this survey as well as through Roe and Johnson's knowledge of known contaminated sites were compiled into a list with summaries for students. Students then read about each and ranked their willingness to work on them. This offered students agency, set a collaborative tone from the outset, and was used by Johnson to assign student project groups.

Box 5: Group Formation

What I'm like when I'm in a group

Complete the following sentences

- In groups, I tend to ...
- In groups, I tend to avoid ...
- I like groups where ...
- I don't like groups where ...
- In this group, I would like to be ...
- How I'd like this group to be for me is ...

Answer the following question

- What are my goals for this group?

The course emphasized explicit skill development crucial for effective community engagement and community engaged research. A primary focus was fostering robust teamwork. Before forming groups, students engaged in a "What I'm Like in a Group" reflection (Box 5). They also articulated their personal goals for the group. This foundational exercise facilitated the setting of clear group agreements and encouraged the assignment of daily roles like facilitator, recorder, summarizer, and taskmaster, promoting accountability and shared responsibility.

Box 6: Meeting Guidelines

Meeting #1 (Today)

If you get through #1–5 below in your first meeting, great! If not, not to worry, we will have plenty of time to continue this conversation throughout the semester.

- **MEET** your group, introduce yourself, and start a conversation together with your answers to the "What I'm Like When I'm in a Group" exercise.
- **ASSIGN** roles for the day.
- **NAVIGATE** to our class GoogleDrive folder (see d21 under "Kokomo—Toxic Action Lab" for link, **CREATE** a New Folder, **GIVE** your folder a name corresponding to the name of your site.

- **CREATE** a new document. **GIVE** it an Appropriate Title (Group Norms, Terms of Engagement, Operation TEAM Flourish, whatever), **ADD** team members' names and **DOCUMENT** your response to #5 below.
- Based on your "what I'm like when I'm in a group" responses, **BEGIN** drafting terms of engagement and expectations with your group.

Johnson provided scaffolding for meeting and project organization to support student learning. She created detailed, do-able lists of tasks and goals, especially for the first few in-class work sessions (Box 6 shows instructions for the students' first meeting). Subsequent in-class work sessions offered more flexibility for student-led agenda and task setting, while also providing an overarching goal so students left each class meeting with a sense of accomplishment. For the project as a whole, students were guided through the following tasks to produce useful knowledge for Kokomo residents and develop their research skills:

- **Cataloging and Analysis** of Indiana Department of Environmental Management (IDEM) archives using Zotero, a free and open source collaborative reference management software.
- **Historical Research** through searches and analysis of historical newspapers for relevant information.
- **Report Drafting and Revision** on the site and in response to peer reviews and detailed instructor feedback.
- **Final Presentations and Team Evaluations**, including publishing their reports publicly and offering self and peer evaluations of team contributions.

The culmination of their efforts was the collective StoryMap: "Historic and Contemporary Contamination in Kokomo, IN." Leveraging the expertise of students with GIS training and those skilled in writing short histories, we collaboratively designed, compiled, published, and circulated an interactive StoryMap using ArcGIS, which included links to each group's full report. Through this process, students gained: (1) valuable, portable tools for effective teamwork; (2) proficiency in accessing and analyzing documents from state and federal environmental management agencies; and (3) skills in organizing and citing information using Zotero. Students who chose to be co-authors were credited for their contributions, further acknowledging their professional development.

The students' research garnered local recognition. A news piece in a Kokomo outlet highlighted this work, underscoring its relevance and impact beyond the university setting. This case study demonstrates that it is indeed possible and practical to teach community engagement, foster critical skill-building, and effectively manage student and community partner needs without necessarily leaving the classroom, offering a scalable model for similar courses.

3 Results and Discussion: Capacities and Processes for Community Engagement

After years of employing the methods of teaching community engagement described above, we have distilled several observations of key challenges and learned strategies to address them. While there is no one-size-fits-all recipe for community engagement, we have identified five capacities and processes crucial to our community engagement teaching. These are: (1) dedicating time to building community partnerships; (2) creating support systems to facilitate student-community relationships; (3) skill development for communication, conflict, and collaboration; (4) balancing and (re)evaluating community, student, and faculty goals; and (5) developing reflexivity and mindfulness practices. In this section, we describe these lessons learned in relation to our practice and discuss them in relationship to relevant academic literature on community engaged postsecondary education. For each section, we include a box of guiding questions for aspiring professors; these questions are intended to help faculty take concrete steps toward developing each of the skills and capacities described below.

3.1 Dedicate Time to Building Community Partnerships

We must be willing and able to build and maintain caring, trusting, and invested relationships with other people to do sustainable community engagement. In their discussion of scholar-activism, Routledge and Derickson (2015) suggest that the impetus for establishing a community—academic relationship is often rooted in our feelings of care, anger, sadness, or hope. To seek out community collaboration, we must "be moved" and "inspired by the collective visions and critiques that we share with our collaborators" (pp. 396–397). Emotional and visionary attunement is a precursor to establishing meaningful partnerships (Hill & MacDonald, 2022). Further, the act of discussing cares, anger, sadness, and hope in one-to-one meetings—to see if that attunement is there—is the stuff that builds relationships and connection in the first place (Chambers, 2003; Ganz, 2024, n.d.; Gecan, 2002; West/Southwest Industrial Areas Foundation, 2014).

Yet, even if we care deeply about the work a community organization does, a partnership can fall apart without an upfront, honest assessment of needs, capacity, and boundaries (Doubeni et al., 2022). Over-promising and under-delivering is common in community work, and it can happen on all sides (Spade, 2020). Community engagement is a responsibility and practice that not all faculty or community partners have capacity for or should be pushed to undertake. To avoid over-promising and under-delivering, direct, clear, and frequent communication is key. First, with oneself: figure out how much time you actually have in the week for more emails, meetings, problem-solving, and set goals accordingly. Second, with each other: state clear expectations of time and commitment from one another, describe the roles and needs each person has without sugar-coating, and give time for questions and consideration. This can be hard to do when approaching overstretched community organizations, but upfront assessments and statements of expectation are essential to building trust and respecting boundaries.

Finally, we find that successful community-university partnerships emerge when organizations are committed to, and have time for, working with students. This is especially essential in community engagement practice that prioritizes collaboration, student leadership, and decentralized power as core goals. Some community groups may need a cadre of volunteers and don't have the capacity to develop leadership with students; they just need the work done! There's nothing wrong with that, but it may not be a good fit for community partnerships built around student learning. Community partners who are invested in student growth and skill formation, who have the time to show a student how to do something, and who will genuinely dialogue with students about the issues central to the collaboration, are ideal.

In sum, building community partnerships requires self and other assessment—both personal and logistical. It can be helpful for faculty who are new at this work to shadow or develop mentoring relationships with faculty who are already well-established and networked in the community to support finding potential partners, assessing capacity and needs alignment, and communication throughout. The same goes for community partners: create spaces where experienced partners can share wisdom and experience with new ones.

Box 7: Guiding Questions for Building and Maintaining Community Partnerships

- Why and how am I "moved" by this issue? How important is this work to me?
- How networked am I in the community already? Do I know what kind of work has already been done related to this issue, who is doing it, and what next steps are? Who does know?
- How much time do I realistically have to dedicate to building and maintaining relationships in the community for the long-term? What is my back-up plan in case my capacity changes?
- For a community partner to meaningfully participate, what are my expectations? How many times would they need to come to meetings? How often do I anticipate emailing? What are the draws on their time and energy that I can anticipate?
- What does my potential community partner need, and where are they under-resourced? Could those needs be met well by students in my program within the time frame we have available?
- Does my past experience with students suggest that they would be passionate about the kind of work that this potential community partner needs someone to do? Would students enjoy this work and learn worthwhile skills from it?

- Does this community partner have the resources, time, energy, and overall capacity to meaningfully collaborate with students in my program?

3.2 Create Support Systems to Facilitate Student-Community Relationships

Community engaged pedagogy often takes more and different kinds of work than traditional classroom-based lecture courses—for faculty, students, and support staff. The work requires extra communication, coordination, and resources (Doubeni et al., 2022). In our experience, setting up support systems and bringing in resources well before the start of the semester has been a best practice to facilitate student relationships with community partners and places.

Even amongst the most eager, students are understandably unsure about how to interact with community members and physically visit the places and communities they are working with. While faculty can get discreet answers to these questions before the start of the semester, they will need to either devote their own time or work collaboratively with support staff to help students negotiate these challenges during the course of the semester. Actively mentoring and facilitating student relationships with community partners often means helping students with the social and logistical support they need to get out of the classroom to build rapport and understand the context of their issue. For many young people, and especially those impacted by the Covid-19 pandemic at a formative age, connecting with new people via phone or in-person can be anxiety-inducing (Bairwa et al., 2024). If we give students vague instructions like: "get to know your community partner," they will not know what to do with that information and often err toward doing nothing at all or sending one email and, when they don't get an immediate response, giving up. Specific instructions, regular check-ins, taking time to help students make contact (email drafts, phone scripts, rides to/from sites), and/or dedicated support staff goes a long way to help bridge the gap.

Practically, one option for structuring the facilitation of community partner relationships can be a "near-peer" mentorship model. In this model, faculty support graduate students, graduate students support undergraduates, and experienced undergraduates support newcomers. Near-peer mentoring uses resources and funding efficiently. Rather than all of the support work falling to faculty, sharing it with graduate students and undergraduate leaders provides those students with valuable learning, leadership, and ideally, paid work, without taking up the otherwise more expensive faculty time. A near-peer mentoring model also supports continuity and the transmission of institutional memory and builds strong leadership skills amongst graduate and undergraduate students.

Regardless of the model, mentoring and facilitating student relationships with community partners must be factored into community engagement teaching. Even the easiest and most successful community-engaged experiences come with communication challenges that require support. Allocating resources and class time to strong stewardship of the relationship and attentive mentorship of students are key ingredients to manage the unknowns and dynamic twists of community-university partnerships.

Box 8: Guiding Questions for Facilitating Students' Relationships with Community Partners

- How much interaction between students and the community partner is ideal?
- What kinds of support do students need if they are going to interact with the community partner in the ways we want?
- Are there near-peers, such as graduate students, who could support the mentoring of the students doing community engagement?
- How do we, as a program/as faculty, mentor the near-peers?
- What plans, values, and practices do we have established for negotiating conflict or misunderstandings?

3.3 Skill Development for Communication, Conflict, and Collaboration

Quality community engagement requires skilled communication, conflict negotiation, and collaborative organization at all levels. No one is born knowing these skills, and while social gender norms often culturally teach them to women, all people have the capacity to learn the art of how to work well with people (Doubeni et al., 2022; Federici, 2008; Lorde, 1984; Merchant, 2005). In higher education programs focused on bringing community engagement to undergraduates, we must not only give students the opportunity to learn through practice, but we (faculty) must take time to metacognitively identify, practice, and reflect upon student development of those skills (Hill & MacDonald, 2022). Pairing the practice with the theory, iteratively, holds tremendous potential to teach our students portable skills that are not only useful for "community engagement," but for all kinds of professional, community-based, and personal contexts (Kezar & Maxey, 2014).

What kinds of skills do students need to learn to do community engagement well? We break the larger concepts of communication, collaboration, and conflict negotiation down into the following discrete skills: visioning and goal-setting; meeting facilitation and agenda creation; project timelines and task management; group dynamics and conflict negotiation; direct communication; accountability and consensus decision making; and joyful community building. This list is not exhaustive and each of us shifts it slightly depending on our context. However, these skills represent foundational skills for working together democratically—where everyone can lead, contribute, and direct.

Although tempting to cut corners on skill development and add more readings on the topic of the course, the best way to ensure success in collaborative student engagement with community members is to dedicate real time to skill building. In community engagement courses, we thus ensure that several learning objectives for the course, assignments, readings, and class sessions are allocated to skill development. Students (and faculty) need support identifying, analyzing, practicing, reflecting on, and refining the arts of working with people, so it is best to incorporate them into the core of the course/program. Organize a syllabus or degree program arc such that skills are introduced at the moment students need them to move their projects forward, giving them space to incorporate and reflect on those skills as they come up.

Box 9: Guiding Questions for Skill Development

- How old are my students and what kinds of professional and/or collaborative work experiences have they had to date?
- What are the most important people skills this group of students need to build to work well with their community partner(s)?
- Who in the class already demonstrates strong skills in this regard and might be willing to help mentor others?
- Who in the community can help me teach the students these skills?
- How can I time lessons about these skills in the arc of the semester so they align well with when students will be needing them most?

3.4 Balance and (Re)Evaluate Community, Student, and Faculty Goals Regularly

Balancing our community partners' goals with academic timelines and learning objectives presents myriad challenges. There are diverse needs and pressures, requiring diligent (re)evaluation. For example, while the community partner needs to get a mailer out, our students need to learn proper citation formatting. Or, the community partner has a big stakeholder meeting in October, but the students just got started in September and will not have enough time to prepare a report by then. Even though there are many places where community-university needs overlap (let's not forget that the people *in* a university are *part of* a community), we can also recognize that the norms, structures, and demonstrations of success often do look different across university/community contexts. How do we negotiate these differences?

Does this partnership continue to serve everyone involved? Why or why not? If not, is it time to transition out of it?

Both the research and our experience show that early, direct, and regular check-ins are key, as is a willingness to let a partnership go in the event that needs do not align (Doubeni et al., 2022; Routledge & Derickson, 2015). When relationship-building happens from the start (taking the time to truly listen, discuss needs, and share limits honestly), it becomes easier to acknowledge what is and is not working and to turn mistakes into learning experiences. Meetings are where the magic happens; have multiple one-on-one meetings from the beginning, ensure you are checking in about the direction of the work at the mid-way point or just before (with enough time to correct if necessary), and hold whole-group reflections at the end of the semester. Make space for all to share: how are the students feeling about the partnership? What about the instructors, partners, and coordinators? Everyone's experience matters.

Box 10: Guiding Questions for Balancing Goals

- Where are the alignments between our community partner and our academic context, and are there any misalignments? Consider needs, capacity, timing, etc.
- How will we collect formative and summative feedback from all stakeholders before, during, and after the semester? Who will process that feedback and share it with everyone involved?
- What mistakes have we made and how can we use those as learning and/or re-adjustment opportunities?
- How is this partnership feeling to the students, partner, coordinators, and faculty?

3.5 Develop Reflexivity and Mindfulness Practices

Sustainability change agents need the capacity to slow down, sit with uncertainty, and adapt to changing circumstances (Carp, 2012; Frisk & Larson, 2011; Hart, 2004; Hull et al., 2020). Many learners study sustainability because they feel deeply connected to the natural world or their communities, relationships that are anchored in feelings of love, care, beauty, and awe. But engaging with communities on sustainability work often demands exposure to the devastating impacts of ecological disasters and social inequities, which can also evoke feelings of grief, fear, frustration, and defeat. Balancing the pull of conflicting emotions and staying with the hard work despite set-backs can lead to apathy and burnout (Brosch, 2021; Vidra, 2015; Wapner, 2016) without deliberate attention to the internal work of developing psychological resilience. Psychological resilience describes the capacity to cope with or transform in response to challenge (APA), and it is a vital component of sustainability learning and practice.

Reflexivity and mindfulness practices are integral for student empowerment and long-term dedication to action. Common approaches to reflection in higher education are assignment-focused, often written, and emphasize criticality. While these are useful, the personal development work required to cultivate agency and psychological resilience can more often come from contemplative practices. Contemplative practices in the classroom can take many shapes. Creative, movement- or stillness-based, activist, ritualistic, and relational practices facilitate self and other awareness, personal growth, and connections to people, place, and ideas. Choosing what and how to include contemplative practice in the learning experience will depend on the institutional and disciplinary context, cultural appropriateness, content focus of the course, and objectives for the contemplative activities.

Contemplative practices support a sense of purpose, self-awareness, empathy, problem-solving capacity, social connection, an internal locus of control, and confidence (Sivilli & Pace, 2014; Zajonc, 2013), in addition to a commitment to social change (Kaufman, 2017). By and large, studies find that contemplative practices can support the internal transformations needed to catalyze the social transformation at the heart of sustainable change (Böhme et al., 2022; Woiwode et al., 2021).

Box 11: Guiding Questions for Practicing Contemplation and Reflection

- How are my students feeling about the state of the world?
- What practices from my own culture, or from the cultures of my students, can we use to sit with our feelings and experiences?
- How can I make space at the start of class, or as needed, to guide my students (and myself) in contemplative practices?

4 Conclusion

In this global moment of climate change, mass extinctions, political and economic uncertainty, and the widening distrust of academia, we argue that community engagement is necessary. Community partnerships help us bridge deep critiques of our current systems and the need for immediate action within the imperfect world in which we live. Our students ask: "why am I getting a degree? Why do I learn about how terrible things are but I do not DO anything?" And community engaged teaching helps give us and them hope in the face of such uncertainty. Community engagement allows us to try new ideas and practices, see what works, and ultimately build the skills we collectively need to disrupt power and build systems that better align with values of justice and sustainability.

Despite our lofty intentions, the methods used and best practices offered in this chapter are not without their limitations. Our models for community engagement in sustainability education are limited by the political norms and ideologies shaping higher education in the United States (and other Anglophone societies) at this moment in time (Marginson, 2025). Further, while we write from our diverse experiences and with the inclusion of relevant research literature, we recognize that there are myriad ways of understanding and practicing community engagement, many of which we do not have the space or ability to explore herein. To move the sustainability education field forward in its theorizing and practicing of community engagement, we encourage more regional and international gatherings of practitioners, cross-disciplinary learning (e.g. with fields that have a long history of community engagement, such as ethnic studies), and edited publications detailing curricular and pedagogical possibilities, such as this one.

In this chapter, we have shared how we help students build skills and capacities through community engagement in multiple pedagogical spaces: in the classroom, via community-partnered and student-led research, and in an undergraduate minor. As we collectively work toward sustainability education that makes a difference, we hope the curricular and programmatic models and insights we have shared offer inspiration, camaraderie, and tangible possibilities for meaningful engagement with community.

References

Alinsky, S. D. (1971). *Rules for radicals: A practical primer for realistic radicals*. Vintage Books.

Bairwa, Y. P., Udayaraj, A., & Manna, S. (2024). The fear of smartphone notifications and calls among medical students: The phone ring phobia syndrome or telephobia. *Journal of Family Medicine and Primary Care, 13*(5), 1850–1855.

Böhme, J., Walsh, Z., & Wamsler, C. (2022). Sustainable lifestyles: Towards a relational approach. *Sustainability Science, 17*(5), 2063–2076.

Brosch, T. (2021). Affect and emotions as drivers of climate change perception and action: A review. *Current Opinion in Behavioral Sciences, 42*, 15–21.

Carnegie Classification of Institutions of Higher Education. (n.d.). *Community engagement*. American Council on Education. Retrieved July 8, 2025, from https://carnegieclassifications.acenet.edu/elective-classifications/community-engagement/

Carp, R. M. (2012). Toward a resilient academy. In L. F. Johnston (Ed.), *Higher education for sustainability: Cases, challenges, and opportunities from across the curriculum*. Routledge.

Chambers, E. T. (2003). *Roots for radicals: Organizing for power, action, and justice*. Bloomsbury Publishing.

Doubeni, C. A., Nelson, D., Cohn, E. G., Paskett, E., Asfaw, S. A., Sumar, M., Ahmed, S. M., McClinton-Brown, R., Wieland, M. L., Kinney, A., Aguilar-Gaxiola, S., Rosas, L. G., & Patino, C. M. (2022). Community engagement education in academic health centers, colleges, and universities. *Journal of Clinical and Translational Science, 6*(1), e109.

Engle, E. W., & Halsell, M. W. (2017). Pursuing reciprocity in engaged scholarship partnerships using community sustainability certification programs. *Journal of Community Engagement and Higher Education, 9*(3), 62–73.

Engler, M., & Engler, P. (2016). *This is an uprising: How nonviolent revolt is shaping the twenty-first century*. Nation Books.

Federici, S. (2008). *Precarious labor: A feminist viewpoint*. The Middle of a Whirlwind. Retrieved August 12, 2025, from https://inthemiddleofthewhirlwind.wordpress.com/precarious-labor-a-feminist-viewpoint/

Frisk, E., & Larson, K. L. (2011). Educating for sustainability: Competencies and practices for transformative action. *Journal of Sustainability Education, 2*(1), 1–20.
Ganz, M. (n.d.). *Shape your world: A college student's political action manual.* Retrieved August 12, 2025 from https://teachingsocialaction.org/social-action-books-articles-videos/shape-your-world-a-college-students-political-action-manual
Ganz, M. (2024). *People, power, change: Organizing for democratic renewal.* Oxford University Press.
Gecan, M. (2002). *Going public: An organizer's guide to citizen action.* Anchor Books.
Hart, T. (2004). Opening the contemplative mind in the classroom. *Journal of Transformative Education, 2*(1), 28–46.
Hill Zuganelli, D. (2017). Chicano studies: Proliferation of the discipline and the formal institutionalization of community engagement, 1965 to present. In *Dissertation abstracts international section A: Humanities and social sciences* (Vol. 78, Issue 1-A(E)).
Hill, C., & MacDonald, M. (2022). Working in a "community-engaged" university during an era of reconciliation. *Australian Educational Researcher, 49*(3), 547–569.
Hull, R. B., Robertson, D. P., & Mortimer, M. (2020). *Leadership for sustainability: Strategies for tackling wicked problems.* Island Press.
Janke, E. M., Quan, M., Jenkins, I., & Saltmarsh, J. (2023). Persistence and proliferation: Integrating community-engaged scholarship language into 59 departments, seven units, and one university's academic promotion and tenure policies. *Michigan Journal of Community Service Learning, 29*(1), 27–48.
Kaufman, P. (2017). Critical contemplative pedagogy. *Radical Pedagogy, 14*(1), 1524–6345.
Kelly, W. B., & Given, L. M. (2024). The "Community Engagement for Impact" (CEFI) framework: An evidence-based strategy to facilitate social change. *Studies in Higher Education, 49*(3), 441–459.
Kezar, A., & Maxey, D. (2014). Collective action on campus toward student development and democratic engagement. *New Directions for Higher Education, 2014*(167), 31–41.
Loew, P. (2014). *Seventh generation earth ethics: Native voices of Wisconsin.* Wisconsin Historical Society Press.
Lorde, A. (1984). Uses of the erotic: The erotic as power. *Sister outsider: Essays and speeches by Audre Lorde* (pp. 53–59). The Crossing Press.
Malin, S. A., & Kallman, M. E. (2022). *Building something better: Environmental crises and the promise of community change.* Rutgers University Press.
Marginson, S. (2025). Problems of the public good in higher education: Building the common amid sovereign individualism, capital and the state. *Higher Education, 89*, 29–52. https://doi.org/10.1007/s10734-024-01341-8
Merchant, C. (2005). *Radical ecology: The search for a livable world* (2nd ed.). Routledge.
Payne, C. M. (1996). *I've got the light of freedom: The organizing tradition and the Mississippi freedom struggle.* University of California Press.
Pellow, D. N. (2012). Activist-scholar alliances for social change: The transformative power of university-community collaborations. In B. Stockdill & M. Y. Danico (Eds.), *Transforming the Ivory tower: Challenging racism, sexism, and homophobia in the academy* (pp. 106–119). University of Hawai'i Press.
Richards, R. (2011). "I could have told you that wouldn't work:" cyberfeminist pedagogy in action. *Feminist Teacher, 22*(1), 5–22.

Roe, S. (2022). *Profits over people: A story of tragedy and corporate malfeasance*. Seven Interactive Press.
Routledge, P., & Derickson, K. D. (2015). Situated solidarities and the practice of scholar-activism. *Environment & Planning D: Society & Space, 33*(3), 391–407.
Sarkissian, W., Hofer, N., Shore, Y., Vajda, S., & Wilkinson, C. (2009). *Kitchen table sustainability: Practical recipes for community engagement with sustainability*. Routledge.
Sheridan, M. P., & Jacobi, T. (2014). Critical feminist practice and campus-community partnerships: A review essay. *Feminist Teacher: A Journal of the Practices, Theories, and Scholarship of Feminist Teaching, 24*(1–2), 138–150.
Sivilli, T. I., & Pace, T. W. (2014). The human dimensions of resilience: A theory of contemplative practices and resilience. Garrison Institute. Retrieved August 12, 2025, from https://www.garrisoninstitute.org/wp-content/uploads/2016/03/The_Human_Dimensions_of_Resilience.pdf
Smolentseva, A. (2022). The contributions of higher education to society: A conceptual approach. *Studies in Higher Education, 48*(1), 232–247.
Spade, D. (2020). *Mutual aid: Building solidarity during this crisis (and the next)*. Verso.
Steneck, N. H. (2007). *ORI introduction to the responsible conduct of research* (updated ed.). Department of Health and Human Services, Office of the Secretary, Office of Public Health and Science, Office of Research Integrity. https://ori.hhs.gov/sites/default/files/2018-04/rcrintro.pdf
Vidra, R. L. (2015). Cultivating hope through contemplative methods. *Journal of Sustainability Education, 10*.
Wapner, P. (2016). Contemplative environmental studies: Pedagogy for self and planet. *Journal of Contemplative Inquiry, 3*(1), 67–83.
West/Southwest Industrial Areas Foundation. Ernesto Cortes on Relational Meetings. (2014). https://www.youtube.com/watch?v=g42mLGATG-c
What is contemplative pedagogy? (2014). *Contemplative pedagogy network*. Retrieved August 12, 2025, from https://contemplativepedagogynetwork.com/what-is-contemplative-pedagogy/
Williams, B., Squire, D., & Tuitt, F. (2021). *Plantation politics and campus rebellions: Power, diversity, and the emancipatory struggle in higher education*. Critical Race Studies in Education. SUNY Press.
Woiwode, C., Schäpke, N., Bina, O., Veciana, S., Kunze, I., Parodi, O., Schweizer-Ries, P., & Wamsler, C. (2021). Inner transformation to sustainability as a deep leverage point: Fostering new avenues for change through dialogue and reflection. *Sustainability Science, 16*, 841–858.
Zajonc, A. (2013). Contemplative pedagogy: A quiet revolution in higher education. *New Directions in Teaching and Learning, 134*, 83–94.

Nora Timmerman teaches and researches environmental justice, sustainability education, ecofeminist philosophy, higher education, and social movement activism and community organizing. She is a Teaching Professor in the Sustainable Communities Program at Northern Arizona University and serves as coordinator for the Sustainable Communities MA and Community University Public Inquiry Programs.

Leah Mundell is Associate Teaching Professor in the Department of Anthropology and the Sustainable Communities Program at Northern Arizona University, where she also coordinates the Community Engagement Minor. Her applied research uses community organizing to understand migration and issues affecting immigrant families in the U.S. and South Africa.

Alder Keleman Saxena is an environmental anthropologist and Assistant Professor at Northern Arizona University in the Department of Anthropology and Sustainable Communities Program. Her research considers the intertwining of biodiversity and food systems, particularly in Latin America, and more recently addresses the more-than-human crises of the Anthropocene.

Lissy Goralnik is an Associate Professor in the Department of Community Sustainability and the Associate Director of Research for the Center for Interdisciplinary Research, Collaboration, Learning, and Engagement (CIRCLE) at Michigan State University. Her transdisciplinary work explores how connections to place catalyze care and responsibility for landscapes and communities.

Lindsey Falkenburg graduated from the Masters in Sustainable Communities program at Northern Arizona University. While there, she was the Community Engagement Minor Graduate Facilitator. Her background is in community-based ecological restoration, facilitation, visionary fiction, and project/program management. She currently works as Parks Planning Manager for the City of Redmond.

Jennifer Lee Johnson is Associate Professor in the Department of Community Sustainability at Michigan State University and Founder/Director of the Toxic Action Lab. Working alongside residents, resource managers, scientists, and the state, Johnson researches the confluence of economic, ecological, and socio-political transformations in Uganda and the US Midwest.

Envisioning Pathways for Building Sustainable Communities: A University Coursework Research Agenda

Dzintra Iliško, Jeļena Badjanova, Anita Pipere, and Wei Zhang

Abstract Sustainability has emerged as a global priority, with higher education institutions playing a crucial role in advancing sustainable development through education, research, and practical initiatives. This chapter aims to demonstrate how university coursework on the development of sustainable communities can enhance the sustainability competencies of master-level students and help them envision practical solutions for local challenges, thereby contributing to curriculum development that promotes sustainable development and active citizenship. This paper examines the efforts of Master level students from the Community Management programme at one of the regional universities in addressing pressing sustainability-related issues within the community and globally. The course provides a theoretical foundation for sustainability models and theories, engages students in research projects that explore unsustainability in their communities, and encourages them to envision sustainable solutions. Most of these students are in leadership roles within municipal organisations and government sectors. As part of the coursework, students have identified critical urban issues within the community and proposed actionable solutions for promoting sustainable city development. These solutions include fostering multi-stakeholder collaboration, incorporating eco-friendly materials, and developing sustainable transportation systems. The study underscores the potential for universities to partner with cities in advancing sustainability agendas, emphasising the importance of collaboration in policy planning and implementing environmentally sustainable practices. The methodology employed is the case study that examines the involvement of students from a regional university in Latvia and their participation in a research-focused course on sustainable design and planning, one of the sustainability-oriented educational courses. The paper elaborates on the usefulness of research-based study courses at the University.

Keywords Leadership · Sustainable urban design · Case study · Stakeholder collaboration · Eco-friendly materials · Sustainable transportation

D. Iliško (✉) · J. Badjanova · A. Pipere · W. Zhang
Daugavpils University, Daugavpils, Latvia
e-mail: dzintra.ilisko@du.lv

W. Leal Filho et al., *Curriculum Development for Sustainable Development*, World Sustainability Series, https://doi.org/10.1007/978-3-032-22146-9_18

1 Introduction

This chapter aims to explore the potential of a research- and action-oriented course for master-level community management students to develop their understanding of sustainability challenges in their communities and plan solutions for these issues in collaboration with stakeholders. The course engages students in transdisciplinary research and applies inquiry-based learning, contributing to the development of a curriculum that supports sustainable development.

The course has been designed for master-level students who hold leadership positions in diverse organisations, such as municipalities, police, state security, prisons, schools, sports-related organisations, as well as private companies and businesses, and for students who plan to apply for management positions in the future. Sustainability has become crucial for businesses, as politics increases demands across all governmental and business sectors to apply a sustainability strategy in practice. This raises managers' awareness of their contribution to Sustainable Development Goal 2030 and the implementation of the Sustainable Development Goals (SDGs). The SDGs' framework allows for flexible and holistic approaches to integrating them into the University curriculum. Abulebdeh et al. (2024) assert that Industry 4.0 requires equipping employees with new skills for implementing the SDGs as relevant to their businesses and organisations, and to undertake responsibility for the social, economic, and ecological well-being of citizens.

Students' leadership positions at the local level are essential for sustainable development; therefore, the course intends to cover theoretical and philosophical aspects, as well as practical aspects on how to build more sustainable communities. The course aims to develop students' competencies and capabilities in solving complex issues by applying interdisciplinary approaches, engaging diverse stakeholders, and fostering transformative action and collective leadership. The society is becoming increasingly divided, extremist, and intolerant towards otherness, necessitating alternative future scenarios. Therefore, participants were asked to identify unsustainability in their immediate environment and to join a common effort to make an in-depth exploration of the problem's roots, and to reimagine better solutions in joint efforts with diverse stakeholders from various fields of work to reimagine a better future. By undertaking autonomy at the local level, students and their organisations are becoming alternative centres of power. By mobilising existing networks, the course participants have identified barriers to sustainable solutions and planned a collaboration between diverse stakeholders to bring forward sustainable solutions for the identified problems. As Vare et al. (2019) assert, higher institutions need to enable students to promote meaningful and sustainable change in their communities, and students must acknowledge their role as change agents (Weinberg et al., 2020). The students were engaged in learning about sustainable urban planning and reimagining a more sustainable city (SDG 11). They have explored the cases of best practice from cities that have made significant progress in enhancing sustainability, energy efficiency, and reducing carbon emissions (Lai, 2024). As Wals et al. (2021) note, education needs to harness its power to reorient people's minds towards sustainability through diverse and innovative approaches. Leal Filho (2020) views universities as living Labs for experimentation with innovative ideas to meet the challenges of sustainable development and implementation of SDGs. Universities need to make a strong commitment to integrating sustainability into their programs and to provide strategies for all level program students to become change agents in their communities (Machado & Davim, 2023). Stubbs (2013) sees the potential of the University curriculum to offer sustainability management programs that prepare students as future leaders to carry out a systemic change leading to attitudinal changes. Wang et al. (2022) emphasise the use of efficient pedagogical approaches that lead to transformations in students' mindsets and attitudes, as well as the development of sustainability competencies, by exploring specific issues in their communities through economic, cultural, social, and environmental aspects.

2 Education for Sustainable Development in Progress

Education for Sustainable Development has been at the centre of debate among many scientists and educational experts (Hopkins et al., 2020; Sterling, 2019, 2024; Vare et al., 2019). By considering the interplay of global and local challenges, Latvia has developed a Sustainable Development Strategy spanning until 2030. According to Saima (2010), this serves as a priority in transforming education, as it fosters innovative approaches in management, emphasising the preservation of Latvia's unique cultural space, while developing the agency of all citizens to promote sustainable changes in society. As stated in the Development Cooperation Policy Guidelines for 2021–2027, Latvia has set its thematic priorities, namely; SDG 16, peace and inclusive institutions, gender equality (SDG 5), quality education (SDG 4), decent work and economic growth (SDG 8), climate action (SDG 13) and partnership to implement those goals (SDG 17).

Due to its interdisciplinary nature, education for Sustainable Development is a broad term that encompasses various fields, including global citizenship education, environmental education, and climate change education. By having its core definition derived from the *Our Common Future report* (Brundtland, 1987), "*the development that meets the needs of the present without compromising the ability of future generations to meet their needs*" (p. 43), the definition has been expanded by new dimensions, determined by new topicalities of the day. *The Agenda for Sustainable Development 2030 outlines 17 Sustainable Development Goals (SDGs) and highlights the essence of sustainable development by emphasising environmental sustainability, social inclusion, and economic efficiency, with a focus on a transformative agenda for education towards* a sustainable future. This vision challenges the purpose of education (Hopkins et al., 2020). The new vision of education requires changes of conscience and "an altered sense of being in the world" (O'Sullivan & Morrell, 2002, p. 18). As such, this requires a transformative education that is holistic, involving cognitive, emotional, and interpersonal aspects of learning.

The report '*Reimagining our Futures Together: A New Social Contract for Education*' (UNESCO, 2021) holds a strong advocacy for education as a common good. The report serves as an invitation to action to achieve this goal. The document underscores the transformational power of education in achieving sustainable collective futures and highlights challenges such as inclusion, cooperation, and solidarity. Therefore, universities need to cooperate to implement this vision. Universities need to create a new social contract that encourages all students to work towards the common good and to develop shared purposes.

It requires repurposing education from anthropocentric to a renewed vision of transformed education. It requires epistemic change towards a more resilient system. As stated in the report, "*education needs to be 'ecological, intercultural and interdisciplinary'*" (UNESCO, 2021, p. 4) *to promote transformations towards 'more just, equitable and sustainable futures*' (p. 4). Higher education has a potential role in fostering education for sustainable development. For the transformations to take place, both changes on individual and institutional levels should take place with the involvement of diverse stakeholders. Therefore, all participants should be engaged in capacity building through the development of systems thinking and collaborative decision-making, aspects that were thoroughly considered in the coursework. As Brundiers et al. (2021) assert, education needs to strengthen system, strategic, and future thinking and problem-solving competencies. Lozano et al. (2017) extend this list of competencies by adding justice, responsibility, change of perspective, empathy, tolerance for ambiguity, and the ability to cope with uncertainty as essential for the current day. According to Brundiers et al. (2021), by integrating all those competencies, we can design viable alternatives for sustainable communities. Thus, education offered to master-level students can be described as a holistic and transformational process that serves as an integral part of lifelong and high-quality education. Reaching sustainable development goals requires changes in thinking and becoming "change makers" (UNESCO, 2017). The primary societal and economic challenges that cities and communities face include rapid population growth in major cities, pollution, new energy demands, and other pressing issues.

On the other hand, advanced countries have found viable technological solutions to increase people's quality of life by creating more sustainable communities. Despite advancements in the community's infrastructure, it is suggested that students explore issues that still need sustainable solutions. Imagination of different, alternative and desirable futures will require the use of life-affirming metaphors (Glasser, 2019), to start with (1) learning about the current state of being, (2) understanding the causes and consequences of some taken for granted assumptions, since global unsustainability patterns are deeply grounded in our assumptions; (3) becoming aware of deep roots of systematic causes of unsustainability, (4) imagining more healthier future by the use of life affirming metanarratives as "a manifold possibility" (p. 50), visons; (5) undertaking responsibility and taking steps in transforming current reality (Glasser, 2019). In the era of the Anthropocene, this requires reevaluating how we build relationships with a larger community and strive for the Common good. Sustainability, being defined as a complex issue, can be measured by several dimensions, including our perceptions of quality of life and what constitutes progress.

As Meadows (2024) suggests, the starting point is honouring life, using life-affirming metaphors, employing creative imagination, making new connections, learning and working with others, thus building resilience in diversity, experimenting with the ideas, prototyping, taking action that contributes to a common good, and acting according to our commitments to a sustainable future. The students were encouraged to explore all aspects of life, energy, quality of life, health issues, and local urbanism.

3 The Design of the Course

The course "Sustainable Community development strategy" covers; the United Nations Sustainability goals (SDGs), theoretical and philosophical aspects of the sustainability concept and models, greening possibilities of industries, eco-friendly policies and actions in organisations, cities by exploring cases of good practices internationally and locally, management of sustainable organisational culture and corporate social responsibility of employees, sustainable human resource management, exploring imperatives set in the Green Deal, as well as sustainability of digital transformations in organisations. The course integrates issues such as climate issues, sustainable consumption, collaborative decision-making, and future thinking. The course offers an interactive, learner-centred environment and action-oriented pedagogy. Then, students were encouraged to explore sustainability issues in their environment and jointly develop scenarios for addressing these issues. The course employs transformative pedagogy that supports collaboration, action research, self-directed learning, and transdisciplinarity, leading to transformative action and new competencies that contribute to more sustainable livelihoods. Participants in the course were asked to explore cases of good practice initiated by their Municipalities, local activists, and organisations that employed bottom-up approaches. The course was based on the principles of socio-constructivist learning theory (Fig. 1).

In the post-pandemic era, organisations are encountering widespread unrest, including rising unemployment and uncertainties about the future. Organisations cannot continue business as usual; instead, they need to seek alternative solutions to the current issues, explore new management models, and devise strategies to build

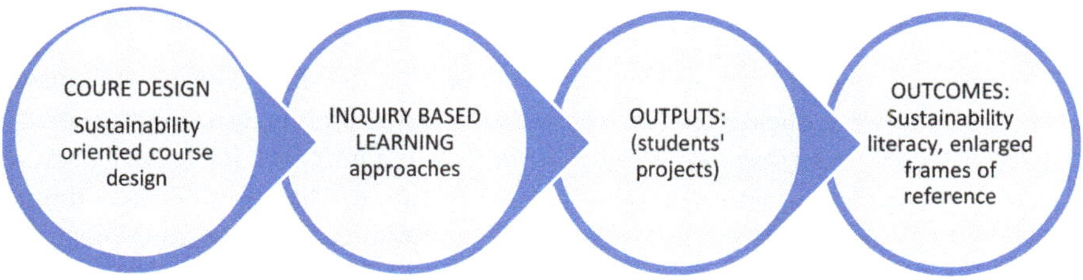

Fig. 1 Sustainability-oriented course design

more inclusive and sustainable cities. The course empowers students to become transformative agents in their cities and organisations, bringing forward an agenda for sustainable and inclusive communities. The aim of the course was not only to equip learners with knowledge about more sustainable models of governance, but also to develop sustainability competencies that foster sustainable transformations in their communities by engaging them in critical thinking and applying transdisciplinary approaches. As many scientists assert, developing sustainability competencies becomes one of the significant challenges in academia (Lozano et al., 2017).

4 Applying Inquiry-Based Pedagogical Approaches

Transformations in higher education necessitate changes in pedagogical approaches towards more inquiry-based methods that foster critical thinking, active engagement, and creative problem-solving (Almulla, 2023). New educational arrangements and models require recontextualising current arrangements. A growing body of literature suggests that students create new knowledge more effectively by applying inquiry and discovery learning, Socratic questioning, reflection, and future envisioning. This approach enables one to view knowledge not as a collection of theories, but as situated in an always-evolving, living field (Davis & Luce-Kapler, 2008).

Management programme students need those skills to solve complex issues in their industries and participate fully in their communities. Equally important, inquiry-based learning enables them to pose their own questions and construct their own meaning (Sam, 2024). This method is grounded in socio-constructivist theories, which enable students to take ownership of their learning. Cooperation with classmates in solving complex societal issues is also among the required 21st-century skills in industries (Cochran-Smith & Lytle, 2009) note that full engagement and immersion in problem-solving can be fostered by promoting not only cognitive thinking but also innovation-driven leadership in the industries and workplaces where students are employed. As leaders, they will (Leal Filho et al., 2020) take responsibility for more sustainable processes and involve others in this co-creation process. As argued by Leal Filho et al. (2020), students who absorb a sustainability culture and engage in sustainability initiatives will become sustainable leaders of tomorrow. By strengthening their visionary and creative capacities, they would contribute to a sustainability agenda.

Problem-solving skills are also listed among the essential skills. During the coursework, students were encouraged to explore real-world problems relevant to them, seeking a deeper understanding (Cochran-Smith & Lytle, 2009). By building future scenarios, the students could reimagine the desired future. There is much uncertainty and turbulence about how the future may unfold. There is no historical evidence to rely on in building the future scenario. Scenario planning, as a collaborative effort, helps students gain new insights into opportunities and challenges for future developments. Moltmann (2009) advocates for envisioning a future by not withdrawing from the world and not holding unrealistic hopes, but by having realistic hopes and by engaging actively in the transformation processes that build a more sustainable future. Moltmann's theology of hope requires one to engage in positive future thinking, leading to the healing of existing injustices, renewing hope for sustainable transformations and renewed relationships, and fostering hope for learning into the sacramentality of creation. Education has been based on a growth-based modernisation paradigm. Now we need to move towards a new ecologically oriented paradigm of education that recognises the limits of growth and hopes for a different kind of future. Therefore, universities need to adapt more effective approaches to developing students' capabilities and skills to work towards desired futures.

5 Research Methodology

The findings presented in the paper are derived from a case study conducted at one of Latvia's regional universities. The case study presents an exploratory case based on descriptive analyses. The case study is designed to provide a deeper understanding of innovative, action-oriented approaches applied in university-level courses. By offering innovation-oriented discourse, the authors have analysed the research projects presented by students on chosen unsustainable issues in the community, as well as by the analysis of students' reflections on their gains, and developed a richer understanding of sustainability. The study employed a qualitative research methodology, which, as noted by Queirós et al. (2017), is related not to numerical representation but to a deeper understanding of the researcher's issue and aims at in-depth knowledge of the usefulness of inquiry-based approaches in understanding community issues and reimagining viable solutions. However, the case study enables researchers to examine complex situations with multiple variables and advanced knowledge bases from various fields, offering an opportunity for innovation. For this study, the authors have employed a case study design focusing on a single regional university. They employ descriptive and explanatory design to explain the benefits of the case study. For a deeper exploration, the authors have employed thematic analyses of students' reflections on their projects by analysing their written feedback. Similar cases could be implemented in the curriculum design of other higher institutions as well as aligned with the competency-based approach.

6 Research Findings

Each group of students has identified an unsustainable issue in their community; they have explored the cause of this issue, the vulnerabilities, and the capacities of all involved members by designing diverse solutions to address a chosen issue. Among the problems identified by the students, building a more inclusive environment in town, implementing in-practice waste management (organic composting) (how to reduce waste by prevention, reduction, and recycling), greening of the city, carrying out a more sustainable hospitality (tourism) business, and teaching youth about sustainable consumption were highlighted. The chosen projects were designed in accordance with the thematic priorities set by the Government of Latvia, specifically SDG 16 (inclusive institutions), SDG 5 (gender equality), SDG 4 (quality education), SDG 8 (decent work), and SDG 13 (climate action). The students were encouraged to work in cross-disciplinary teams by exploring political, economic, social, and cultural aspects of their issue of study. The course aimed at deepening students' knowledge about sustainability issues and current research, as well as developing students' competencies to carry out exploration and investigation of complex climate and societal, economic, and sustainable issues in their communities, by empowering them to engage in envisioning more sustainable solutions to unsustainable current problems in more interdisciplinary partnerships and cooperation. They were asked to reflect on how their initiatives align with the communities long-term planning strategies and policies, and whether they have a positive impact on local communities.

6.1 Envisioning Sustainability in the Community

The students were invited to explore aspects of unsustainable development in a small peripheral, "paradigmatically shrinking town", the most distant from the capital, the poorest, and "the most neglected financially and politically" (Mallach, 2024, p. 18) that global forces and trends, such as many cities in Latvia, have influenced. The town has gained its status as "the other" due to its poor connection to the rest of Latvia and its economic backwardness. To prevent negative growth, recession, and an alarming number of young people leaving the town, the students were asked to envision more viable scenarios for development. Due to economic marginality in the city, characterised by a lack of jobs, many young people are leaving to other EU countries, resulting in the city's social and linguistic status being defined as "the Other." Based on the statistics, Mallach (2024) describes Latvia as "one of the world's fastest-shrinking countries whose population since 1991 has dropped from 2.679 million to 1.882 million" (Eurostat, 2024), and IHME (2022) predicts it to be 1.253 million in 2030.

The students in this course have developed a vision of a sustainable city that encompasses renewable energy, high-quality education, sustainable transportation, inclusive infrastructure for all, effective waste management, and the adoption of sustainable initiatives, including accessible transportation and the greening of the city. They were asked to identify unsustainable practices in their environment and seek viable solutions.

Selected cases for exploration differ in terms of type and context. Evaluation of the presented cases was based on how well students went beyond reporting pure information about the selected case. It also addressed whether all sustainability dimensions (political, economic, cultural, and ecological) were covered, and if their interconnectivity had been acknowledged, as well as students' agency in reimagining a better future and developing more innovative solutions. Furthermore, the evaluation assesses whether suggested initiatives lead to positive transformations, including whether students become aware of the necessary systemic changes and the changes required at both the individual and societal levels.

By engaging with the communities of learning, students had the opportunity to identify issues for deeper exploration and develop a shared understanding. Senge (1990) emphasizes the value of such learning communities in developing shared and alternative future visions through the active participation of all students, including involving them in a continuous process of learning from one another. Sharing of visions was carried out in a dialogue where all students were invited to participate in collective planning and decision-making on the most viable future scenarios of the chosen issues. Inquiry-based learning allowed the course participants to undertake an ownership. As one of the participants commented in the reflection: "*Together with other participants of the course, I engaged in discussion in an informal and friendly atmosphere until we agreed on some aspects of how to investigate a chosen issue*" (Table 1).

One group of students examined the issue of enhancing quality education (SDG 4), highlighting the fragmented efforts to integrate sustainability into school settings. They opted to investigate the application of sustainability as a comprehensive institutional approach to more effectively address the diverse aspects of sustainability in current practices. The identified aspects included sustainable waste management, the use of sustainable learning materials, building a sustainable school culture, non-hierarchical leadership, and fostering stronger communities of practice where teachers were empowered to take ownership of small sustainable initiatives and practices. They reported that "*All is changing, students are changing, therefore, schools and approaches towards teaching need to change as well to be more sustainable.*"

The students who were engaged in exploring issues related to waste of food and sustainable consumption (SDG 12) concluded that, for the transformative change to take place, "*there is a need for engagement and cooperation among many stakeholders, changing behaviours and attitudes as well as spreading the cases of good practice among youth*" The other participants of this research group commented that joint planning with participants from different field of work allowed "*to come up with innovative and alternative solutions to societal problem.*" Envisioning solutions in collaboration with diverse stakeholders enabled course participants to plan collective action and take ownership.

The group that had explored the environment in organisations (SDG 16; SDG 5) concluded that "*it was far from what we call inclusive to people with diverse needs.*" They made careful observations and photo documentation of the settings that were not inclusive for all people. They explored the policy and practice from the perspective of inclusivity, examining whether it had flexible arrangements and

Table 1 The projects explored by the groups of students were as follows

SDG	Vision	Type of activities	Stakeholders involved
SDG 4 Quality education	Building a sustainable school environment) (waste management, sustainable use of materials, involvement of stakeholders in co-creation)	Experiential learning, co-creation, co-production, joint learning, sharing ideas, and examples of good practice	School administration, NGO's, municipality, school council, parents, students' council
SDG 16, SDG 5 Inclusive environment in the institutions	Building an inclusive environment (gender inclusive policy, infrastructure)	Co-creating solutions for inclusive practices, undertaking ownership towards inclusive behaviour, and participating in a professional community of learning	Municipality, employees, citizens
SDG 13, SDG 11 Climate action, greening the cities	Making the city a more sustainable and green place involves greening cities, expanding urban green spaces, and supporting and organising climate actions	Re-imagining a greener city, advocating for sustainability in the city, strategic planning, reimagining ecologically more sustainable ways of living, and green markets	Municipality, NGO's, businesses and industries, citizens, government, universities, community groups, innovators
SDG 13, SDG 12 Climate action, sustainable consumption	Promoting sustainable food consumption among youth, raising awareness on responsible consumption, and encouraging sustainable lifestyles (minimalism, sharing economy, local products)	Building awareness, developing knowledge and attitudes, behaviours, and a sustainability mindset among youth. The use of digital tools, game-based learning, and evaluating one's footprint	Educators, NGO's, mass media, and citizens

(continued)

Table 1 (continued)

SDG	Vision	Type of activities	Stakeholders involved
SDG 7 Transition to a renewable energy	Encourage private sector investment in renewable projects, promote public–private partnerships, integrate renewable energy topics into school curricula, and encourage energy-efficient behaviour	Exploring different aspects of the issue, examining costs and long-term benefits and climate impact, prototyping for finding better local solutions, and learning from scientists from other disciplines	Government, municipality, business companies, NGO's
SDG 11 Smart cities are more sustainable and inclusive venues	Ensuring inclusive urban planning, promoting green buildings with innovative energy systems, and developing participatory digital platforms that enable residents to contribute to the co-design of city projects	Optimised resource consumption, faster and more efficient public services, optimisation of public services, sustainable infrastructure, and sustainable modes of transportation	Government, non-profit organisations, municipality, business incubators, civil society organisations

an anti-discriminatory policy, if all participants had equal access to training and advancement, accessibility, psychological safety, training opportunities, and how diversity, inclusion, and equity had been prioritised in their institution. The group working on this issue concluded that, even if the policy is inclusive, "*much work needs to be done to ensure inclusivity in practice.*" After completing their inquiry, they have commented that "*deep exploration of the issue allowed them to discover many hidden aspects that need to be raised to ensure more inclusive everyday practice in the organisation.*"

The group of participants who chose to work on exploring a renewable energy project (SDG 7) concluded that small initiatives have been implemented in the city. Regarding renewable energy, such as solar power panels, it was highlighted that some research initiatives are conducted within Nordic-Baltic programmes. Since Latvia aligns with the EU's clean energy goals, several initiatives are fostering a gradual transition to renewable power. They acknowledged a lack of strategic thinking and action in intensifying such initiatives in the city. More collaborative efforts are needed at all levels to pursue this goal.

One group of students discussed the model of smart cities (SDG 11) as more inclusive and sustainable due to intensive use of innovative technologies. They referred to the UNICE (2021) concept of a holistic approach to building sustainable cities, which includes the notion of harmonious societies. This reflects a city that is modern yet preserves its heritage and culture, making people's lives "more liveable" (UNICE, 2021).

The group that focused on SDG 7 (a transition to renewable energy) explored aspects such as the use of low-carbon energy sources. Additionally, they discussed the pros and cons of using solar panels, the concept of degrowth, reducing energy use, and strategies for minimising one's ecological footprint. Historically, most societal transformations have been driven by revolutions, wars, and natural disasters. However, this group of students explored viable solutions for changes at the individual citizen's level, focusing on reducing energy use and seeking alternative energy sources (Table 2).

Inquiry learning is a practical approach that enables students to recognise problems in their surroundings or within companies, formulate testable questions, analyse, evaluate, and interpret findings, and propose solutions to identified problems, communicating them to their classmates. The engagement of students is significantly higher since they choose the issue for exploration that is more relevant to them (Cents-Boonstra et al., 2020). Inquiry methods allow students to become active participants in the learning process by considering alternative solutions to the issues. By integrating sustainability issues into the coursework, universities can equip future graduates to manage sustainability issues in their work settings. Students of all majors should be aware of sustainability issues. By involving them in inquiry learning, they can learn through hands-on, real-life experiences and case studies (Abo-Khalli, 2024). A comprehensive strategy at the University level is needed to integrate sustainability issues not only in coursework but also in research and management processes for sustainability to become a whole institutional approach.

Table 2 Sustainability competences developed during coursework

Competences	Description	Activities	Outcomes
System thinking	Understanding the interrelation among political, cultural, economic and environmental aspects	Mapping sustainability challenges	Analyses of communal issues
Critical thinking	Questioning current assumptions, evaluating current practices	Group discussions, reflections	Evidence-based decision making
Collaboration	Collaborating with diverse stakeholders	Projects with multiple stakeholders	Social learning
Action	Searching solutions	Scenario planning, design thinking	Undertaking leadership
Future thinking	Envisioning futures	Inquiry-based projects	Long-term planning

7 Challenges and Limitations of the Study

Due to its qualitative character, the study's findings cannot be generalised internationally, as it is a case study conducted at a single regional University. Qualitative inquiry facilitated a deeper understanding of teaching sustainability issues through the inquiry-based approach in the university setting, as well as the gains, potentialities, and limitations of using inquiry-based approaches while exploring sustainability issues in the community. Case studies challenge current assumptions and practices, offering a foundation for innovation.

Regarding the limitations of the present study, it is noteworthy that the research was conducted at a single university, which has a relatively small sample size. As a qualitative study, it has limited generalisability and may contain student self-reporting bias. For the future perspective, digital sustainability tools might be integrated, and comparative studies across regions can be employed.

8 Conclusions

This chapter showcases the potential of action-oriented research-based study course design for a better understanding of unsustainable issues in the community and possibilities for working towards viable solutions by engaging stakeholders.

The course, designed for master's-level students, helped them become aware of urgent issues in society and the city, and to address these issues in a transdisciplinary way and to reimagine how they can become agents of change and commit themselves to sustainable societal transformations. The course unleashed their potential to engage with future thinking and collaborative action planning, together with multiple stakeholders.

Offering a sustainability course for Master's-level management programme students, and exposing them to sustainability issues topical in their communities and globally, makes it possible to bridge the gap between theory and practice and mobilise students to become agents of change in their communities. The opportunity to train current and future leaders in sustainability enables them to enhance their capacity to become influential transformational leaders in their organisations and communities, thereby implementing sustainable practices within their organisations.

The course fostered historically low levels of university-business and NVO cooperation in discussing urgent sustainability issues by developing cross-sectoral partnerships. As Whitehouse (2015) suggests, cross-sectoral partnerships between institutions and individuals can foster a more efficient sustainability agenda in communities. This enabled a shared vision and leadership, as well as initiative and open-mindedness, among all actors involved. Cooperation with diverse partners, including businesses, local municipalities, networks, alliances, and NGOs, would enable students to learn from and about real-life challenges, benefiting from discussions with various stakeholders.

Sustainable development was viewed by the students as focusing not only on environmental aspects but also on integrating all dimensions, including social, political, economic, and cultural aspects, by involving students in an inquiry-based process for exploring their cases. The task was to educate and raise awareness not only of theoretical models and notions of sustainable management and governance, but also to engage students in identifying unsustainability in their neighbourhood and modelling viable solutions. Linking theory to practical cases will help students develop the competencies necessary to be change agents in their communities.

Inquiry-based learning and a case study approach were used as an effective pedagogical tool for sustainability teaching, developing students' skills, knowledge, and attitudes by engaging them in the investigation of real-life community issues. The coursework provided students the opportunity to engage with real-life situations. By applying transformative pedagogy, students were encouraged to ask critical questions and offer solutions for community problems. By engaging in collaborative problem-solving efforts, the course fostered students' knowledge construction through brainstorming ideas and reimagining future scenarios. By proposing solutions to chosen community issues, they applied principles of sustainability.

There is a strong need for a curriculum framework that promotes sustainability literacy among students in management programs. As graduates of the program, they will pursue careers in municipalities, governmental organisations, and communities, and will engage in implementing the Agenda 2030 and SDG goals in their communities. Such a curriculum framework will be transformational in shaping students' attitudes towards acting sustainably. Action- and inquiry-oriented pedagogical approaches are the most suitable for transformative experiences and practices. These approaches are more viable in both teacher training and management programs.

Solving unsustainability issues requires cooperation between diverse sub-systems and stakeholders in the community, leading to the co-creation of more sustainable visions for society. By approaching real-world problems, the students have demonstrated action-oriented competence and chosen sustainability-related issues that are of the most concern to them and their community. The course strengthened their transversal competences, such as negotiation, problem-solving, decision-making, and finding consensus. Inquiry-based learning developed the students' action competence and built their self-reflectivity.

Education for sustainable development involves not only the transmission of knowledge about emerging climate, social, and economic issues but also a need to focus on transformative and transdisciplinary teaching to highlight the interconnectivity with socio-economic and political systems. This would be promoted by engaging students in an active exploration and inquiry process of the issue from multiple perspectives and engaging in undertaking 'ownership' of initiating positive change in their communities.

References

Abo-Khalli, A. G. (2024). Integrating sustainability into higher education: Challenges and opportunities for universities worldwide. *Heliyon, 10*(3), e29946. https://doi.org/10.1016/j.heliyon.2024.e29946

Abulebdeh, A., Zaidan, E., & Abulibdeh, R. (2024). Navigating the confluence of artificial intelligence and education for sustainable development in the era of industry 4.0: Challenges, opportunities, and ethical dimensions. *Journal of Cleaner Production, 437*, 140527. https://doi.org/10.1016/j.jclepro.2023.140527

Almulla, M. A. (2023). Constructivism learning theory: A paradigm for students' critical thinking, creativity, and problem solving to affect academic performance in higher education. *Cogent Education, 10*(1), 1–25. https://doi.org/10.1080/2331186X.2023.2172929

Brundiers, K., Barth, M., Cebrián, G., et al. (2021). Key competencies in sustainability in higher education—Toward an agreed-upon reference framework. *Sustainability Science, 16*, 13–29. https://doi.org/10.1007/s11625-020-00838-2

Brundtland, G. H. (1987). *Our common future: Report of the World Commission on Environment and Development*. Geneva, UN-Dokument A/42/427. http://www.un-documents.net/ocf-ov.htm. viours

Cents-Boonstra, M., Lichtwarck-Aschoff, A., Denessen, E., et al. (2020). Fostering student engagement with motivating teaching: An observation study of teachers and students' behaviours. *Research Papers in Education, 36*(6), 754–779. https://doi.org/10.1080/02671522.2020.1767184

Cochran-Smith, M., & Lytle, S. L. (2009). *Inquiry as stance: Practitioner research for the next generation*. Teachers College Press.

Davis, S., & Luce-Kapler. (2008). *Engaging minds: Changing teaching in complex times*. Lawrence Erlbaum Associates.

Development Cooperation Policy Guidelines for 2021–2027. (2021, April 16). The Ministry of Foreign Affairs: Cabinet of Ministers. *Latvijas Vēstnesis, 73*.

Eurostat. (2024). *Population projections to 2100* (updated March 1, 2024). European Union.

Glasser, H. (2019). Toward robust foundations for sustainable well-being societies: Learning to change by changing how we learn. In J. Cook (Ed.), *Sustainability, human well-being, and the future of education*. Palgrave Macmillan. https://doi.org/10.1007/978-3-319-78580-6

Hopkins, C. A., Michelsen, G., & Salīte, I. et al. (2020). Sustainability as a purpose and the new path of learning for the future. In *Humanistic futures of learning*. UNESCO.

Humanistic futures of learning: Perspectives from UNESCO Chairs and UNITWIN Networks (2020). https://unesdoc.unesco.org/ark:/48223/pf0000372577. https://doi.org/10.54675/AYFL2310

Institute for Health Metrics and Evaluation (IHME). (2022). University of Washington Population Forecasts, Reference Scenario: 2018–2100.

Lai, O. (2024). *Top 7 smart cities in the world in 2022*. https://earth.org/top-7-smart-cities-in-the-world/

Leal Filho, W. (2020). Living labs for sustainable development: The role of the European school of sustainability sciences and research: In W. Leal Filho et al. (Ed.), *Universities as labs for sustainable development. World sustainability series*. Springer. https://doi.org/10.1007/978-3-030-15604-6_1

Leal Filho, W., Eustachio, J., Caldana, A. C., Will, M., Lange Salvia, A., Rampasso, I., Anholon, R., Platje, J., & Kovaleva, M. (2020). Sustainability leadership in higher education institutions: An overview of challenges. *Sustainability, 12*, 3761. https://doi.org/10.3390/su12093761

Lozano, R., Merrill, M., Sammalisto, K., Ceulemans, K, & Lozano, F. (2017). Connecting competences and pedagogical approaches for sustainable development in higher education: A literature review and framework proposal. *Sustainability, 9*, 1889. https://doi.org/10.3390/su9101889

Machado, C. F., & Davim, J. P. (2023). Sustainability in the modernisation of higher education: Curricular transformations and sustainable campus—A literature review. *Sustainability, 15*(11), 8615. https://doi.org/10.3390/su15118615

Mallach, A. (2024). Peripherality, marginality and global change: Daugavpils and the future of peripheral cities. *Journal of Comparative Studies, 17*(46), 1–47. https://doi.org/10.59893/jcs.17(46).001

Meadows, D. (2024). Envisioning a sustainable world. In R. Costanza & I. Kubiszewski (Eds.), *Creating sustainable and desirable future: Insights from 45 global theologian leaders* (pp. 9–14). World Scientific.

Moltmann, J. (2009). *Theology of hope: On the ground and the implications of Christian eschatology*. Translated from German by J. Leitch. Augsburg Fortress.

O'Sullivan, E., & Morrell, A. (Eds.). (2002). *Expanding the boundaries of transformative learning: Essays on theory and praxis*. Palgrave Press.

Queirós, A., Faria, D., & Fernando, A. (2017). Strengths and limitations of qualitative and quantitative research methods. *European Journal of Education Studies, 3*(9), 369–387. https://doi.org/10.5281/zenodo.887089

Saima. (2010). *Sustainable development strategy of Latvia until 2030*. Saima. https://www.mk.gov.lv/lv/media/15129/download?attachment

Sam, R. (2024). Systematic review of inquiry-based learning: Assessing impact and best practices in education. *F1000Research, 13*, 1045. https://doi.org/10.12688/f1000research.155367.1

Senge, P.M. (1990). The fifth discipline: The art and practice of the learning organization. Doubleday.

Sterling, S. (2019). *Becoming 'learner drivers' for the future: Re-thinking learning and education in a troubled world*. Routledge.

Sterling, S. (2024). *Learning sustainability in dangerous times*. Norwgen Publishing.

Stubbs, W. (2013). Addressing the business-sustainability nexus in postgraduate education. *International Journal of Sustainability in Higher Education, 14*, 25–41. https://doi.org/10.1108/14676371311288930

UN. (2015). Transforming our world: 2030 Agenda for Sustainable Development. https://sustainabledevelopment.un.org/content/documents/21252030%20Agenda%20for%20Sustainable%20Development%20web.pdf?ref

UNESCO. (2017). Education for sustainable development: Learning objectives. https://unesdoc.unesco.org/ark:/48223/pf0000247444. https://doi.org/10.54675/CGBA9153

UNESCO. (2021). *Reimagining our futures together: A new social contract for education*. International Commission on the Futures of Education. https://unesdoc.unesco.org/ark:/48223/pf0000379707?posInSet=1&queryId=54ba8493-2ba4-41bb-8510-683267361b42

UNICE. (2021). *People-smart sustainable cities*. https://unece.org/housing/publications/people-smart-sustainable-cities-E

Vare, P., et al. (2019). Devising a competence-based program for educators of sustainable development: Lessons learned. *Sustainability, 11*(7), 18–90. https://doi.org/10.3390/su11071890

Wals, A. E. J., Tassone, V. C., Hampton, G. P., & Reams, J. (2021). Eco-social innovation through sustainability-oriented higher education. In M. Barth, G. Michelsen, M. Rieckmann, & I. Thomas (Eds.), *Handbook of higher education for sustainable development*. Routledge. https://doi.org/10.4324/9781315852249

Wang, Y., Sommier, M., & Vasques, A. (2022). Sustainability education at higher education: Pedagogies and students' competences. *International Journal of Sustainability in Higher Education, 23*(8), 174–193. https://doi.org/10.1108/IJSHE-11-2021-0465

Weinberg, A., Trott, C., Wakefield, W., et al. (2020). Looking inward. Outward, and forward: Exploring the process of transformative learning in teacher education for a sustainable future. *Sustainability Science, 15*(2), 1767–1787. https://doi.org/10.1007/s11625-020-00831-9

Whitehouse, H. (2015). Cross-sectoral relationships for education for sustainability. In G. Moscardo, & P. Benckendorff (Eds.), *Education for sustainability in tourism, CSR, sustainability, ethics and governance*. Springer. https://doi.org/10.1007/978-3-662-47470-9_8

From Theory to Practice: Rethinking the Sustainable Design Curriculum in Mexico Through Complexity and Participation

Francisco Platas-López

Abstract Traditional pedagogical models in Mexican architecture and design programs are often characterized by nonjudgmental, object-centered stances. These models are inadequate for addressing complex sustainability challenges. This chapter presents an alternative interdisciplinary curriculum framework developed at the National Autonomous University of Mexico (UNAM). The framework is designed to provide architects and designers with the systemic and ethical competencies necessary to address today's complex challenges. The primary objective is to validate this pedagogical model by detailing its theoretical foundations, demonstrating its practical application, and documenting its real-world impact. The proposed framework integrates constructivist epistemology, participatory design methodologies, and the sciences of complexity. This shifts the educational focus from creating objects to making systemic interventions. Two case studies demonstrate its practical application in a rural indigenous community and an informal urban settlement. These studies confirm the framework's adaptability and effectiveness. These conclusions are directly linked to the project's practical application. The project received a United Nations Special Prize for its significant contribution to Sustainable Development Goal 10, which aims to reduce inequalities. This award validates the framework's value, which lies not only in its physical outcomes but also in its process. The process promotes procedural and epistemic justice by rebalancing the power dynamics between experts and communities. In conclusion, this pedagogical model is a relevant, replicable approach to reforming sustainable design education, especially in the Global South.

Keywords Curriculum studies in Mexico · Sustainable design education · Complex design for vulnerability and risk · Sustainable development goals (SDGs) · Curriculum framework

---

F. Platas-López (✉)
Universidad Autónoma del Estado de México, UAEMéx, Casa de Cultura La UAEMéx en Tlalpan, Mexico City, México
e-mail: fplatasl@uaemex.mx

W. Leal Filho et al., *Curriculum Development for Sustainable Development*, World Sustainability Series, https://doi.org/10.1007/978-3-032-22146-9_19

1 Introduction

Latin American societies currently navigate an environment defined by profound uncertainty and complexity. These societies confront a dual set of challenges: those shared with developed nations, such as environmental pollution, resource scarcity, and urban overcrowding, alongside persistent issues largely eradicated in the Global North, including endemic diseases, food insecurity, and high rates of inequality and corruption (Platas & Carreto, 2025). In this context, sustainability transcends its role as an adjective or a specialization to become the central, defining problem of design. Consequently, the training of architects and designers in Mexico has arrived at a critical juncture. Inherited pedagogical models, generally anchored in a modernist vision that privileges the creation of aesthetic and functional objects in isolation, are proving profoundly inadequate for addressing the systemic nature of contemporary socio-ecological crises.

This traditional approach, which often treats complex problems as discrete technical challenges, fails to equip practitioners with the necessary tools to understand or intervene in interconnected systems. The simple aggregation of knowledge, frequently drawn from dissimilar or even antagonistic epistemic frameworks, does not yield an integrated understanding. Instead, it produces an accumulation of disconnected perspectives that cannot capture the feedback loops, emergent properties, and non-linear dynamics inherent in sustainability problems. The unique challenges of the region thus demand more than a simple update to course content; they necessitate a fundamental departure from pedagogical models that are often uncritically imported and ill-suited to the realities of the Global South. An effective educational response must arise from and be tailored to its specific context, revaluing local knowledge systems and community-led solutions.

This chapter presents and argues for an alternative theoretical-methodological framework that invites a rethinking of the sustainable design curriculum in Mexico. It proposes complexity and participation as a robust scaffolding for coherent disciplinary convergence. This framework moves beyond a passive understanding of complexity to actively apply its principles, recognizing and empowering disadvantaged communities as active agents in the construction of their own habitats and futures. To demonstrate the validity and applicability of this approach, the chapter is structured in three parts. First, it presents the current state of curriculum development for sustainable design in Mexican architectural education. Second, it details the innovative theoretical, epistemic, and methodological foundations of the curriculum for the subject "Sustainability in Architecture and the City" at the National Autonomous University of Mexico (UNAM), followed by an analysis of two practical case studies where this framework has been applied. Finally, it discusses the international recognition this work has received, particularly its contribution to the UN Sustainable Development Goals, and reflects on its broader implications, scope, and limitations.

To address this critical gap between traditional pedagogy and contemporary needs, this chapter introduces and validates an alternative curriculum framework developed and implemented at the National Autonomous University of Mexico (UNAM). This framework integrates the science of complexity with participatory design methodologies to equip emerging architects and designers with the systemic and ethical competencies required to tackle socio-ecological challenges. We argue that by shifting the pedagogical focus from object-creation to system-level intervention, it is possible to cultivate a new generation of professionals capable of designing more sustainable and resilient futures. This chapter documents the theoretical foundations, practical application, and demonstrable impact of this innovative educational model, positioning it as a replicable approach for sustainable design education, particularly within the context of the Global South.

Accordingly, the primary objectives of this chapter are threefold:

1. **To present** the theoretical and epistemological foundations of a novel interdisciplinary curriculum for sustainable design, grounded in constructivism, complexity science, and participatory methods
2. **To demonstrate** the framework's practical effectiveness and adaptability through an analysis of its implementation in two distinct case studies—a rural indigenous community and an informal urban settlement in Mexico
3. **To validate** the model's contribution to achieving global sustainability targets, particularly the UN Sustainable Development Goals, thereby establishing its broader relevance for higher education institutions seeking to reform their design curricula.

2 Methodology: Theoretical Foundations for an Integrated Curriculum

The proposed curriculum is built upon a coherent synthesis of three theoretical pillars: a constructivist and participatory epistemology, which redefines the roles of designer and community; and the sciences of complexity, which provide the analytical tools to understand and engage with socio-ecological systems.

2.1 A Constructivist and Participatory Stance

The framework is grounded in a constructivist epistemology, which posits that knowledge is not a passive reflection of reality but is actively constructed through an individual's interaction with their environment. This perspective, drawing on the work of Jean Piaget (1950) as interpreted by Rolando García (2006), understands the city not as a static object but as a complex system whose elements can only be understood through their mutual relationships. This epistemological stance necessitates a pedagogical shift away from knowledge transmission and toward capacity building through experiential learning.

This foundation naturally leads to a participatory design methodology. If knowledge is co-constructed, then the traditional hierarchy of the "architect-genius" imposing solutions upon a community becomes untenable. The framework reframes the designer's role from that of a solitary author to a facilitator of "collective intelligence". The architect becomes a technical companion, while the community becomes the active protagonist in the design process, providing irreplaceable knowledge of their own lived experiences. This approach aligns with the work of Romero and Mesías (2004), who argues for understanding habitat production not as a technical problem to be solved, but as an integral and interdisciplinary process driven by the community itself.

To bridge the ethical and the experiential, the curriculum incorporates Katya Mandoki's concept of "aesthetics for everyday life" (prosaic aesthetics) (Mandoki, 2016). This moves aesthetics beyond the traditional study of art and beauty to focus on esthesis: the sensory and cognitive opening of the subject to the world. By analyzing social interactions through visual, auditory, verbal, and bodily registers, students learn to perceive buildings and cities as living, interconnected systems. This fosters an awareness of the environment that is essential for developing a holistic design consciousness—one that is at once personal (self-ethical), communal (socio-ethical), and planetary (anthropo-ethical).

2.2 Navigating Complexity in Socio-ecological Systems

The framework's unique analytical power derives from its synthesis of three distinct but complementary schools of complexity theory that have been influential in Latin America. It combines the philosophical-humanistic approach of Edgar Morin (2024), the structural-epistemological framework of Rolando García (Castorina & Barreiro, 2023), and the computational-modeling tools of the broader sciences of complexity. This integration creates a robust meta-methodology capable of addressing both the qualitative (cultural, ethical) and quantitative (network analysis, simulation) dimensions of sustainability challenges.

2.2.1 Edgar Morin's Complex Thought

Developed by the French thinker Edgar Morin, complex thought provides a way of thinking that is capable of weaving together seemingly contradictory elements (Morin, 2024). The curriculum applies three of Morin's guiding principles to urban and architectural problems.

Dialogical thinking: This principle addresses the need to maintain the duality of contradictory yet complementary concepts. For example, the modern city is a product of the dynamic tension between order (urban planning, infrastructure) and disorder (informal growth, pollution). An attempt to eliminate one pole inevitably alters the other; understanding this dynamic is crucial for effective intervention.

Recursive organization: This principle describes cyclical causality. A society's ideology produces its architecture and urban forms, but these built environments, in turn, modify the environment and reshape societal norms and demands. For instance, a lack of urban green space contributes to "heat islands" and health problems, which generates social demand for more sustainable policies, creating a continuous feedback loop.

Hologrammatic principle: This principle posits that the part is in the whole, and the whole is in the part. A local environmental problem, such as a single street with little vegetation, is not an isolated issue but a reflection of the global dynamics and urban policies of the entire metropolis. Conversely, a small-scale intervention, what Morin calls an "ecology of action" like a green roof, contains the potential to influence the entire system (Distler, 2021).

2.2.2 Rolando García's Complex Systems

Rolando García's genetic constructivist epistemology provides a structured analytical framework to "decompose" complex problems into their constituent parts and scales of influence, making them researchable (Becerra & Amozurrutia, 2015). Key principles include:

Principle of Stratification: Urban reality is analyzed across three interconnected levels: the global/international (e.g., financing policies of international organizations), the national/territorial (e.g., land use laws, building regulations), and the local (e.g., the design of a specific building, community organization). The framework emphasizes understanding the interactions between these levels, moving beyond traditional architectural studies that often focus solely on the local.

Principle of Internal Articulation: Within each level, there are subsystems (economic, political, technological, environmental) whose elements are more tightly interconnected with each other than with elements of other subsystems. For example, at the local level, a new environmental regulation (political subsystem) mandating sustainable materials (technological subsystem) can increase construction costs (economic subsystem), affecting housing affordability.

Boundary Conditions: These are the flows of energy, capital, information, and influence that occur between the different levels and subsystems. For example, global capital flows (third-level processes) directly influence national urban development projects (second-level processes), which in turn determine the viability of local architectural designs (first-level processes).

2.2.3 Complex Systems Science

Drawing on the work of international research centers such as the Centre for Advanced Spatial Analysis (CASA), the Santa Fe Institute (SFI), and UNAM's

Center for Complexity Sciences (C-3), the curriculum introduces students to a range of analytical and modeling tools.

Cellular Automata (CA) and Agent-Based Models (ABM): These are computational models used to simulate the evolution of complex systems from simple local rules. CA can model urban sprawl or deforestation, while ABM can simulate the actions of autonomous agents (e.g., people choosing transport routes) to understand emergent system-level behavior.

Complex Networks and Space Syntax: Network theory is used to analyze the structure of urban systems (transport, social, ecological networks) to identify vulnerabilities and opportunities for enhancing connectivity. When combined with Space Syntax, which analyzes how spatial configuration affects human movement, it allows for the design of more intuitive, functional, and socially vibrant public spaces.

Fractal Principles and "Urban Acupuncture": The curriculum also explores concepts like fractal geometry to design more organic and efficient urban systems, and "urban acupuncture" (Elshinawy et al., 2023).

3 The Curriculum in Practice

The theoretical foundations described above are translated into a concrete pedagogical structure within the Master's in Architecture program at UNAM as shown in Fig. 1.

The curriculum is designed as a cognitive scaffold, deliberately building student capacity layer by layer—from historical context to philosophical grounding, ethical

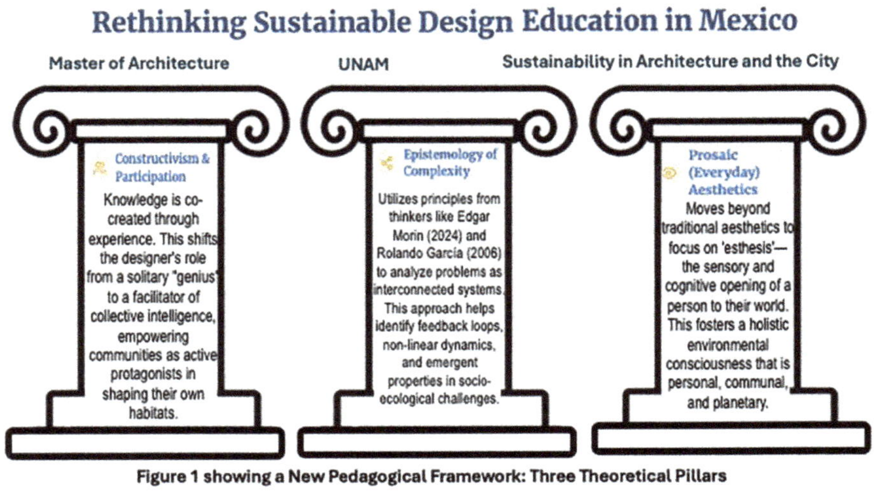

Fig. 1 Showing a new pedagogical framework: three theoretical pillars

commitment, analytical tools, and finally to integrated action. This structure is a direct reflection of the constructivist principle that knowledge must be built sequentially to be meaningful and applicable.

3.1 *The "Sustainability in Architecture and the City" Course Framework*

The one-and-a-half-hour weekly course is structured over a 16-week semester, organized around five interconnected goals that guide students on a journey of knowledge construction.

Goal 1: Genetics and History (Weeks 1–2): The course begins by establishing the historical origins (genesis) of contemporary problems. It studies the evolution of the pre-Hispanic city of Tenochtitlan, revaluing ancestral knowledge systems like the chinampa agricultural method to provide a deep contextual grounding for current environmental challenges in Mexico City.

Goal 2: Ontology, Teleology, and Epistemology (Weeks 3–4): This phase prompts students to question the fundamental nature (ontology) and purpose (teleology) of design. It introduces core sustainability concepts like ecological footprint and carrying capacity, framing them through different ideological lenses (e.g., design for profit, for social need, or for self-expression) and establishing the inhabitant as the central actor in the design process.

Goal 3: Ethics, Axiology, and Aesthetics (Weeks 5–6): This goal focuses on incorporating complex ethical principles. For Edgar Morin (2023), the essential task of the human sciences is to shed light on the complex nature of humanity, which is often obscured by the boundaries between academic disciplines. Participatory design is introduced as the practical method for enacting this ethical shift, transferring power from the expert to the collective (Romero & Mesías, 2004).

Goal 4: Complexity Methodologies (Weeks 7–9): This is a focused module where students are formally introduced to the analytical tools of Morin (2024), Rolando García's Epistemic Framework (Castorina & Barreiro, 2023), and the sciences of complexity (Batty, 2024), preparing them for practical application.

Goal 5: Synthesis and Practical Application (Weeks 9–16): The final and longest phase of the course is dedicated to applying the integrated knowledge through a capstone transdisciplinary methodology known as "Complex Design for Vulnerability and Risk" (Platas, 2020).

3.2 A Methodological Framework: Complex Design for Vulnerability and Risk

To translate the theoretical foundations of complexity science and participatory epistemology into a replicable and rigorous research-design process, the curriculum employs the "Complex Design for Vulnerability and Risk" methodology. Developed by Francisco Platas, this framework serves as the primary practical tool of the course, providing a structured yet flexible guide for students to analyze and intervene in complex socio-ecological systems. The overarching goal is to move beyond superficial problem-solving to cultivate a deep, systemic understanding that can inform targeted, effective, and ethically grounded design interventions.

The methodology consists of five iterative and interconnected phases:

Phase 1: Delimitation. The initial phase, Delimitation, establishes the boundaries and context of the research problem. This process begins with an exercise in critical self-reflection, wherein the designer or researcher explicitly maps their own worldview (weltanschauung) and epistemic framework. This step is crucial for acknowledging inherent biases and positioning the researcher within the system of study. Concurrently, a "genetic" analysis is conducted to trace the historical evolution and origins of the problem at hand. By combining this internal self-contextualization with an external historical analysis, this phase facilitates a clear and well-justified delimitation of the specific system under investigation. The objective is to define what is inside and outside the analytical frame, preventing the scope from becoming unmanageably broad while ensuring that critical historical determinants are not overlooked.

Phase 2: Structuring. The Structuring phase aims to deconstruct the delimited system into its constituent parts and map their relationships. This involves identifying the key elements and subsystems—such as economic, political, social, environmental, and technological components—and, most importantly, articulating their dynamic interactions. The analysis moves beyond a simple inventory of parts to focus on uncovering the emergent structural properties of the system. Researchers are guided to identify fundamental characteristics like stability, fragility, resilience, or robustness, which arise not from individual elements but from the pattern of their interconnectedness. This phase provides the architectural blueprint of the problem, revealing the underlying logic that governs the system's behavior.

Phase 3: Complex System Configuration. In this phase, a descriptive model of the system is constructed to visualize its multi-scale dynamics. The Complex System Configuration synthesizes the findings from the previous phase into a coherent representation. This is not necessarily a computational model but rather a conceptual one, often created using tools such as mind maps, graphs, or Checkland's Soft Systems Methodology (SSM) (Stowell, 2024). The model's primary function is to illustrate the intricate web of causality within the system, showing how processes at global and national scales (e.g., economic policies, climate change) create boundary conditions that influence territorial and local realities (e.g., housing availability, community organization). This visualization is a critical analytical step that helps generate initial working hypotheses about the causal relationships driving the system's behavior and producing specific vulnerabilities.

Phase 4: Empirical Validation. The Empirical Validation phase is designed to rigorously test the hypotheses generated during the configuration stage. This is achieved through a mixed-methods approach that integrates diverse forms of evidence. Quantitative tools, qualitative analysis, and computational models drawn from the sciences of complexity may be employed. Critically, this phase mandates social participation as an essential validation too. The constructed academic arguments and models are presented to the community, whose lived experience and local knowledge serve to verify, refute, or refine the researchers' understanding. This participatory element ensures that the analysis is grounded and aligns with the framework's constructivist and ethical commitments, transforming community members into active partners in the knowledge-creation process.

Phase 5: Synthesis. The final phase, Synthesis, culminates in the integration of all preceding findings into a comprehensive and deep understanding of the problem. This is not merely a summary of results but a holistic interpretation of the system's structure, dynamics, and vulnerabilities, informed and validated by both academic analysis and community knowledge. This profound, systemic understanding becomes the robust foundation upon which well-informed and targeted interventions can be designed. The output of this phase is a diagnosis or a set of design proposals—often developed in collaboration with a multidisciplinary team—that are precisely tailored to the unique complexities of the situation, thereby maximizing their potential for positive impact and minimizing the risk of unintended negative consequences.

4 Case Studies in Diverse Mexican Contexts

The robustness and flexibility of the curricular framework are demonstrated through its application in two postgraduate research projects conducted in radically different Mexican contexts. These case studies provide concrete evidence of the framework's ability to adapt to diverse scales and challenges while generating relevant, context-specific knowledge.

4.1 The Rural Indigenous Context: Habitat Transformation in El Vithé, Hidalgo

This master's thesis by Rosa Isabel Gallardo applied the framework at a micro-scale to explore the transformation of habitat in the Hñähñu indigenous community of El Vithé, a mountainous region marked by the socio-cultural impacts of international migration and transculturation (Gallardo, 2019). The analysis mapped a system that included interacting subsystems such as Hñähñu culture (including collective work "Rä Nfáxâ´befi"), construction technology (the transition from traditional maguey stalk to modern concrete block), the local economy (driven by migrant remittances), and migratory processes (the influence of North American lifestyle models).

The central validation tool was a community participation workshop. Through maps and game dynamics, inhabitants collectively identified their community's boundaries, problems, desires, and priorities. This process revealed that the adoption of "modern" housing was not a simple loss of identity but a complex adaptive mechanism. It was a form of self-regulation by which the community system responded to new aspirations for social status and prestige, mediated by the migratory experience, while simultaneously seeking to preserve the underlying communal logic of Hñähñu inhabitation (Gallardo, 2019).

4.2 The Urban Informal Context: Self-production of Habitat in Barranca de Huentitán, Jalisco

In a contrasting case, Alfonso Arias's master's thesis applied the framework to an informal urban settlement in the Barranca de Huentitán, an area of high socio-environmental vulnerability, topographical complexity, and institutional marginalization (Arias, 2021). The research focused on how inhabitants, in the absence of the state and professional designers, generate their own habitability strategies that respond simultaneously to contextual pressures (real estate speculation, lack of services) and the physical conditions of the site (steep slopes, water runoff).

Arias introduced a methodological innovation within the framework: a comprehensive multi-scale spatial analysis during the validation phase. This integrated three distinct methods:

Basin Analysis (Territorial Scale): Using hydrological principles to map the micro-basin and understand natural risk areas that condition the settlement.

Urban Morphological Analysis (Community Scale): Analyzing the settlement's structure to reveal an organic morphology that adapts intelligently to the topography.

Analysis of Habitability Indicators (Architectural Scale): Evaluating dwellings to conclude they reached a "Sufficient" level of habitability despite precarious services.

To integrate these analyses, Arias developed the PUAAT (Urban-Architectural Processes of Territorial Appropriation) model (Arias, 2021). This synthesis conceptualized habitat production as a function of the interaction between the subject, context, constructive tradition, and spatial intuition. The research demonstrated that inhabitants, through an intuitive reading of the territory and adapted building traditions, generate resilient and spatially intelligent design solutions. This case highlights the framework's capacity to decode and value the implicit knowledge embedded in community-led design processes, which is often ignored by formal planning.

5 Results: Broader Impacts and Contribution to Sustainable Development Goals

The academic framework and its practical applications have achieved significant real-world impact, connecting directly to global sustainability agendas (Platas & Carreto, 2025). In 2020, the project team—Francisco Platas López, Rosa Isabel Gallardo Martínez, and Alfonso Arias Martínez—received the United Nations Special Prize as part of the National Design Award: Diseña México, the highest recognition of its kind in the country (Frías Peña & Chávez López, 2022). This award, supported by international bodies including the UNEP—United Nations Environment Programme and UNESCO, specifically recognized the project for its contribution to achieving the Sustainable Development Goals of the 2030 Agenda.

Crucially, the award was granted under the axis of SDG 10: Reduced Inequalities. This recognition is significant because it validates that the process of the design framework is as important as its physical outcomes. The academic curriculum that the students followed was very important:

> This was because of the "ethical commitment to society" that the students were expected to have, given that they were attending a public university with a strong social commitment. The least the teachers could do was try to change their students' perspective on design. (México Desconocido, 2021)

It reduces inequality not merely by providing improved housing for vulnerable populations, but by rebalancing power dynamics between experts and communities, thereby promoting procedural justice and epistemic equality. This aligns perfectly with the social inclusion dimension of sustainable development that is central to the 2030 Agenda.

Media and institutional responses to the award underscored this contribution. Reports highlighted the project's "ethical commitment to societies," (Milenio, 2021) its use of "disciplinary convergence to address inequality problems," (INFOBAE, 2021) and its success in valuing "constructive traditions" (DGCS-UNAM, 2021) to achieve "social welfare" (UDG, 2020). The UNAM Gazette noted that the work empowers inhabitants to "make decisions about the construction of the space they will inhabit," which is a "fundamental task for the designer of the twenty-first century" (UNAM, 2021).

Beyond SDG 10, the curriculum framework makes substantial contributions to other key goals (United Nations, 2015), as shown in Fig. 2.

SDG 4 (Quality Education): The framework offers an innovative and replicable pedagogy for sustainability in higher education, directly addressing the need to equip learners with the skills to "promote sustainable development". It fosters critical and systemic thinking, moving beyond rote learning to build genuine problem-solving capacities.

SDG 11 (Sustainable Cities and Communities): The methodology is purpose-built to address the challenges of making human settlements "inclusive, safe, resilient and

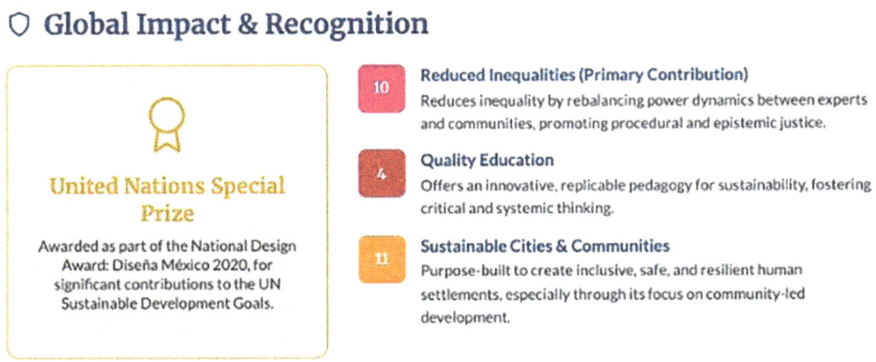

Fig. 2 Showing a global impact and recognition

sustainable," particularly through its focus on informal settlements and community-led development, which are key targets within SDG 11.6.

6 Conclusion: Synthesis, Synthesis, Implications, and Future Directions

6.1 Contextualization and Methodological Approach

This chapter addressed the critical inadequacy of traditional, object-centered design education in Mexico to contend with complex, systemic sustainability challenges. In response, we presented a robust pedagogical alternative: an interdisciplinary curriculum framework developed at the National Autonomous University of Mexico (UNAM). The framework's theoretical scaffolding rests on a coherent synthesis of constructivist epistemology, participatory design methodologies, and the sciences of complexity. Its core practical tool, the "Complex Design for Vulnerability and Risk" methodology, guides students through an iterative process of delimitation, structuring, modeling, validation, and synthesis to intervene in complex socio-ecological systems.

6.2 Key Findings and Implications

The implementation of this framework demonstrated its effectiveness in fostering advanced cognitive skills, such as systems thinking and an understanding of non-linear causality. The case studies confirmed its adaptability in radically different Mexican contexts. In the rural indigenous community of El Vithé, the analysis revealed that the adoption of "modern" housing was a complex adaptive response to migration and new social aspirations, rather than a simple loss of identity. In the informal urban settlement of Barranca de Huentitán, the research uncovered how inhabitants generate resilient and spatially intelligent design solutions through an intuitive reading of the territory, a form of knowledge often ignored by formal planning.

The broader implications of this work are significant. The project's recognition with a United Nations Special Prize for its contribution to

SDG 10 (Reduced Inequalities) validates that the framework's true value lies not just in its physical outcomes but in its process. By rebalancing power dynamics between experts and communities, it promotes procedural justice and epistemic equality, a core tenet of sustainable development. This suggests the framework serves as a replicable and highly relevant model for design education beyond Mexico, particularly in other Global South regions facing similar pressures of inequality, rapid urbanization, and the need to integrate local knowledge into planning.

6.3 *Limitations and Future Prospects*

Despite its success, the broader implementation of this curriculum faces notable challenges. A primary limitation is the need for intensive faculty training in complexity science and participatory facilitation, skills not standard in most architectural programs. Furthermore, institutional inertia within departments anchored in traditional, modernist paradigms can present considerable resistance to such fundamental change. Finally, the resource-intensive nature of authentic community engagement, which requires significant time for trust-building, is often at odds with the rigid calendars and budgets of academic institutions.

Future research should focus on addressing these limitations. Key prospects include developing scalable and effective models for faculty training in complexity and participation.

Longitudinal studies are also needed to assess the long-term impact of this curriculum on the professional practices and career trajectories of its graduates. Finally, further work is essential to explore how the principles of this master's-level framework can be adapted for integration into undergraduate design education, ensuring that the next generation of professionals is equipped with a systemic, ethical, and participatory mindset from the outset of their training.

References

Arias, A. (2021). *Mecanismos de atención a la demanda habitable autogenerados por pobladores de la barranca de Huentitán, Jalisco: una propuesta de análisis metodológico complejo [A proposal for complex methodological analysis: Self-generated housing demand response mechanisms for residents of the Huentitán Ravine, Jalisco]* [Master's thesis, UNAM]. TESIUNAM.

Batty, M. (2024). Digital twins in city planning. *Nature Computational Science, 4*, 192–199.

Becerra, G., & Amozurrutia, J. A. (2015). Rolando García's "complex systems theory" and its relevance to sociocybernetics. *Journal of Sociocybernetics, 13*(1), 18–30.

Castorina, J. A., & Barreiro, A. (Eds.). (2023). *The development of social knowledge: Towards a cultural-individual dialectic*. Information Age Publishing.

DGCS UNAM. (2021). *Premio especial de la ONU a universitarios por su compromiso social [The university students received a special UN award for their social commitment]*. UNAM. Retrieved September 5, 2025, from https://www.dgcs.unam.mx/boletin/bdboletin/2021_042.html

Distler, F., & Rasolofo-Distler, F. (2021). Dans les pas d'Edgar Morin: vers une écologie de l'action complexe *[Following in the footsteps of Edgar Morin: Toward an ecology of complex action]. Projectics/Proyéctica/Projectique, 30*(3), 111–121.

Elshinawy, E., Serag, Y., Elshater, A., & Afifi, S. (2023). Urban acupuncture theory for reclaiming urban spaces during the COVID-19 lockdown in New Cairo City. *HBRC Journal, 19*(1), 301–318.

Frías Peña, J., & Chávez López, C. (Eds.). (2022). *Premio Nacional de Diseño: Diseña México 2020. El diseño y los objetivos de desarrollo sostenible [National design award: Diseña México 2020 design and the sustainable development goals]*. UNAM.

Gallardo, R. I. (2019). *Procesos de transformación en la construcción social de lo espacial habitable, en la sociedad hñähñu, desde la perspectiva de la complejidad. [A complexity perspective on transformation processes in the social construction of habitable space in Hñähñu Society]* [Master's thesis, UNAM]. TESIUNAM.

García, R. (2006). *Sistemas complejos. Conceptos, método y fundamentación epistemológica de la investigación interdisciplinaria [Complex systems: Concepts, methods, and epistemological foundations of interdisciplinary research]*. Gedisa.

INFOBAE. (2021). *Estudiantes de Arquitectura de la UNAM recibieron un premio de la ONU por proyecto de diseño sostenible [The sustainable design project of architecture students from UNAM was recognized with a UN award*]. INFOABE. Retrieved September 5, 2025, from https://www.infobae.com/america/mexico/2021/01/18/estudiantes-de-arquitectura-de-la-unam-reciben-premio-de-la-onu-por-proyecto-de-diseno-sostenible/

Mandoki, K. (2016). *Everyday aesthetics: Prosaics, the play of culture and social identities*. Routledge.

México Desconocido. (2021). *ONU premia a estudiantes de la UNAM por investigación sustentable [The UN awards UNAM students for their sustainable research*]. México Desconocido. Retrieved September 5, 2025, from https://www.mexicodesconocido.com.mx/onu-premia-a-estudiantes-de-la-unam-por-investigacion-sustentable.html

Milenio. (2021). *¡Orgullo nacional! ONU premia a estudiantes de la UNAM por proyecto sustentable [National pride! The UN awards UNAM students for their sustainable project*]. Milenio. Retrieved September 5, 2025, from https://www.milenio.com/ciencia-y-salud/unam-onu-premia-estudiantes-proyecto-sustentable

Morin, E. (2023). *Encore un moment… Textes personnels, politiques, sociologiques, philosophiques et littéraires [One more moment. Personal, political, sociological, philosophical, and literary texts*]. Denoël.

Morin, E. (2024). *La méthode de la méthode [The method of the method]*. Actes Sud.

Piaget, J. (1950). *Introduction à l'épistémologie génétique [Introduction to genetic epistemology]*. PUF.

Platas, F., & Carreto, F. (2025). Transformative interdisciplinary research initiatives by autonomous university of the state of Mexico: Advancing education and diminishing inequalities in Latin America. In *Green metric in higher education—Measuring and reporting on sustainability initiatives at higher education institutions* (pp. 1–21).

Platas, F. (2020). Diseño complejo para la vulnerabilidad y el riesgo en megaciudades [Complex design for vulnerability and risk in megacities]. *Academia XXII, UNAM, 11*(20), 114–130.

Romero, G., & Mesías, R. (2004). *La participación en el diseño urbano y arquitectónico en la pro-ducción social del hábitat [Participation in the social production of habitat through urban and architectural design*]. Cyted/Habyted.

Stowell, F. (2024). The art of soft systems inquiry: Retracing the footsteps of Churchman and Checkland. *Systemic Practice and Action Research, 37*, 1123–1140.

UDG. (2020). *Gana egresado del CUAAD el Premio Nacional de Diseño 2020 [A graduate won the 2020 national design award]*. UDG. Retrieved September 5, 2025, from https://www.gaceta.udg.mx/gana-egresado-del-cuaad-el-premio-nacional-de-diseno-2020/

UNAM. (2021). *Premio Especial de la ONU a diseñadores universitarios [UN special award for university designers*]. UNAM. Retrieved September 5, 2025, from https://www.gaceta.unam.mx/premio-especial-de-la-onu-a-disenadores-universitarios/

United Nations. (2015). *Transforming our world: The 2030 agenda for sustainable development unit*. United Nations.

Francisco Platas-López is a Professor at the Autonomous University of the State of Mexico (UAEMéx) and an Associate Professor at the National Autonomous University of Mexico (UNAM). He holds a Bachelor of Engineering (B. Eng.), a Master of Design (M. Des.), and a Doctorate in Design (D. Des.). Dr. Platas is a member of Mexico's National System of Researchers (SNII) and an International Scholar with the Society for the History of Technology (SHOT). In 2020, he was a Design Facilitator for the Generation Equality Asia Pacific Design Challenge, a joint initiative of UN Women Asia and the Pacific and the World Design Organization (WDO). His research interests lie at the intersection of aesthetics and design in everyday life, the history of infrastructure, and design-based approaches to vulnerability and risk. Dr. Platas is the creator of the Complex Design for Vulnerability and Risk, a novel epistemology developed to analyze and address these multifaceted challenges.

Teaching Sustainability in Interior Design: Interdisciplinary Approaches and Pedagogical Strategies from a Portuguese Higher Education Context

Maria Constança Simões Rigueiro and Maria Alzira Pimenta Dinis

Abstract The Anthropocene marks a new geological era characterized by pressing global challenges, particularly climate change, environmental degradation, and resource scarcity. Within this context, Education for Sustainable Development (ESD) plays a key role in equipping future professionals with the knowledge, values, and skills necessary to design more sustainable and responsible solutions. This chapter examines how sustainability principles are integrated into the undergraduate and master's programmes in "Interior Design and Equipment" and "Interior Design and Furniture" at the Polytechnic Institute of Castelo Branco, Portugal, focusing on the curricular units of "Materials" and "Sustainability and Innovation". In the curriculum of these curricular units, sustainability is positioned as a central educational objective, offering students both a theoretical foundation and practical tools such as Environmental Product Declarations (EPD), Life Cycle Assessment (LCA), and criteria-based assessment tools (CBT). The curriculum development is grounded in real-world case studies, interdisciplinary methods, and project-based learning, promotes critical thinking, ethical responsibility, informed decision-making and sustainable development. By exploring material selection processes and sustainable design methodologies, students are encouraged to develop creative and ecologically responsible solutions that reflect the realities of Interior Design (ID) practice today.

M. C. S. Rigueiro (✉)
Institute for Sustainability and Innovation in Structural Engineering (ISISE), Instituto Politécnico de Castelo Branco, Castelo Branco, Portugal
e-mail: constanca@ipcb.pt

M. A. P. Dinis
Fernando Pessoa Research, Innovation and Development Institute (FP-I3ID), Porto, Portugal

Marine and Environmental Sciences Centre (MARE), University of Coimbra, Coimbra, Portugal

M. A. P. Dinis
e-mail: madinis@ufp.edu.pt

W. Leal Filho et al., *Curriculum Development for Sustainable Development*, World Sustainability Series, https://doi.org/10.1007/978-3-032-22146-9_20

This chapter contributes to the broader discussion on how Higher Education Institutions (HEIs), in the curriculum of the courses, can support the integration of sustainability into design curricula, fostering both technical competence and environmental awareness.

Keywords Higher education institutions (HEIs) · Education for sustainable development (ESD) · Interior design (ID) · Portugal · Teaching strategies

1 Introduction

Higher education institutions (HEIs) must significantly intensify their efforts to tackle sustainability challenges by promoting institutional progress across three interconnected dimensions: social, environmental, and economic. This involves integrating sustainability into curricula and research agendas, adopting environmentally responsible campus operations, and fostering inclusive and equitable communities (Husaini et al., 2018; Leal Filho et al., 2025; Lozano et al., 2015; Rigueiro et al., 2025). HEIs play a crucial role in generating knowledge, fostering innovation, and preparing future professionals to effectively address climate change, resource scarcity, and environmental challenges. By actively engaging students, they contribute to transforming the systems and structures that enable practical and effective responses to sustainability challenges and sustainable development. Students themselves can become powerful agents of change, helping to shift the underlying beliefs, values, and paradigms that shape society's understanding and approach to sustainability (Leal Filho et al., 2024; UNEP, 2021). It is in this context that the present chapter is introduced. Its purpose is to highlight the methodologies employed with students at the Polytechnic Institute of Castelo Branco to foster knowledge of sustainability within the practice of interior design (ID). This chapter details how sustainable design has been promoted within the curriculum of the "Materials" and "Sustainability and Innovation" courses in the undergraduate degree in "Interior Design and Equipment" and the Master's degree in "Interior Design and Furniture" at the Polytechnic Institute of Castelo Branco, in order to support sustainable development.

2 Theoretical Background

This section outlines the key concepts, tools, and methodologies that inform the integration of sustainability into ID education.

2.1 Interior Design Sustainable Education Development

The interior components of buildings, which Brand (1994) refers to as "services, space plan, and stuff", tend to cycle in and out much more rapidly than the structural and exterior elements of a building. ID, broadly defined, involves furniture, finishes, materials, and equipment to enhance human comfort and well-being (CIDQ, 2019). The scope of ID projects can range from a single residential room to large-scale commercial and institutional campuses. Interior designers specify products sourced from across the globe, and the design process itself is energy-intensive. It often involves advanced software for spatial planning and visualization, the global shipment of material samples, virtual team collaboration, frequent site visits, and frequent redesigns driven by changing trends. Many interiors created by professional designers, and the products they include, are shaped more by market demands and client preferences than by thoughtful consideration of their environmental and social impacts (Irwin, 2015).

In his influential 1970 work (Papanek, 1991), Design for the Real World, Victor Papanek offers a practical guide for designers aiming to make a meaningful contribution in the face of resource scarcity and environmental challenges. Translated into more than twenty languages, the book remains one of the most widely read and cited texts in the field of design, inspiring generations of designers to look beyond aesthetics and profit. Papanek advocates for design that is not only functional and visually appealing, but also promotes reuse and recycling, while avoiding harmful practices such as addictive design and consumer lock-in. He underscores the importance of holistic, student-centred learning, encouraging the use of learners' strengths and resources to drive real-world transformation through cross-disciplinary and interdisciplinary collaboration.

Providing ID students with sustainability skills enables them to drive meaningful change by promoting design practices that support environmental responsibility and contribute to societal transformation. However, this requires more than simply adopting concepts that go beyond traditional education. Effectively developing the cross-cutting skills needed to address sustainability challenges demands both the integration of sustainability-focused content into ID curricula and the reinforcement of higher education institutions through active, transformative pedagogical approaches (Astles & Matthews, 2025).

2.2 Sustainable Construction Tools

At present, the construction sector stands as the leading driver of climate change (UNEP, 2023). It is responsible for 40% of electricity consumption, over 35% of the European Union total waste produced, 50% of materials usage (Cabeza et al., 2022; Carvalho et al., 2023; IPCC, 2022) and 37% of greenhouse gas global emissions (UNEP, 2023). Moreover, several critical challenges, such as rapid population growth, industrial concentration, excessive resource consumption, and increasing social inequality, although not directly attributable to the construction sector, exacerbate the broader systemic pressures contributing to environmental and social instability (IPCC, 2018). Significant changes are essential to reduce the environmental impact of buildings, construction practices, and human behaviour (Reyes Nieto et al., 2024; WMO-No. 1368, 2025). The field of sustainable construction science has emerged as a promising response to these challenges. In this context, sustainable construction has introduced a range of strategies and methodologies aimed at advancing sustainable development on two key scales: the building and the urban. These strategies and methodologies support strategic planning, marketing, decision–making, enable the identification of a project disadvantages and provide guidance to improve them, promoting a more suitable construction environment Reyes (2019). In this study, only the first one is highlighted, the building level, where the environmental, social and economic impacts and efficiency of buildings can be assessed. These strategies and methodologies can be aggregated in two different groups: The Criteria-Based-Tools (CBT), which are based on a criteria system and scores; The Life Cycle Assessment (LCA) tools where the evaluation techniques enable the knowledge of the environmental performance of the analysed system, i.e., materials and or buildings, through its life.

All over the world, at building level, different CBT has been developed, for example: Building Research Establishment Environmental Assessment Method (BREEAM) (2025), Leadership in Energy and Environmental Design (LEED) (2025), Sustainable Build environment (SBTool) (2022), Comprehensive Assessment System for Built Environment Efficiency (CASSBEE) (2001), German Sustainable Building Council (DNGB) (2015), Green Star (2025), Haute Qualité Environnementale (HQE) (2025) and BEAM (2024) (He et al., 2018; Sanhudo et al., 2018). Each approach emphasizes different indicators and perspectives on sustainability requirements, yet all share the common goal of evaluating sustainability and guiding actions toward the creation of sustainable buildings (Ade & Rehm, 2019).

The LCA tools cover principles and structure presented in ISO 14040:2006 (2006) according to the requirements and guidelines on ISO 14044:2006 (2006). The assessment of the environmental performance comprises four different phases: the goal and scope definition phase, the life cycle inventory (LCI) phase, the life cycle impact assessment (LCIA) phase, and the interpretation phase. The LCA tools provide better understanding and increase the awareness of the importance of environmental protection and the possible impacts associated with products (EC-JRC, 2011; Saouter et al., 2020). Some of these tools are available in diverse platforms and have been developed to support the development of more sustainable products and services, as example the software SimaPro (2025), GaBi (now Sphera LCA for Experts) (2025), OpenLCA (2025) and One Click LCA (2025), and many others. The application of CBT and LCA tools constitute a voluntary approach without mandatory regulation to promote standards and the development of sustainable buildings.

2.3 Sustainable Design Tools

The earliest concerns regarding resource limitations and the environmental impact of material use in production can be traced back to Fuller (1969), who warned that without a conscious human effort to care for and preserve the Earth, its systems could be severely compromised or even face collapse. A notable milestone in advancing sustainable thinking within the design field was the publication of Green Design by the Design Council in the United Kingdom (Burall, 1991), which marked the beginning of a broader and more systematic engagement with sustainability in design.

During the 1990s, design practices increasingly embraced a redesign approach aimed at minimizing environmental impacts and improving product efficiency. This shift was closely linked to the application of LCA to evaluate the entire life cycle of products, i.e., Eco Design. Since the 2000s, a variety of design strategies and approaches have emerged: product-service systems, sustainable design, emotionally durable design, cradle-to-cradle design, biomimicry, and design for the base of the pyramid focusing on extending product life cycles through durability, reparability, and adaptability (Ceschin & Gaziulusoy, 2020). Today, the field of Design for Sustainability (DfS) or sustainable design has evolved to encompass approaches such as product–service systems and design for social innovation, reflecting a significant expansion of research and practice in the design area (Cole & Hamilton, 2024). The development and application of DfS have become important topics for companies, consumers, communities, and decision-makers. To support this, design professionals have access to a variety of tools, both qualitative and quantitative, that vary in complexity, cost, focus, and the type of results they produce. These DfS tools can also be categorized by their purpose, such as universal tools for general guidance, participatory tools for stakeholder engagement, and assessment tools for evaluating impacts (Camocho, 2022; Kozlowski et al., 2019).

The use of these DfS tools allow for the integration of sustainable principles into design practices and structures the development process, guiding teams towards more efficient sustainable solutions and address society's challenges and needs (Camocho, 2022). However, their adoption remains neither systematic nor fully effective, as many designers and professionals involved in product and service development often lack the knowledge and expertise required for their consistent and comprehensive application or lack of integration into standard workflows due to complexity, or insufficient alignment with industry needs (Faludi et al., 2020).

In the last decade, research emphasized the need for clearer definitions, better classification, and more user-friendly interfaces to increase tools uptake (Vicente & Camocho, 2023). For example, tools such as KATCH-e (2020), SinnDesign (2015), and InEDiC (2011) were developed within the Portuguese context to enhance understanding of sustainability concepts in the design of products. They allow users to evaluate the implications of sustainable strategies on the product design, business model and its environmental effects. Aimed at design students in higher education institutions and professionals, these tools provide both theoretical foundations (as circular economy strategies, eco-design, life cycle assessment, material selection databases) and practical resources (Excel files, games, apps and software for process optimization and synthesis of the products), as well as guidance on where to access relevant information (Vicente & Camocho, 2024). SinnDesign (2015) and InEDiC (2011) are qualitative tools specialized in certain fields such as ID (furniture, textiles for the habitat and building materials) and ceramic, respectively. Tailored design tools have to address sector-specific challenges, such as material resources or process efficiency. A systems approach is recommended, combining the use of multiple tools, either qualitative or quantitative, to address the full spectrum of sustainability challenges and avoid contradictions (Robèrt et al., 2002).

3 Methods

This section outlines the teaching approach and the implementation of sustainability into ID education courses.

3.1 Case Studies in Sustainable Interior Design Education

The case studies were conducted in the undergraduate and master's programmes in "Interior Design and Equipment" and "Interior Design and Furniture" at the Polytechnic Institute of Castelo Branco, Portugal, with a particular focus on the curricular units of "Materials" and "Sustainability and Innovation".

The "Materials" course takes place in the second of three years of 'Interior Design and Equipment' degree. It usually has around 40–50 students. Students are divided in groups of three in order to keep up with the practical work developed in the classroom context. This course unit aims to develop a clear understanding of how to apply sustainable development principles in the ID and equipment sector, particularly in the selection of materials. In this way, students are guided to understand the importance of material characteristics for successful ID. For that, initially, concepts of sustainable development, sustainability and sustainable design are presented. Various materials are presented, namely their physical, mechanical, thermal, electrical and sustainability characteristics. Concerning the sustainable characteristics, they are analysed by the EPD. The EPD provides the environmental performance of the product, described from a life cycle perspective, by conducting a LCA of the product in accordance with ISO 14025:2006 (2006). The results of the LCA study comply with the rules, requirements and guidelines for developing an EPD for a specific product category. These requirements are defined by the reference Product Category Rules (PCR). The EPD is verified by an approved independent verifier before being registered and published in the International EPD System through our EPD Portal, EN15804:2012 (2012), EN15942:2011 (2011) and CEN/TR 15941:2010 (2010).

The "Sustainability and Innovation" course takes place in the first of two years of the master in 'Interior Design and Furniture'. It usually has around 25 students.

The objectives of this course are to contribute to the development and creation of an innovative and sustainable design. Students are guided in the area of sustainable and innovative design and motivated to understand the key concepts, theories and methodologies essential to the products they create, for example: Sustainable Development (SD), LCA methodologies and boundaries of the analysed system, "Cradle to Grave", i.e., a linear model where the products are discarded at the end of their life, or "Cradle to Cradle", i.e., a model that prioritizes the circulation of materials aiming to eliminate the waste, Ecological footprint, Greenhouse Gas Protocol, Low carbon emission products, Recycling, Reduce, Reuse, Readapt, Refurbishment, Renewable energies. Since this is a course unit for a master's degree in interior and furniture design, different tools for assessing sustainability are presented. That is, the most common CBT (or the basis) for assessing the sustainability of buildings in the construction sector, BREEAM, LEED and the Portuguese SBTool^{PT}-H, are presented (Martinho et al., 2013). The SBTool^{PT}-H (Mateus & Bragança, 2011), is a Portuguese adaptation of SBTool, where each sustainability criterion is assigned a quantitative score based on a comparison between two benchmarks: national best practice and conventional practice. The score is normalized and turned in a qualitative score from A+ to E. The goal is to establish a dimensionless value that allows the building's sustainability score to be easily understood by non-experts. On the qualitative scale, grade D represents national conventional practice, while grade A reflects national best practice. After all parameters are weighted and aggregated, the building receives a final sustainability rating from A+ to E.

With regard to product design sustainability assessment, the Katch-e and SinnDesign tools are presented and explained. The use of these two tools has the potential to influence both the process and the outcomes of design, contributing to more responsible, ethical, and impactful practices across the sector. It is widely recognized that the integration and effective application of these tools can play a transformative and foundational role in supporting the evolution of the design profession.

3.2 *Learning Activities and Assessment Strategies*

During the classes of the material course, the necessary elements for understanding and practical application are provided, i.e., for the different groups of materials (metals, wood and its derivatives, ceramics, polymers, mortars and concretes, and natural materials of plant origin), the characteristics, processes, and production techniques are presented. Digital media and samples from manufacturers are used so that students can get a real sense of the materials, their textures, colours and finishes. LCA methodologies are presented, along with examples of EPD corresponding to different types of materials, as illustrated in Fig. 1. In this way, students must complete several assignments, including one involving the selection of an EPD for a material specified by the lecturer. The aim of this work is to search existing databases of environmental product declarations (EPD) and analyse the environmental impacts inherent in the

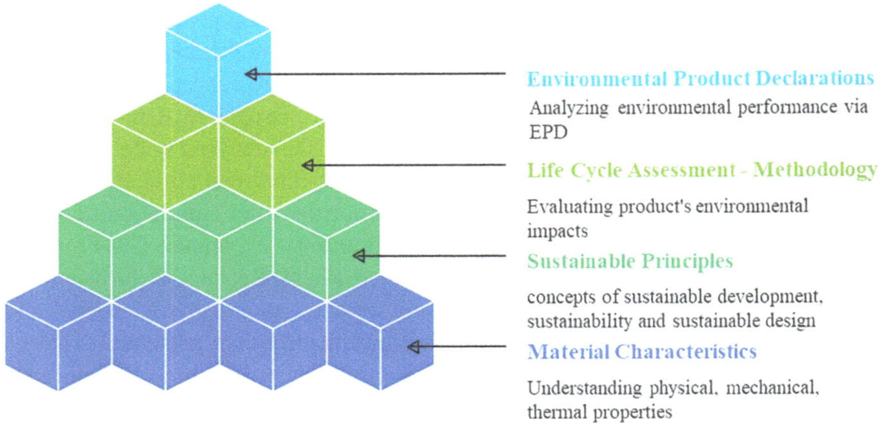

Fig. 1 Learning process in sustainable interior design education on "Materials"

production of material. For the development of this work some guidance is given, the students must answer the following questions:

1. EPD—what is it?
2. What is the purpose of the EPD?
3. Identify the unit declared in the EPD analysed;
4. Identify the manufacturing process of the material presented in the EPD;
5. Explain the production steps of the material, what is in and what is out;
6. What are the parameters that describe the environmental impacts?
7. Taking into account the environmental performance of the product, chosen, quantify the potential environmental impacts associated with greenhouse gas emissions and energy consumption. To do this, the students must to define the functional unit of the material.

For the assessment of the "Sustainability and Innovation" course, several practical works are given to the students. Two assignments are set on the area of ID—construction, and the third is given in the domain of the product/equipment. One of them (in the domain of the ID—construction) has the goal to provide skills in the field of sustainability assessment using CBT. Using the SBTool^{PT}-H the students have to access to an ID project already developed in the master course of "Interior Design and Furniture". This is important because the use of CBTs such as the SBTool^{PT}-H is highly time-intensive, requiring data from multiple sources related to the building—for example, material specifications, bill of materials, costs, and a set of predefined calculations concerning both thermal and acoustic performance. Therefore, the building project should be as complete as possible in the design phase to be assessed.

When using SBTool^{PT}-H, the students have to analyse the following dimensions, categories, indicators and parameters:

1. **Environmental Dimension**

Category C2—Land use and biodiversity—Parameters: P4—Utilised percentage of net available use index; P5—Percentage of green areas occupied by native plants; P6—Percentage of plant area with reflectance equal to or greater than 60%.

Category C4—Materials and solid waste—Parameters: P9—Reused materials and products; P10—Use of materials with recycled content; P11—Use of certified organic materials.

2. **Social Dimension**

Category C6—Comfort and health of users—Parameters: P16—Efficiency of natural ventilation; P17—Toxicity of coating materials; P19—Visual comfort with natural light; Category C7—Accessibility—Parameters: P21—Public transport accessibility index.

3. **Economic Dimension**

Category C9—Initial investment costs—Parameters: P24—Value of initial investment cost per square metre of usable area; P25—Present value of utilisation costs per square metre of usable area.

The dimensions, categories, and parameters selected for students to analyse are defined according to their profile and the objectives of their previously conducted study, which serves as the basis for the case study illustrated in Fig. 2.

At the end of this assignment, the students must present an appropriate presentation of the project design used including the bill of materials and costs; a descriptive and justificative report covering the results obtained using the SBTool^{PT}-H evaluation guide for the indicators analysed; The parameters for which an improvement in

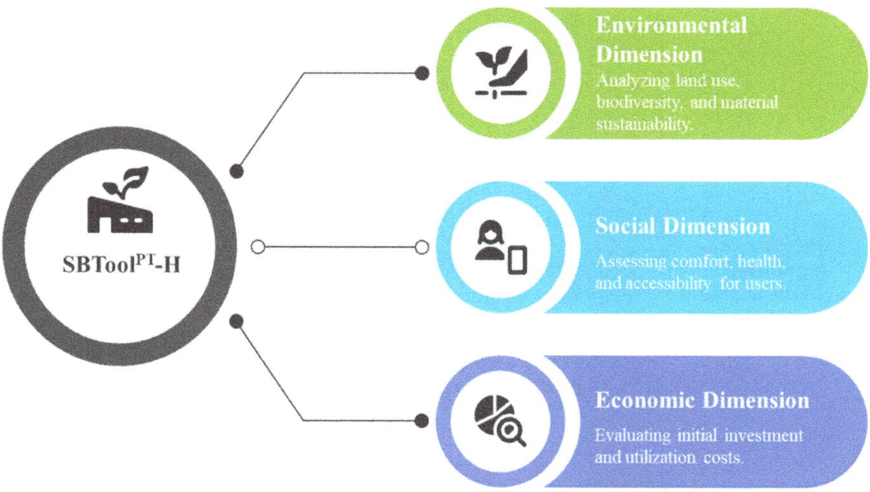

Fig. 2 Using the SBTool^{PT}-H in interior design

sustainability is needed must be justified. If necessary, a sensitivity analysis of the parameters must be conducted. Parameters that exhibit some degree of randomness and have a significant influence on the final classification should be subjected to a variation study to assess how their changes affect the final result.

The second assignment, focused on ID and construction, is developed using the LCA methodology. Considering the quantities of material associated with the established solution, the students must quantify the environmental impacts (in terms of global warming and embodied energy—renewable and non-renewable) of the design interior project using appropriate software or literature. The "Cradle to Grave" or "Cradle to Cradle" approach must be defined, where applicable, and, if necessary, students are required to carry out a sensitivity study of the critical materials, as shown in Fig. 3. The aim of these two assignments is to assess the sustainability of buildings, raise students' awareness of the choices they have made and how to proceed in order to achieve more sustainable solutions.

With a view to minimising ecological, economics and social impacts, the third assignment has the goal to provide skills in the field of sustainability assessment of a product (or equipment) and a LCA assessment of the material is also completed. For this purpose, the DfS methodology—SinnDesign is used. This tool promotes innovative and competitive solutions for the habitat and building construction. To do this, and according the methodology, a reference product must be analysed, and that process of the assignment is detailed in Fig. 4.

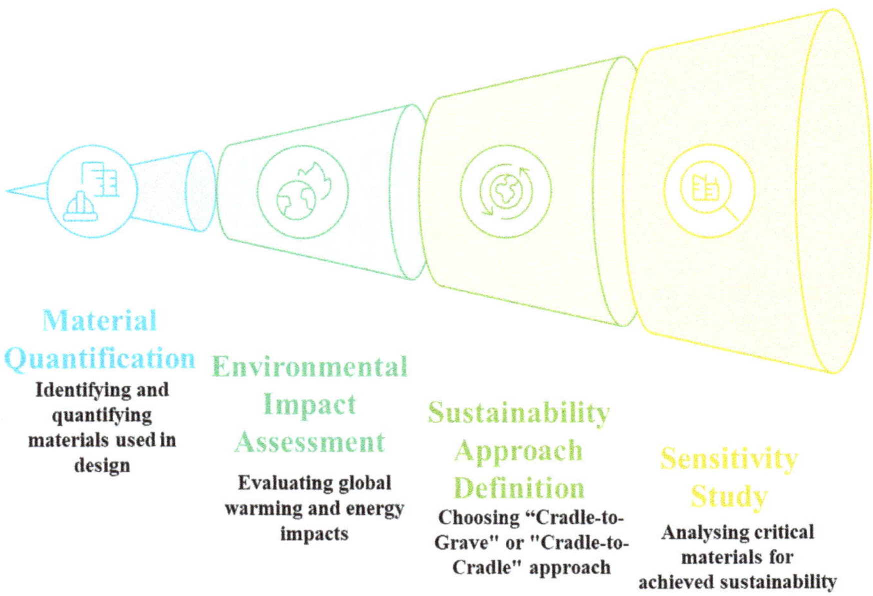

Fig. 3 Process conducted in interior design projects—construction

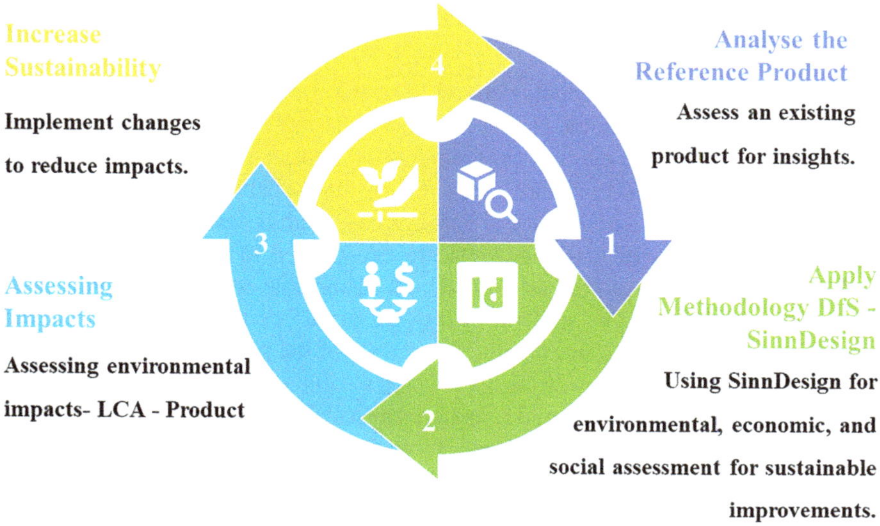

Fig. 4 SinnDesign tool to assess sustainability in interior design products and equipment

It can be stated that these methodologies were chosen because they provide adequate guidance on where to access relevant information, which is particularly important for design master students, as presented in Fig. 5.

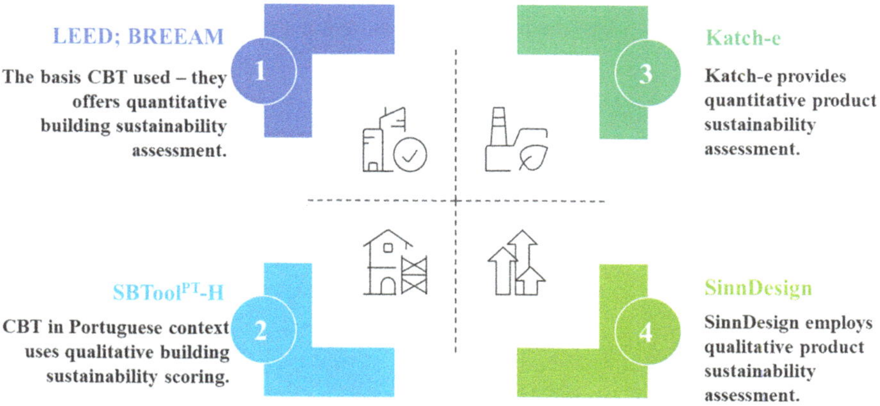

Fig. 5 Sustainability assessment tools in sustainable interior design education

4 Student Learning Outcomes and Reflections

The main insights of the students derived from the pedagogical approach that are employed by the lecturer in the classes of "Materials" and "Sustainability and Innovation" in the degree and master's programmes in "Interior Design and Equipment" and "Interior Design and Furniture" at the Polytechnic Institute of Castelo Branco, Portugal, are presented and discussed. Through case studies, project-based learning, and collaborative design tasks, students are encouraged to develop creative and ecologically responsible solutions aligned with sustainable development principles.

The outputs produced by students in the "Materials" course are closely aligned with the teaching methods and learning objectives of the program. Throughout the progress of the practical work, students develop a solid understanding of the fundamental concepts related to material selection and sustainability. They are introduced to various analytical methodologies that enable them to critically assess the environmental performance of different materials. A significant emphasis is placed on the use of EPD as a key tool for evaluating environmental impacts, particularly in terms of global warming potential and embodied energy. By the end of the course, students are able to apply these methods in a systematic and informed manner to support responsible and sustainable material selection in their design projects, as shown in Fig. 6.

Students are required to search for existing EPD of materials. They search EPD on national and international websites. Then they quantify the environmental impacts of the selected material specifically focusing on global warming potential and embodied energy—both renewable and non-renewable. Each working group presents the outcomes of their work to the rest of the class, using a presentation format which

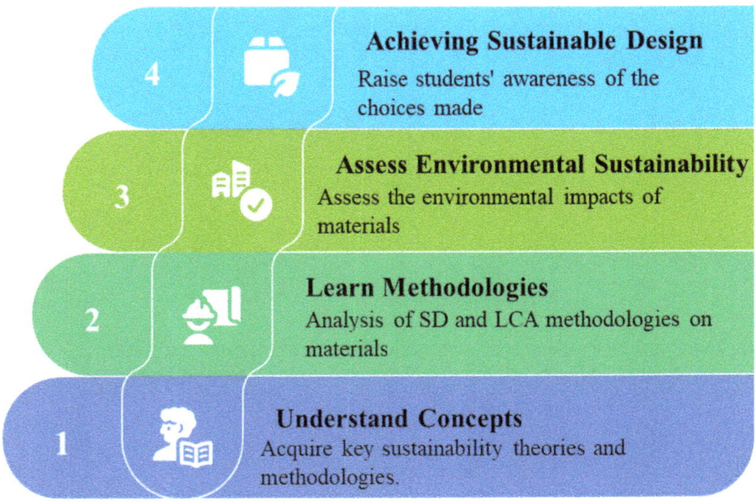

Fig. 6 Developing sustainability awareness in material selection

they consider most appropriate. This allows students to communicate their findings creatively and effectively to their peers while developing product lifecycle awareness, as shown in Fig. 7.

Students enrolled in the "Interior Design and Furniture" master course demonstrate a higher degree of maturity and complexity in their approach. Consequently, the results of their projects demonstrate greater depth, detail, and refinement, reflecting their advanced level of knowledge and skills in key concepts as well as their familiarity with tools applicable in the field of construction and product development, as evidenced in Fig. 8.

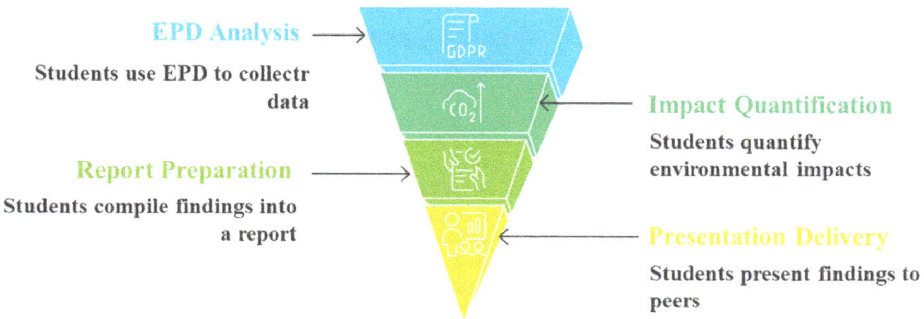

Fig. 7 Key learnings on sustainability and communication in material production

Fig. 8 Building sustainability awareness in interior design projects

In their analysis and conclusions on the application of the SBTool^{PT}-H methodology, the students identified both strengths and weaknesses in their projects, along with suggestions for improvement. The use of this methodology encouraged them to reconsider material reuse—for example, recognizing how the choice of surface finishes can impact indoor air quality. It also highlighted the importance of windows in providing natural lighting and emphasized the need for efficient use of both natural and financial resources in design projects.

In relation to the use of sustainability assessment tools in the context of product development, it is possible to identify several relevant aspects emphasized by the students in their final reflections. These points reflect not only their understanding of the tools themselves but also their critical evaluation of the tools applicability, advantages, and limitations when integrated into the design process:

i. Design Thinking with a Focus on Stakeholder Engagement;
ii. Evaluation frameworks, like SinnDesign, help translate sustainability from theory into practice;
iii. Recognize the impact of design on multiple actors: consumers, manufacturers, recyclers, regulators, etc.;
iv. Allow holistic sustainability evaluation;
v. Explore how compliance with ecolabels and international standards shapes design choices and marketing strategies;
vi. A great impact of communication in sustainable products;
vii. Inclusive and Universal Design Considerations;
viii. Gain the ability to assess the pros and cons of materials and production methods from energy, cost, and environmental perspectives;
ix. Learn to identify where sustainable options may reduce or increase costs—and how these trade-offs influence stakeholder decisions;
x. Circular Economy Integration, develop skills in mapping how products can transition from linear to circular systems through strategies like reuse, recycling, remanufacturing;
xi. Design for Durability and Longevity—Understand how good design can extend product life and create emotional or aesthetic value that discourages premature disposal.

Together, these insights reflect the students' growing ability to approach product development with a strategic, informed, and multidimensional perspective on sustainability, combining environmental, economic, and social considerations within ID practice.

5 Conclusions

Although the application of sustainability is fundamental in HEIs, its inherent complexity requires the careful selection of tools aligned with educational objectives and course curricula. Teachers must therefore be familiar with a wide range of available methodologies and conduct a thorough analysis to choose those that best fit their pedagogical goals. The sustainability indicators analysed must be clearly aligned with the objectives defined for the course, playing an essential role in guiding this tool selection process. This research aimed to present the methodologies and tools applied within the bachelor's and master's degree programmes in ID at the Polytechnic Institute of Castelo Branco. To frame these course units, a comprehensive review of tools for construction and product development was conducted. These tools not only provide designers with structured, data-driven methods for making informed decisions, but also foster a more critical and reflective design practice. By incorporating these resources into the learning process, students not only question conventional practices but also demonstrate the ability to adopt sustainable strategies and design innovative solutions. They emerge with a clearer understanding of how to respond effectively to current environmental, social, and economic challenges, showing measurable growth in critical thinking, creativity, and sustainability-driven decision-making.

Furthermore, this pedagogical approach promotes a shift in mindset, encouraging students to embrace greater accountability and sensitivity to the needs of communities, ecosystems, and future generations. Through the application of these tools and methodologies in real-world projects, students come to understand that assessing sustainability involves not only evaluating environmental impacts but also considering economic viability and social consequences. They recognize that there is no universal solution; each product, material, and context demands a carefully adapted sustainable strategy. Circularity is understood as an integral aspect of design, requiring deliberate planning from the outset rather than being treated as a corrective measure after a product's use. Students learn that sustainable design adds value by extending product lifespans, building consumer trust, achieving cost savings, and offering market differentiation. They also develop an appreciation for the importance of fair, factual, and transparent communication regarding environmental performance, recognizing its role in fostering credibility and trust. Additionally, they highlight the significance of designing for disassembly and end-of-life, understanding how modularity, clear labelling, and material selection can facilitate efficient recycling and reuse. Finally, they acknowledge the practical importance of integrating eco-labels and sustainability standards into real-world design applications, thereby reinforcing the professional relevance and impact of their learning.

The methodology presented here and its results should be considered in light of certain limitations. These include the limited time allocated within the classroom for these curricular units, when compared to others, as the students' varying levels of knowledge in conducting qualitative analysis, using software tools beyond those already familiar with, and interpreting numerical values and resulting quantitative classifications. Additionally, the constraints and complexity associated with autonomous work and the application of sustainability principles across other course units further impact the depth and consistency of learning. One way to mitigate these limitations and constraints would be to establish stronger integration between the different curricular units, along with better organization of course objectives, ensuring that the implementation of sustainability development becomes a concrete and cohesive component of the academic program, to be considered in the future.

References

Ade, R., & Rehm, M. (2019). The unwritten history of green building rating tools: A personal view from some of the 'founding fathers.' *Building Research & Information, 48*(1), 1–17. https://doi.org/10.1080/09613218.2019.1627179

Astles, P., & Matthews, C. (2025). Communicating sustainability competencies: A case study of sustainability and learning design. In W. Leal Filho, L. V. Trevisan, P. J. Pace, M. Mifsud (Eds.), *Education for sustainable development: The contribution of universities. World sustainability series.* Springer. https://doi.org/10.1007/978-3-031-86985-3_6

Brand, S. (1994). *How buildings learn: What happens after they're built* (2nd ed.). Phoenix Illustrated.

BREEAM, Building Research Establishment Environmental Assessment Method. (2025). Retrieved July 15, 2025, from https://breeam.com/

Burall, P. (1991). Green design. Design Council.

Cabeza, L. F., Bai, Q., Bertoldi, P., Kihila, J. M., Lucena, A. F. P., Mata, É., Mirasgedis, S., Novikova, A., & Saheb, Y. (2022). Buildings. In P. R. Shukla, J. Skea, R. Slade, A. Al Khourdajie, R. van Diemen, D. McCollum, M. Pathak, S. Some, P. Vyas, R. Fradera, M. Belkacemi, A. Hasija, G. Lisboa, S. Luz, & J. Malley (Eds.), *IPCC, 2022: Climate change 2022: Mitigation of climate change. Contribution of working group III to the sixth assessment report of the intergovernmental panel on climate change.* Cambridge University Press. https://doi.org/10.1017/9781009157926.011

Camocho, D. J. G. (2022). *Transition to circular and sustainable economy through design* [Ph.D. thesis, Faculdade de Design, Tecnologia e Comunicação, Universidade Europeia].

Carvalho, J. P., Bragança, L., & Mateus, R. (2023). Automating building sustainability assessment using building information modelling: A case study. *Journal of Building Engineering, 76*, 107228. ISSN: 2352-7102. https://doi.org/10.1016/j.jobe.2023.107228

CASSBEE - Comprehensive Assessment System for Built Environment Efficiency. (2001). Retrieved July 15, 2025, from https://www.ibecs.or.jp/CASBEE/english/index.htm

CEN/TR 15941:2010. (2010). *Environmental product declarations—Methodology for the selection of generic data.* European Committee for Standardisation.

Ceschin, F., & Gaziulusoy, I. (2020). *Design for sustainability: A multi-level framework from products to socio-technical systems.* Routledge, Taylor & Francis Group. https://doi.org/10.4324/9780429456510

CIDQ. Council for Interior Design Qualification. (2019). *Definition of interior design—Abbreviated definition of interior design.* Retrieved July 15, 2025, from https://www.cidq.org/

Cole, L. B., & Hamilton, E. M. (2024). Transition interior design: Reimagining practice and education for a post-carbon society. *Journal of Sustainability Research, 6*(2), e240018. https://doi.org/10.20900/jsr20240018

DNGB—German Sustainable Building Council. (2015). Retrieved July 15, 2025 from https://www.dgnb.de/en

EN15804:2012. (2012). *Environmental product declarations—Rules for product categories.* European Committee for Standardisation.

EN15942:2011. (2011). *Environmental product declarations—Format for communication.* European Committee for Standardisation.

European Commission-Joint Research Centre (EC-JRC) - Institute for Environment and Sustainability: International Reference Life Cycle Data System (ILCD). (2011, November). *Recommendations for life cycle impact assessment in the European context* (1st ed.). EUR 24571 EN. Publications Office of the European Union.

European Commission, Buildings and Construction. (n.d.). Retrieved July 08, 2025, from https://single-market-economy.ec.europa.eu/industry/sustainability/buildings-and-construction_en

Faludi, J., Hoffenson, S., Kwok, S. Y., Saidani, M., Hallstedt, S. I., Telenko, C., & Martinez, V. (2020). A research roadmap for sustainable design methods and tools. *Sustainability, 12*(19), 8174. https://doi.org/10.3390/su12198174

Fuller, R. B. (1969). *Operating manual for spaceship earth.* Southern Illinois University Press.

Green Building – BEAM. (2024). Retrieved July 15, 2025, from https://www.hkgbc.org.hk/eng/main/index

Green Star. (2025). Retrieved July 15, 2025, from https://new.gbca.org.au/green-star/rating-system/

He, Y., Kvan, T., Liu, M., & Li, B. (2018). How green building rating systems affect designing green. *Building and Environment*, *133*, 19–31. ISSN: 0360-1323. https://doi.org/10.1016/j.buildenv.2018.02.007

HQE - Haute Qualité Environnementale. (2025). Retrieved July 15, 2025, from https://www.hqegbc.org/en/qui-sommes-nous-alliance-hqe-gbc/la-certification-hqe/

Husaini, M. Z., Jusoh, A., & Kassim, M. S. (2018). Analysis of sustainability assessment tools (SATs) for higher education institutions (HEIs). *International Journal of Academic Research in Business and Social Sciences, 8*(8), 558–571.

InEDIC. (2011). Innovation and eco-design in the ceramic industry, an overview of knowledge needs in Portugal, Spain and Greece. EU life long learning—Leonardo da Vinci funded project, which aims at developing training materials and eco-design tools to support the integration of environmental considerations in the design of ceramic products, thus contributing to the innovation and competitiveness of this longstanding industry in Portugal, Spain and Greece.

IPCC. (2018). Summary for policymakers. In V. Masson-Delmotte, P. Zhai, H.-O. Pörtner, D. Roberts, J. Skea, P. R. Shukla, A. Pirani, W. Moufouma-Okia, C. Péan, R. Pidcock, S. Connors, J. B. R. Matthews, Y. Chen, X. Zhou, M. I. Gomis, E. Lonnoy, T. Maycock, M. Tignor, & T. Waterfeld (Eds.), *Global warming of 1.5 °C. An IPCC special report on the impacts of global warming of 1.5 °C above pre-industrial levels and related global greenhouse gas emission pathways, in the context of strengthening the global response to the threat of climate change, sustainable development, and efforts to eradicate poverty* (pp. 3–24). Cambridge University Press. https://doi.org/10.1017/9781009157940.001

IPCC. (2022). Summary for policymakers [P. R. Shukla, J. Skea, A. Reisinger, R. Slade, R. Fradera, M. Pathak, A. Al Khourdajie, M. Belkacemi, R. van Diemen, A. Hasija, G. Lisboa, S. Luz, J. Malley, D. McCollum, S. Some, & P. Vyas (Eds.)]. In P. R. Shukla, J. Skea, R. Slade, A. Al Khourdajie, R. van Diemen, D. McCollum, M. Pathak, S. Some, P. Vyas, R. Fradera, M. Belkacemi, A. Hasija, G. Lisboa, S. Luz, & J. Malley (Eds.), *Climate change 2022: Mitigation of climate change. Contribution of working group III to the sixth assessment report of the intergovernmental panel on climate change.* Cambridge University Press. https://doi.org/10.1017/9781009157926.001

Irwin, T. (2015). Transition design: A proposal for a new area of design practice, study, and research. *Design and Culture.*

ISO 14025:2006. (2006). *Environmental labels and declarations—Type III environmental declarations—Principles and procedures.*

ISO 14044:2006. (2006). *Environmental management—Life cycle assessment—Requirements and guidelines.*

ISO 14040:2006. (2006). *Environmental management—Life cycle assessment—Principles and framework.*

KATCH_e Project. (2020). *Knowledge alliance on product-service development towards circular economy and sustainability in higher education* (project no. 575793-EPP-1-2016-1-PT-EPPKA2-KA).

Kozlowski, A., Bardecki, M., & Searcy, C. (2019). Tools for sustainable fashion design: An analysis of their fitness for purpose. *Sustainability, 11*(13), 3581. https://doi.org/10.3390/su11133581

Leal Filho, W., Trevisan, L. V., Dinis, M. A. P., Ulmer, N., Paço, A. M. Fd., Borsari, B., Sierra, J., & Salvia, A. L. (2024). Fostering students' participation in the implementation of the sustainable development goals at higher education institutions. *Discover Sustainability*, *5*(1), 1–24. https://doi.org/10.1007/s43621-024-00204-7

Leal Filho, W., Trevisan, L. V., Sivapalan, S., Mazhar, M., Kounani, A., Mbah, M. F., Abubakar, I. R., Matandirotya, N. R., Dinis, M. A. P., Borsari, B., & Abzug, R. (2025). Assessing the impacts of sustainability teaching at higher education institutions. *Discover Sustainability, 6*(1), 1–22. https://doi.org/10.1007/s43621-025-01024-z

LEED - Leadership in Energy and Environmental Design. (2025). Retrieved July 15, 2025, from https://www.usgbc.org/leed

Lozano, R., Ceulemans, K., Alonso-Almeida, M., Huisingh, D., Lozano, F. J., Waas, T., Lambrechts, W., Lukman, R., & Hugé, J. (2015). A review of commitment and implementation of sustainable development in higher education: Results from a worldwide survey. *Journal of Cleaner Production, 108*, 1–18. https://doi.org/10.1016/j.jclepro.2014.09.048

Martinho, S., Rigueiro, C., & Mateus, R. (2013). Cost/benefit analysis in the implementation of sustainable construction principles in a residential building. Portugal SB13, Contributions of sustainable building for EU-20-20-20 targets, Guimarães. http://hdl.handle.net/10400.11/2777

Mateus, R., & Bragança, L. (2011). Sustainability assessment and rating of buildings: Developing the methodology SBTool^{PT}-H. *Building and Environment*, *46*(10), 1962–1971. ISSN: 0360-1323. https://doi.org/10.1016/j.buildenv.2011.04.023

One Click LCA. (2025). Retrieved July 15, 2025, from https://oneclicklca.com/

OpenLCA. (2025). Retrieved July 15, 2025, from https://www.openlca.org/

Papanek, V. (1991). *Design for the real world: Human ecology and social change* (revised 2nd ed.). Thames & Hudson.

Reyes, J. E. (2019, July 8). Development of sustainable solutions for urban rehabilitation. Ph.D. degree in Steel and Composite Construction from the University of Coimbra.

Reyes Nieto, J. E., Rigueiro, C., da Silva, L. S., & Murtinho, V. (2024). Urban rehabilitation model: A blueprint for efficient, livable, smart and sustainable urban environments. In *2024 IEEE international smart cities conference (ISC2), Pattaya, Thailand* (pp. 1–6). https://doi.org/10.1109/ISC260477.2024.11004267

Rigueiro, M. C. S., Caeiro, S., Câmara, P., Fonseca, A., Godinho, T., Mapar, M., Moreira, A. M., Dinis, M. A. P., Farinha, C. S., Simão, J., & Silva, M. V. d. (2025). A critical analysis of sustainability indicators for education and curricula in higher education institutions and their adaptation to the Portuguese context. In W. L. Filho, L. V. Trevisan, P. J. Pace, & M. Mifsud (Eds.), *Education for sustainable development: The contribution of universities* (pp. 465–488). Springer Cham. https://doi.org/10.1007/978-3-031-86985-3_28

Robèrt, K. H., Schmidt-Bleek, B., Aloisi de Larderel, J., Basile, G., Jansen, J. L., Kuehr, R., Thomas, P. P., Suzuki, M., Hawken, P., & Wackernagel, M. (2002). Strategic sustainable development—Selection, design and synergies of applied tools. *Journal of Cleaner Production, 10*(3), 197–214. ISSN: 0959-6526. https://doi.org/10.1016/S0959-6526(01)00061-0

Sanhudo, L. P. N., & Martins, J. P. d. S. P. (2018). Building information modelling for an automated building sustainability assessment. *Civil Engineering and Environmental Systems, 35*(1–4), 99–116. https://doi.org/10.1080/10286608.2018.1521393

Saouter, E., Biganzoli, F., Ceriani, L., Versteeg, D., Crenna, E., Zampori, L., Sala, S, & Pant, R. (2020). *Environmental footprint: Update of life cycle impact assessment methods—Ecotoxicity freshwater, human toxicity cancer, and non-cancer*. EUR 29495 EN. Publications Office of the European Union. ISBN: 978-92-76-17143-0. https://doi.org/10.2760/300987, JRC114227.

SBTool - Sustainable Build Environment. (2022). Retrieved July 15, 2025, from https://www.iisbe.org/sbmethod

SimaPro. (2025). Retrieved July 15, 2025, from https://simapro.com/

SinnDesign. (2015). *Sustainable innovation through design, tools and training materials to integrate sustainability in design.* Conducted by a consortium from Portugal, Spain, Austria and Denmark. Lifelong Learning Programme, European Commission.
Sphera LCA for Experts. (2025). Retrieved July 15, 2025, from https://sphera.com/?utm_source=google&utm_medium=paid&gclid=CjwKCAjw4efDBhATEiwAaDBpbnzVJzePf3mxK1Xx5uS3B1K2cLzthoQvaDykpamWN0l3XlBbMfl-CRoCNToQAvD_BwE
UNEP. (2021). *The UNEP sustainability university framework.* United Nations Environmental Program. https://www.unep.org/resources/toolkits-manuals-and-guides/uneps-sustainable-university-framework
United Nations Environment Programme (UNEP), & Yale Center for Ecosystems + Architecture. (2023). *Building materials and the climate: Constructing a new future.* https://wedocs.unep.org/20.500.11822/43293
Vicente, J., & Camocho, D. (2023). Design for sustainability tools: Categories of classification towards practical use. In *Human-centered design and user experiences* (Vol. 114, pp. 41–49). https://doi.org/10.54941/ahfe1004220
Vicente, J., & Camocho, D. (2024). Design for sustainability tools: Definition and criteria towards practical use. *Journal of Cleaner Production, 479*, 144041. ISSN: 0959-6526. https://doi.org/10.1016/j.jclepro.2024.144041
World Meteorological Organization WMO-No. 1368. (2025). *State of the global climate 2024.* World Meteorological Organization (WMO). ISBN: 978-92-63-11368-5.

Maria Constança Simões Rigueiro Ph.D. in Civil Engineering in 2009 from the University of Coimbra. She is a full member of the ISISE research centre, where she is a member of the SMCT—Steel and Mixed Construction Technology research team. She is a Coordinating Professor at the Castelo Branco Polytechnic Institute. Since February 2025, she has been the coordinator of the Degree Course in Civil Engineering. In her curriculum vitae, the most frequent terms used to contextualise her scientific, technological production are: Structural Dynamics; Bridges; Extreme Actions; Design; Sustainability; Sustainable Construction; Sustainability Assessment; Sustainability in Teaching and Curricula.

Maria Alzira Pimenta Dinis Habilitation, Ph.D., M.Sc. is Associate Professor at the Faculty of Science and Technology, University Fernando Pessoa, Porto, Portugal. Graduated in Metallurgical Engineering and Master Science in Materials Engineering at the Faculty of Engineering, University of Porto, Portugal. Ph.D. in Earth Sciences and Habilitation in Ecology and Environmental Health at the Faculty of Science and Technology, University Fernando Pessoa, Portugal. Co-Coordinator of the Earth Sciences doctoral program. Researcher at Fernando Pessoa Research, Innovation and Development Institute (FP-I3ID) and Centro de Ciências do Mar e do Ambiente/Marine and Environmental Sciences Centre (MARE), University of Coimbra in environment-related topics. Experienced author, editor and reviewer. Sustainability expertise.

Integrating Design, Project Management, and People Management in Sustainable and Innovative Learning Processes

Patrícia Borba Vilar Guimarães, Cynara Carvalho de Abreu, José Guilherme da Silva Santa Rosa, and Josué Vitor de Medeiros Júnior

Abstract This chapter explores how interdisciplinary integration in a professional master's program advances SDG-oriented curriculum development with measurable institutional impact. Higher education institutions must prepare professionals to address climate change and sustainability challenges. In this program, students engaged in collaborative, practice-based analyses of institutional processes, applying design thinking and agile project management to create innovative solutions. This approach strengthened problem-solving, teamwork, and systemic thinking, resulting in nine Technical Products (PTTs) aligned with United Nations Sustainable Development Goals (SDGs) 4, 9, 11, 12, and 16. The model shows how master's programs in the Global South can link curriculum innovation to the climate and sustainability agenda, supporting leadership in public sector sustainable change.

Keywords Sustainable curriculum · Interdisciplinary learning · Project-based learning · Climate change education · SDGs · Innovation in higher education

In a world grappling with the impacts of climate change, imagine a university classroom where tomorrow's leaders are equipped to design sustainable solutions today. The urgent need to address climate change and foster sustainable development calls for higher education institutions to take proactive roles in preparing future professionals. This chapter focuses on how universities can cultivate critical and systemic thinkers ready to design practical solutions for environmental, social, and economic

P. B. V. Guimarães (✉) · C. C. de Abreu · J. G. da Silva Santa Rosa · J. V. de Medeiros Júnior
Postgraduate Program in Institutional Process Management (PPGPI), Federal University of Rio Grande do Norte (UFRN), Natal, Brazil
e-mail: patricia.borba@ufrn.br

C. C. de Abreu
e-mail: carvalho@ufrn.br

J. G. da Silva Santa Rosa
e-mail: jose.guilherme@ufrn.br

J. V. de Medeiros Júnior
e-mail: josue.medeiros@ufrn.br

W. Leal Filho et al., *Curriculum Development for Sustainable Development*, World Sustainability Series, https://doi.org/10.1007/978-3-032-22146-9_21

challenges. International frameworks, such as the United Nations Sustainable Development Goals (SDGs), place education at the heart of transformation and emphasize the importance of curriculum innovation for inclusive, sustainable change.

Interdisciplinary and integrated curricular strategies are effective for linking theory with practice in higher education. These approaches foster collaborative, contextualized learning and help develop technical competencies and transversal skills. However, implementation often faces institutional resistance and rigid structures. Deliberate planning and faculty engagement are essential to realize the benefits of curricular integration for sustainability.

Higher education institutions are increasingly seen as 'living laboratories' for sustainability. Building on the need for curricular reform, international examples reinforce this idea. Studies show university campuses can foster skills and institutional transformation. They can implement energy-efficient operations (Leal Filho et al., 2018) and redesign curricula to integrate sustainability across disciplines (Leal Filho & Portela de Vasconcelos, 2022). In India, higher education institutions have promoted the adoption of renewable energy through curricular and training strategies (Prasath et al., 2015). These examples highlight the importance of linking curriculum development to sustainability practices and aligning with global experiences and the SDGs.

This chapter presenThis chapter presents an interdisciplinary curriculum integration initiative in the Graduate Program in Institutional Process Management (PPGPI) at the Federal University of Rio Grande do Norte (UFRN), Brazil. The program connects Product and Process Design, Project Management, and People Management within a unified learning framework. Public servants in this master's program developed Technical Products (PTTs) to address institutional and societal needs. Nes the stages of this curricular integration. Section 1 explores foundational concepts, Sect. 2 details the interdisciplinary implementation, Sect. 3 discusses outcomes and challenges, Sect. 4 addresses scaling strategies, and the conclusion reviews the initiative's contribution to the SDGs, specifically SDGs 4, 9, 11, 12, and 16.

By connecting theory, practice, and policy, this work examines the role of postgraduate education in public sector transformation and innovation. The analysis centers on how curriculum integration in a professional master's program advances sustainable development through interdisciplinary, project-based learning and the creation of technical products aligned with SDGs.

The integration of sustainability into educational frameworks is an urgent priority as the world faces pressing challenges like climate change, resource depletion, and social inequality. Building upon the need for curriculum integration identified earlier, scholars contend that education—particularly at universities—must evolve. Students need tools to understand and tackle these complex issues holistically. Integrating sustainability into curricula remains central to reform efforts. Interdisciplinary approaches are especially important because they blend the environmental, social, and economic dimensions of sustainability.

1 Transformative and Interdisciplinary Education

Sterling (2001) and Tilbury (2011) have long argued for a shift from traditional educational paradigms toward more transformative approaches. Transformative learning is characterized by critical thinking and deep engagement, in which students are not merely passive recipients of knowledge but active participants in reshaping their understanding of sustainability. This approach, which emphasizes the development of new skills, perspectives, and behaviors, is increasingly seen as necessary for addressing the challenges posed by climate change, biodiversity loss, and social inequalities.

This shift often occurs through interdisciplinary education, which helps students draw on knowledge from diverse fields. A multidisciplinary approach equips individuals to analyze complex, interconnected problems. With this preparation, they can view sustainability issues through multiple disciplinary lenses such as economics, environmental science, politics, and sociology (Horn et al., 2023). By engaging a range of disciplines, students develop more comprehensive solutions to sustainable development challenges.

2 The Role of Project-Based Learning in Sustainability Education

Project-based learning (PBL), an instructional method in which students actively explore real-world problems over extended periods, is another key approach for fostering a deep understanding of sustainability in higher education. PBL allows students to engage with authentic challenges and apply their knowledge in meaningful ways. This methodology has been shown to significantly enhance students' problem-solving abilities, teamwork, and leadership skills, which are essential for addressing sustainability challenges (Gupta et al., 2022). By engaging in hands-on projects, students can work with communities, businesses, and governments to implement sustainable solutions, which enhances their sense of responsibility and agency.

In practice, PBL is evident in several international programs. For example, the Erasmus+ PECT Project (Protecting the Environment through Conservation Techniques) exemplifies how PBL can be used to teach students about environmental sustainability. The project involved students from Malta and Romania who collaborated with local communities to protect natural resources. Through hands-on activities such as tree planting, habitat restoration, and environmental monitoring, students applied theoretical knowledge in real-world contexts, gaining valuable experience in environmental conservation and management (Hazard, 2023).

Furthermore, Tilbury (2011) emphasizes the need for participatory, action-oriented learning approaches that foster critical thinking and problem-solving skills. PBL, when implemented in the context of sustainability, aligns well with these goals by encouraging students to engage directly with the issues they are studying.

3 Sustainability and People Management in Education

Another critical aspect of sustainability education is integrating people management skills. As sustainability challenges require coordinated action among various stakeholders, leadership, collaboration, and effective management are essential for implementing sustainable practices. Northouse (2018) argues that effective leadership in sustainability contexts requires not only technical knowledge but also strong interpersonal skills, such as emotional intelligence, conflict resolution, and team management.

Incorporating people management into sustainability curricula helps students develop the necessary skills to lead teams, navigate organizational challenges, and manage relationships among diverse stakeholders. Horn et al. (2023) underscore the importance of leadership training in sustainability education, noting that future leaders must be able to manage people in ways that encourage collaboration and long-term commitment to sustainable practices.

Educational programs, such as the Sustainability Management and Leadership Program at UC Berkeley, have successfully integrated people management into their curriculum by focusing on leadership in sustainability contexts. These programs prepare students to lead sustainability initiatives across sectors, ensuring that they are equipped not only with technical expertise but also with the people skills needed to drive change.

A significant aspect of sustainability education that is often overlooked is the importance of context—both local and global. The challenges of sustainability are not only international but are also profoundly influenced by local cultural, political, and economic conditions. As Tilbury (2011) notes, sustainability education must be contextualized to be practical. This means that educational approaches should consider the specific needs, values, and challenges of the communities in which they are implemented.

Global frameworks, such as the United Nations Sustainable Development Goals (SDGs), provide a broad outline for sustainable development. Still, it is the local context that determines how these goals are achieved. For example, in Brazil, federal universities have created postgraduate programs that link sustainability with local development goals, particularly in rural areas, where issues such as deforestation, land degradation, and water scarcity are pressing concerns. These programs integrate local knowledge and traditional practices with modern scientific approaches to develop sustainable solutions tailored to the region's unique needs.

In contrast, European initiatives, such as the Erasmus+ projects, foster international cooperation and learning across borders, enabling students to gain a global perspective on sustainability while addressing specific regional challenges. By integrating local and global perspectives, these initiatives promote the exchange of knowledge and ideas, which enhances the effectiveness of sustainability education.

Despite the growing emphasis on sustainability education, several challenges remain in its integration into higher education curricula. Watson (2003) notes that while many educational institutions recognize the importance of sustainability, implementing effective sustainability programs is often hindered by traditional teaching methods, limited resources, and a lack of interdisciplinary collaboration.

Moreover, Sachs (2015) argues that sustainability education often focuses on environmental issues at the expense of social and economic dimensions. To address this, universities must adopt a more holistic approach that encompasses all three pillars of sustainability—environmental, social, and financial. This integrated approach not only provides students with a broader understanding of sustainability but also equips them to tackle complex, multifaceted problems in their professional lives.

Despite these challenges, there are numerous opportunities for expanding sustainability education. The growing global focus on sustainability, along with the increasing demand for professionals with sustainability expertise, presents an opportunity for higher education institutions to lead the way in preparing the next generation of leaders. As UNESCO (2023a, 2023b) highlights, there is a growing recognition that sustainability is a key driver of innovation and economic growth, making sustainability education not only a moral imperative but also a financial necessity.

3.1 Positioning of UFRN Governance on Sustainability and SDGs: Based on the PDI and Management Plan

The Federal University of Rio Grande do Norte (UFRN) plays a significant role in promoting sustainability and aligning its institutional strategies with the United Nations Sustainable Development Goals (SDGs). As one of the leading public universities in Brazil, UFRN's commitment to sustainability is evident through its institutional policies, particularly in the Institutional Development Plan (PDI) (UFRN, 2020) and the Management Plan (UFRN, 2022), which outline the university's strategic priorities for the coming years. These documents reflect a comprehensive approach to integrating sustainability into the university's core operations, academic activities, and research initiatives, aligning with both national and global sustainability agendas.

3.2 Institutional Development Plan (PDI) and Sustainability

The Institutional Development Plan (PDI) of UFRN is a fundamental document that guides the university's activities over a specific period, typically every five years. The most recent version of the PDI integrates sustainability as one of its central pillars, ensuring that sustainability is woven into the fabric of UFRN's academic, research, and extension efforts. It provides a strategic framework for integrating the SDGs into the university's goals, ensuring that UFRN contributes to the global sustainable development agenda while addressing local challenges specific to Rio Grande do Norte.

The PDI explicitly identifies the need to advance sustainability education by encouraging interdisciplinary approaches that address its environmental, social, and economic dimensions. This is seen in UFRN's emphasis on integrating sustainability-related topics across various academic programs, fostering innovation through sustainable design and development, and creating research opportunities focused on sustainability challenges. Moreover, the PDI highlights the importance of community engagement and the application of knowledge to tackle local and regional sustainability issues.

The PDI also underscores UFRN's commitment to environmental responsibility in its physical operations. This includes promoting energy efficiency, waste management, and the sustainable use of resources across the campus, aligning with SDG 12 (Responsible Consumption and Production). UFRN's initiatives, such as reducing carbon emissions, adopting green technologies, and developing sustainable infrastructure, reflect its broader commitment to the SDGs.

3.3 UFRN'S Management Plan: Operationalizing Sustainability

Complementing the PDI, the Management Plan outlines the specific actions, resources, and institutional structures required to implement the university's goals. The Management Plan of UFRN identifies key priorities for integrating sustainability, linking operational and academic strategies to the SDGs in practical ways. This includes capacity-building initiatives for staff and students, institutional sustainability policies, and the development of a sustainability action plan with measurable goals.

A significant element of UFRN's Management Plan is the interdisciplinary research and projects related to sustainability. For instance, the university has established research centers focused on environmental sustainability, energy, and sustainable urban planning that collaborate with national and international institutions. These initiatives promote a comprehensive understanding of sustainability that goes beyond traditional academic boundaries in environmental science to encompass social, economic, and governance-related issues.

The Management Plan also emphasizes institutional governance structures that ensure sustainability principles are integrated at all levels of decision-making. This includes the creation of dedicated committees and sustainability officers to oversee the implementation of sustainability practices and policies within the university. These governance bodies help establish clear lines of accountability, ensuring sustainability goals are met across departments and initiatives.

3.4 UFRN'S Contribution to SDGs: Integration of Global and Local Perspectives

UFRN's sustainability governance is not only about internal policies but also about how it engages with the broader community, both locally and globally. By aligning with the SDGs, UFRN strengthens its international role in higher education for sustainability. For example, UFRN's involvement in international projects, UNESCO's ESD for 2030, facilitates cross-cultural exchange and the sharing of best practices for sustainability in higher education.

At the local level, UFRN's actions align with the regional sustainability challenges faced by the state of Rio Grande do Norte, including issues like water scarcity, deforestation, and social inequality. The university is actively engaged in projects that address these issues by promoting social inclusion, local economic development, and environmental protection. Initiatives such as community-based environmental monitoring and sustainable agricultural practices exemplify how UFRN applies sustainability concepts to real-world contexts, fostering both local impact and global relevance.

In line with SDG 11 (Sustainable Cities and Communities), UFRN's urban sustainability efforts focus on creating green spaces on campus, promoting sustainable mobility, and improving infrastructure to enhance the quality of life for students, faculty, and the local community. The university is also a key partner in developing sustainable urban planning models in Natal, the state capital, aimed at reducing urban sprawl, promoting public transportation, and incorporating green areas into urban planning.

3.5 Challenges and Opportunities for UFRN'S Sustainability Governance

Despite these advancements, UFRN faces several challenges in fully operationalizing its sustainability goals. One of the primary obstacles is the financial constraints that often limit the implementation of large-scale sustainability initiatives. While the university is committed to sustainability, insufficient funding for some projects means that specific goals are achieved more slowly than desired. Additionally, there is a need for greater integration of sustainability across all academic and administrative functions, as it should not be limited to research and extension projects but also be embedded in the day-to-day functioning of the institution.

Another challenge lies in engaging and involving the entire university community—from students and faculty to staff and management—in sustainability initiatives. Although UFRN has made strides in this area, more work is needed to ensure that sustainability becomes part of the institutional culture, with active involvement from all stakeholders.

However, these challenges also present opportunities for UFRN to enhance its leadership in sustainability education and research. The growing global demand for professionals with sustainability expertise presents an opportunity for UFRN to expand its academic offerings, create new interdisciplinary programs, and strengthen its leadership in sustainability education. Furthermore, UFRN's experience in managing sustainability initiatives at both local and global levels positions it to influence policy development and shape the future of sustainability education in Brazil and beyond.

In light of this institutional framework, the following section presents the methodological approach that operationalizes UFRN's sustainability strategies through an integrated, practice-oriented curriculum design.

4 Methodology

4.1 Integrating Design, Project Management, and People Management

The methodological approach of the integrated curricular experience involves a structured step-by-step process focusing on the context, the diverse background of participants, and specific evaluation criteria. This approach ensures clarity, interdisciplinarity, and practical application, all while aligning with good research practice in curriculum studies.

The methodology integrates three complementary components—Product and Process Design, Project Management, and People Management—delivered through interconnected modules featuring joint challenges and shared deliverables. Rooted in project-based learning (PBL), the curriculum organizes students into interdisciplinary teams to address authentic or simulated institutional issues, using applied research, stakeholder involvement, and systems thinking to develop context-specific solutions.

The Product and Process Design component helps students develop innovative, user-centered, and sustainability-oriented solutions. By leveraging engineering principles, service design, and innovation management, teams are encouraged to prototype enhancements to public services, institutional workflows, or digital tools, while critically evaluating the social and environmental implications of their design choices. As posited by Brown (2009), design thinking is an iterative, collaborative, and problem-solving methodology that prioritizes the human being in decision-making. When applied to global governance, particularly within the domain of climate change, this paradigm can foster the creation of more adaptable, inclusive, and innovative institutional frameworks capable of addressing the multifaceted array of actors and interests involved in establishing a stable climate. According to Santa Rosa and Moraes (2012), participatory design comprises a set of techniques that include integrating users into the design process, thereby ensuring that their perspectives, needs, and experiences inform the development of interfaces and systems. In the context of global climate governance, this approach can be adapted to enhance stakeholder engagement, fostering co-created solutions that are both technically effective and socially legitimate in pursuit of a stable climate.

Project Management is approached not only as a technical discipline, but as a strategic competence for driving innovation in contexts of uncertainty and resource constraints. Students are introduced to traditional and agile approaches, and they develop work plans, risk matrices, and performance indicators aligned with institutional and sustainability objectives.

The People Management component serves as the human connector within the integrated methodology. Drawing upon organizational psychology, leadership studies, and public administration, students delve into motivational strategies, conflict resolution, collaborative governance, and inclusive management practices. This human-centered approach ensures that sustainability extends beyond products and processes to encompass the ethical, emotional, and organizational facets of institutional existence.

To enhance the experiential and cooperative dimensions of the program, it integrates mentorship, structured peer evaluation, and engagement with public-sector stakeholders, thereby fostering an authentic environment for innovation. Graduates are prepared as professionals adept at spearheading sustainable transformation, addressing intricate societal and institutional imperatives, and cultivating transformative governance methodologies. The outcomes were evaluated against the CAPES criteria for professional master's programs, specifically: achieved applicability, prospective applicability, transferability, and demonstrable impact. Furthermore, each PTT was correlated with at least one United Nations Sustainable Development Goal (SDG) to ascertain its congruence with international sustainability frameworks.

This integrated methodology culminates in the development of Technical Products (PTTs), concrete applications of interdisciplinary learning. Each PTT is conceived to solve a real institutional problem, frequently accompanied by a distinctive visual identity that strengthens its usability and professional recognition. This seamless transition from methodology to tangible products exemplifies the transformative potential of integrated curricula in professional master's programs.

4.2 Case Study: Context and Objectives

The Professional Master's Program in Institutional Process Management (PPGPI) at the Federal University of Rio Grande do Norte (UFRN) undertook a significant curricular restructuring in 2024, leading to the implementation of an innovative integrated approach to teaching and learning. This restructuring aimed to enhance the curriculum's interdisciplinary nature and foster a more practical, collaborative, and sustainability-focused learning environment. The integrated approach involved the simultaneous offering of three core components—Product and Process

Design, People Management, and Project Management—designed to complement and reinforce one another.

The decision to implement this integration arose from a thorough self-assessment and an external evaluation conducted by CAPES (the Brazilian higher education regulatory agency) of the doctoral-level training offered by PPGPI. These evaluations highlighted the need for a more cohesive, integrated curriculum that better aligns with the interdisciplinary guidelines. As a result, the curriculum integration was seen as an essential step towards improving both the academic quality and practical relevance of the program.

The primary objective of the integrated course offering was to enhance the program's innovative potential through the teaching methodologies employed. By bringing together Product and Process Design, People Management, and Project Management, the PPGPI sought to provide students with a holistic understanding of organizational processes, emphasizing sustainability and real-world applications. This integrated model aimed not only to equip students with the theoretical knowledge required in each of these areas but also to foster a collaborative learning environment where students could apply their knowledge in a practical, project-based context.

Furthermore, the integration of these components was intended to align with UFRN's Institutional Development Plan (PDI), which emphasizes interdisciplinary teaching, innovation, and sustainability in shaping graduates' academic and professional profiles. By focusing on the relationships among design, management, and people, the program sought to create professionals capable of managing complex processes and leading teams to achieve sustainable, innovative outcomes. This aligns with UFRN's broader commitment to addressing global challenges through higher education, as outlined in its PDI and the United Nations' Sustainable Development Goals (SDGs), particularly SDG 4 (Quality Education), SDG 9 (Industry, Innovation, and Infrastructure), SDG 11 (Sustainable Cities and Communities), and SDG 12 (Responsible Consumption and Production).

Through this integrated approach, the course also aimed to promote a shift from traditional, siloed teaching methods to a more collaborative and project-driven model of learning. This change was designed to foster greater engagement and motivation among students, as well as to develop essential transversal skills, such as critical thinking, problem-solving, and teamwork, which are increasingly valued in the contemporary workforce.

In summary, the integrated Project Management course within the PPGPI was designed to offer an innovative, interdisciplinary educational experience that would better prepare students for the challenges of managing organizational processes in a rapidly changing world. By emphasizing sustainability, collaboration, and real-world applications, this initiative sought to create a curriculum aligned not only with labor-market needs but also with the global imperatives of sustainable development.

4.3 Strategy

The strategy for implementing the integrated curriculum in the Professional Master's Program in Institutional Process Management at UFRN was designed to ensure alignment among the three core components—Product and Process Design, People Management, and Project Management—into a cohesive, practical learning experience. The implementation began with an initial meeting among the faculty members involved in the course. Each faculty member presented a brief overview of the content and objectives of their respective disciplines. This collaborative approach enabled the development of a strategy to integrate these disciplines into the professional master's program's overall curriculum.

Following the faculty presentations, a detailed analysis was conducted of students' professional backgrounds, areas of expertise, and sectors in which they worked. This analysis helped in defining the "Guiding Questions," which would later serve as a foundation for the students' projects. These guiding questions were intended to drive interdisciplinary collaboration and ensure that the students' projects addressed real-world issues within their respective institutions, with a particular focus on social impact.

The first encounter with the students began with a presentation by the Program Coordinator, outlining the innovative approach being introduced in the course and emphasizing the integration of the three disciplines. After this, the faculty members introduced their respective courses, including the structure, content, and assessment methods. Following the introduction of the courses, the guiding questions were presented and discussed with the students. The next step was to explain the criteria for team formation. Once the teams were defined, the project development process officially began, with each team tasked with creating a technical product to solve concrete institutional problems with social impact.

In the first cohort, three teams were formed; in the second, six, all working on the same methodological approach. The repetition of the experiment provided an opportunity to compare the outcomes of the two cohorts and refine certain teaching practices. This iterative process allowed the faculty to evaluate the strengths and weaknesses of their approach, ensuring continuous improvement in the delivery of the interdisciplinary curriculum.

Before proceeding with the project work, students were given time to define the specific details of their proposals within their teams and then present them to the larger group. Subsequently, they began using the "Life Cycle Canvas" (LCC) to track and monitor the progress of their projects (Medeiros et al., 2018). The LCC provided a structured framework for students to visualize and manage the various stages of their projects, from conception to implementation.

Throughout the course, particularly during periodic meetings to monitor project progress, a noticeable increase in student engagement, participation, and interaction was observed. By the second meeting, the faculty members noticed that students were becoming more active in their participation. This shift was attributed to students' transition from a defensive posture—initially apprehensive about the course's interdisciplinary nature—to a more open and proactive stance. The faculty's enthusiasm and availability to support students through mentorship helped foster this shift in mindset, as students began to view the faculty not as critics but as partners in their learning journey.

The faculty members played an active role in encouraging critical thinking and student proactivity, particularly in identifying problems and proposing solutions, while respecting each team's thematic choices. This collaborative approach, which integrated students' diverse backgrounds and expertise, was a central element of the course's design.

However, this approach also posed a challenge for students, as they were accustomed to traditional, siloed learning environments where each discipline was taught in isolation. The interdisciplinary model, in which four professors collaboratively guided students' learning, was a new and unfamiliar experience for many. This initial discomfort and skepticism were evident in the students' lack of verbal participation and their tendency to keep their webcams off during the first remote sessions, suggesting their hesitance to engage with the new format fully.

At the conclusion of the formal lessons, an Integrated Panel was convened to provide students with feedback on their project progress. This session was followed by a series of feedback meetings where the faculty provided individualized guidance to each team. The feedback process was extensive and interactive, designed to help students refine their projects and ensure alignment with the guiding questions and the course's overall objectives.

In summary, the implementation strategy for the integrated curriculum was a dynamic process that involved continuous reflection, adjustment, and collaboration among faculty and students. Despite the initial challenges, the plan succeeded in fostering a collaborative, interdisciplinary learning environment that encouraged critical thinking, problem-solving, and practical application of knowledge. The integration of these three components—Product and Process Design, People Management, and Project Management—provided students with a unique opportunity to develop their skills and competencies in a real-world, socially responsible context.

4.4 Applied Cases and Technical Products (PTTs)

In Brazil, the "Coordenação de Aperfeiçoamento de Pessoal de Nível Superior" [Coordination for the Improvement of Higher Education Personnel] (CAPES) is the federal agency responsible for evaluating and accrediting graduate programs, including professional master's and doctoral degrees. CAPES plays a pivotal role in ensuring that postgraduate education contributes to scientific advancement, technological innovation, and societal impact. Its assessment methodology combines quantitative indicators, such as scientific production and graduate outcomes, with qualitative evaluations, including the relevance and originality of technical and technological products (PTTs).

For professional master's programs, CAPES places special emphasis on the applicability, social relevance, and innovation of outputs, as these programs are expected to generate solutions that address real institutional or societal challenges. The evaluation of each PTT, therefore, considers multiple criteria, including:

1. Actual applicability: the extent to which the product was successfully implemented to meet its intended goals, considering the mobilization of internal and external resources when necessary.
2. Potential applicability: the degree to which the product can be transferred or adapted to other institutional contexts with thematic or conceptual alignment.
3. Replicability: the scalability of the product, reflecting its ability to be reproduced or adapted across multiple settings.
4. Impact: the potential of the PTT to produce significant organizational, social, or policy-related changes, for instance, by restructuring programs, processes, or services.

Applying these criteria allows the program to demonstrate compliance with CAPES guidelines and to evidence the social, institutional, and educational impact of each product. Table 1 summarizes the PTTs developed within the integrated course, linking them to their respective SDGs and evaluation criteria.

The implementation of the integrated curriculum in the Professional Master's Program in Institutional Process Management (PPGPI/UFRN) led to the development of 9 interdisciplinary Technical Products (PTTs). Each PTT was designed by student teams in response to concrete institutional challenges, leveraging the

Table 1 Methodological approach adopted in the experience

Context definition	Participants
The integrated curricular experience was implemented within the professional master's program in institutional process management (PPGPI) at the Federal University of Rio Grande do Norte (UFRN) during the 2024 curricular restructuring. The program adopted an interdisciplinary approach, in line with CAPES guidelines and UFRN'S institutional development plan (PDI), to bolster innovation and sustainability in professional education	The experiment involved two cohorts of professional master's students from diverse academic and professional backgrounds, including law, administration, design, and engineering. The teaching staff comprised four professors specializing in product and process design, Project Management, and people management, who collaborated to design and deliver the course
Implementation	*Project development and outputs*
Faculty members convened initial meetings to harmonize objectives, delineate the content of their respective disciplines, and devise shared pedagogical approaches. This collaborative effort yielded the guiding questions that directed the students' projects, thereby ensuring congruence with institutional and societal imperatives	Students were arranged into interdisciplinary teams and familiarized with the integrated methodology via introductory lectures and workshops. The life cycle canvas (LCC) and collaborative tools such as Trello were used to oversee project progress. Consistent mentoring sessions and feedback panels were conducted to provide uninterrupted guidance

Source Elaborated by the authors (2025)

combined knowledge of Design, Project Management, and People Management. This applied approach embodies the principles of project-based learning, fostering organizational innovation and sustainable development in real-world contexts.

Table 2 summarizes the main PTTs developed, their institutional focus, and their alignment with the United Nations Sustainable Development Goals (SDGs).

The nine PTTs illustrate the practical application of interdisciplinary learning to institutional problem-solving. Students were encouraged to use tools such as the Life Cycle Canvas (LCC) and Trello boards to plan, monitor, and evaluate their projects, enhancing both project management competencies and the reflective dimension of professional learning. Beyond addressing concrete operational or social problems, each PTT was accompanied by the creation of a distinctive visual identity, including a logo or branding element that represented its functional focus and target audience. This dual outcome—technical solution and identity design—reinforced students' competencies in design thinking, communication strategy, and innovation

Table 2 Technical products (PTTs) developed and Their SDG alignment

PTT/Case	Problem addressed	Main deliverable/ Product	SDG alignment
1. Workflow optimization for public service	Inefficient internal administrative processes	Redesigned workflow and procedural manual	SDG 16: Peace, justice and strong institutions
2. Institutional communication protocol	Misaligned and slow communication channels	Communication flowchart and stakeholder map	SDG 9, SDG 16
3. Sustainability in procurement	Wasteful procurement and absence of green criteria	Sustainable procurement guideline	SDG 12: Responsible consumption and production
4. Employee engagement toolkit	Low engagement and team cohesion	Digital and physical engagement toolkit	SDG 8: Decent work and economic growth
5. Lifecycle canvas-based project plan	Projects are delayed and unmonitored	LCC planning board with Trello integration	SDG 9: Industry, innovation and infrastructure
6. Internal training module on agile & design	Teams lacking innovative management methods	Structured training module (PTT)	SDG 4: Quality education
7. Interdisciplinary service redesign	Fragmented public services for end users	Service blueprint with ergonomic improvements	SDG 9, SDG 11: Sustainable cities
8. Mental health and workload awareness campaign	High stress and low awareness of Well-being	Internal multimedia campaign and toolkit	SDG 3: Good health and Well-being, SDG 8
9. Institutional podcast for knowledge sharing	Limited dissemination of innovative practices	Podcast series: "Innovation in public management"	SDG 4, SDG 9

Source Elaborated by the authors (2025)

management, while reflecting a professional standard often required in public-sector initiatives.

Among the Technical Products (PTTs) developed through the integrated curriculum, the "HABITABA" project stands out as a model of social innovation aligned with sustainability principles. Designed as a digital platform (app/website), "HABITABA" connects low-income university students seeking affordable housing with local homeowners offering underutilized living spaces. The project addresses key challenges in student retention, urban density, and community engagement by promoting a coliving model that fosters inclusion and reduces housing-related dropout rates. Its development followed the integrated methodology combining Product and Process Design, Project Management, and People Management, progressing from immersion and ideation to prototyping with tools such as the Life Cycle Canvas (LCC) and user-centered design techniques. The initiative's applicability, replicability, and potential institutional impact were systematically assessed according to CAPES evaluation criteria, as summarized in Table 3.

Among the nine PTTs developed, "HABITABA" stood out for its focus on sustainable student housing and collective spaces. Its distinctive visual identity reinforces the interdisciplinary and community-oriented approach. The name "HABITABA" evokes *taba*, the term used by Indigenous peoples to designate a traditional dwelling, which also inspired the demonym *Potiguar*—the name historically associated with the Indigenous peoples of the Rio Grande do Norte region. The visual identity symbolizes community-based, sustainable housing initiatives aligned with local heritage.

Several innovation pathways emerged from these cases:

1. Organizational Efficiency and Process Optimization
2. Projects such as the Workflow Optimization Manual and the Lifecycle Canvas-based Project Plan enabled public institutions to streamline internal processes, improve resource allocation, and establish clear operational standards.
3. Sustainability and Responsible Practices

Table 3 Evaluation of the HABITABA technical product (PTT) according to CAPES criteria

Evaluation criteria	Description applied to "HABITABA"
Realized applicability	The PTT was feasible with moderate effort, requiring the mobilization of the interdisciplinary team and support from the graduate program, without preventing its execution
Potential applicability	Highly, the model can be applied in other university contexts facing housing and student retention challenges
Replicability	High; the PTT is scalable to other public institutions and urban regions with similar socioeconomic dynamics
Impact	Strong; the project can influence university housing policies, improve community engagement, and reduce student dropout rates due to housing costs

Source Elaborated by the authors (2025)

4. The Sustainable Procurement Guideline and the Interdisciplinary Service Redesign demonstrated how professional master's programs can integrate environmental and social responsibility into institutional routines, directly supporting SDG 12 and SDG 11.
5. Human-Centered Innovation and Well-being
6. Initiatives such as the Employee Engagement Toolkit and the Mental Health Campaign addressed the human dimension of organizational sustainability, reflecting the interaction between People Management and Design in creating inclusive, motivating work environments.
7. Knowledge Dissemination and Learning Ecosystems
8. The Institutional Podcast and the Internal Training Module fostered knowledge sharing and continuous learning, positioning the program as an agent of institutional transformation and promoter of SDG 4.

By integrating real-world challenges, systemic thinking, and SDG-oriented innovation, these cases substantiate the effectiveness of interdisciplinary, project-based curricula in professional education. They also illustrate the transferability and scalability of this model to other institutions seeking to align higher education with the Sustainable Development Goals.

The development of interdisciplinary Technical Products (PTTs) represented the culmination of this integrated pedagogical approach, translating the program's methodological principles into tangible institutional outcomes. Each PTT emerged from a cycle of collaborative ideation, prototyping, and iterative feedback, integrating design thinking, project management tools, and people-centered strategies. By addressing real institutional challenges—such as workflow inefficiencies, sustainable procurement, social housing, and employee well-being—the projects generated solutions with both immediate applicability and high potential for replication in other public-sector contexts.

Beyond their technical deliverables, many PTTs were accompanied by distinctive visual identities that enhanced their recognition, usability, and institutional integration, reflecting a professional standard common to public innovation initiatives. This dual outcome—functional solutions and branded products—illustrates how interdisciplinary curricula can simultaneously foster organizational innovation and the development of sustainability-oriented professional competencies.

This linkage between methodology and product development provides the foundation for the program's educational and institutional results, presented in the next section, highlighting student engagement, interdisciplinary learning gains, and measurable contributions to the United Nations Sustainable Development Goals (SDGs).

5 Results

These results confirm the importance of interdisciplinary and project-based learning approaches in sustainability education, as emphasized by Sterling (2001), Tilbury (2011), and Horn et al. (2023). The initiative achieved three headline outcomes: a noticeable increase in student engagement and autonomy, the generation of impactful Technical Products, and the alignment of these products with Sustainable Development Goals (SDGs). The progressive increase in student engagement and autonomy reflects the potential of transformative learning processes, which move students from passive recipients of knowledge to active co-creators of solutions for real-world challenges.

While the PPGPI/UFRN experience emphasized digital platforms, collaborative tools, and project-based methodologies, other contexts have focused on operational infrastructures and cross-disciplinary strategies. For example, *Towards Green Campus Operations* describes low-carbon campus buildings and regenerative design solutions (Leal Filho et al., 2018), while *Sustainability on University Campuses* highlights institutional transformations driven by student-centered learning. In the Indian context, renewable energy integration through higher education initiatives provides an additional perspective (Prasath et al., 2015). Collectively, these experiences illustrate the diversity of pathways through which universities can contribute to sustainable development agendas.

The implementation of the integrated curriculum within the PPGPI/UFRN produced multidimensional outcomes across three levels: student engagement and competency development, institutional innovation, and societal impact. By merging design thinking, project-based learning, and interdisciplinary collaboration, the experience fostered the creation of nine Technical Products (PTTs) aligned with CAPES evaluation criteria and the United Nations Sustainable Development Goals (SDGs).

The combination of technical, transversal, and socio-environmental skills demonstrates the holistic perspective advocated by Gupta et al. (2022) in project-based learning for sustainability. In line with Tilbury (2011), this experience fostered critical thinking and collaborative action, strengthening students' ability to connect theory and practice.

A primary educational achievement was the progressive increase in students' engagement and autonomy. Initially, participants demonstrated hesitance in the interdisciplinary setting, reflecting a traditional, siloed academic background. Over time, and particularly after successive team interactions and mentoring sessions, students transitioned to active collaboration, critical reflection, and proactive problem-solving.

Using the *Life Cycle Canvas* (LCC) and *Trello* boards for project planning, students developed: Technical skills: process mapping, agile project planning, sustainable service redesign, and visual identity creation; Transversal competencies: teamwork, leadership, negotiation, and systemic thinking; Socio-environmental awareness: the ability to embed sustainability principles into institutional practices. By the end of the course, student teams demonstrated a holistic understanding of design, management, and people dynamics, producing outputs that addressed real institutional demands while reflecting environmental and social responsibility.

The nine PTTs represent tangible applications of the integrated methodology, each designed to address concrete institutional challenges while advancing sustainable development. A representative example is HABITABA, a digital platform that connects low-income students with local homeowners offering underutilized housing. By fostering a coliving model, the project promoted social inclusion by mitigating student dropout caused by housing shortages, supported urban sustainability by optimizing the use of existing housing stock, and enhanced community engagement by strengthening ties between the university and surrounding neighborhoods. A similar initiative by the University of Cape Town in South Africa demonstrates the broader applicability of this approach; their 'Green Housing Connect' program shares similar goals of linking students with local housing resources, further emphasizing the global relevance and adaptability of such innovative solutions in the Global South context.

Evaluated under the CAPES criteria for professional master's technical products, "HABITABA" demonstrated realized applicability with moderate implementation effort, high potential applicability in universities facing similar housing challenges, high replicability to other institutions and urban regions, and significant impact, with the potential to influence student retention strategies and institutional housing policies. Projects such as HABITABA illustrate how design thinking (Brown, 2009) and participatory design (Santa Rosa & Moraes, 2012) can be successfully integrated into public sector innovation. By aligning digital tools with social inclusion, the initiative echoes international experiences such as the Erasmus+ projects (Hazard, 2023), confirming that project-based learning can generate socially relevant and scalable solutions.

Other PTTs complemented this focus by targeting different but interconnected dimensions of institutional sustainability. Some projects enhanced organizational efficiency, such as workflow optimization initiatives and Lifecycle Canvas project plans, directly contributing to SDG 16 (Peace, Justice, and Strong Institutions). Others prioritized sustainable practices, exemplified by the development of procurement guidelines with environmental criteria and the redesign of institutional services, aligning with SDG 11 (Sustainable Cities and Communities) and SDG 12 (Responsible Consumption and Production). Initiatives focused on human-centered innovation included employee engagement toolkits and mental health awareness campaigns, advancing SDG 3 (Good Health and Well-being) and SDG 8 (Decent Work and Economic Growth). Finally, projects dedicated to knowledge dissemination, such as institutional podcasts and internal training modules, fostered continuous learning and innovation, supporting SDG 4 (Quality Education) and SDG 9 (Industry, Innovation, and Infrastructure).

Together, these outputs illustrate the capacity of interdisciplinary, project-based learning to generate practical, scalable, and socially relevant solutions. By connecting design thinking, project management, and people management with sustainability-oriented objectives, the integrated curriculum not only transformed students into reflective and innovative professionals but also delivered measurable institutional benefits aligned with the United Nations Sustainable Development Goals (SDGs).

Societal impact was particularly evident in projects such as "HABITABA", which addressed both urban and social challenges, demonstrating how higher education can generate concrete solutions aligned with community needs and the United Nations Sustainable Development Goals (SDGs). The initiative advanced SDG 4 (Quality Education) by implementing a participatory, project-based learning model in which students engaged with real institutional problems and developed technical products (PTTs) that bridged the gap between academic knowledge and practical application. SDG 9 (Industry, Innovation, and Infrastructure) was promoted through the creation of innovative digital tools, such as the "HABITABA" platform and process redesign projects, which increased institutional efficiency and introduced scalable technological solutions. SDG 11 (Sustainable Cities and Communities) was directly addressed by initiatives that optimized the use of urban housing stock, promoted coliving models, and fostered closer connections between universities and their surrounding neighborhoods, contributing to more inclusive and resilient urban environments. SDG 12 (Responsible Consumption and Production) was advanced by PTTs that redesigned procurement and internal workflows to minimize waste, enhance resource efficiency, and improve transparency in public sector practices. Finally, SDG 16 (Peace, Justice, and Strong Institutions) was reinforced by products that strengthened institutional governance, improved communication protocols, and embedded ethical and participatory practices into organizational routines, supporting more collaborative and accountable public management. Collectively, these outcomes illustrate how interdisciplinary, project-based learning can generate measurable social and institutional benefits while fostering professional competencies aligned with sustainability.

These findings reinforce Sachs's (2015) claim that sustainability education must address not only environmental but also social and economic dimensions. The institutional products developed in this experience embraced this holistic view by integrating efficiency, governance, well-being, and community engagement. Moreover, the challenges reported—such as initial student resistance and limited resources—corroborate Watson's (2003) findings, which highlight barriers to implementing sustainability initiatives within higher education institutions.

These outcomes demonstrate that integrating design, project management, and people management in professional education meets CAPES expectations for professional master's programs while contributing to sustainable institutional transformation.

6 Conclusion

This chapter examined the implementation of an interdisciplinary curriculum integration within the Professional Master's Program in Institutional Process Management (PPGPI/UFRN), highlighting its capacity to advance sustainable curriculum development in higher education. By combining Product and Process Design, Project Management, and People Management, the initiative provided a coherent framework that connected theory with practice, institutional priorities with societal needs, and local challenges with global sustainability agendas.

The results demonstrated three key contributions. First, at the educational level, students developed technical, transversal, and socio-environmental competencies, transitioning from passive learners to active co-creators of solutions. Second, at the institutional level, nine Technical Products (PTTs) were generated, addressing concrete challenges in public management and aligning with CAPES evaluation criteria for professional master's programs. Third, at the societal level, the projects contributed to multiple Sustainable Development Goals (SDGs), with *HABITABA* serving as a notable example of social innovation for student housing and community engagement.

Despite its achievements, the intervention faced constraints. The short semester limited the full consolidation of some products; a suggested solution is to extend mentoring into the inter-session period to provide students with additional time and support. Students initially resisted the unfamiliar interdisciplinary format; introducing preparatory workshops or orientation sessions could ease this transition. Moreover, the dual role of faculty as both instructors and evaluators introduced the potential for bias. To mitigate this, separating the roles by involving external evaluators or peer assessments could ensure a more objective evaluation process. These challenges echo broader concerns in the literature (Watson, 2003; Sachs, 2015) regarding barriers to embedding sustainability in higher education.

Nonetheless, the integrative model demonstrates that professional master's programs—particularly in the Global South—can act as catalysts for institutional transformation and sustainable innovation. By embedding project-based, interdisciplinary methodologies into postgraduate education, such programs enhance both academic relevance and societal impact.

Looking forward, future developments should prioritize three directions:

1. Faculty development to expand the integrative methodology across other courses and programs.
2. Longitudinal assessment of sustainability competencies and institutional impacts, supported by more systematic evaluation tools.
3. Inter-institutional collaboration to scale the replicability of technical products and strengthen partnerships between universities, public organizations, and communities. In doing so, higher education institutions can reinforce their role as laboratories for sustainability and innovation, advancing the global sustainable development agenda while producing context-sensitive solutions with tangible public value. Let us co-create these systemic solutions together, ensuring that the

impact of our collaborative efforts resonates across educational and institutional boundaries.

References

Brown, T. (2009). *Change by design: How design thinking creates new alternatives for business and society*. Harvard Business Press.

Gupta, R., Jain, S., & Sharma, R. (2022). Enhancing environmental competencies through project-based learning: A study on sustainability education. *Environmental Education Research, 28*(4), 524–541. https://doi.org/10.1080/13504622.2022.2030583

Hazard, M. (2023). Project-based learning for sustainability: A framework for education. *Journal of Sustainability Education, 14*(2), 83–96. https://doi.org/10.4018/ijse.2023.09.03.07

Horn, A., Scheffelaar, A., Urias, E., & Zweekhorst, M. B. M. (2023). Training students for complex sustainability issues: A literature review on the design of inter- and transdisciplinary higher education. *International Journal of Sustainability in Higher Education, 24*(1), 1–27. https://hdl.handle.net/1871.1/ad8f4315-1c8f-44d8-99c6-b0c97e38c42a

Leal Filho, W., Frankenberger, F., Iglecias, P., & Mülfarth, R. C. K. (Eds.). (2018). *Towards green campus operations: Energy, climate, and sustainable development initiatives at universities. World Sustainability Series*. Springer.

Leal Filho, W., & Portela de Vasconcelos, C. R. (2022). Pursuing sustainable development across disciplines. In Leal Filho, W., & Vasconcelos, C. R. P. (Eds.), *Handbook of best practices in sustainable development at the university level. World Sustainability Series*. Springer.

Medeiros, J. D. F., Araújo, T. A., & Oliveira, L. S. (2018). Life Cycle Canvas: a tool for project management in the context of professional master's programs [Life Cycle Canvas: uma ferramenta para gestão de projetos no contexto de programas de mestrado profissional]. *Revista Brasileira de Gestão e Desenvolvimento Regional, 14*(1), 45–58.

Northouse, P. G. (2018). *Leadership: Theory and practice* (8th ed.). Sage Publications.

Prasath, R. A., Poyyamoli, G., Boruah, D., Nandhivarman, M., & Edwin, G. A. (2015). The role of higher educational institutions and other training organizations to promote renewable energy in India. In Leal Filho,W., Muthu, N. Edwin, G., & Sima, M. (Eds.), *Implementing campus greening initiatives: Approaches, methods and perspectives. World Sustainability Series* (pp. 37–52). Springer International Publishing.

Sachs, J. D. (2015). *The age of sustainable development*. Columbia University Press.

Santa Rosa, J. G., & Moraes, A. (2012). *Design participativo: Técnicas de inclusão de usuários no processo de ergodesign de interfaces [participatory design: User inclusion techniques in the ergodesign process of interfaces]*. Rio Books.

Sterling, S. (2001). *Sustainable education: Re-visioning learning and change*. Green Books.

Tilbury, D. (2011). *Education for sustainable development: An expert review of processes and learning*. UNESCO. https://unesdoc.unesco.org/ark:/48223/pf0000190735

UNESCO. (2023a). *Education for sustainable development (ESD) for 2030: A roadmap*. UNESCO. https://www.unesco.org/en/sustainable-development/education

UNESCO. (2023b). *Sustainable development goals and higher education: A guide for universities*. UNESCO. https://www.unesco.org/en/sustainable-development/education

Universidade Federal do Rio Grande do Norte – UFRN. (2020). Plano de Desenvolvimento Institucional (PDI) 2020–2029 [institutional development plan 2020–2029]. https://www.ufrn.br/resources/documentos/pdi/PDI-2020-2029.pdf

Universidade Federal do Rio Grande do Norte – UFRN. (2022). *Management plan for sustainability and SDGs implementation*. UFRN.

Watson, R. (2003). *Sustainability and governance in higher education*. Oxford University Press.

Patrícia Borba Vilar Guimarães is a lawyer and Associate Professor at the Federal University of Rio Grande do Norte (UFRN), working in the Department of Procedural Law and Propaedeutics (DEPRO). She leads the "Law and Development" research group (CNPq), is active in both the Academic Law Graduate Program and the Professional Master's in Institutional Process Management. Dr. Guimarães earned her PhD in Natural Resources from Universidade Federal de Campina Grande in 2010.

Cynara Carvalho de Abreu is a Psychologist, a Master in Administration, and earned her PhD in Education. She is currently a postdoctoral researcher in Clinical Psychology at PUC-SP and a full Professor at UFRN since 2010, with expertise in phenomenological-existential clinical psychology and educational sustainability.

José Guilherme da Silva Santa Rosa is an Associate Professor in the Department of Design (Arts) at UFRN. His research focuses on design, user interfaces, multimedia, ergonomics, and human–computer interaction. He coordinates Ergodesign and related research initiatives.

Josué Vítor de Medeiros Júnior has been a Professor at UFRN since 2009. He holds a PhD in Administration and has experience in process and project management. He founded and led UFRN's Project Management Office from 2016 to 2019 and currently serves as the University's Pro-Rector of Planning.

Advancing Sustainability Education Through Capstone Curriculum

Nataliya Apanovich and Alice Reznickova

Abstract Capstone courses are widely recognized as high-impact educational practices that support sustainability learning through applied, interdisciplinary, and community-engaged experiences. This chapter critically examines two contrasting capstone models—an undergraduate research-based capstone at the University of Arizona and a graduate-level, applied project-based capstone at the University of Colorado Boulder. Through analysis of reflective practice grounded in the authors' instructional and administrative experiences, we explore how capstone courses foster sustainability competencies while grappling with persistent challenges. Drawing on existing literature and our own practice, we identify four interrelated challenge areas: managing stakeholder relationships and expectations, conceptual ambiguities in sustainability education, career preparation and professional alignment, and structural and resource constraints. These challenges are shaped not only by pedagogical complexity, but also by broader trends in higher education, including increased reliance on contingent faculty, growing teaching loads, and limited institutional investment in experiential learning. We conclude with practical insights for designing capstone programs as part of a broader curriculum for sustainable development, offering guidance on how to embed experiential, systems-oriented learning into curricular structures that prepare students for sustainability challenges across diverse institutional contexts.

Keywords Capstone courses · Sustainability education · Community-engaged learning

---

N. Apanovich (✉)
University of Arizona, Tucson, AZ, USA
e-mail: apanovich@arizona.edu

A. Reznickova
University of Colorado Boulder, Boulder, CO, USA
e-mail: alice.reznickova@colorado.edu

W. Leal Filho et al., *Curriculum Development for Sustainable Development*, World Sustainability Series, https://doi.org/10.1007/978-3-032-22146-9_22

1 Introduction

Capstone courses are increasingly recognized as powerful culminating experiences in sustainability curriculum, designed to integrate students' interdisciplinary knowledge and apply it to real-world problems (Ho, 2023). Grounded in experiential learning theory, capstones help bridge the gap between abstract classroom concepts and practical application by engaging students in projects with community, industry, or government partners (Brundiers & Wiek, 2017). These courses are widely praised for fostering critical thinking, systems understanding, and ethical reasoning—skills that are essential to sustainability education (Kovaric & Gingell, 2024; Chang et al., 2018). When well-designed, capstones can cultivate the core competencies for sustainability, including anticipatory foresight, normative reasoning, strategic intervention, and collaboration (Levesque & Blackstone, 2020; Brundiers et al., 2021). Through structured engagement with complex environmental and social challenges, students synthesize diverse disciplinary insights into applied solutions and practice the professional skills of project management, interdisciplinary teamwork, and stakeholder communication (LeChasseur et al., 2024). In this sense, capstone courses operationalize the aspirations of sustainability education, preparing graduates to apply their knowledge and skills in service of broader societal goals. These expectations have led many institutions to place capstones at the center of their sustainability curricula, anticipating that real-world projects will promote deeper learning and support students' development as effective change agents (Budowle et al., 2021).

However, while capstone courses offer significant pedagogical value, they also encounter persistent challenges in design and implementation. Increasingly, researchers note that capstone models often struggle to reconcile the tension between educational goals and institutional limitations (Evans, 2019; Barth & Rieckmann, 2015). Faculty must balance academic rigor with applied relevance, foster student independence while providing adequate structure, and accommodate interdisciplinary breadth within constrained timelines (Evans, 2019; Brundiers et al., 2021). Without intentional course design, scaffolding, and sustained mentorship, capstone projects can result in superficial engagement that emphasizes completion of deliverables over meaningful learning (Watson et al., 2020). Outcomes vary widely depending on how projects are framed, the quality of advising, and the availability of institutional resources (Shephard et al., 2015). Resource-intensive by nature, capstones often demand close faculty supervision, flexible scheduling, and administrative support—elements that are not universally accessible in all academic programs. If such needs go unmet, capstones risk being reduced to narrowly scoped or overly simplified exercises that fail to capture the complexity of sustainability issues (Watson et al., 2020). Empirical studies show that while these experiences enhance student learning, they also place a substantial burden on faculty and require clear institutional commitment to succeed (Minshew et al., 2020; Bhagwan, 2020).

A growing body of scholarship categorizes these challenges into several thematic areas: time and resource constraints, stakeholder alignment, conceptual ambiguity, inequities in partnerships, student preparedness, difficulties in assessing learning outcomes, and career preparedness and professional alignment (Landfried et al., 2023; Knox & Nairn, 2025; Watson et al., 2020; Fraustino et al., 2019). These thematic groupings help educators analyze and respond to common pitfalls in capstone delivery, and they provide a framework for identifying areas in need of reform.

This chapter focuses on four core categories of challenges found consistently across the literature: managing stakeholder relationships and expectations, conceptual ambiguities in sustainability education, career preparation and professional alignment, and structural and resource constraints. Below, we describe each category in greater detail.

Managing Stakeholder Relationships and Expectations. A defining characteristic of sustainability capstones is engagement with external partners who have their own organizational missions, constraints, and expectations. Managing these relationships involves navigating mismatched timelines, goals, and quality standards. Partners may expect usable deliverables, while instructors must prioritize student development. Students can find themselves caught between these pressures, particularly when asked to operate in professional-like roles before they are adequately trained (Park et al., 2025). Additionally, inequities can emerge when academic institutions extract value from partnerships without offering reciprocal benefits, especially in collaborations with under-resourced community organizations (Sasson, 2019). Faculty must invest time in building trust, clarifying expectations, and ensuring transparency—all of which require careful coordination and communication.

Conceptual Ambiguities in Sustainability Education. Capstone students often struggle to articulate a coherent understanding of sustainability. Sustainability's normative and interdisciplinary nature—encompassing ecological, economic, and social dimensions—makes it difficult to define project boundaries, goals, and measures of success. Without sufficient theoretical grounding or critical reflection, students may adopt superficial sustainability frames or default to well-worn narratives without interrogating their assumptions (Brundiers et al., 2021; Annelin & Boström, 2024). Faculty must help students navigate these complexities, encouraging them to think systemically, critically, and ethically. However, embedding this reflective work within the limited instructional time of a capstone can be difficult, particularly when project timelines are tight, and partners emphasize technical outputs.

Career Preparation and Professional Alignment. Capstones are often positioned as stepping stones into professional roles, yet this objective is not always met. Students may enter capstone courses without the necessary preparation in project management, research methods, or stakeholder engagement. Furthermore, projects may not always align with individual career goals, making it difficult for students to showcase relevant skills in their portfolios or job applications (Ferrante et al., 2025; Brundiers & Wiek, 2017). Programs that do not explicitly integrate career development into the capstone process may miss opportunities to support students' transition into the workforce. Moreover, programs sometimes struggle to systematically measure the long-term professional impacts of capstone experiences (Evans, 2019).

Structural and Resource Constraints. Sustainability capstones are labor-intensive courses that often outpace the logistical capacity of the departments that house them. They require low student–instructor ratios, sustained mentoring, and extended planning, yet are frequently confined by standard academic calendars, credit limitations, and faculty workloads. Programs lacking sufficient institutional support may find it difficult to maintain the depth and quality of the learning experience (Gresham et al., 2023; Evans, 2019). Even when faculty are committed, the absence of financial compensation, administrative assistance, or professional development opportunities may undermine their ability to fully support students. Graduate and undergraduate programs face distinct but overlapping pressures: the former may need to meet professional standards in deliverables, while the latter often contend with uneven student readiness and require additional scaffolding (Konrad et al., 2018).

These four themes capture some of the most persistent and impactful challenges facing sustainability capstones. While these categories are not exhaustive, they reflect key leverage points where thoughtful design and intentional pedagogy can significantly improve the student experience and the broader value of capstone work.

Drawing on these insights, this chapter critically examines how these four categories of challenges manifest in two institutional settings: an undergraduate research-based capstone in the Sustainable Built Environments program at the University of Arizona, and a graduate-level, applied project-based capstone in the Masters of the Environment program at the University of Colorado Boulder. By analyzing these contrasting models, we aim to illuminate how capstone instructors and administrators navigate the practical tensions and pedagogical opportunities embedded in sustainability-focused experiential learning. These case studies provide not only contextualized illustrations of common challenges, but also a basis for informing curriculum development in sustainability education. The chapter contributes to the broader conversation on how capstone courses can be designed as integrative, high-impact components of sustainability curricula.

2 Methodology

This chapter is based on a qualitative inquiry that draws on two distinct case studies to examine how university capstone experiences can advance sustainability education. Specifically, we draw on personal experiences in the roles of educator (Nataliya) and administrator/educator (Alice) to critically assess the design, implementation, and management of capstone programs. As of 2025, Nataliya has completed three years of teaching the Senior Capstone course in the Sustainable Built Environments (SBE) program across multiple modalities, including in-person, online, and micro-campus formats. Alice has completed two years of leading the Capstone program in the Masters of the Environment (MENV) graduate program.

The analysis is informed by reflective practice and draws on a variety of data sources, including informal conversations with faculty, staff, students, and capstone partners, as well as formal and informal course evaluations. We employed thematic analysis to identify recurring challenges across the two cases, focusing on patterns that may inform broader discussions in sustainability curriculum design. This work is grounded in reflective practitioner research, with the goal of offering practical, transferable insights for educators and administrators designing or refining sustainability capstone programs. Using a reflective practitioner approach is particularly appropriate for this study because it centers the lived experiences, situated knowledge, and critical reflections of educators directly involved in capstone design and delivery (Schön, 2017). Additionally, reflective inquiry aligns with the goals of transformative learning by emphasizing iterative refinement, collaboration, and responsiveness to context (Bergmark, 2022). Rather than seeking generalizability, this method prioritizes insight, transferability, and relevance to practitioners working in similarly complex educational environments.

Case Study 1: Independent Research-Based Sustainability Capstone

University's Setting

The Bachelor of Sustainable Built Environments (SBE) Program is an interdisciplinary program located in the College of Architecture, Planning, and Landscape Architecture at the University of Arizona (UA), Tucson. UA is a land-grant, Hispanic-Serving Institution (HSI) with the Tucson campus situated in the borderlands region, characterized by proximity to the Mexican border and the confluence of American, Hispanic, and Indigenous cultures. The region is part of the Sonoran Desert, with extreme (above 100 °F) summertime temperatures and low annual precipitation (~10″) and is one of the fastest warming places in the country (NOAA, 2025). Tucson is home to more than half a million people (approximately half of whom are Hispanic) and has a poverty rate above the national average (U.S. Census Bureau, 2023). The Tucson community represents a very active civil society, with dozens of non-profits addressing issues that directly impact the region: water shortages, rising temperatures, lack of affordable housing, homelessness, ecological degradation, food insecurity, and energy poverty.

Program's Background.

The SBE program, created in 2017, is an interdisciplinary undergraduate program designed to provide students with a comprehensive understanding of sustainability principles and equip them with the skills to enhance the sustainability of communities, buildings, and open spaces. The motivation behind establishing the program was to address the growing need for professionals who can integrate sustainability into the built environment. The program aims to cultivate leaders capable of creating sustainable solutions that balance environmental, social, and economic factors. It is available in-person on the main campus, fully online, and in a micro-campus format.

Design of the Capstone Course

Each graduating class consists of 40–50 students. Students take most of their courses from existing programs within the college: Architecture, Landscape Architecture,

Urban Planning, Real Estate Development, and Heritage Conservation. The purpose of the Senior Capstone is to serve as a culminating experience, demonstrating the ability to synthesize the knowledge gained throughout the major. The capstone requires students to write a thesis based on their work. Some theses are focused on data analysis and theory, while others are more design-oriented.

Students begin preparing for their capstone projects in the fall semester while taking the Research Methods course. This course is designed to teach students how to conduct academic research using mixed methods. Specifically, students work on crafting a problem statement, conducting a literature review, formulating a research question, collecting and analyzing different types of data, and writing a discussion and conclusions. Students are encouraged to focus on sustainability-related issues that are local in scale, personally meaningful, and feasible to complete in one semester. As a result, many students work on topics relevant to the campus, the city in which they live, or a specific neighborhood. Students are generally discouraged from pursuing national or international projects, as these typically require more time and resources than available. Common areas of research include active transportation (e.g., walking and biking), affordable housing (e.g., infill development), sustainable food systems (e.g., food forests), access to green spaces (e.g., parks), energy efficiency in buildings (e.g., LEED certification), and inclusive design (in architecture). At the end of the course, students produce a pilot study that they then carry over into their capstone course in the spring.

In the capstone course, students expand their pilot studies to produce a final thesis. Expansion takes different forms: some students collect additional data for a more robust analysis, while others edit their research questions to narrow or change their focus. The instructors' role is two-fold: to ensure students are advancing their projects in a timely manner and to connect them to necessary resources and mentors on- and off-campus. In addition to regularly meeting in a classroom (available only to in-person students), instructors highly encourage students to schedule individual meetings for personalized mentorship. Only approximately half of the students choose to meet with instructors regularly on a one-on-one basis.

All students are encouraged to work with a capstone mentor, who can help them locate appropriate sources, revise their work, or provide feedback on research methodology. Students are encouraged to work with faculty or professionals who share similar interests. They are required to present their final work at a poster showcase event at the end of the semester. The college and outside members are invited to the poster session, with typically five to seven reviewers commenting on and scoring each poster using a standardized rubric. Students are graded on both the final report and poster presentation, with several smaller assignments to monitor progress throughout the semester. The capstone allows students to produce an original project related to sustainability that can be added to their portfolio or graduate school applications.

The course is typically co-taught by two instructors, each advising approximately 15 students at a time. This structure allows for more individualized guidance but also places considerable mentoring demands on each instructor, especially when student preparation levels vary widely.

Challenges & opportunities.

While the capstone project offers an exciting experience for students, it can also be demanding and, at times, disappointing. One of the biggest challenges is that students are often underprepared to engage in academic-level research. One semester of learning research methods is insufficient preparation for the rigor expected in a final capstone project. Students need much more time and practice conducting literature reviews, identifying gaps in existing knowledge, and formulating appropriate research questions. As a result, it is not uncommon for students to spend half the semester refining their questions and literature review, leaving limited time for data collection, analysis, and writing the thesis. This is especially reflected in the quality of their final papers where students often struggle to connect their research to broader themes in sustainability.

Another challenge stems from the mentorship model, where mentorship is encouraged but not required. While this structure benefits students who are motivated and well-connected within the faculty or community, it disadvantages those who are less motivated, introverted, or lack existing networks. Instructors leverage their personal networks to connect students with mentors whenever possible, but this only occurs when a good match exists between student needs and available resources. As a result, most students conduct their projects independently without the involvement of community partners. This represents a missed opportunity, as Tucson offers a rich landscape of sustainability-focused work and networks that could greatly enhance student learning and project impact.

Students also face technical challenges when performing desired analyses. Although all students complete one GIS course and several design-focused courses, these offerings are sometimes insufficient to meet project needs. When possible, instructors pair students with faculty who have specific technical skills, allowing students to increase their proficiency and produce higher-quality work.

Another challenge relates to connecting capstone projects to post-graduation employment outcomes. While the program does not systematically track employment following capstone completion, in the author's three years of teaching the course, she was unaware of any capstone project directly leading to employment. This may partly be due to many students choosing to continue into graduate education.

From the instructors' perspective, additional challenges arise. First, the time commitment is substantial, particularly when students are underprepared for independent research. Rather than focusing on project-specific mentoring, instructors often spend significant time guiding students on conducting literature reviews, designing unbiased survey or interview questions, and improving academic writing. Second, one of the greatest frustrations is the inability—due to time or professional network limitations—to connect every student to a mentor. Without mentorship, students work in silos and risk conducting research that lacks meaningful academic or community impact.

Importantly, all of these challenges are amplified for online students. It is often unrealistic to expect the same level of engagement, mentoring, and project integration from students in the fully online modality. As a result, online students often have a different capstone experience compared to their in-person peers.

Community stakeholders also face challenges, particularly regarding availability and funding. Non-profit organizations, which are common sources of mentorship, often operate with limited staff and budgets, making sustained engagement difficult. Since the capstone course lacks a dedicated budget, mentors cannot be compensated for their time. Without a structured mentorship program, students independently determine how and how often to engage with community partners, creating uncertainty and undermining the program's ability to retain mentors for future cohorts (see Table 1). Similar challenges exist for faculty mentors, most of whom carry heavy teaching loads of up to five courses per semester. Asking faculty to take on additional mentorship responsibilities compromises their regular work and places undue pressure on them.

Despite these challenges, the capstone course also creates important opportunities. It allows students to build closer relationships with faculty and community members, sometimes leading to internship and job opportunities. It provides space for students to strengthen existing skills and/or learn new ones. The deep exploration of sustainability issues through the capstone also incentivizes many students to pursue graduate education as they seek to further develop their expertise. Finally, it incentivizes students to learn about their local settings, issues, and environments, fostering a greater appreciation for place-based learning and local sustainability solutions. As a result of this appreciation, after graduation, some students choose to remain in the Tucson area to continue learning about sustainability in a region highly impacted by climate change and related environmental and social challenges. These positive outcomes happen for some students while additional resources could provide the same robust experience for all students and even community partners.

Case Study 2: Team-Based Sustainability Capstone

University's Setting

The Masters of the Environment (MENV) is a professional sustainability graduate program within the Environmental Studies (ENVS) Department at the University of Colorado– Boulder (CU Boulder). CU Boulder is the flagship campus of the CU public university system. CU Boulder is located on the Front Range of the Rocky Mountains within Boulder County. Boulder County experienced rapid population growth (over 16% since the 2010 census). The majority of Boulder County population is White (not Hispanic or Latino) (76.2%), more people received a Bachelor's degree or higher than in the U.S. on average (63.9% vs. 35.0%), and median household income is much higher than the U.S. average ($102,772 vs. $78.538) (U.S. Census Bureau, 2024a, 2024b). The Front Range area is emerging as a hub for innovation and sustainability with both established companies and start-ups concentrating there, offering a variety of industry partnerships. In addition, proximity to many non-profit organizations and state and local government agencies in the Denver metropolitan area allow for partnerships with public entities.

Table 1 Key challenges in capstone courses, with examples from the SBE and MENV case studies, and connection to curriculum

Challenge	Explanation	Example	Connection to Curriculum
Managing stakeholder relationships & expectations	Complexity of aligning the different expectations and roles of students, faculty, and external partners involved in capstone projects.	Case study 1: The lack of structured mentorship program creates confusion and uncertainty among community partners and faculty mentors about their responsibilities and undermines the program's ability to retain them for future cohorts. Case study 2: Students work as "consultants" for partner organizations, creating inherent tension between fulfilling academic expectations of faculty advisors and producing professional deliverables. Partners sometimes shift project scopes midstream, challenging students' ability to maintain academic rigor while adapting to client demands.	Include faculty and external partners as a part of the curriculum design team—Their feedback informs curriculum design for future cohorts to help align expectations. Navigate the differences and expectations from faculty and partners as well as adapting to shifting project scopes is embedded in the coursework, learning outcomes, and assessment.
Conceptual ambiguities in sustainability education	Varying perceptions of what sustainability entails and how it should be addressed through experiential learning.	Case study 1: Students often pursue projects on topics like bike lanes or green buildings but struggled to articulate how their projects address broader sustainability systems (social equity, economic resilience, etc.). Faculty frequently have to push students to critically reflect on sustainability dimensions rather than assume surface-level "green" actions sufficed. Case study 2: Partners sometimes propose projects narrowly focused on compliance or technical fixes (e.g., GHG inventories) without broader reflections on sustainability competencies such as systems, anticipatory, and normative thinking. Students are more invested in learning and practicing technical skills and believe they had enough critical reflection in other coursework.	Add new learning outcomes (and accordingly, more time for these learning outcomes) to the curriculum to practice sustainability competencies such as systems, anticipatory, and normative thinking. Add or redesign existing assessment for critical reflection. Offer training to advisors and partners to incorporate critical reflection with students throughout the project.

(continued)

Table 1 (continued)

Challenge	Explanation	Example	Connection to Curriculum
Career preparation & professional alignment	Challenge of connecting capstone experiences meaningfully to students' career pathways and professional development.	Case study 1: Although students produce independent research projects, these rarely translate into direct employment opportunities. Many projects remain academic exercises rather than producing portfolios or networks valued by employers. Case study 2: Career development is woven into the capstone through skill workshops, resume development, networking events with partners, and explicit listing of skills and opportunities on capstone project porposals. Nevertheless, some capstones may in the end not offer them during the course of the project and students may not take full advantage of them fully while concentrating on the project deliverables. Students may not identify how opportunities offered through the program benefit them.	Build better alignment between student career aspirations and their projects (source a variety of projects based on the cohort's characteristics) and frame projects explicitly in terms of skill and professional network development. Design curriculum to include career and professional development, both in class and through other faculty and partner organizations. Introduce evaluations to measure short- and long-term career outcomes and evolve curriculum based on this learning. Source projects that are current and require skills sought by the industry. Leverage alumni networks in capstone courses to emphasize how experiential learning supported their careers.
Structural & resource constraints	Limitations related to time, funding, and administrative support necessary to design, deliver, and sustain effective capstone experiences.	Case study 1: Faculty experience heavy time burdens mentoring students who lack research skills, often needing to teach basic research methods during capstone rather than focusing on project-specific guidance. Case study 2: MENV allocates substantial funding and course equivalencies for capstone management (~34 course equivalents per cohort), but this is a new development, secured through advocacy of faculty and administrators and may still not be enough to address all the challenges listed above while keeping an up-to-date curriculum and building individual mentoring relationships with students.	Identify the right structure for capstone courses based on the available resources. Appropriate resources should be allocated to design and facilitate effective curriculum.

Program Background

MENV was founded in 2014 as a part of a university-wide initiative to support professional education. The program was incubated in the Environmental Studies Department and through CU Boulder Graduate School. The first enrolled MENV class of ~30 students matriculated in 2016. From 2021, each entering class numbers roughly 80–100 students.

MENV is an innovative program that focuses on high-impact, experiential learning. The coursework is centered on professional, applied skills with the goal of developing leaders–and workforce–in the environment and sustainability. From the start to this day, the centerpiece of the program was the Capstone project.

Design of the Capstone Course

Currently, the Capstone project is administered in a four-semester sequence–three semesters in the classroom (Year 1 Fall–1 credit, Year 1 Spring–2 credits, and Year 2 Fall–1 credit) and Year 1–Summer semester of only project work (5 credits). The purpose of the three accompanying courses, Capstone Innovation Labs, is to: 1. Guide students through the Capstone process, 2. Establish "soft" skills (e.g., leadership, project management, communication) and technical skills (e.g., GIS, data visualization, qualitative and quantitative data collection and analysis), and 3. Support students' professional development. Capstone is extensive in scope: the projects are designed to be completed by a team of students in ~900–1200 team hours during the summer. The students also spend ~450–600 team hours preparing the scope of work in the spring semester and another ~225–600 team hours in the fall on editing and design. In the fall, MENV hosts poster session and symposium, during which students present their projects to the program and to the public respectively.

Capstone projects are developed by capstone partner organizations in collaboration with the capstone administrator and program faculty. Capstone partners are for profit, not-for-profit, and government organizations, many of which have offices in Boulder County or on the Front Range. The projects focus on sustainability and advance capstone partner organization's mission while being academically and professionally rigorous. The capstone process is competitive for partner organizations as not all projects are selected by students. Some years, fewer than half of the projects are selected.

Year 1–Fall semester serves as a preparation for students so that they are able to pick a project that best aligns with their interests and career aspirations. Students review proposals, meet capstone partners in a networking session, and submit their preferences. The students are then assigned to a project based on their preference with input from capstone partners and program faculty. Skill alignment, team diversity, and team dynamics are additional considerations.

The MENV Capstone model is student-as-a-consultant–students plan, design, execute, and report on the project. The students work closely with the capstone partner, meeting weekly during the year to report to them on progress and ask for feedback. Capstone partners ensure that the deliverables respond to their needs and that the final product is useful to them. Capstone teams also report to their faculty advisor weekly or every other week to update them on progress and to discuss any questions and issues. Students can ask capstone advisors for advice on group conflict, partner communication, project management, and other issues.

Career and professional development are further woven through the capstone process. This includes mentoring students on project selection in alignment with their career, refining their professional job application materials, and in-class skill development. Apart from the in-class experiences, capstone faculty advisors are an important resource as they coach students individually and in teams. Finally, many capstone partners facilitate other learning and professional development opportunities, including interviews with subject matter experts and other employees in their organization, attendance in organizational meetings and conferences, and technical skill workshops.

Challenges and Opportunities

MENV Capstone is a large-scale, complex process. With nearly 150 stakeholders (80–100 students, 25–40 Partners, and 10–13 Advisors) collaborating over four semesters each year, aligning expectations is a key challenge (see Table 1). While students are assigned to their preferred capstone, projects often evolve between selection and launch. At times, projects cannot be completed as planned, and pivots are necessary later in the process. Despite clearly stated expectations for students' professionalism and project work, capstone partners, advisors, and the students themselves interpret these expectations differently (for instance, quality standards may vary between students on a single team, partners, and advisors). The model of student-consultant is often new for capstone partners given the prevalence of traditional internships, leading to heavier management of student teams. On the other hand, some partners also view students as professional consultants and hold them to professional, paid-consultant standards. On both ends of the spectrum, capstone partners may not be fully content with their teams' performance.

The second challenge is teaching sustainability through a large-scale applied professional project. Each project is different, and therefore capstone faculty advisors and partners are best placed to guide the students to examine sustainability from a critical, systems-based perspective. Despite the capstone sequence offering three semesters of classroom time, time in the classroom is limited as most of the time is dedicated to skill building, professional development, and the complex capstone process itself. The projects are fast-paced, and student attention can be split with other courses and employment outside of MENV. Students therefore largely focus on the challenges ahead of them: technical aspects of the project, project management, team dynamics, or client communication. Critical and systems-focused sustainability education takes time and in these fast-paced projects, may not be prioritized.

While not many students receive—or accept—employment offers directly with their partners, we believe that the capstone project leads to full-time employment through skill development, integrated work-experience in the year-long applied project, and professional network building. Offering so many different potential projects leads to high satisfaction among students in terms of career alignment. Capstone partners and faculty advisors are important mentors for students' career development, but not every student may receive the same level of support. Continued encouragement of partners and advisors is essential. MENV also innovates in this space and incorporates career development content in the capstone curriculum directly. This includes a newly introduced Alumni Engagement Day during which program graduates host panels to discuss their careers, including how their capstone experience prepared them for success. Some students may regard this content as easy and not necessary to prioritize alongside their other coursework and alumni can offer stories of how they leveraged experiential learning in their careers. This year,

MENV also incorporated capstone-specific questions into our short-term and long-term career outcome surveys. This will help with curriculum evaluation to ensure that we are aligning our capstone with careers.

Finally, complex, large-scale programs require time to design, execute, and maintain. This in turn can be expensive for the department and university. Major tasks include finding capstone partners, developing proposals, maintaining industry relationships, mentoring students, providing effective programming for students' professional growth, and continued evaluation and redesign of the overall process. Each year, MENV dedicates six course equivalents for program administration, four course equivalents for classroom instruction, and ~ 27 course equivalents for Capstone mentoring (three teams per semester is equal to one course equivalent for a faculty member). This figure does not include the time contributed by others in the department, such as the Career and Professional Development Manager and the 2–3 Graduate Student Assistants who support the program. It also does not include the time capstone partners dedicate to the students and their projects. This level of funding and time commitment allows MENV to successfully execute the program and to continue to proactively address the challenges described above through the curriculum. Yet this level of support was not always available, it was only made possible through MENV faculty and administrators' advocacy for the program and the university's willingness to provide it. And even this level of support does not cover everything we would want to do had we had more time and resources. Our ability to continue to address our challenges in the curriculum leads to many successful outcomes. Capstone teams often build close relationships with each other, their capstone advisor, and their capstone partner, as well as with many other MENV faculty. The students approach challenges maturely and proactively. Many of the students work in our local communities and advance sustainability on the Front Range and beyond, while practicing new skills and expanding their professional networks. In the end, many students return to MENV as capstone partners to mentor the new generation of sustainability professionals. Our career outcomes show the effectiveness of our overall program: 87% of our graduating students have a job within six months of graduation with most employed in their field of study in 2024.

3 Discussion

Capstone projects are a critical tool for advancing sustainability competencies and preparing students to address complex societal and environmental challenges. They provide opportunities to apply classroom learning in authentic, interdisciplinary contexts through collaboration with faculty and community or industry partners. These programs offer considerable value—not only for students, who develop practical skills and professional networks, but also for institutions seeking to fulfill public service and sustainability missions through community-engaged learning.

Our two case studies demonstrate the versatility of capstone courses across institutional contexts. The undergraduate case centers on a single-semester, research-based model with volunteer mentorship, while the graduate case features a multi-semester, team-based format with structured advising and external partnerships. Despite differences in scope and structure, both programs encountered recurring challenges with implications for curriculum design and implementation.

A persistent challenge is the divergence of expectations among students, faculty, and community partners. Each stakeholder group may hold differing views on project goals, deliverables, and measures of success. While these differences can enhance authenticity and prepare students for real-world negotiation, they must be addressed through thoughtful curricular scaffolding. For example, including faculty and external partners in curriculum design ensures their perspectives inform learning goals and assessment strategies. Expectations for new partners and faculty should be clearly communicated at the outset of the capstone. Meanwhile, students receive instruction on how to navigate divergent perspectives using competencies such as normative reasoning, strategic action, and interpersonal communication.

Another challenge is balancing technical skill development with opportunities for systems thinking and critical reflection. Capstones are often constrained by timelines, partner demands, and student workloads, which can limit space for reflective learning. When mentorship is unpaid or spread thin, reflection on deeper sustainability themes may be sidelined. Even in well-resourced models, mentorship time is often consumed by logistical coordination and technical support. Programs may choose to scaffold certain sustainability competencies in earlier coursework, using the capstone to reinforce rather than introduce them. To strengthen these learning goals, programs can revise assessments, introduce structured debriefing, and leverage collaborative models such as communities of practice where student teams exchange peer feedback and navigate complex decisions. Students are also evaluated throughout the process on their adaptability, communication, and collaboration, reinforcing key interpersonal and strategic competencies essential to sustainability leadership.

Measuring capstone outcomes, particularly career impacts, remains difficult. While students often gain valuable experience and networks, these outcomes are not always tracked or made visible to students or administrators. Career development activities—such as resume workshops, alumni panels, and skill-mapping exercises—should be built into the curriculum and reinforced by faculty and partner engagement. Programs can also improve alignment between student aspirations and capstone opportunities by sourcing a diversity of projects and explicitly framing each one in terms of skill-building and professional relevance. Introducing evaluation mechanisms to assess short-and long-term career outcomes will help programs refine their approach and demonstrate their value to students and institutions.

A final challenge is the limited institutional infrastructure to support capstone delivery. Faculty require time to build and revise curriculum, cultivate relationships with external partners, and mentor students effectively. Yet increasing workloads, reliance on contingent faculty, and public disinvestment in higher education have reduced institutional capacity for this work (AFT, 2020). Programs must carefully structure their capstone models based on available resources and ensure that faculty receive adequate time, compensation, and support. Sustainable implementation also depends on consistent funding, administrative support, and alignment between capstone logistics and broader departmental strategies. Investing in this infrastructure is critical not only for course delivery but also for maintaining up-to-date, effective sustainability curricula over time.

4 Conclusions

Capstone projects are a critical pedagogical tool for advancing sustainability competencies and preparing students to address complex, real-world challenges. Through two contrasting case studies—one undergraduate, one graduate—this chapter demonstrates how capstones can be implemented across diverse institutional and curricular contexts. Grounded in reflective practitioner research and informed by instructional experience, student feedback, and engagement with capstone partners, the analysis offers insights into how capstones operate within institutional systems and contribute to sustainability curriculum development.

Across both cases, we identify four persistent challenge areas: managing stakeholder relationships and expectations, addressing conceptual ambiguities in sustainability, aligning capstone work with career development, and navigating structural and resource constraints. These challenges are not isolated but intertwined, often shaped by broader dynamics in higher education such as increased reliance on contingent faculty, heavier teaching loads, and insufficient funding for community-engaged learning. Addressing these challenges requires not only thoughtful course design but also institutional investment and faculty support.

Implications for practice include the need to formalize capstone leadership within faculty workload models, create structures for consistent mentorship, and embed space for critical reflection—especially on issues of justice, systems thinking, and long-term impact. The findings suggest that capstone courses can play a vital role in developing sustainability competencies, but only when supported by intentional, equity-focused curricular and institutional frameworks.

More broadly, this chapter contributes to the field of curriculum development for sustainable development by showing how capstone courses can function as integrative, applied components of sustainability curricula that can include and model innovative practices. By identifying recurring challenges, we aim to inform how capstones can be better embedded into curricular structures that promote systems thinking, stakeholder engagement, and real-world problem solving.

While the two case studies offer valuable insights into capstone design and implementation, the analysis is based primarily on the authors' institutional experiences and informal evaluations. As such, the findings may not be fully generalizable to other contexts. Future research could expand on this work by systematically comparing capstone models across institutions, assessing long-term student outcomes, and evaluating how different curriculum structures influence sustainability competency development. There is also a need to investigate how equity, justice, and inclusion can be more deeply embedded into capstone frameworks.

References

American Federation of Teachers. (2020). An army of temps: AFT 2020 adjunct faculty quality of work/life report. Retrieved June 16, 2025, from https://www.aft.org/sites/default/files/adjuncts_qualityworklife2020.pdf.

Annelin, A., & Boström, G.-O. (2024). Interdisciplinary perspectives on sustainability in higher education: A sustainability competence support model. *Frontiers in Sustainability, 5, 1416498.*

Barth, M., & Rieckmann, M. (2015). *State of the art in research on higher education for sustainable development*. Routledge.

Bergmark, U. (2022). The role of action research in teachers' efforts to develop research-based education in Sweden: Intentions, outcomes, and prerequisite conditions. *Educational Action Research, 30*(3), 427–444.

Bhagwan, R. (2020). Towards the institutionalisation of community engagement in higher education in South Africa. *Perspectives in Education, 38*(2), 36–55.

Brundiers, K., Barth, M., Cebrián, G., Cohen, M., Diaz, L., et al. (2021). Key competencies in sustainability in higher education—Toward an agreed-upon reference framework. *Sustainability Science, 16*(1), 13–29.

Brundiers, K., & Wiek, A. (2017). Beyond interpersonal competence: Teaching and learning professional skills in sustainability. *Education Sciences, 7*(1), 39.

Budowle, R., Krszjzaniek, E., & Taylor, C. (2021). Students as change agents for community–university sustainability transition partnerships. *Sustainability, 13*(11), 6036.

U.S. Census Bureau. (2024a). Quick facts: Tucson city, Arizona. Retrieved April 18th 2025, from https://www.census.gov/quickfacts/fact/table/tucsoncityarizona

U.S. Census Bureau. (2024b). Quick facts: .. Retrieved May 16th 2025, from https: //www.census.gov/quickfacts/fact/table/bouldercountycolorado/PST045224

Chang, C.-C., Kuo, C.-G., & Chang, Y.-H. (2018). An assessment tool predicts learning effectiveness for project-based learning in enhancing education of sustainability. *Sustainability, 10*(10), 3595.

Evans, T. L. (2019). Competencies and pedagogies for sustainability education: A roadmap for sustainability studies program development in colleges and universities. *Sustainability, 11*(19), 5526.

Ferrante, C. J., Heppard, K. A., DePorres, D. P., & Green, S. G. (2025). The evolution of capstone sequences for undergraduate programs. *Journal of Higher Education Theory and Practice, 25*(1), 32–45.

Fraustino, J. D., Pressgrove, G., & Colistra, R. (2019). Extending understanding of service-learning projects: Implementing place-based branding for capstone courses. *Communication Teacher, 33*(1), 45–62.

Gresham, M., Boswell, C., Anderson, O. S., Laye, M. J., & Smith-Sherwood, D. (2023). Understanding faculty needs in capstone experiences. In *Cultivating capstones* (pp. 167–178). Routledge.

Ho, C. T. (2023). Making career development sustainable: A senior capstone course case study. In *Educating the sustainability leaders of the future* (pp. 297–309). Switzerland.

Knox, R. C., & Nairn, R. W. (2025). Twenty-five years of domestic service learning and community engagement projects: A framework for state-funded university environmental capstone classes. *Journal of Civil Engineering Education, 151*(3), 05025002.

Konrad, R., Hall-Phillips, A., & Vila-Parrish, A. R. (2018). Are our students prepared? The impact of capstone design pedagogical approaches on student skill development during industry-sponsored fieldwork. *INFORMS Transactions on Education, 18*(3), 183–193.
Kovaric, K., & Gingell, G. (2024). Effects of an experiential learning curriculum on systems-thinking. *Journal of Medical Education and Curricular Development, 11.*
Landfried, M., Chen, E., Savelli, L. B., Cooper, M., Price, B. N., & Emmerling, D. (2023). MPH capstone experiences: Promising practices and lessons learned. *Frontiers in Public Health, 11,* 1129330.
LeChasseur, K., Levey, F., Sabuncu, A. C., Ebadi, A., & McNeill, J. (2024). Capstone projects for self-efficacy, skills, and successful careers. In *In 2024 ASEE Annual Conference & Exposition.*
Levesque, V. R., & Blackstone, N. T. (2020). Exploring undergraduate attainment of sustainability competencies. Sustainability: *The journal of. Record, 13*(1), 32–38.
Minshew, L. M., Yi, J., Morbitzer, K. A., & McLaughlin, J. E. (2020). Use of capstone experiences in pharmacy education to synthesize and apply students' knowledge and skills. *American Journal of Pharmaceutical Education, 84*(11), 8060.
Oceanic, N., & Administration, A. (2025). *Tucson monthly and daily normals and records.* U.S. Department of Commerce. Retrieved April 18th 2025, from https: //www.weather.gov/twc/ Tucson Monthly NormalExtremes
Park, H. Y., Palazzo, D., & Hollstein, L. (2025). Case study for planning education: Lessons from incorporating an interdisciplinary teaching approach and APA trend reports in capstone planning studios. *Sustainability, 17*(3), 1294.
Sasson, I. (2019). Building a sustainable university–community partnership: Case study in science education. *Studies in Higher Education, 44*(12), 2318–2332.
Schön, D. A. (2017). *The reflective practitioner: How professionals think in action.* Routledge.
Shephard, K., Harraway, J., Lovelock, B., Mirosa, M., Skeaff, S., et al. (2015). Seeking learning outcomes appropriate for "education for sustainable development" and for higher education. *Assessment & Evaluation in Higher Education, 41*(6), 825–842.
Watson, M. K., Barrella, E., Wall, T., Noyes, C., & Rodgers, M. (2020). Comparing measures of student sustainable design skills using a project-level rubric and surveys. *Sustainability, 12*(18), 7308.

Nataliya is an educator and interdisciplinary researcher at the University of Arizona. Her areas of expertise include sustainability education, urban agriculture, food systems, food security, and biochar for soil amelioration. She holds a Ph.D. in Sustainable Agriculture and Biorenewable Resources and Technology from Iowa State University and an M.A. in Environmental Law and Policy.

Alice Reznickova is an Associate Teaching Professor and Capstone Lead in the Masters of the Environment Program at the University of Colorado Boulder. Alice is an educator, mentor, and researcher in sustainability with a particular interest in human behavior, policy, food systems, urban design, and sustainability education. Alice received her M.S. in Chemistry and Ph.D. in Environment and Resources, both from the University of Wisconsin Madison.

Integrating Sustainability and Practice into Engineering Education: An Interdisciplinary Curriculum Model in Côte D'Ivoire

Lisa Schmitt, Charlotte Newiadomsky, Arne Graßmann, Ahoutou Paul Kouakou, and N'guessan Raymond Kre

Abstract This chapter presents a successful curriculum model that was implemented at a university in Côte d'Ivoire. The model emphasises sustainability and practical experience through an interdisciplinary dual-study engineering programme. It was developed under the project "Industry-Integrated Dual Engineering Study in a North–South Collaboration" (IIDES-NSC). The curriculum combines academic training with practical experience through demanding semester-long study projects and multiple industry internships, addressing the urgent need for energy and water supply solutions in rural Côte d'Ivoire by educating and training students to develop and implement new solutions. A central objective is to integrate sustainability not as a standalone subject but as a cross-cutting principle throughout the curriculum. This is exemplified by the development of a solar-powered prototype for electricity generation, water supply and water treatment. This approach directly supports the United Nations Sustainable Development Goals (SDGs), particularly clean energy (SDG 7), clean water (SDG 6), quality education (SDG 4), gender equality (SDG 5), decent work and economic growth (SDG 8), and industry, innovation, and infrastructure (SDG 9). An interdisciplinary approach was central to the curriculum development. Stakeholders, including academic institutions, ministries, local industries,

L. Schmitt (✉) · C. Newiadomsky · A. Graßmann · A. P. Kouakou
SWK E2—Institute for Energy Technology and Energy Management, Hochschule Niederrhein, Krefeld, Germany
e-mail: lisa.schmitt@hs-niederrhein.de

C. Newiadomsky
e-mail: charlotte.newiadomsky@hs-niederrhein.de

A. Graßmann
e-mail: arne.grassmann@hs-niederrhein.de

A. P. Kouakou
e-mail: kouakoupaul.sfa@univ-na.ci

N. R. Kre
Unité de Formation et de Recherche Sciences Fondamentales Appliquées, Université Nangui Abrogoua, Abidjan, Côte d'Ivoire
e-mail: kreray.sfa@univ-na.ci

W. Leal Filho et al., *Curriculum Development for Sustainable Development*, World Sustainability Series, https://doi.org/10.1007/978-3-032-22146-9_23

and international organisations such as German Society for International Cooperation (GIZ), collaborated through round tables, workshops, and surveys to align the curriculum with the industry's needs and the country's national development plan. The implementation faced challenges, including bureaucratic and cultural barriers, language differences, and a tight schedule. Acceptance of the curriculum's practical orientation required the inclusion of industry-based lecturers and a focus on applied learning. Nevertheless, these challenges presented opportunities to enhance the university's reputation, improve teaching quality, and better prepare graduates for the labour market. The chapter's objective is to show how an industry-integrated, sustainability-focused curriculum can enhance employability, foster innovation, and strengthen higher education's contribution to **sustainable development**. The curriculum demonstrates how a practice-oriented education model can empower students by addressing real-world challenges and leveraging collaborative development. The case study illustrates the transformative potential of embedding sustainability into curriculum design, offering a replicable model for similar contexts in sub-Saharan Africa.

Keywords Curriculum development · Dual-study programme · Practice-oriented learning · Sustainable development · Industry collaboration · Employability · Energy and water supply · Rural area

1 Introduction

Efforts to better align engineering education with sustainable development objectives have gained international momentum. The United Nations Educational, Scientific and Cultural Organization (UNESCO) formally shifted educational agendas toward sustainability with initiatives such as "Regional Centres of Expertise" and the "Global Universities Partnership on Environment and Sustainability" to embed sustainability across curricula and higher-education networks (UNEP, 2025; UNESCO, 2020). More recently, the International Decade of Sciences for Sustainable Development (2024–2033) underscores a renewed global commitment to transdisciplinary, problem-solving-oriented science and education, supported by international networks and institutions like the World Federation of Engineering Organizations and UNESCO (UNESCO, 2024a, 2024b).

Sustainability has become essential in engineering education as societies worldwide face environmental degradation, resource scarcity, and infrastructure inadequacies. These systemic challenges are particularly pressing in Sub-Saharan Africa, where over 600 million people still lack reliable access to electricity (International Energy Agency, 2022; IRENA, 2024), and around 300 million are without access to clean water (UNICEF and WHO, 2022), severely impeding progress in critical societal domains such as education, healthcare, and economic development (Sule et al., 2022).

At the regional level, the engineering sector in Sub-Saharan Africa continues to grapple with a mismatch between graduate skills and labour-market needs. The Royal Academy of Engineering's, 2025 report highlights the urgent need for updated curricula, stronger industry-academia partnerships, and scalable education models to build local engineering capacity tailored to development challenges aligned with the United Nations Sustainable Development Goals (SDGs) (Royal Academy of Engineering, 2025). Studies across the region, including those in Ghana (Manteaw et al., 2025), Nigeria (Ukhurebor et al., 2025) and Zimbabwe (Kumba et al., 2024), echo this sentiment, highlighting persistent issues such as outdated teaching models, colonial legacies in pedagogy, limited practice-oriented learning and underfunded infrastructure. These barriers hinder the implementation of sustainability-focused education (Mashiyane et al., 2023).

In this context, Côte d'Ivoire exemplifies where educational interventions can make a deep impact. Despite its leadership in West Africa, the country faces persistent infrastructure deficits in rural areas, with only 43% of rural households being connected to the power grid as of 2020 (World Bank, 2024). This creates disproportionate burdens on vulnerable groups (e.g., women and children) for accessing clean water (Carr et al., 2024). Such conditions accentuate the need for a curriculum that equips students with practical skills, systems thinking, and interdisciplinary collaboration competencies that are hallmarks of a development-oriented engineering education aligned with the SDGs.

To meet this demand, the "Industry-Integrated Dual Engineering Study in a North–South Collaboration" (IIDES-NSC) project was initiated to provide a dual-study Master's programme that integrates classroom instruction with project-based modules and industry internships. This approach ensures that students gain hands-on experience while learning theoretical concepts, enabling them to effectively address pressing issues in renewable energy and water management. Fostering interdisciplinary collaboration and aligning with Côte d'Ivoire's national development strategy, the curriculum prepares graduates to lead a sustainable transformation in their communities.

This chapter aims to present and critically reflect on the design, implementation, and outcomes of the IIDES-NSC dual-study Master's programme, with a focus on sustainable development. Specifically, it seeks to demonstrate how sustainability can be embedded holistically across a curriculum without requiring a dedicated "sustainability" course. It also shows how combining academic study with real-world industry experience can prepare graduates to address urgent infrastructure and environmental challenges. Lastly, the curriculum offers a model for aligning curricula in higher education with national development strategies and the SDGs.

To achieve these aims, the curriculum integrates classroom learning with a semester-spanning interdisciplinary project and industry placements, ensuring that students apply theoretical knowledge to pressing real-world problems in renewable energy and water management. The curriculum was co-developed with stakeholders from academia, industry, and government, ensuring relevance to both, the labour market demands and the national priorities.

By documenting the curriculum development process, challenges, and outcomes, this chapter contributes to the broader discussion on how higher education can be a driver of sustainable transformation in sub-Saharan Africa.

2 Background

The curriculum developed within the IIDES-NSC project represents a partnership between Côte d'Ivoire's Université NANGUI ABROGOUA (UNA), Germany's Hochschule Niederrhein University of Applied Sciences, and local and international industry stakeholders. Together, these partners designed a curriculum that combines advanced academic training with real-world applications. This collaborative effort aimed to bridge the gap between theoretical knowledge and practical skills while addressing local energy and water challenges.

The curriculum focuses on three core areas:

1. **Renewable Energy**: Students are trained to design, implement, and maintain systems such as solar power plants and hybrid solutions tailored to local conditions. By utilising renewable energy resources, the curriculum contributes to reducing dependence on fossil fuels and enhancing energy security.
2. **Water Management**: The curriculum addresses the critical need for clean and sustainable water solutions. Students learn techniques for water purification, distribution, and conservation, ensuring they are prepared to implement effective systems in resource-scarce environments.
3. **Entrepreneurial and Soft Skills**: Beyond technical expertise, students develop skills in project management, intercultural communication, business planning, and innovation. These competencies empower graduates to establish enterprises that address regional infrastructure deficits while creating economic opportunities.

The integration of these elements ensures that graduates are well-prepared to meet the demands of both the labour market and their communities, making them key contributors to sustainable development. As a result, the main pillars of the Master Ingénierie Eau et Énergie (MIEE; Master Water and Energy Engineering) study programme consists of courses with a focus on technical, economic, and social content, addressing all three core areas (Fig. 1).

Furthermore, the combination of these elements, which are embedded in a student project of three semesters, increases motivation as the students experience that solid knowledge is essential to develop technical solutions and gives confidence about regular technical or financial decisions. Students experience that decision-making is difficult if you have to account for the consequences and when the decisions have an immediate influence on the project budget or the achievability of project goals.

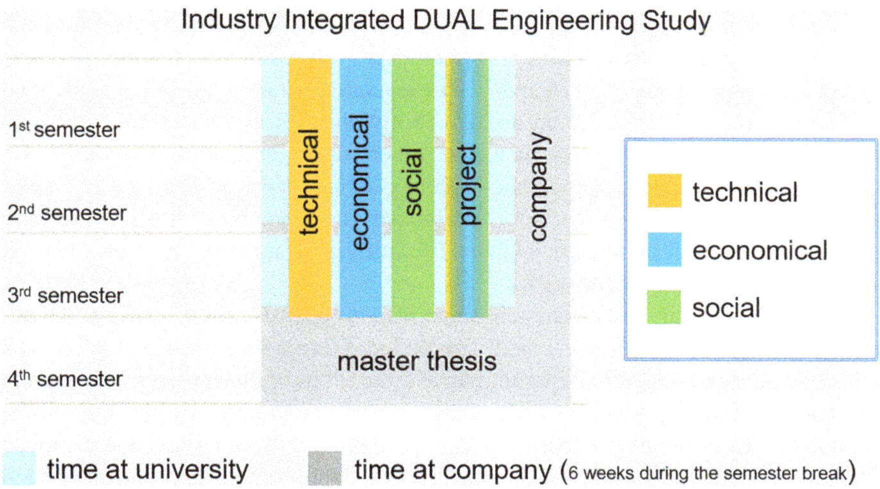

Fig. 1 Structure of the MIEE curriculum

3 Methodology

The curriculum development followed a structured, participatory, and iterative process, ensuring it met both rigorous academic standards and the needs of the local labour market (SDGs 8 and 9). The approach combined stakeholder engagement, co-design of a dual-study model, integration of prototype-based learning, continuous monitoring, and deliberate inclusion of gender and disciplinary diversity.

The process began with an extensive needs assessment involving representatives from academia, industry, government ministries, non-governmental organisations (NGOs), and international organisations such as GIZ. **Stakeholder engagement** took place through multi-day workshops where participants defined the skills graduates would require to address pressing sectoral challenges, as well as through targeted surveys of industry partners, graduates from related study programmes, and community representatives to quantify skill gaps and training needs. Round table discussions on themes such as renewable energy, water management, gender equality, and sustainability provided additional insights. This participatory process ensured that the curriculum was aligned with Côte d'Ivoire's national development plan, the SDGs, and the realities of the local socio-economic environment.

The core of the curriculum is a **dual-study model** (SDG 4), adapted from European cooperative education systems but tailored to Côte d'Ivoire's context. The academic calendar alternates between 12 and 14 weeks of classroom instruction and six weeks of industry internships. Formal agreements with local companies, utilities, and NGOs guarantee internship availability, define learning outcomes, and ensure that academic modules are sequenced so students can immediately apply newly acquired theoretical knowledge in real-world settings.

A distinctive pedagogical element of the curriculum is the requirement for each cohort to **design, construct, and test prototypes** addressing community needs in energy and water supply (SDG 6 and 7). These semester-long projects follow defined milestones, from concept design and feasibility analysis to construction and field testing. Examples include solar-powered water supply and purification units for rural villages and hybrid photovoltaic–battery systems for off-grid electricity. Supervised jointly by university lecturers and industry mentors, these projects not only strengthen technical competencies but also cultivate teamwork, problem-solving, and community engagement skills.

To maintain quality and ensure relevance, a comprehensive **monitoring and evaluation framework** was implemented. This includes formative assessments after each academic module and internship, summative evaluations of final projects and reports, and graduate tracking surveys conducted six months and one year after programme completion. Community feedback from areas where prototypes have been deployed provide an additional measure of impact, capturing both functional performance and social acceptance.

From the outset, the curriculum sought to foster inclusivity and diversity. A **gender balance** (SDG 5) target of at least 40% female participation was established, supported by targeted outreach campaigns and orientation workshops promoting respectful collaboration across differences in religion, age, culture, and ethnicity. Respect for all voices, particularly those of female students, was considered a core selection criterion. Equally important was the recruitment of students from **varied academic backgrounds**, including electrical engineering, electronics, water management, and theoretical physics. This deliberate mix allowed teams to benefit from complementary expertise, such as physics graduates contributing to system modelling, electronics specialists managing circuit design, and water engineers overseeing the treatment system development.

By combining collaborative curriculum design, integrated industry engagement, project-based learning, and inclusive recruitment, this methodology offers a replicable model for developing sustainability-oriented engineering study programmes in contexts facing similar socio-economic and infrastructural challenges. Further information on the curriculum development process can be found in Newiadomsky et al., 2023.

4 Results and Discussion

The implementation of the MIEE curriculum under the IIDES-NSC project has yielded significant results, demonstrating the value of integrating sustainability, interdisciplinarity, and practice-oriented learning into engineering education.

Curriculum development: Various new academic modules have been implemented, embedding sustainability principles and strengthening technical training. The combination of theoretical instruction, applied laboratory work, and field-based activities has enabled students to develop competencies directly aligned with industry

Fig. 2 Prototype system in the model village

requirements in renewable energy and water management. By linking each module to the semester-spanning interdisciplinary project, the curriculum ensures that theoretical knowledge is immediately applied to real-world problems, reinforcing learning outcomes and fostering problem-solving skills.

Prototype achievements and practical impact: Two flagship prototypes emerged as tangible outputs of the students' work. The first prototype was deployed and installed in a rural community (Fig. 2). It provides reliable access and supply to electricity and clean water to areas outside both the electrical grid and water distribution systems. This installation demonstrates the scalability and replicability of decentralised infrastructure solutions in similar contexts.

The second prototype, installed on the university campus, serves as a hands-on demonstrator for training and experimentation (Fig. 3). Students can test and optimise system configurations, while the pumped water supports aquaculture tanks in a neighbouring faculty, being an example of interdisciplinary synergies. In Fig. 4, on of the MIEE students is showcasing the prototype's functionalities to stakeholders, reinforcing its role as a learning and outreach tool.

Strengthened partnerships and stakeholder engagement: Collaboration with industry partners such as euregio-solarzentrum GmbH, WILO SE, and FIACER (Fédération Ivoirienne des Associations en Efficacité Energétique, Energies Renouvelables et Climat, engl. Ivorian Federation of Energy Efficiency, Renewable Energy and Climate Associations) has been instrumental in keeping the curriculum industry-relevant. These partnerships facilitated internships, mentoring, and exposure to professional environments, bridging the gap between academics and industry. Furthermore, it ensured that prototypes were designed to meet real operational needs, increasing the likelihood of adoption and long-term use. An equally important

Fig. 3 Exterior view of the prototype system installed at UNA

Fig. 4 MIEE student showcasing the prototype at the inauguration event at UNA

success factor was the involvement of local intermediary organisations and individuals, particularly from the Association of Cameroonian Engineers and Computer Scientists, Ruhrbezirk e.V. (VKII e.V.) and TOUKAM Consulting. Their ability to navigate and connect both Ivorian and German institutional and cultural contexts proved essential for building trust, facilitating communication, and resolving challenges throughout the project. The experience highlighted the crucial role of such bridge-builders in the successful implementation and effective collaboration between all partners in such complex international projects.

Community empowerment and social outcomes: Beyond technical outputs, the curriculum has delivered community benefits by providing local workshops and training sessions that raise awareness of sustainable practices. This knowledge transfer empowers local populations to adopt renewable energy solutions and implement water-saving measures, contributing to resilience and self-reliance.

Critical reflections and lessons learned: The success of the MIEE Master's study programme underscores the effectiveness of embedding sustainability as a cross-cutting principle rather than a stand-alone subject. The dual-study model's alternating structure between academic semesters and industry placements allows students to contextualise learning, enhancing both technical proficiency and employability. However, the process has not been without challenges. Navigating bureaucratic requirements, managing cultural differences, and ensuring timely prototype completion demanded adaptive strategies and continuous stakeholder communication. Additionally, securing adequate financing for hands-on projects emerged as a critical enabler, since the authenticity and educational value of practical components would be compromised without dedicated budgets.

Implications for replication and scaling: The adaptability of the dual-study model makes it suitable for replication in other regions with similar infrastructure deficits. Success factors include close industry-university collaboration, early integration of practical components, and stakeholder involvement in curriculum co-design. For maximum impact, scaling efforts should also focus on institutionalising funding mechanisms for applied projects and maintaining strong alumni-industry linkages.

To summarize, the MIEE curriculum demonstrates how a carefully designed, context-adapted curriculum can serve as a catalyst for both educational innovation and sustainable development. Its graduates are not only technically competent but also equipped with the entrepreneurial mindset and collaborative skills necessary to tackle pressing energy and water challenges in Côte d'Ivoire and beyond.

5 Conclusion

The IIDES-NSC project responds to the urgent need for practice-oriented, sustainability-focused engineering education in Côte d'Ivoire. This need being characterised by persistent rural infrastructure deficits, limited access to reliable energy and clean water, and a shortage of highly skilled professionals capable of addressing these challenges.

Contextualisation: By embedding sustainability as a transversal principle, the MIEE programme aligns with both Côte d'Ivoire's national development strategies and the SDGs. It addresses key regional challenges such as decentralised energy provision, water supply and purification, and community empowerment, offering a replicable framework for other sub-Saharan African contexts facing similar development constraints (SDGs 6 and 7).

Methods: The curriculum was developed through a structured, participatory process involving academia, industry, government, and community stakeholders. It adopts a dual-study model, which combines classroom-based learning with industry internships, while integrating a semester-spanning interdisciplinary project focused on the design and implementation of community-oriented prototypes. Continuous monitoring and evaluation, as well as deliberate gender and disciplinary diversity in student selection (SDG 5), ensured both an academic rigor as well as a socio-economic relevance.

Key Findings: The analysis presented demonstrates that a sustainability-oriented, dual-study curriculum model can be effectively adapted to the socio-economic and infrastructural realities of a Sub-Saharan African context (SDGs 4 and 9). The participatory co-design process, that integrates input from academia, industry, government, and community stakeholders, proved to be essential for ensuring both labour-market relevance and alignment with national development priorities (SDG 8). Embedding sustainability as a transversal principle across modules, rather than isolating it in a stand-alone course, emerged as a viable and replicable approach to fostering interdisciplinary competencies (SDGs 6 and 7). Furthermore, the integration of industry internships with an extended interdisciplinary project provided a coherent link between theoretical knowledge and applied problem-solving. The findings also highlight the importance of deliberate inclusivity strategies, particularly in gender balance and disciplinary diversity (SDG 5), as key factors in fostering innovation, collaboration, and real-world applicability of student work.

Limitations: Despite these achievements, challenges remain. The project's reliance on external funding for prototype development poses sustainability risks if resources are not institutionalised. Bureaucratic procedures and cultural differences among stakeholders can slow or possibly halt implementation altogether. Finally, the relatively small initial cohort size limits the immediate reach and scalability of this approach.

Impact and Outlook: The dual-study approach contributes to SDG 4 by bridging the gap between theoretical knowledge and practical application. In addition, it addresses SDGs 8 and 9 by improving graduate employability and supporting technology transfer to underserved communities. Partnerships with industry and international collaborators emerged as critical to ensuring real-world applicability and long-term sustainability of project outcomes. Looking forward, expanding enrolment, diversifying funding streams, and scaling prototype deployment to additional rural communities could amplify the impact of both the project and the new curriculum. Strengthening alumni networks, deepening South-South collaborations, and integrating emerging topics like climate resilience and circular economy will further future-proof the curriculum. The experience and lessons learned from the IIDES-NSC project provide a strong foundation for developing similar curricula in other regions where sustainable infrastructure solutions and skilled professionals are urgently needed.

In conclusion, the Master's study programme illustrates how context-sensitive, practice-oriented engineering education can equip graduates with the competencies to lead sustainable transformation in their own communities while addressing pressing development needs.

References

Carr, R., Kotz, M., Pichler, P.-P., Weisz, H., Balmin, C., & Wenz, L. (2024, July). Climate change to exacerbate the burden of water collection on women's welfare globally. *Nature Climate Change, 14*, 700–706.

International Energy Agency (2022). *Africa energy outlook 2022—English version.* World Energy Outlook Special Report. Retrieved 11 Aug 2025, https://iea.blob.core.windows.net/assets/6fa5a6c0-ca73-4a7f-a243-fb5e83ecfb94/AfricaEnergyOutlook2022.pdf

IRENA (2024). *World energy transitions outlook 2024: 1.5°C Pathway, international renewable energy agency,* Abu Dhabi. Retrieved 11 Aug 2025, https://www.irena.org/-/media/Files/IRENA/Agency/Publication/2024/Nov/IRENA_World_energy_transitions_outlook_2024.pdf

Kumba, H., Olanrewaju, O., Akpan, J., Mafuratidze, F., & Ndiyamba, D. (2024). *Engineering education and industry synergy in Zimbabwe: A divide and redirection.* Innovation and Evolution in Higher Education. Retrieved 13 Aug 2025, https://www.intechopen.com/chapters/1188736

Manteaw, B. O., Boafo, Y. A., Owusu, E. H., Enu, K. B., & Amoah, A. B. (2025). Transgressive pedagogies in a climate-constrained Africa: Grounding climate change education and sustainability learning in local communities. *Frontiers in Education*, Vol. 10, Article 1519914. Retrieved August 13th, 2025, from https://doi.org/10.3389/feduc.2025.1519914

Mashiyane, T., Salifu, S., Ogunbiyi, O., & Oketola, A. (2023). Challenges, strategies and recommendations for reengineering engineering education in Sub-Saharan Africa: A review. *Science, Engineering and Technology, 4*(1). https://doi.org/10.54327/set2024/v4.i1.102, Retrieved 13 Aug 2025.

Newiadomsky, C., Saars, L., Graßmann, A., & Kouakou, A.P. (2023). Dual engineering study program with focus on sustainability in Côte d'Ivoire. In U. Bode (Ed.), *Universities, Entrepreneurship and Enterprise Development in Africa—Conference Proceedings 2022* (pp. 231–242). Hochschule Bonn-Rhein-Sieg.

Royal Academy of Engineering (2025). *Engineers for Africa 2025 report: Engineering capacity in Sub-Saharan Africa.* London: Royal Academy of Engineering. Retrieved 13 Aug 2025, https://raeng.org.uk/media/rz0gr5xb/engineers-for-africa-2025-report.pdf

Sule, I. K., Yusuf, A. M., & Salihu, M.-K. (2022). Impact of energy poverty on education inequality and infant mortality in some selected African countries. *Energy Nexus, 5*, Article 100034.

The World Bank (2024). *World Development Indicators.* Retrieved 11 Aug 2025, https://databank.worldbank.org/reports.aspx?source=2&country=CIV

Ukhurebor, K. E., Egenti, G., Uduehi, I., Odezuligbo, I., Aliu, D., Abiodun, A. O., Adama, K. K., Eragbe, B. S., Talabi, A. A., Jokthan, G., Opateye, J., Nwankwo, U. C., & Eneche, B. M. (2025). The role of continuous engineering and scientific education for sustainable development in Sub-Saharan Africa: Insights from Nigeria. *Journal of Infrastructure, Policy and Development, 9*(1), Article 8989. Retrieved 13 Aug 2025. https://doi.org/10.24294/jipd8989

UNDP (2024). *Annual Report 2023*. One United Nations Plaza, New York, MY, Retrieved 11 Aug 2025, https://annualreport.undp.org/assets/Annual-Report-2023.pdf

UNEP (2025). *Global Universities Partnership on Environment and Sustainability (GUPES)*. Retrieved 13 Aug 2025, https://www.unep.org/explore-topics/education-environment/why-does-education-and-environment-matter/global-universities

UNESCO (2020). *Education for sustainable development: A roadmap*. In S. Giannini (Ed.) *Foreword*. UNESCO. Retrieved 13 Aug 2025. https://doi.org/10.54675/YFRE1448

UNESCO (2024a). *International Decade of Sciences for Sustainable Development 2024–2033: Strategic Plan*. UNESCO. Retrieved 13 Aug 2025, https://unesdoc.unesco.org/ark:/48223/pf0000391869

UNESCO (2024b). *International Decade of Sciences for Sustainable Development 2024–2033: Fostering Science for All*. UNESCO. Retrieved 13 Aug 2025, https://unesdoc.unesco.org/ark:/48223/pf0000391158

UNICEF, WHO (2022). *Progress on drinking water, sanitation and hygiene in Africa 2000–2020: Five years into the SDGs*. United Nations Children's Fund (UNICEF) and World Health Organization (WHO), New York. Retrieved 11 Aug 2025, https://www.unicef.org/media/117726/file/JMP%20Snapshot%20on%20Progress%20on%20Drinking%20Water,%20Sanitation%20and%20Hygiene%20in%20Africa%202000-2020.pdf

Lisa Schmitt M. Eng., is a scientific researcher at SWK E2 - Institute of Energy Technology & Energy Management at the University of Applied Science in Krefeld. She completed her dual bachelor studies in electrical engineering with a focus on energy technology in 2018 and her master's studies in energy management engineering in 2021. In 2022, she worked on a project on renewable energies for the German NGO 'Ingenieure ohne Grenzen e. V.' designing and implementing a solar system for a school in Malawi. Since 2023, she has been working on the project "Growing Rwandan Energy Awareness Through highER eduation" (GREATER) in Rwanda to develop new study programmes and design so-called Living Labs as well as on the project "Industry-Integrated Dual Engineering Study in a North-South Collaboration" (IIDES-NSC) in Côte d'Ivoire.

Charlotte Newiadomsky is working at the SWK E2 – Institute of Energy Technology and Energy Management at Hochschule Niederrhein University of Applied Sciences. She works on electricity price modelling, energy efficiency in the industrial sector and climate change impacts on electricity prices. She is a PhD Student at Europa University in Flensburg and deals with "Assessing the Resilience of Electricity Generation Technologies to Climate Change Impacts in Developing Countries." She studied chemical engineering at the Aachen University of Applied Sciences and finished in 2011 her master studies in Technology and Resources Management in the Tropics and Sub-tropics with major in energy management at the TH Köln – University of Applied Sciences.

Arne Graßmann is a lecturer at Hochschule Niederrhein University of Applied Sciences in the Department of Industrial Engineering. His teaching and research field is energy technology with a focus on aero-dynamic optimisation of components, high-energy (transonic) flows, energy storage and damage analysis (thermo-aero-dynamic) in steam power plants. He is the Deputy Head of SWK E^2 – Institute of Energy Technology & Energy Management. Prior to this, he was involved in the development of simulation and optimisation tools for steam power plants at Siemens AG in the field of steam turbine development. He has worked as project manager of two international cooperation project in Côte d'Ivoire and Rwanda.

Ahoutou Paul Kouakou is a professor of physics and lecturer at UNA in Côte d'Ivoire. His fields of expertise are plasma physics, material science, photovoltaics and energy efficiency. He has experience working abroad since he did his bachelor's and master's degrees as well as his doctorate in physics at the University of Lorraine in France. He worked as project manager of the project "IIDES-NSC" from 2021 to 2024 and is responsible for the dual study master programme developed in that project. Since 2021, he has been a trainer for Energy Efficiency trained by the GIZ project "Projekt Berufliche Bildung im Bereich Erneuerbare Energien und Energieeffizienz" (ProFERE).

N'guessan Raymond Kre is a lecturer at UNA in Côte d'Ivoire and the director of the Department of Fundamental and Applied Science. This department consists of the fields of mathematics-informatics and physics-chemistry.

Subnational Climate Planning Education: Meeting the Needs of the Coalition for High Ambition Multilevel Partnerships Through Penn State's Local Climate Action Program

Peter Buck and Brandi Robinson

Abstract Anthropogenic climate change presents a unique challenge to planning educators as an ever-present and invisible threat as well as a multi-scalar "wicked" or "superwicked" problem. Capacity-building to support subnational and local climate action has been recognized in global policy frameworks since at least the Rio Summit, most recently in the Coalition for High-level Ambition and Multi-level Partnerships (CHAMP) created at the 28th Conference of Parties (COP 28). Capacity-building requires focused and systematic curriculum and praxis in universities to educate the climate-planning workforce. This chapter explores the global to local warrant, context, structure, mission, values, goals, and practices of Penn State's Local Climate Action Program (LCAP), an immersive capacity-building climate planning educational program at one of the United States's land grant universities. The authors intend to provide a view of the LCAP so university faculty, administrators, and student advocates can see commonalities and recognize what makes their own national, subnational, and institutional contexts distinct and potentially replicate or refine the program.

Keywords Capacity building · Climate planning · Experiential learning · Greenhouse gas inventory · Subnational government

P. Buck (✉) · B. Robinson
Penn State's Local Climate Action Program, Pennsylvania, USA
e-mail: PeterBuck@psu.edu

John and Willie Leone Family Department of Energy and Mineral Engineering, The Pennsylvania State University, Pennsylvania, USA

Penn State Sustainability, Pennsylvania, USA

B. Robinson
e-mail: brobinson@psu.edu; bjn151@psu.edu

W. Leal Filho et al., *Curriculum Development for Sustainable Development*, World Sustainability Series, https://doi.org/10.1007/978-3-032-22146-9_24

1 Introduction

Anthropogenic climate change presents a unique challenge to planning educators as an ever-present and invisible threat as well as a multi-scalar "wicked" or "super-wicked" problem. Capacity-building to support subnational and local climate action has been recognized in global policy frameworks since at least the Rio Summit, most recently in the Coalition for High-level Ambition and Multi-level Partnerships (CHAMP) created at the 28th Conference of Parties (COP 28). Capacity-building requires focused and systematic praxis in universities to educate the climate-planning workforce. This chapter explores the global to local warrant, context, structure, mission, values, goals, and practices of Penn State's Local Climate Action Program (LCAP), an immersive capacity-building climate planning program at one of the United States's land grant universities. The authors intend to provide a view of the LCAP so university faculty, administrators, and student advocates can see commonalities and recognize what makes their own national, subnational, and institutional contexts distinct and potentially replicate or refine the program.

Climate change's cause is both straightforward and bedevilingly complicated. On one hand, the physical fact is that humans have added greenhouse gases to the atmosphere at a scale and pace such that 2024 broke multiple records: the global average temperature clearly exceeded the Paris Agreement's 1.5 °C pre-industrial level, atmospheric carbon dioxide went over 422 ppm, and extreme events, including floods, heatwaves, wildfires and other climate violence wreaked havoc across the planet (Copernicus, 2025). But the problem is not simple.

Following Rittel and Webber's "Dilemmas in a General Theory of Planning" (1973), Incropera (2016) describes climate change's myriad and layered repercussions across the physical, biological, economic, technological, social, political, and moral spheres, classifying it as a "wicked" problem. Climate change is difficult to define, lacks a stopping rule, has no immediate or ultimate solution, requires trial-and-error to manage all or some of it, has limitless potential solutions, is unique, symptomatic of other problems, has numerous explanations, and does not permit a "wrong" answer because answers alter lives (Rittel and Webber, pp. 160–167). Sun and Yang (2016) write, "The impossibility of understanding the full repercussions of climate change and the impact of each decision at both an organizational and a national level make many stakeholders reluctant to move forward toward proposing or implementing potential solutions." Levin et al (2012) place climate change in another category—"super wicked"—with four more features: "time is running out; those who cause the problem also seek to provide a solution; the central authority needed to address them is weak or non-existent; and irrational discounting occurs that pushes responses into the future" (p. 124; see also Levin et al, 2007; Lazarus, 2009). More recently, climate-strategic and ecophilosophical discourses have merged in Elizabeth Boulton's conception of climate change as a "hyperthreat" (2018).

The hyperthreat has warlike destructive capabilities that are so diffuse that it is hard to see the enormity of the destruction coherently nor who is responsible for its hostile actions. It defies existing human thought and institutional constructs. It is powered and energized by three key enablers: its invisibility, its ability to evade all existing human threat-response mechanisms, and by human hesitancy—the slower humans are to act, the stronger it becomes (p. 84).

The Guardian announced in 2019 that it would use the terms "climate emergency, crisis or breakdown" and "global heating" instead "global warming" to convey the urgency of the situation (Carrington, 2019). Climate change, however framed, demands nuanced understanding and multi-scalar action.

2 Multi-Scalar Climate Policy

Calls for multi-scalar action on climate change has been present in global policy for decades. The Local Governments and Municipal Authorities (LGMA) constituency has officially spoken for local and subnational governments in the United Nations Framework Convention on Climate Change since the first Conference of Parties (COP) in 1995, with ICLEI: Local Governments for Sustainability serving as the focal point organization (LGMA, 2023; see also ICLEI, 2025). Articles 7 and 11 of the Paris Agreement (COP, 2015) recognize the necessity for climate change adaptation and capacity-building at the subnational and local levels. Following Paris, the LGMA advocated for systematic local and subnational governments' recognition, engagement, and empowerment. The Glasgow Climate Pact included the term "multilevel action" for the first time (COP, 2022).

At COP 28, the Coalition for High-level Ambition and Multi-level Partnerships (CHAMP) was formed. The CHAMP Pledge recognizes the importance of engagement by all levels of government, reaffirms that collaborative, coordinated, and science-based climate action between nations and subnational governments is necessary to reduce emissions in line with global goals, and acknowledges that national laws and circumstances and the UNFCCC and the Paris Agreement itself are the primary vehicle for global climate response (COP28 UAE, 2023). Governments that endorse CHAMP also have the discretion to deliver CHAMP commitments, to determine the time, place, and manner for subnational input on climate commitments in light of their own Nationally Determined Contributions (NDC), and whether and how to fund local subnational projects. In summer 2024, Katja Dörner, Mayor of Bonn and chair of the Climate Action Governance Portfolio for ICLEI's Global Executive Committee (ICLEI, 2024), Frente Nacional de Prefeitas e Prefeitos (2024), and the United States Conference of Mayors (2024) made calls to action to support CHAMP to achieve NDC goals in line with the Paris Agreement. The United States was an original signatory of both the Paris Agreement and CHAMP.

The United States of America 2021 NDC pledged to reduce overall emissions by 50–52% by 2030 (p. 2). It provides emissions reduction pathways for the electricity, transportation, building, industry, agriculture and lands, and non-carbon dioxide GHG emissions including methane, hydrofluorocarbons, and other potent short-lived climate pollutants (pp. 3–5). Legislation was passed during the Biden administration that enabled decarbonization, especially the Inflation Reduction Act (IRA). The

IRA created the Climate Pollution Reduction Grants (CPRG) program to provide "$5 billion in grants to states, local governments, tribes, and territories to develop and implement ambitious plans for reducing greenhouse gas emissions and other harmful air pollution" (EPA, 2025a). Section 60,114 of the Inflation Reduction Act created a two-phase program that included a $250 million block for noncompetitive planning grants to create state, city, tribal, and other subnational Priority Climate Action Plans (PCAP) (EPA, 2025b).

In 2008, the Pennsylvania legislature passed its Climate Change Act (2008) requiring the creation of a climate impacts assessment (CIA) (Sect. 24.3), a statewide greenhouse gas (GHG) inventory (Sect. 24.4), and the creation of a climate action plan (CAP) (Sect. 5). There have been multiple CIAs, GHG inventories, and CAPs. In 2019, former Pennsylvania Governor Tom Wolf signed Executive Order 2019–01 (Commonwealth of Pennsylvania, Governor's Office, 2019) committing the Commonwealth to a 26 percent reduction of net greenhouse gas emissions statewide by 2025 from 2005 levels, an 80 percent reduction of net greenhouse gas emissions by 2050, and guidance to state agencies for their own actions to mitigate greenhouse gas emissions. Pennsylvania's gross GHG emissions in 2021 were 258.42 MMTCO2e with net emissions at 234.87 MMTCO2e after accounting for forestry and land use. 2021 net emissions represents an increase of 18.08 MMTCO2e or 8.3% from 2020. The largest contributing sectors were the industrial, electricity production, and transportation sectors, accounting for 83% of 2021's gross emissions (Pennsylvania Department of Environmental Protection, 2024, p. 6). The United States Energy Information Agency (2023) ranks Pennsylvania to be the fourth largest total GHG emitter in the U.S.

Following the passage of the IRA and the implementation of the CPRG, Pennsylvania Department of Environmental Protection released its own PCAP (2024) that includes nine sector-based action areas and goals. As instrumentalities of the Commonwealth, municipal and county governments are subject to state law and can access funding through state and federal programs. Additionally, Pennsylvania mayors are members of the United States Conference of Mayors and signatories of the 2024 Resolution in support of CHAMP.

3 Higher Education for Multi-Scalar Climate Action

Higher education's role in climate action has been refined from the Rio Summit through the Paris Agreement and beyond. *Agenda 21* recognized the role the scientific and technological community and universities would play to build capacity, specifically for the protection of the atmosphere including the intensification of anthropogenic climate change through energy, transportation, and land use planning (UNCED, 1992, Sects. 9, 10, 31, and 34). Education for All: Global Monitoring Report argued that mitigation of and adaptation to climate change requires education to develop the requisite transferrable skills (UNESCO, 2014), a sentiment reaffirmed by Sustainable Development Goal 4.7 to "ensure all learners acquire knowledge and skills needed to promote sustainable development" and Goal 13.3 to "[i]mprove education, awareness-raising and human and institutional capacity on climate change mitigation, adaptation, impact reduction and early warning" (UNDESA, 2015). Section 12 of the Paris Agreement states, "Parties shall cooperate in taking measures, as appropriate, to enhance climate change education, training, public awareness, public participation and public access to information, recognizing the importance of these steps with respect to enhancing actions under this Agreement (COP, 2015). Higher education needs to meet a growing demand for "environmental auditors, community organizers, corporate managers, engineers, practitioners, technical professionals, policymakers and, most significantly, the community about actions that can be taken to mitigate and adapt to climate change, while concurrently propagating social and governance measures" (Leal Filhio et al., 2021, p. 2).

A survey of literature and programs indicates that LCAP is at least unusual. In their bibliometric research, a worldwide online survey and case studies of climate change education programs across the globe, Leal Filhio et al. (2021) provided evidence of no programs as immersive and integrated as the LCAP, though they report on aspects such as GHG accounting, GHG reporting, or other public-facing programs featuring in some programs (pp. 13–15). The Stockholm Resilience Center offers an occasional five-day asynchronous course that introduces resilience assessments and related social-ecological systems methods (2025). According to EPIC-N, Chinhoyi University of Technology and Harare City Council in Zambia, the University of Lusaka in Zambia, and Bandung Institute of Technology in Bandung, Indonesia are each collaborating on climate resilience strategies (EPIC-N, 2025). Hahn et al. (2020) studied a project between the University of Alaska, the City of Anchorage, and the broader community. Their experience suggests that when support from local leaders, local experts, a steering committee, and diverse stakeholders have good working relationships, there is a better chance for successful community climate mitigation and adaptation planning. Engagement with a university to develop this plan can bolster this success. Additionally, Indiana University (2019), University of Nebraska at Lincoln (2021), and Chapman University (Penn State Sustainability, 2024) have all had experiential courses, externships, or student projects that have partnered with municipalities to conduct greenhouse gas inventories and, in some cases, recommend mitigation actions. None of them appears to be an ongoing program.

Drexel University's Environmental Collaboratory "facilitates the co-creation and implementation of climate justice centered solutions and best practices with the leadership of community partners in the Philadelphia region" (2025a). Students and faculty provided residential energy education materials for seven area governments through the Pennsylvania Climate Initiative (2025b) and supported planning activities through the Philadelphia Climate Justice Initiative. The University of Connecticut (2025) offers the Climate Corps through its Environmental Corps. This academic program is a collaboration between the University's disciplinary units, Center for Land Use Education and Research, and the Connecticut Sea Grant Program. It focuses on community-scale climate change adaptation service learning and work force development. Students take a one-semester academic course on climate change adaptation in Connecticut. They can then engage in a one-semester independent study working with a municipal or non-governmental organization on climate adaptation project. Arizona State University's Sustainable Cities Network (2024) delivers climate solutions to Arizona's cities, towns, and Tribal communities. Students conduct original research and project solutions through SOS 498/594 Urban Sustainability Best Practices Application. None of ASU's projects focus on greenhouse gas mitigation. It is in this broad policy window (See Kingdon, 1995) that Penn State's Local Climate Action Program operates.

4 Penn State's Local Climate Action Program

The Pennsylvania Department of Environmental Protection's Energy Programs Office created the Local Climate Action Program (LCAP) in 2019. It was established "to help local governments lead by example on critical climate change mitigation actions. LCAP provides free technical and personnel assistance to local governments for planning efforts to reduce GHG emissions and address climate change" (Pennsylvania Department of Environmental Protection, 2024, p. 13). The LCAP initially accomplished greenhouse gas mitigation planning education through special topics courses to multiple Pennsylvania colleges and universities. Students attended webinars on GHG inventorying, climate plans, community engagement, environmental justice, and other topics. Since the 2022–23 academic year, it has been administered through Penn State University's John and Willie Leone Family Department of Energy and Mineral Engineering and Penn State Sustainability in partnership with the Pennsylvania Department of Environmental Protection's Energy Programs Office and ICLEI USA. As the Commonwealth's land grant university, founded by the Morril Act in 1862 (Penn State University, 2004), this was a sensible decision. This chapter now reports on Penn State's version of the program.

The Penn State LCAP's mission is to assist Pennsylvania's communities to draw down greenhouse gas emissions through immersive education and workforce development. Its vision is to deliver its students and Pennsylvania's communities the ability to secure and maintain just and sustainable climate futures. The program's Directors believe all people deserve fair treatment and equal involvement in the decisions affecting their lives, access to reliable and valid information, the pursuit of right livelihood, and a safe and secure home. The Co-Directors also believe they have a responsibility as expert educators to influence public policy (see Buckland, Robinson, & Mann, 2020). It has three goals:

1. To educate students and communities in the methods of greenhouse gas accounting and climate planning, policy, and implementation.
2. To provide technical assistance to communities and organizations across Pennsylvania, the United States, and the world.

3. To coordinate, cooperate, and collaborate with governments, educational organizations, non-governmental organizations, and business to advance community-scale climate solutions.

Its learning objectives are consistent with the sustainability meta-competencies literature, including systems thinking, temporal thinking, ethical and normative literacy, interpersonal competence, and strategic competence (Brundiers et al., 2021; Engle et al., 2017; Wiek et al., 2011).

Additionally, the LCAP approaches the entire context as a complete environment, a place, wherein we practice Experiential Learning (EL). Since its inception, research on Experiential Learning Theory (Kolb, 1984) has demonstrated that faculty mentors must assist students in their navigation of multiple dynamics in community-based EL. Therefore, the program is aligned with the National Society for Experiential Education's eight principles (1998):

1. *Intention*: Clarity from the beginning on why the LCAP uses experience as the educational vehicle.
2. *Preparedness and Planning*: The Directors, collaborating staff, and municipal partners are experienced in their relevant technical, policy, and pedagogical duties.
3. *Authenticity*: The program's experiences are co-designed and implemented with community partners.
4. *Reflection*: Students, directors, and government staff are provided with situationally appropriate time to intentionally reflect on their experiences, including in writing, group meetings, and one-on-one discussions.
5. *Orientation and Training*: Students are prepared to accomplish technical tasks through training (ex: GHG inventorying, policy memos) before completing and delivering materials.
6. *Monitoring and Continuous Improvement*: Within a program year and between them, Directors maintain notes on data, sequence, communication, collaborator, student, and their success, challenges, and failures. New processes are updated in a timely fashion.
7. *Assessment and Evaluation*: All delivered documents are kept by the LCAP and by partner governments as part of open records. All deliverable types have been reviewed for their ability to be templatized and analyzed regarding their alignment with stated learning objectives, goals, values, vision, and mission.
8. *Acknowledgements*: At the end of the year, deliverables are reviewed, graded, and approved for transfer and all parties thanked for their participation. Directors have also nominated LCAP students for numerous awards, taken a former student to COP 28, invited former students to be part of a research Accelerator, and been repaid with LCAP alumni forming a supporting Board.

This intentionally designed immersive experience should "simultaneously support students, employers, and the broader economy, catalyzing both individual and collective impacts" (Huber, Jabot, & Heath, 2024, p. 5).

LCAP includes two three-credit courses offered in consecutive fall and spring semesters. Students accepted into the program commit to participating in both courses. After initial offerings under temporary course numbers, both courses now carry permanent course designations: Energy and Sustainability Policy 481 - Local Climate Action Program I: GHG Inventories and Energy and Sustainability Policy 482 – Local Climate Action Program: Climate Action, approved by the Penn State University Faculty Senate. The courses are offered in an online, largely asynchronous format to accommodate returning adult learners and residential students alike.

Third- and fourth-year undergraduates as well as graduate students from any major at any Penn State campus are invited to apply for admission to the program. Most students come from degree programs focused on energy, sustainability, policy, environmental management, and community development. Of the students who have completed the program in its first three years, approximately half have been World Campus students, completing their degrees entirely online. Each of the three annual cohorts has taught between 16–26 students and a total of 57 students. The annual cohorts have served between 10–13 governmental partners. Table 24.1 shows the program's governmental partners.

The fall semester develops students' and community partners' understanding of climate science, multi-scalar climate policy, the development of a community-wide greenhouse gas inventory, and the delivery of that inventory in written and spoken fashion. The fall has three key learning objectives:

1. To understand the historical and contemporary context of the ongoing climate crisis and how policy at various scales of governance contributes to solutions and management;
2. To apply principles of emissions calculations to develop an accurate and complete community-scale greenhouse gas emissions inventory, and
3. To communicate the findings of the emissions inventory in both written and oral formats to local government officials and community members in a public setting.

Table 24.1 Government types and names

Government type	Name
Council of governments	Centre region council of governments
Boroughs	Churchill, Conshohocken, Doylestown, homestead, Mechanicsburg, Selinsgrove, state college
Townships	Buckingham, Doylestown, Easttown, Haverford, Kennett, Lower Macungie, Lower Merion, Palmer, Solebury, Tinicum, Uwchlan
Cities	Allentown, Erie, Harrisburg, Johnstown, Pittston, Reading, Scranton
Counties	Bucks, centre
State agencies	Department of conservation and natural resources
Other	House of representatives climate caucus secretary, Paul Takac (HD-82)

The first month of the fall semester centers on establishing the student-community partner relationship while building the students' understanding of climate science and policy from the global down to the local level. The first assignment is called "A Place You Love." It turns students' attention to the fact that all human beings have a place and people they love, something they can have in common with their community partners. From there, they learn about their community, their partnering staff, and establish a schedule for meeting, preferred communication styles, and any ancillaries. The students complete readings and assignments to familiarize themselves with the Paris Agreement, the United Nations Sustainable Development Goals 11: Sustainable Cities and Communities and 13: Climate Action, the US Nationally Determined Contribution, and Pennsylvania's climate goals as articulated in Executive Order 2019–01 (Commonwealth of Pennsylvania, Governor's Office, 2019), and climate action plans. Students also learn science on climate impacts at the global, northeastern U.S., and Pennsylvania scales as well as Pennsylvania's contributions to the carbon budget. Additionally, students use publicly available cumulative health mapping tools to investigate, understand, and communicate current states of environmental justice (Pennsylvania Department of Environmental Protection, 2025).

After building this foundational knowledge, the class performs a community-wide greenhouse gas emissions inventory. The inventory meets the requirements of the United States Community Protocol, developed by ICLEI USA to guide United States communities to account for emissions generated by activities carried out within their jurisdictions (2019). The inventory follows six principles: Relevance, Accuracy, Completeness, Measurability, Consistency and Comparability, and Transparency (2019, pp. 22–23). It is distinct from but works with the CDG (formerly Carbon Disclosure Project) and the Global Covenant of Mayors (p. 52). Students use ICLEI USA's ClearPath 2.0 software to develop the inventory. Students, faculty, and partners often face challenges in boundary setting, data availability and quality, and calculations as they calculate or estimate emissions in stationary energy, transportation, water, wastewater, solid waste, and agriculture, forest, and other land use for their assigned community (see Diedrich, 2023, pp. 6–20). Program co-directors and a technical adviser from ICLEI troubleshoot along the way and review inventories for accuracy and completeness prior to students presenting their findings to their community partners. Once the inventorying work is finished, students learn to construct an emissions inventory report (building from an ICLEI template) and prepare for their public presentation.

The inventory serves as a critical baseline for climate action for local government partners as they develop plans, policies, and programs to address emissions. Providing our partners with comprehensive, replicable inventories ensures they can use this initial inventory in subsequent years to evaluate the success of measures implemented. ICLEI has placed the conduct of a greenhouse gas inventory at the front of the planning cycle. Krause et al (2025) find evidence that completing a community-wide emissions inventory "leads to a modest but significantly significant reduction in onsite residential emissions" (p. 2).

In the spring semester, the cohort turns its attention to climate action. While the inventorying experience is somewhat standardized for the cohort with all students moving through each sector for their respective communities, the spring work is highly individualized to a community's preferences and needs. It is distinctly more like an internship or independent study. The spring course carries five learning objectives:

1. To articulate specific strategies for implementing climate planning into local scale governance;
2. To identify moral and practical dilemmas to consider how these will impact future climate action decisions before the local government;
3. To communicate the types of projects, programs, and policies local government can leverage to advance climate goals,
4. To produce agreed-upon deliverables to advance the community partner's climate policy planning objectives, and
5. To translate topics from the Hope in Action speaker series into concise, locality-specific information for community partners.

A key design feature of the Local Climate Action Program is its ability to meet a community where they are in climate action and construct climate action planning for the spring semester that honors their circumstance and advances their progress. The instructors' years of practical experience, acquaintance with literature-informed and -guided possibilities, their government partners level of expertise and knowledge of their own community, and student curiosity guide the decisions. Combined, the team assesses the technical potential for GHG emissions reductions from a mitigation opportunity, the initiative feasibility or likelihood that the community will adopt a mitigation initiative, and what the community or stakeholder's behavioral plasticity could be in response to a mitigation initiative (Nielsen et al, 2020, p. 326). The range of climate policies, policy frames, and policy instruments, includes: framing through visions, goals, and planning; governance of municipal operations, in jurisdiction by authority through regulation, by enabling legislation, through partnership, and through experimentation; learning from experience through implementation, through trans-city/municipal networks, and through capacity-building practices and programs (Neij & Heiskanen, 2021, pp. 4–11).

Table 24.2 shows spring semester policy deliverables. All but one of the policies reside in visions, goals, and planning. This makes sense because the LCAP is a consultative capacity-building and technical service educational program with no jurisdictional authority. Should the community partners implement recommended policies, communities could change governmental operations, enter partnerships, acquire services, and change ordinances. The exceptions to the rule came from a serendipitous interaction where students interned with a solar cooperative active within one jurisdiction and provide materials for an ongoing youth engagement program with local non-governmental organizations in another.

The spring semester speaker series, Hope In Action, features speakers from government, private industry, non-governmental organizations, and academia to offer perspectives and proven climate solutions. Entirely web-based, this weekly speaker series is open to all current and past LCAP participant communities and the public. Students are required to attend (or watch the recording) and complete reflective

Table 24.2 Policy intervention types and quantities of each intervention

Policy intervention	Quantity
Climate action, energy transition plan, or sustainability plan guidance	11
Climate action plan, energy transition plan, or sustainability plan	8
Community engagement or education materials	8
Comparative case studies	7
Emissions forecasts	6
Community survey	5
Local government operations inventory	4
Solar energy feasibility study	4
Zoning or other policy review	4
Newsletter, website, or other communication	3
Science-based target guidance	3
Community program support	2
Prioritization matrix	2
Electric vehicle fleet transition guidance	2
Funding guidance document	2
Comprehensive plan recommendations	1
Request for proposal for professional services	1
Legislative report	1

assignments about the talks, sometimes with their community partner as the audience. Example speakers from the series have included representation from chapter leaders for the United States National Climate Assessment, Pennsylvania Department of Environmental Protection's Energy Programs Office, Sustainable Pennsylvania, energy and energy services companies and non-governmental organizations including ICLEI, and local government staff from Pennsylvania, New York, and Colorado.

Having completed three full cohorts, the Directors have identified areas that are working well and areas of ongoing challenge to consider in adapting this model to other universities/local governments. The top three program strengths are complementary instructor expertise, high-quality student applicants, and strong partnerships. The top four challenges are data timing and availability, data accuracy and fit, LCAP's time-intensive nature, and turnover in local governments.

4.1 Strengths

- *Complementary instructor expertise*: The Penn State LCAP currently has two co-directors. One co-director has extensive experience in greenhouse gas emissions inventorying and online course pedagogy and delivery. The other co-director has served in multiple elected and appointed positions of local government and therefore has a strong working knowledge of the jurisdictional authority of Pennsylvania local governments to instigate and implement climate action. Both co-directors have served in voluntary positions supporting the Centre Region's Climate Action and Adaptation Plan from development to implementation.
- *High quality student applicants lead to higher quality deliverables*: The program has attracted highly effective, ambitious, passionate students who engage professionally with their local government partners and their classmates. While the work produced is still student work, they perform at a high level, providing exceptional products their local government partners can use. For example, students who worked with the City of Scranton drafted the "Scope of work" section for the City's Request for Proposals to hire a consultant for longer term climate planning. The students partnered with Lower Merion Township outside of Philadelphia produced emissions reduction scenarios that were published in their adopted Sustainability Plan. The students who worked with Mechanicsburg Borough and Upper Providence Township drafted climate action plans the Township will integrate into their Comprehensive Plans.
- *Strong partnerships*: The LCAP would not be possible without several key partnerships. The first is with the Pennsylvania Department of Environmental Protection's Energy Programs Office who not only manages electric and gas utility data requests for our communities, but also provides the funding for their ClearPath software licenses. ICLEI USA: Local Governments for Sustainability provides our cohort training in the software and accompanying technical support to ensure accurate, complete inventories. The Centre Region Planning Agency's Sustainability Planner and staff at the Pittsburgh area's Congress of Neighboring Communities (CONNECT) are also technical support partners by memoranda of understanding. Other less formal partners include the participants in the Hope in Action series.

4.2 Challenges

- *Data timing*: Unlike most traditional courses, the LCAP relies on external factors, with utility data availability for the inventory chief among them. Directors continue to attempt to access data ahead of when it is needed, but sometimes this is not within program control. Utility companies can be slow to respond. Each year, the Directors have implemented further guardrails to attempt to have the data on time. Curating all data in advance and handing over packages because data collection prevents a real-world learning opportunity. However, the constraints of completing the work within a semester and ensuring students receive their grades in a timely manner necessitates some preemptive data acquisition work.

- *Data accuracy and fit*: Sometimes, utility companies cannot or do not always provide requested electricity or natural gas data by municipal boundary, instead providing them in zip codes. Some utilities have also provided data without indicating units such as kWh or MMBtu. Together with ICLEI, Directors are working to establish methodological downscaling approaches to address problem. These challenges, however, provide learning opportunities including the documenting of assumptions and describing methodological approaches clearly to create reliability between inventories.
- *Time-intensive nature of high-touch experience*: That we frame these as 'courses' is somewhat deceptive because the invisible program management occurring in conjunction with the course delivery itself is considerable. The LCAP is predicated on strong, trusting relationships with community partners and their stakeholders, with students, and with external agencies. Building and maintaining these relationships takes time and attention. While data challenges can consume time, the work students and Directors complete is more iterative in nature than traditional courses. Students do not simply write a paper and submit it for a grade and then move on to the next assignment. Instead, Directors work closely with the students to revise and refine their work until it is the absolute best it can be before turning it over to the community partner. For example, the Directors rehearse presentations with students, sometimes several times, in advance of their public presentations.
- *Personnel turnover and bandwidth*: Community partner engagement throughout the process is key to providing them with actionable climate solutions for their community. And while a local government may be eager to join the program at the outset, unforeseen challenges can compromise their engagement. Personnel or leadership changes and partner bandwidth can negatively impact a community's immersion in the program and make it challenging for students and faculty to complete the work. To some extent these problems are unavoidable and luckily rare. At the start of the LCAP year, we encourage new community partners to scenario plan for the community's continued participation in the program should any individual points of contact move on.
- *Hostility from the Trump administration and Congress*: Actions by the President of the United States and Congress have set out a broad array of challenges to climate action in the United States. These include executive orders attacking renewable energy and renewable energy credits, state laws supporting climate action, and H.B. 1's sunsetting tax credits for solar and wind energy, electric vehicles, batteries, and changing rules related to foreign entities of concern. These are combined with sustained assaults on higher education itself, including research on climate, renewable energy, and environmental justice (see Natural Resource Defense Council, 2025).

4.3 The Future

After a preliminary self-study and program evaluation with peers and collaborators, the LCAP will be taking the following steps. They are intended to make LCAP a more effective educational and capacity-building program, create stickier policy, and disseminate the model to other colleges and universities.

First, the Directors are co-developing a practical research program to determine past effectiveness and better support mitigation policy development. Four questions, programmatic expansion through replication, and a fundraising plan are under consideration:

1. *How can communities be assessed for mitigation pathway policy windows vis a vis the existing policy streams (*Kingdon, 1995*) as well as initiative feasibility and behavioral plasticity (*Nielsen et al, 2020*)?* As the United States political and economic landscape changes under the Trump administration, this question will encounter new challenges. However, state policy and politics, an active private sector, longstanding NGOs, and professional networks are creating dynamic spaces.
2. *Is there longitudinal evidence of greenhouse gas mitigation policy adoption in our cohorts and of what kind? Following adoption, is there evidence through repeated GHG inventories that they are accomplishing their goals? How can the LCAP enhance sticky mitigation policy?* As with the previous question, changing political landscapes at multiple scales will influence the answers to these questions.
3. *Do LCAP participants report efficacy related to mitigation planning, policy, program, and project development after participation in the LCAP? If efficacy is reported, what evidence of that efficacy is shown?* These questions can be answered through surveys, semi-structured interviews, as well as document analysis.
4. *What jobs are LCAP alumni taking after they graduate and what climate solutions are they impacting, if any?* Anecdotally, the LCAP appears to be preparing its graduates to work in sustainability consulting, solar energy, energy services, and sustainability planning. Surveys and follow-up interviews will be required.
5. *How do we replicate the LCAP in different places?* Based on discussions with faculty at universities from Europe, New Zealand, Africa, and across the United States, the Directors see a demand for more programs like the LCAP. Directors are developing a "train the trainer" workshop so university leaders, faculty, governments, and other partners can make their own place-based and context-appropriate climate change mitigation programs. Workshops are anticipated to be executed in summer 2026.
6. *The LCAP must find a long-term sustainable funding model.* The nature of the program means it is not possible to simply scale up student enrollments and generate more revenue back to the John and Willie Leone Family Department of Energy and Mineral Engineering. Therefore, funding through philanthropy, corporate sponsorship, or government grants are being explored.

5 Conclusion

In an interview on *Frontline* (n.d.), Harvard President Derek Bok said, "I think any self-respecting educational institution ought to judge its policies by its best estimate of what their long-term consequences for their students and for the society will be." Today, there is almost no policy issue of greater importance than the superwicked hyperthreat of the climate crisis. As member states to CHAMP have shown, we need subnational mitigation action. To accomplish that, we need educational models that teach students, build capacity, and develop the future workforce. Penn State's Local Climate Action Program has taken that challenge on, setting up a curricular model to be adopted and adapted in institutions of higher education around the world.

References

Arizona State University Sustainable Cities Network (2024) *Anthology of student-led research for local communities*. Retrieved 27 Jul 2025. https://www.dropbox.com/scl/fi/zuuyb3yb2jnw0ficr6guz/Anthology-of-Student-Led-Research_FINAL.pdf?rlkey=q4m3ll4fmi3q9mixovdw8a838&e=1&st=5wxh98hr&dl=0

Boulton, E. (2018). Climate change as a hyperthreat. In S. Pearson (Ed.), *Australian contributions to strategic and military geography, advances in military geosciences*. https://doi.org/10.1007/978-3-319-73408-8_5

Brundiers, K., Barth, M., Cebrián, G., Cohen, M., Diaz, L., Doucette-Remington, S., Dripps, W., et al. (2021). Key competencies in sustainability in higher education—toward an agreed-upon reference framework. *Sustainability Science, 16*(1), 13–29. https://doi.org/10.1007/s11625-020-00838-2

Buckland, P., Robinson, B., & Mann, M. (2020). Science Alone Will Not Save Us. Engaged Civic Life Might. In J. Henderson, A.Drewes (Eds.), *Teaching Climate Change in the United States*. Routledge.

Carrington, D. (2019, May 17). Why the Guardian is changing the language it uses about the environment. *The Guardian*. Retrieved 27 Jul 2025. https://www.theguardian.com/environment/2019/may/17/why-the-guardian-is-changing-the-language-it-uses-about-the-environment?CMP=share_btn_tw

Commonwealth of Pennsylvania, Governor's Office (2019, January 8) *Executive Order 2019–01: Commonwealth leadership in addressing climate change and promoting energy conservation and sustainable governance*. Retrieved 27 Jul 2025. https://www.pa.gov/content/dam/copapwp-pagov/en/oa/documents/policies/eo/2019-01.pdf

Conference of the Parties: Twenty first session. (2015, December 12). *Adoption of the Paris Agreement*. U.N. Doc. FCCC/CP/2015/L.9/Rev/1

Conference of the Parties serving as the meeting of the Parties to the Paris Agreement (2022, Mar 8). *Report of the Conference of the Parties serving as the meeting of the Parties to the Paris Agreement on its third session, held in Glasgow from 31 October to 13 November 2021*. Retrieved 27 Jul 2025. https://unfccc.int/documents/460950

Copernicus (2025, January 10) *The 2024 annual climate summary: Global climate highlights 2024*. Retrieved 27 Jul 2025. https://climate.copernicus.eu/global-climate-highlights-2024

COP 28 UAE (2023) Coalition for high ambition multilevel partnerships for climate action. Retrieved 15 June 2026: https://www.cities-and-regions.org/wp-content/uploads/cop28-champ_pledge.pdf

Diedrich, G. (2023). *From data to action: Harnessing GHG inventories for local sustainability initiatives.* Retrieved 27 Jul 2025. https://graham.umich.edu/media/files/CLCF-2023-Diedrich-Report.pdf

Drexel University (2025a). *The environmental collaboratory: community-based climate justice solutions.* Retrieved 27 Jul 2025. https://drexel.edu/environmental-collaboratory

Drexel University (2025b). *The Environmental Collaboratory: Community-Partnered Initiatives.* Retrieved 27 Jul 2025. https://drexel.edu/environmental-collaboratory/community-partnered-initiatives

Engle, E. W., Barsom, S. H., Vandenbergh, L., Sterner, G. E., & Alter, T. R. (2017). Developing a framework for sustainability meta-competencies. *International Journal of Higher Education and Sustainability, 1*(4). https://doi.org/10.1504/IJHES.2017.090204

Environmental Protection Agency (2025a, February 13). *Carbon Pollution Reduction Grants.* Retrieved on 27 Jul 2025. https://www.epa.gov/inflation-reduction-act/climate-pollution-reduction-grants

Environmental Protection Agency (2025b, January 10) *Priority Climate Action Plan Directory.* Retrieved 27 Jul 2025. https://www.epa.gov/inflation-reduction-act/priority-climate-action-plan-directory

EPIC-N (2025) *Stories: Sustainable development goal 13: Climate action.* Retrieved 27 Jul 2025. https://www.epicn.org/stories/?_page=1&tx_sustainable_development_goals=13-climate-action

Frente Nacional de Prefeitas e Prefeitos (2024, June 17). *Letter Of commitment from the national front of mayors (FNP) with the coalition for high ambition multi-level partnerships (CHAMP).* Retrieved 27 Jul 2025. https://multimidia.fnp.org.br/biblioteca/documentos/item/1213-carta-compromisso-da-fnp-com-a-coalizao-por-parcerias-multiniveis-de-alta-ambicao-champ

Hahn, M. B., Kemp, C., Ward-Waller, C., Donovan, S., Schmidt, J. I., & Bauer, S. (2020). Collaborative climate mitigation and adaptation planning with university, community, and municipal partners: a case study in Anchorage, Alaska. *Local Environment, 25*(9), 648–665. https://doi.org/10.1080/13549839.2020.1811655

Huber, M., Jabot, M., & Heath, C. (2024). *Experiential Learning and Community Partnerships for Sustainable Development: A Foundational Model for Climate Action* (1st edn.). Routledge. https://doi.org/10.4324/9781003489337

Frontline (n.d.) *Secrets of the SAT: An interview with Derek Bok.* Retrieved 27 Jul 2025. https://www.pbs.org/wgbh/pages/frontline/shows/sats/interviews/bok.html

ICLEI (2025) *Who We Are.* Retrieved 27 Jul 2025. https://icleiusa.org/about/who-we-are/

ICLEI—Local Governments for Sustainability USA (2019, July) *U.S. community protocol for accounting and reporting of greenhouse gas emissions, Version 1.2.* Retrieved 27 Jul 2025. https://drive.google.com/drive/folders/1azqkizTz0rXyWJeBmULBxHPFvfTnn5mR

ICLEI World Congress (2024, Jun 23). *Call to support multilevel climate action.* Retrieved 27 Jul 2025. https://www.youtube.com/watch?v=FFyQEKqf1VQ]

Incropera, F. P. (2016). *Climate change a wicked problem: Complexity and uncertainty at the intersection of science, economics, politics and human behavior*. Cambridge University Press.

Indiana University (2019, June 6) Students working with Indiana cities to inventory their greenhouse gas emissions. Retrieved 27 Jul 2025. https://news.iu.edu/live/news/26130-students-working-with-indiana-cities-to-inventory

Kingdon, J. W. (1995). *Agendas, alternatives, and public policies* (2nd ed.). Addison Wesley Longman.

Kolb, D. A. (1984). *Experiential learning: Experience as the source of learning and development* (Vol. 1). Prentice- Hall.

Krause, R., et al. (2025). Do municipal efforts 'move the needle' in reducing local greenhouse gas emissions? An initial assessment of US cities. *Cities, 159*, 1–9. https://doi.org/10.1016/j.cities.2025.105769

Lazarus, R. J. (2009). Super wicked problems and climate change: Restraining the present to liberate the future. *Cornell Law Review 94*, 1153–1234. Retrieved 27 Jul 2025. https://scholarship.law.georgetown.edu/facpub/159

Leal Filho, W., Sima, M., Sharifi, A., Luetz, J. M., Salvia, A. L., Mifsud, M., Olooto, F. M., Djekic, I., Snholon, R., Rampasso, I., Donkor, F. K., Dinis, M., Klavins, M., Finnveden, G., Chari, M., Molthan-Hill, P., Mifsud, A., Sen, S., & Lokupitiya, E. (2021). Handling climate change education at universities: An overview. *Environmental Sciences Europe, 33*, 1–19. https://doi.org/10.1186/s12302-021-00552-5

Levin, K., Bernstein, S., Cashore, B., & Auld, G. (2007, March 3) Playing it forward: Path dependency, progressive incrementalism, and the 'Super Wicked' problem of global climate change. In *Proceedings of the International Studies Association Convention, Chicago, IL, USA 28* (pp. 1–26). https://doi.org/10.1088/1755-1307/6/50/502002

Levin, K., Cashore, B., Bernstein, S., & Aud, G. (2012). Overcoming the tragedy of super wicked-problems: Constraining our future selves to ameliorate global climate change. *Policy Science, 45*, 123–152. https://doi.org/10.1007/s11077-012-9151-0

Local Governments and Municipal Authorities (2023). *History of the LGMA*. Retrieved 27 Jul 2025. https://www.cities-and-regions.org/about-the-lgma/history-of-the-lgma/

Natural Resource Defense Council (2025, July 18). *White house watch: Tracking attacks on our environment & health*. Retrieved 27 Jul 2025. https://www.nrdc.org/resources/white-house-watch-tracking-attacks-our-environment-health

Neij, L. & Heiskanen, E. (2021) Municipal climate mitigation policy and policy learning—A review. *Journal of Cleaner Production, 317*. https://doi.org/10.1016/j.jclepro.2021.128348.

National Society for Experiential Education (1998). *Standards of Practice: Eight Principles of Good Practice for All Experiential Learning Activities*. Norfolk, VA. Retrieved 27 Jul 2025. http://www.naceweb.org/principles/?referal=knowledgecenter&menuID=203

Nielsen, K., Stern, P., Dietz, T., Gilligan, J. M., van Vuuren, D. P., Figueroa, M. J., Folke, C., Gwozdz, W., Ivanova, D., Reisch, L. A., Vandenbergh, M. P., Wolske, K. S., & Wood, R. (2020). Improving climate change mitigation analysis: A framework for examining feasibility. *One Earth, 3*(3), 325–336. https://doi.org/10.1016/j.oneear.2020.08.007

Penn State Sustainability (2024, November 1) *Local climate action program and world campus alum honored in Orange County*. Retrieved 27 Jul 2025. https://www.psu.edu/news/penn-state-sustainability/story/local-climate-action-program-and-world-campus-alum-honored-orange

Penn State University (2004, March 15). Background information about the Morril Act. Retrieved 27 Jul 2025. Retrieved 27 Jul 2025. https://www.psu.edu/news/story/background-information-about-morrill-act

Pennsylvania Department of Environmental Protection (2024, March 1). *Pennsylvania's priority climate action plan*. Retrieved 27 Jul 2025. https://www.pa.gov/agencies/dep/residents/climate-change/pa-climate-action-plan.html

Pennsylvania Department of Environmental Protection (2024, June 24). *Pennsylvania's greenhouse gas inventory report 2024*. Retrieved 27 Jul 2025. https://greenport.pa.gov/elibrary/GetDocument?docId=8731732&DocName=PENNSYLVANIA%20GREENHOUSE%20GAS%20INVENTORY%20REPORT.PDF%20%20%3cspan%20style%3D%22color:green%3b%22%3e%3c/span%3e%20%3cspan%20style%3D%22color:blue%3b%22%3e%28NEW%29%3c/span%3e%209/9/2026

Pennsylvania Department of Environmental Protection (2025). Penn EnviroScreen. Retrieved 27 July 2025. From https://gis.dep.pa.gov/PennEnviroScreen/

Rittel, H., & Webber, M. (1973). Dilemmas in a general theory of planning. *Policy Sciences, 155*, 160–169. https://doi.org/10.1007/BF01405730

Stockholm Resilience Center (2025) *Transdisciplinary education*. Retrieved 27 Jul 2025. https://www.stockholmresilience.org/education/transdisciplinary-education.html

Sun, J., & Yang, K. (2016). The wicked problem of climate change: A new approach based on social mess and fragmentation. *Sustainability, 8*(12), 1312. https://doi.org/10.3390/su8121312

UNESCO. (2014b.) *UNESCO Roadmap for implementing the global action programme on education for sustainable development*. UNESCO. Retrieved 27 Jul 2025. http://unesdoc.unesco.org/images/0023/002305/230514e.pdf
United Nations Commission on Sustainable Development. (1992). *Agenda 21*. Retrieved 27 Jul 2025. http://sustainabledevelopment.un.org/content/documents/Agenda21.pdf
United States Conference of Mayors (2024, June). *In support of United States endorsement*. Retrieved 27 Jul 2025. https://www.usmayors.org/the-conference/resolutions/?category=a0FKY000000sZyJ2AU&meeting=92nd%20Annual%20Meeting
United States Energy Information Agency (2023). *Total CO2 emissions estimates from energy consumption, per capita CO2 emissions, and carbon intensities, ranked by state*. Retrieved 27 Jul 2025. https://www.eia.gov/state/seds/sep_sum/html/pdf/rank_co2_capita.pdf
United States of America (2021, June 21). *The United States' Nationally Determined Contribution: Reducing Greenhouse Gases in the United States: A 2030 Emissions Target*. Retrieved from https://unfccc.int/sites/default/files/NDC/2022-06/United%20States%20NDC%20April%2021%202021%20Final.pdf
University of Connecticut. (2025) Climate Corps. Retrieved 27 July 2025. From https://climate.uconn.edu/caa/
University of Nebraska—Lincoln (2021, March 18). *ICLEI partners with UNL and Conservation Nebraska in student service-learning project*. Retrieved 27 Jul 2025. https://ianrnews.unl.edu/iclei-partners-unl-and-conservation-nebraska-student-service-learning-project
Wiek, A., Withycombe, L., & Redman, C. L. (2011). Key competencies in sustainability: A reference framework for academic program development. *Sustainability Science, 6*, 203–218. https://doi.org/10.1007/s11625-011-0132-6

Peter Buck is the Director of Education at Penn State Sustainability, Co-Director of Penn State's Local Climate Action Program, and a Courtesy Appointment as Associate Teaching Professor in the John and Willie Leone Family Department of Energy and Mineral Engineering. His praxis sits at the nexus of democratic process and ethics, climate change and systemic risk, and educational systems. Dr. Buck has an extensive sustainability-related policy leadership portfolio through elected and appointed offices, testimony to local and state government commissions, non-governmental organization leadership, and public press communication.

Brandi Robinson is an Associate Teaching Professor in the John and Willie Leone Family Department of Energy and Mineral Engineering. Robinson's teaching, service, and research focus on the efficacy of utilizing local policy mechanisms and structures of governance to address issues surrounding global climate change. Specifically, her interest is in understanding how to connect the causes and impacts of climate change with areas of existing concern and interest in local communities to build partnerships for meaningful change.

Educating for Environmental Character: Findings from a University Course in Sustainability Studies

Ryan Juskus, Heather M. Maranges, and Kate Allman

Abstract In recent years, several scholars have argued that an education for sustainable development (ESD) ought to educate students' character in addition to their cognition. What pedagogical practices should be considered while developing curricula for sustainable development that includes character goals? Can character goals be achieved in the context of a traditional classroom-based course? The authors of this chapter answer these questions through empirical research. They use a mixed methods approach to analyze data collected from two semesters of an undergraduate course in environmental leadership and character at an American university to assess the course's impact on students' character compared to a control group. This study contributes to theoretical and empirical scholarship that bridges the fields of ESD and character education to examine whole-person, transformative approaches to environmental and sustainability education. Discovering significant growth in the character traits of wonder (appreciation of beauty) and humility, as well as growth in moral awareness, the authors conclude that character education methods in a traditional environmental studies classroom setting can cultivate persons who are not only knowledgeable about environmental topics but who are also defined by deep and enduring dispositions to foster environments that sustain many forms of life. Finally, they discuss the significance of these findings for whole-person approaches to ESD and propose promising pedagogical practices to implement in traditional classroom contexts.

R. Juskus (✉)
Institute for Social Concerns, University of Notre Dame, Notre Dame, USA
e-mail: rjuskus@nd.edu

H. M. Maranges
Department of Psychology, Florida State University, Tallahassee, USA
e-mail: maranges@psy.fsu.edu

K. Allman
Winston-Salem TEACH, Winston-Salem, USA

Department of Education and Research Fellow, Program for Leadership and Character, Wake Forest University, Winston-Salem, USA

K. Allman
e-mail: allmank@wfu.edu

W. Leal Filho et al., *Curriculum Development for Sustainable Development*, World Sustainability Series, https://doi.org/10.1007/978-3-032-22146-9_25

Keywords Environmental education · Education for sustainable development · Character education · Wonder · Humility · Moral awareness

1 Introduction

Undergraduate students in American universities today enter sustainable development studies classrooms with a complex mix of emotions, values, and identities. For example, they might feel anxious about climate change and the economy but excited about their career prospects and sustainable technologies and business models. Perhaps they embrace conflicting values, such as self-expression, sustainability, justice, and self-interest. Academic institutions are also becoming increasingly diverse in terms of students' cultural, political, and religious background. In short, students both carry and enter into a potentially dizzying degree of ethically significant human plurality.

Given the nature and urgency of environmental problems and the need ultimately for behavioral changes at the level of individuals, communities, and collectives, it is within this context of human plurality that teachers of undergraduate environmental courses often address questions of emotion, meaning, purpose, and ethics in their syllabi, classroom discussions, and assignments. A growing body of theoretical and empirical scholarship on environmental education examines and advocates efforts like these that take a whole-person approach to education, by which we mean an education that combines traditional intellectual goals and skill development with character goals, psycho-emotional and social learning, and reflection on purpose and worldview. Yet, little attention in this literature has been given to efforts to navigate tensions between normativity and plurality, particularly as they both intersect with power dynamics. What pedagogical practices should be considered while developing curricula for sustainable development that includes whole-person learning goals? Can character goals be achieved in a traditional classroom-based course with students representing a plurality of backgrounds and moral commitments? In this article, we respond to these questions through analyzing evidence from two semesters of an undergraduate course at Wake Forest University that included character goals among their learning objectives and sought to present them in a critical and pluralistic manner. We draw the conclusion from our analysis that a classroom-based, semester-long course that is attentive to power and moral plurality can successfully develop in students the character traits of wonder and humility as well as grow their overall moral attentiveness, which are qualities that are associated with sustainable identities, values, and behaviors. We close by proposing promising pedagogical practices to consider incorporating in curricula for sustainable development.

2 Literature Review

In recent decades, some scholars have argued that an education for sustainable development (herein ESD) is inherently a transformative and ethical project that not only engages, but *ought* to engage the whole person (Ingman & McConnell, 2021; Jordan & Kristjánsson, 2017; Maiteny, 2005; Orr, 2004; Podger et al, 2010; Sterling, 2021).[1] Philosopher Matt Ferkany reflects this view when writing that ESD "aims to shape learners' beliefs, values, attitudes, and behavior—in other words, what sort of person they will be—not just teach them about sustainability or impart merely academic skills which they may or may not go on to deploy later in life." For Ferkany, ESD should incorporate the cultivation of character virtues, since "virtues define the sort of person one is and might overlap certain (sets) of ESD's transformative aims" (Ferkany, 2021, p. 70). This call to more directly engage virtue and character in the classroom reflects the mainstreaming of environmental virtue ethics in recent years (Bouma-Prediger, 2020; Cafaro, 2004; Hursthouse, 2007; Kawall, 2021; Sandler, 2007; Wensveen, 2000).

If ESD aims to shape the sort of person learners will be, then what character of person might it aim at? Travis York and Christian Becker propose one answer: a "sustainable person." In their estimation, Sustainability suggests a new self-understanding as a fundamentally threefold-related being: a being related to nature, contemporaries, and future generations." This relational understanding of the human person "represents a substantial paradigm shift in human self-identity": from the "dominant paradigm" of the modern age, which idealizes the autonomous individual who is independent from and gains control over nature through science and technology, to a new paradigm of the self that is in relation to others, nature, and future generations. "The excellent cultivation of this self-identity is called the sustainable person" (York & Becker, 2012, p. 5). Other scholars have formulated something analogous in terms of "nature relatedness," "interconnectedness," "nature connectedness," "sustainable wellbeing," and other comparable concepts (Barrable, 2019; McKeon, 2012; Merino et al, 2020; Ronen & Kerret, 2020).

---

[1] In this article, we take a very broad approach to education for sustainable development (ESD) that includes the frameworks of education for sustainability (EfS), environmental and sustainability education (ESE), and ecological/environmental education (EEE). For further evidence of environmental studies as an ethical project, one can point to the purpose statement of the Association for Environmental Studies and Sciences, which reads as follows: "Fundamental to our embrace of higher education is the notion that broad advances in environmental knowledge that benefit both humans and the rest of nature require disciplinary, interdisciplinary, and transdisciplinary approaches to research and learning. Moreover, they require humility about what we know and don't know, both as individuals and as representatives of disciplines. For it is only through learning communities of the type proposed for AESS that we can achieve 'whole system' environmental education and the creative synthesis of new knowledge that promotes a healthier Earth. Such a synthesis would truly mark the coming of age of ES&S as a professional society." In this case, "a healthier Earth" is the substantive good envisioned as the goal of environmental studies as an interdisciplinary field, a goal that will be advanced through cultivating humility, creativity, systems thinking. Furthermore, the statement orients environmental studies toward the benefit of humans and the rest of nature, thus rejecting at least some forms of anthropocentric moral ontology.

It is not necessary to agree either on the exact shape of this "sustainable person" or the character of the transformation required to affirm these scholars' more basic point about the transformative dimensions of ESD. Moreover, the kind of character education framework proposed by some scholars of ESD does not insert character education where it previously was absent; rather, it draws on philosophical, theoretical, and empirical resources to intentionally consider how and toward what an education in environment and sustainability might shape students. York and Becker thus echo theorists of environmental education that have countered the idea that ESD should be only technical and scientific in nature.[2] Because ESD seeks to shape persons, and because character education centers the identity and formation of persons, these scholars advocate incorporating character and virtue into environmental studies curricula.

Environmental philosopher Brian Treanor argues for character in environmental studies when he writes, "We will need both intelligence and creativity *and* the wisdom to use them well as we move forward: good scientists, engineers, economists, politicians, and activists to be sure, but only if they are also good people." (Treanor, 2014, p. 89). Whether using the language of "sustainable persons," "virtuous humans," or an analogous concept, the idea is that ESD should seek to cultivate persons who possess praiseworthy character traits that are conducive to just sustainability, traits which may include, for example, justice, empathy, courage, honesty, humility, curiosity, simplicity, creativity, care, forgiveness, openness, and good judgment.

Recent empirical studies have complemented the theoretical literature by testing the relationship between measures of character and environmental sustainability. One study of 900 university students in Spain found that character strengths, including appreciation of beauty, love of learning, creativity, and zest, were associated with nature relatedness (Merino et al, 2020). Notably, appreciation of beauty was the strongest predictor. Another study of university students found that sustainable consumption practices were related to the character strengths of appreciation of beauty, humility, love of learning, kindness, leadership, and perspective (Valor et al, 2020). Similar patterns emerge in other works that find VIA character strengths associated with sustainable behavior in Mexico (Corral-Verdugo et al, 2015) and environmental responsibility in Ghana (Kabir and Wiium, 2021). Though still in a nascent phase, empirical studies on the relationship between character and environmental sustainability suggest a promising link that can benefit from more research in this area.

Among the character traits that have emerged from theoretical and empirical studies as being particularly important for sustainability commitments are the appreciation of beauty, wonder, and humility (Brady, 2018; Diessner & Niemiec, 2022; Ferkany & Whyte, 2012; Johnson, 2014; Merino et al, 2020; Pulkki, 2022; Taylor et al, 2020; Valor et al, 2020; Washington, 2018). These are also some of the traits

[2] See, for example, the work of David Orr, Stephen Sterling, and Paul Maiteny, each of whom has developed research, frameworks, and curricula for a whole-person approach to environmental education that includes attention to worldview, psycho-emotional factors, spirituality, and the cultivation of environmental virtues.

centered in the course studied here, with appreciation of beauty considered a proxy for wonder in nature. Indeed, Merino et al. suggest "appreciation of beauty is the strongest covariate of nature relatedness" and, since it is the latter's top "signature strength," presents an opportunity for environmental educators to aim at cultivating it in their methods and pedagogies (Merino et al, 2020, p. 1707, 1721). Diessner and Niemiec's study also found appreciation of beauty to be the strongest predictor of pro-environmental behavior, intention for pro-environmental behavior, and susceptibility to elevation, the latter of which is associated with engaging moral beauty (Diessner & Niemiec, 2022). Jennifer Cole Wright, a psychologist and leading scholar of the virtue of humility, has also collaborated on empirical research that links "environmental humility" with sustainable behaviors (Taylor et al, 2020). Based on these links between sustainable behaviors and views and beauty, humility, and wonder, our study focuses on the degree to which these traits can be cultivated in an undergraduate classroom.

The question we sought to answer in this study was: *Can a pluralistic, undergraduate, classroom-based, semester-long course in environmental studies cultivate in students environmental character traits, particularly humility, wonder, and attentiveness to the moral dimensions of life?* We answered this question by conducting a mixed-methods, pre-post study of two sections of an undergraduate course entitled "Environmental Leadership and Character" (herein Environmental L&C) taught at a private university in the southern United States during the 2021–22 academic year. In addition to intellectual goals and assignments, this course required students to engage in various character exercises intended to build traits of humility, wonder, and justice, and to reflect on students' personal vocation, character, and moral commitments. Based on findings from the study, this paper suggests that environmental studies and sustainable development courses that supplement traditional intellectual goals with character goals can help students discover purpose and strategies for ethical leadership in sustainable development.

3 The Course: Environmental Leadership and Character

The course that was researched is an environmental and sustainability studies course that aims to support students in exploring questions of identity, purpose, history, character, and meaning so that they are equipped to apply their cognitive and technical environmental learning toward ethical leadership in diverse sustainable development contexts. To achieve these ends, students integrate their environmental knowledge, moral and political commitments, religious or philosophical traditions, and self-understanding. It does this by examining Black, liberal, religious, and Indigenous traditions of ethical environmental leadership in the United States as well as the particular environmental character virtues of humility, wonder, and justice. It asks students to articulate their own approach and moral influences through comparison and contrast with other approaches, thus highlighting likeness and difference as sources of clarity about students' personal commitments and aspirations.

Informing the course design was research in character studies about effective strategies for cultivating character in university students. In particular, it used the framework of seven strategies for cultivating character advocated by Lamb, Brandt, and Brooks: habituation through practice; reflection on personal experience; engagement with virtuous exemplars; dialogue that increases virtue literacy; awareness of situational variables; moral reminders; and friendships of mutual accountability (Lamb et al, 2021).

Among its student learning objectives, the course includes both traditional intellectual goals and character goals. For example, one of the intellectual goals of the course is to differentiate between contemporary frameworks of ethical environmental leadership. The primary character-related goal is: Develop specific character traits for ethical and effective environmental leadership through articulating a personal approach, understanding and cultivating key virtues, identifying orienting narratives, and articulating a vision of how to integrate science, moral commitments, and a common life with diverse others.

Assignments throughout the course give students opportunities to work toward these intellectual and character goals. The intellectual goals are achieved through readings, a book review, a key concepts exam, a character profile of an exemplary environmental leader, and a personal vision for ethical environmental leadership. The character goals are achieved through a series of character building exercises that are integrated with the three main units of the course, each one focused on a particular tradition of ethical leadership: Indigenous, mainstream, and Black and popular environmental leadership. The course opens by introducing key concepts and then proceeds to address each of these traditions in the context of a journey in search of environmental practical wisdom. Paired with each main course unit is an environmental virtue that is central to that tradition: Indigenous environmental leadership is paired with the virtue of humility, understood ecologically; mainstream environmental leadership is paired with the virtue of wonder in nature; and Black and popular environmental leadership is paired with the virtue of justice.

As for the character building exercises, during the first unit on humility, students must interview or have a conversation with two people who hold environmental views on key topics, like climate change, that differ from their own. The goal is not to debate but to listen and learn the other person's perspective. Students reflect on this experience in a written reflection. During the unit on wonder, students must spend two sessions each week in a place not primarily inhabited by humans, unplugged and attentive to what is around them. They respond to targeted prompts, some directly from Rachel Carson and some designed by the instructor, and keep a journal of their experiences and responses. For the unit on justice, students create a moral reminder to themselves and display it in a place where they will see it every day. They keep a journal with reflections on living with that moral reminder for a few weeks. They must also attend and reflect on a justice-related event on campus.

Having chosen depth over breadth by focusing on only three traditions of ethical environmental leadership, the course also includes a number of assignments, like a book review assignment and a profile and presentation of an exemplary environmental leader, in which students have freedom to explore other traditions and frameworks. In their book reviews students engage with environmental virtue in, for example, popular culture, Buddhism, Christianity, Black feminism, and Enlightenment humanism. In their profile of an exemplary environmental leader, students must give a character profile and presentation on someone from business, grassroots activism, politics, academia, or governance who has received a major environmental award. These tend to be very diverse. The final paper asks students to develop a personal vision of environmental leadership and character that also synthesizes their learning throughout the course.

4 Study Design

The current study entailed a mixed-methods, pre-post approach in which we tested whether students in the Environmental L&C course grew in appreciation of beauty, humility, and moral attentiveness from the beginning to the end of the course compared to a control group across two semesters. This research was approved by Wake Forest University's Institutional Review Board.[3]

Participants. In Fall 2021, 50 students provided consent and completed measures of appreciation of beauty, humility, and moral attentiveness as part of a larger survey at the beginning and end of the semester. In Spring 2022, 56 students provided consent and completed that same survey at the beginning and end of the semester. Students also completed a final reflection assignment at the course that asked students to synthesize their course learning by bringing it into conversation with their personal, vocational, educational, and career aspirations. All of the eight fall students (100%) and twelve of the 14 (85.7%) spring students consented for their course reflection to be used for research purposes. See table below for demographics.

Fall 2021		*Environmental leadership and character (n = 8)*	*Control group (n = 42)*
	Age *M (SD)*	19.50(1.20)	18.71(1.15)
	Gender	75% women, 25% men	76% women, 24% men
	Race and ethnicity	6 White, 1 Asian, 1 White and Asian	25 White, 4 Asian, 3 Hispanic/Latinx, 2 Black, 7 Multiracial, 1 chose not to respond
Spring 2022		*Environmental leadership and character (n = 14)*	*Control group (n = 42)*
	Age *M (SD)*	20.64(1.01)	19.00(1.36)

(continued)

[3] The IRB number assigned to this study by Wake Forest University is IRB00023552.

(continued)

Fall 2021		*Environmental leadership and character (n = 8)*	*Control group (n = 42)*
	Age *M (SD)*	19.50(1.20)	18.71(1.15)
	Gender	75% women, 25% men	76% women, 24% men
	Race and ethnicity	6 White, 1 Asian, 1 White and Asian	25 White, 4 Asian, 3 Hispanic/Latinx, 2 Black, 7 Multiracial, 1 chose not to respond
	Gender	64% Women, 36% Men	79% Women, 21% Men
	Race and ethnicity	16 White	26 White, 6 Asian, 6 Multiracial (5 identified in part as Hispanic/Latinx; 1 identified in part as Asian), 3 Black, 1 chose not to respond

Quantitative Method

Participants responded to questions about their Appreciation of Beauty and Humility as assessed by the Virtues-in-Action (VIA) questionnaire and questions about their moral attentiveness (Moral Attentiveness Scale, Reynolds, 2008) as part of a larger survey. For means, please see tables in results section.

Appreciation of Beauty. Participants responded to the 3-item virtue in action (VIA) subscale for appreciation of beauty and excellence. Participants read the definition ("You notice the beauty and excellence around you; you are often awe-struck by beauty, greatness, and/or the moral goodness you witness; you are often filled with wonder") and respond with the extent to which the virtue is an essential part of who they are, is natural and effortless to express, and is uplifting or energizing to express on a scale from 1 (strongly disagree) to 7 (strongly agree).

Humility. Participants responded to the 3-item virtue in action (VIA) subscale for humility. Participants read the definition ("You let your accomplishments speak for themselves; you see your own goodness but prefer to focus the attention on others; you do not see yourself as more special than others; you admit your imperfections") and respond with the extent to which the virtue is an essential part of who they are, is natural and effortless to express, and is uplifting or energizing to express on a scale from 1 (strongly disagree) to 7 (strongly agree).

Moral Attentiveness. Participants responded to the 5-item moral attentiveness scale (MAS; Reynolds, 2008), with response options from 1 (strongly disagree) to 7 (strongly agree). Sample items include "I regularly think about the ethical implication of my decisions" and "I like to think about ethics."

Qualitative Method

Quantitative data was triangulated with a culminating reflection assignment that was completed at the end of the course. The purpose of this assignment was for students to synthesize their course learning by bringing it into conversation with their personal, vocational, educational, and career aspirations. The assignment prompt was:

> In this summative written assignment, you should bring together all of the aspects of our course—your research throughout the course, your character building exercises and reflections, our readings and discussions, the videos and podcasts, your book review, your colleagues' presentations, and your profile of an environmental leader—to author a personal vision that (a) describes the kind of environmental character you imagine yourself to be, (b) names and explores the frameworks and traditions that orient you, and (c) examines how your approach relates to, differs from, and/or is challenged by other frameworks of environmental leadership and character. This paper should include elements of both self-reflection and critical examination and analysis through close engagement with course texts and themes. Additional research may be necessary for you to carry out this assignment, especially in light of the fact that our course explored in depth only a small number of traditions and individuals. Because this assignment is unique, I will include guiding questions for you to organize your thoughts and a list of required items that you must incorporate in some fashion. How you weave together these various threads is up to you—i.e. whether you write in an essay format that creatively ties all of the pieces together into a unified narrative/argument or you break it up into several sections that are only loosely related to one another.

To better understand how the course supported the development of students' appreciation of beauty, humility, and moral awareness, the written assignments were analyzed using directed content analysis method (Hseih & Shannon, 2005). Direct content analysis is a useful analytic approach when attempting to identify and categorize all instances of a particular phenomenon. First, the entire text was read and text was highlighted or annotated to indicate first impressions of the text. The next step in the analysis was to code all of the passages using the predetermined parent codes. Child codes were assigned based on emergent themes related to the predetermined parent codes. Text excerpts that could not be categorized with the initial coding scheme were assigned a new code. Codes were then collapsed into categories and then refined into themes.

5 Results

For each outcome, we conducted repeated-measure ANOVAs to assess whether students in the Environmental L&C (vs. control) group demonstrated an increase from the beginning to the end of the course in Appreciation of Beauty, Humility, and Moral Attentiveness. Specifically, we modeled how Group (Environmental L&C or Control), Time (Pre- to Post-semester), and the Group × Time interaction predicted each outcome in separate analyses for both Fall and Spring semesters.

Appreciation of Beauty. For Fall descriptive statistics, see Table 25.1; for results of the repeated measure ANOVA, see Table 25.2; and for graphical representation of the interactions, see Fig. 25.1. As expected, students in the Environmental L&C group grew significantly in Appreciation of Beauty from the beginning to the end of the semester relative to the control group, the students of which did not change in Appreciation of Beauty.

For Spring descriptive statistics, see Table 25.3; for results of the repeated measure ANOVA, see Table 25.4; and for graphical representation of the interactions, see Fig. 25.2. As expected, students in the Environmental L&C group grew significantly

Table 25.1 Appreciation of beauty descriptive statistics, fall

Group	Time	Mean	Std. error	95% Confidence interval	
				Lower bound	Upper bound
Environmental leadership	pre	4.875	0.412	4.046	5.704
	post	6.917	0.415	6.083	7.750
Control	pre	5.317	0.180	4.956	5.679
	post	5.310	0.181	4.946	5.673

Table 25.2 Appreciation of beauty ANOVA statistics, fall

Predictor	df	F	p-value
Group	1, 48	2.42	0.126
Time	1, 48	16.23	<0.001
Group × Time	1, 48	16.48	<0.001

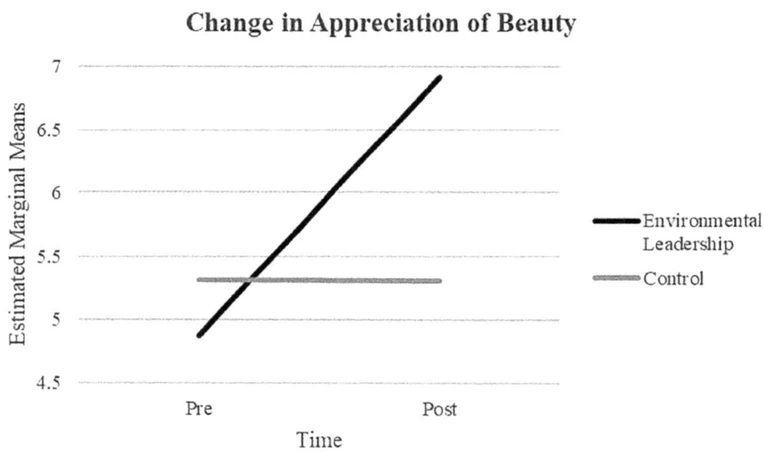

Fig. 25.1 Change in appreciation of beauty from the beginning to the end of the semester for both the environmental leadership and control group, fall

in Appreciation of Beauty from the beginning to the end of the semester relative to the control group, in which the increase of Appreciation of Beauty was not significant.

These findings were triangulated by the qualitative analysis, with almost all participants ($n = 17$) referencing how the course increased their appreciation of beauty. One student wrote:

> The biggest lesson I've learned regarding the environment is to appreciate its beauty and fight for its preservation. As Joy Harjo stated in her poem, it's important to remember how we are all born from the earth. We owe the earth our lives. Her poem compels the reader to think about all the beauty the earth provides: the sun, moon, sky, stars, plants, animals. It asks us to remember our families and their histories and the histories across the world.

Table 25.3 Appreciation of beauty descriptive statistics, spring

Group	Time	Mean	Std. error	95% Confidence interval	
				Lower bound	Upper bound
Environmental leadership	pre	5.762	0.302	5.157	6.367
	post	6.143	0.261	5.619	6.666
Control	pre	5.040	0.174	4.690	5.389
	post	5.167	0.151	4.864	5.469

Table 25.4 Appreciation of beauty ANOVA statistics, spring

Predictor	df	F	p-value
Group	1, 54	10.274	0.002
Time	1, 54	1.795	0.186
Group × Time	1, 54	0.449	0.506

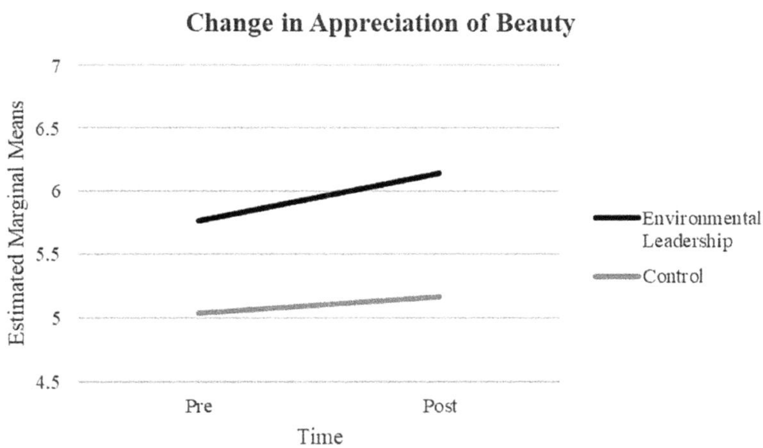

Fig. 25.2 Change in appreciation of beauty from the beginning to the end of the semester for both the environmental leadership and control group, spring

> 'Remember you are all people, and all people are you.' Humans share this earth with each other and with the living things around us.

Many participants ($n = 9$) closely linked their appreciation of beauty with the development of the virtue of wonder discussed in the course. Wonder is an intellectual virtue that arouses the intellect and directs it towards objects that are seen as intellectually understandable and decipherable (Kristjanson, 2019). In the course, following Kathleen Dean Moore's scholarship on Rachel Carson, the virtue of wonder was defined as "the propensity to respond with delight, awe, or yearning to what is beautiful and mysterious in the natural world when it unexpectedly reveals itself" in such a way that it "leads [one] to celebrate and honor the more-than-human world, to care for it, to protect its thriving" (Moore, 2005, p. 265). One participant explained the connection they perceived between wonder and appreciation of beauty:

> In exercising my cultivation of wonder throughout this course, I chose to also tap into my virtue of the appreciation of beauty and excellence through being present in the environment and appreciating it in its most raw form. The virtue of wonder relates to a fresh and child-like awe of the world, involving sensory and emotional excitement in nature. I found wonder to relate to the appreciation of beauty and excellence as I believe you can not feel wonder without it...Being able to exercise my sense of wonder was something emphasized throughout the course, and became a virtue I found almost essential for both environmental leadership and general happiness.

Overall, the qualitative data supported the growth in the appreciation of beauty reflected in the quantitative data.

Humility. For the Fall, descriptive statistics are in Table 25.5; for results of the repeated measure ANOVA, see Table 25.6; and for graphical representation of the interactions, see Fig. 25.3. Relative to students in the control group, which began high but decreased in Humility, students in the Environmental L&C group significantly increased in Humility from the beginning to the end of the semester.

For the Spring, descriptive statistics are in Table 25.7; for results of the repeated measure ANOVA, see Table 25.8; and for graphical representation of the interactions, see Fig. 25.4. As predicted, students in the Environmental L&C group significantly increased in Humility from the beginning to the end of the semester compared to students in the control group, which decreased in Humility on average.

The participants' final reflections triangulated this finding, with 11 students highlighting how the course readings supported their valuing of humility. One student wrote:

> I regard humility as the most important virtue that I have come to understand in our course. Specifically, I found our reading 'Humility as a Foundational Virtue' to be very insightful, and in particular, Jennfier Cole Wright's statement that 'humility's ethical alignment orients

Table 25.5 Humility descriptive statistics, fall

Group	Time	Mean	Std. Error	95% Confidence interval	
				Lower bound	Upper bound
Environmental leadership	pre	4.667	0.353	3.957	5.376
	post	5.625	0.349	4.923	6.327
Control	pre	5.603	0.154	5.293	5.913
	post	5.389	0.152	5.083	5.695

Table 25.6 Humility ANOVA statistics, fall

Predictor	df	F	p-value
Group	1, 48	1.14	291
Time	1, 48	3.55	0.066
Group × Time	1, 48	2.46	0.005

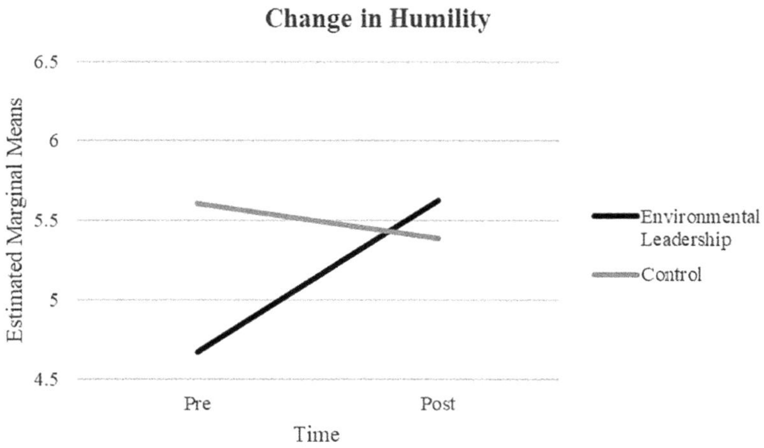

Fig. 25.3 Change in humility from the beginning to the end of the semester for both the environmental leadership and control group, fall

Table 25.7 Humility descriptive statistics, spring

Group	Time	Mean	Std. error	95% Confidence interval	
				Lower bound	Upper bound
Environmental leadership	pre	5.548	0.257	5.033	6.062
	post	6.071	0.289	5.493	6.650
Control	pre	5.437	0.148	5.139	5.734
	post	5.246	0.167	4.912	5.580

Table 25.8 Humility ANOVA statistics, spring

Predictor	df	F	p-value
Group	1, 54	3.336	0.073
Time	1, 54	0.824	0.368
Group × Time	1, 54	3.786	0.057

toward others, allowing us to understand and experience ourselves as only one among a host of other morally relevant beings, whose interests are as real, legitimate, and as worthy of attention and concern as our own' stood out in the reading (156). This appealed to me because it highlighted the importance of *humility* in the practice of recognizing the importance of other 'morally relevant beings,' a sentiment that I held before this class. In this sense, learning about humility in this context helped me become aware that this is a virtue that I need to focus on in order to better improve myself in accordance with my own values.

Five participants also referenced how the readings written by indigenous or authors of color specifically supported the development of humility in the course. One student explained:

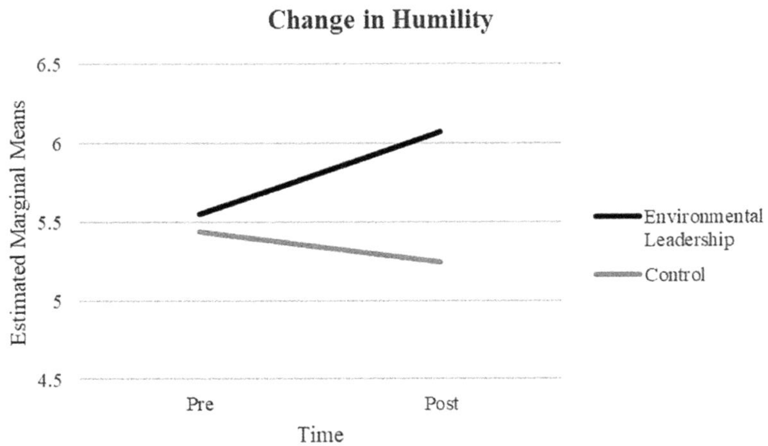

Fig. 25.4 Change in humility from the beginning to the end of the semester for both the environmental leadership and control group, spring

> When reading about indigenous environmental leadership in *As Long as Grass Grows*, one of the main takeaways that I brought with me was the uniqueness of traditional indigenous relationships with the environment. Because of their place-based identity and their spiritual and cultural practices, many indigenous people did (and do) regard themselves as a small piece in the greater force of nature. Actions are taken to act in harmony with nature rather than trying to conquer it. I think that these values and actions are exemplary of the cultivation of humility. Especially in comparison to settler-colonial perspectives and attitudes towards the environment as something to conquer and domesticate, an understanding of a sustainable attitude towards nature is crucial in addressing the climate crisis.While settler-colonialists approach the environment and nature as something unruly and in need of human intervention, indigenous people have historically cultivated mutually beneficial relationships with nature.

Overall, the qualitative data supported the growth in humility reflected in the quantitative data.

Moral Attentiveness. For Fall, descriptive statistics, see Table 25.9; for results of the repeated measure ANOVA, see Table 25.10; and for graphical representation of the interactions, see Fig. 25.5. Consistent with expectations, the Environmental Leadership students increased in their moral attentiveness from the beginning to the end of the semester, whereas the control group students' moral attentiveness decreased.

For Spring, descriptive statistics, see Table 25.11; for results of the repeated measure ANOVA, see Table 25.12; and for graphical representation of the interactions, see Fig. 25.6. Descriptively, as predicted, the Environmental Leadership group grew in moral attentiveness over the course of the semester compared to the control group, who decreased in moral attentiveness, but the interaction did not reach statistical significance.

The qualitative data also triangulated this finding, with six students highlighting how the course fostered their sense of awareness. One student wrote:

Table 25.9 Moral attentiveness descriptive statistics, fall

Group	Time	Mean	Std. error	95% Confidence interval	
				Lower bound	Upper bound
Environmental leadership	pre	5.300	0.339	4.619	5.981
	post	6.000	0.327	5.343	6.657
Control	pre	5.457	0.148	5.160	5.755
	post	5.310	0.143	5.023	5.596

Table 25.10 Moral attentiveness ANOVA statistics, fall

Predictor	df	F	p-value
Group	1, 48	0.698	0.408
Time	1, 48	2.54	0.118
Group × Time	1, 48	5.97	0.018

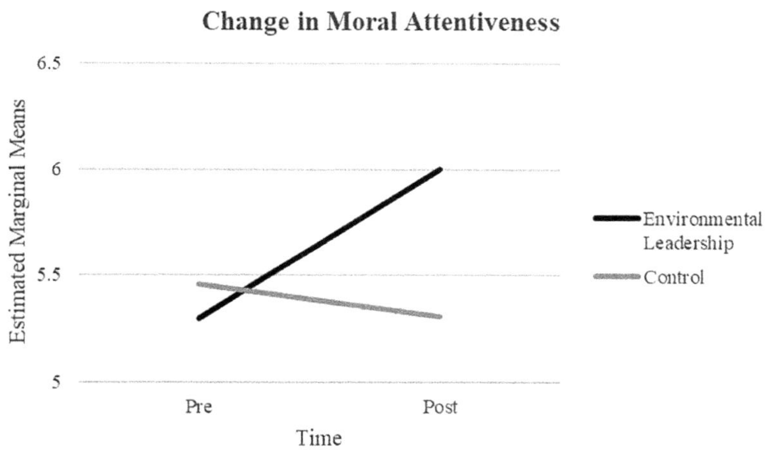

Fig. 25.5 Change in moral attentiveness from the beginning to the end of the semester for both the environmental leadership and control group, fall

Table 25.11 Moral attentiveness descriptive statistics, spring

Group	Time	Mean	Std. error	95% Confidence interval	
				Lower bound	Upper bound
Environmental leadership	pre	5.929	0.308	5.310	6.547
	post	6.386	0.289	5.807	6.965
Control	pre	5.338	0.178	4.981	5.695
	post	5.233	0.167	4.899	5.568

Table 25.12 Moral attentiveness ANOVA statistics, spring

Predictor	df	F	p-value
Group	1, 54	7.842	0.007
Time	1, 54	1.398	0.242
Group × Time	1, 54	3.554	0.065

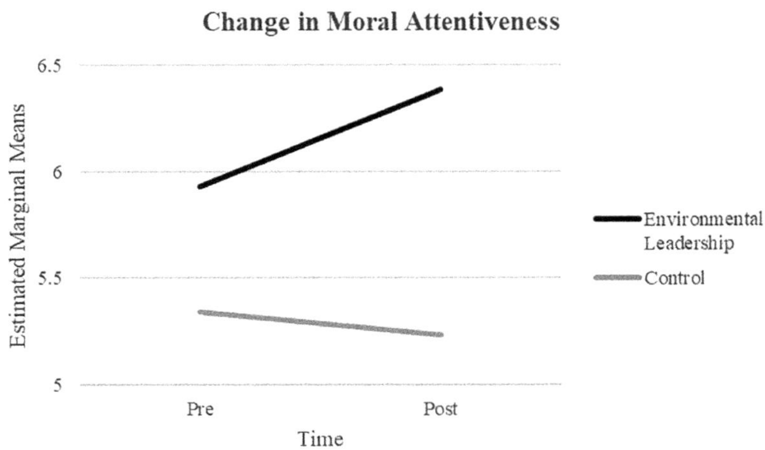

Fig. 25.6 Change in moral attentiveness from the beginning to the end of the semester for both the environmental leadership and control group, spring

> The primary takeaway from this class within my sense of character is a sense of awareness and intentionality. Given that I spend a lot of my time pondering elements of environmental leadership and character, and educating myself on related topics, the values and subject matter of this course were not new to me. However, this was the first time in my life in which moral values have been taught to me in a formal education setting—apart from occasional activities with my elementary school guidance counselor. It is an odd phenomenon that so often we are told that our character is what is most important about us, yet there is little time designated to cultivating what is supposedly the most important thing about us in academic settings...

Another student wrote:

> I came into this class unsure of my purpose in life and how I wanted to get better at life. This course has helped me understand both myself and what I wish to look like in the future. My values have changed due to this course because I am so much more educated in topics I know nothing about. […] ENV305 was the first class in a long time where I have been actively interested in every topic and all of the readings. This course was a new experience for me, and my character has been positively changed.

This change supported the growth in moral attentiveness reflected in the quantitative data.

6 Discussion and Conclusions

The current work tested whether pedagogy that integrated assignments aimed at intellectual and character development through use of the seven strategies of character education in a traditional classroom context supported university students' development in appreciation of beauty, humility, and moral awareness across a semester in two studies. Overall, the current empirical investigation provides evidence for the efficacy of such pedagogical strategies in a course on environmental leadership and character compared to a control group. In Fall, across the semester, students in the Environmental L&C class grew in the character virtue of appreciation of beauty—i.e., noticing and being affected by beauty, particularly via feelings of wonder and awe. In Spring, although students increased in appreciation of beauty, the interaction did not reach statistical significance. We also observed consistent significant changes in students' humility in the intervention class, with students growing in their tendency to focus attention outside of themselves, over the course of both semesters. Finally, students developed in their moral attentiveness, more often and deeply thinking about the moral implications of their behaviors, across both semesters compared to the control group. These quantitative patterns were triangulated via qualitative analyses. In sum, quantitative and qualitative results suggest that character has a place and can be grown in a class that focuses on environmental leadership from the lens of individual moral responsibility.

A number of leading theorists of education for sustainable development (ESD) have argued that ESD should be transformative, not just teaching students about environmental and sustainability topics but also seeking to surface or shape their values, attitudes, purposes, and behaviors vis-a-vis pressing environmental challenges. An explicitly and reflectively transformative approach like this is meant to counteract the often unreflective yet nevertheless character-forming power of what York and Becker call "the dominant paradigm" that predominates in American universities (York & Becker, 2012). It is essential to investigate whether this type of whole-person education can be done within the context of a traditional classroom-based, semester-long course in a pluralistic university environment.

We conclude that by inviting students to engage in whole-person learning through character-focused pedagogical strategies, environmental and sustainability studies courses can successfully integrate intellectual and character goals in a way that helps students cultivate dispositions, particularly wonder and humility, that are positively associated with environmentally sustainable motivations and behaviors. Student writing and feedback from the course suggests that this integrative and invitational approach to ESD also meets students' desire for more holistic forms of education that provide them with knowledge, skills, dispositions, and a sense of purpose, especially given the existential stakes of planetary sustainability challenges like climate change. What is more, ESD courses can integrate cognitive knowledge and character in ways that help them become more attuned to the moral dimensions of life in general, thus breaking down the idea that sustainability and social concerns are inherently at odds.

Our study, therefore, suggests one pathway through which research on the positive association between measures of character and commitment to sustainability can be translated into a practical program for ESD. Empirical research suggests that the particular traits of appreciation of beauty—which we take as a proxy for the virtue of wonder—and humility have strong positive associations with sustainable motivations and practices. We have shown that an ESD course and curriculum can effectively target and develop these traits in students through assignments that invite students to reflectively engage in practices that they might otherwise never engage in, such as listening to someone with whom they disagree with an eye toward understanding rather than debating or spending time in an attentive and reflective posture in a natural landscape.

A second implication of our study is more conjectural in nature, since there is no control group that participated in a version of the Environmental L&C course that was inattentive to moral plurality and power asymmetries. Nevertheless, we speculate that attention to human difference and power asymmetries is a factor in the success of the course and should therefore also be central to any attempt to combine intellectual learning with character. The course studied here does not impose one conceptualization of environmental virtue onto students. Rather, it invites them to consider and practice relevant virtues carried by three different traditions of environmental thought and practice and to bring them into mutually critical and constructive dialogue with one another. By beginning the course with an Indigenous environmental history that sustains a critique of settler-colonialism as a form of environmental deprivation, for example, students are invited to pay close attention to the interplay of power, environmentalism, and epistemology. Under the rubric of humility, they are then invited to look critically at themselves and their own tradition(s) even as they are invited to learn from others' traditions. In other words, the course assumes that no one tradition carries some kind of "pure" environmental virtue; rather, it assumes that the only way to develop environmental character is through openness to the (human and ecological) other.

Based on the transformative aims of environmental and sustainability education, the structure of the course studied here, and the findings from our study of its impact on students, we propose the following practices for teachers who would like to complement their intellectual goals with character goals in an environmental studies course:

- Use one or more of the seven strategies for cultivating virtue suggested by Lamb et al. "How is Virtue Cultivated?" (2021)
- Promote reflective writing on character-related experiences and exercises (graded for thoroughness and completion, not for content)
- Create assignments that help students integrate scientific knowledge of nature with personal experience of and appreciation for the natural world
- Develop assignments that help students make meaning of sustainability issues in a way that invites them to consider their existing and emerging moral commitments
- Present environmental virtues as general concepts that can be filled out differently according to diverse moral traditions and commitments

- Create opportunities for students to research and present on diverse exemplars of ethical environmental leadership

7 Limitations and Future Directions

Notwithstanding the strengths of the current research, it is important to note several aspects which may limit generalizability of conclusions. First, students and control groups were largely White and women. Although 32% students were people of color, it will be important to apply the pedagogical practices employed in the current work in contexts with more diverse students. Relatedly, the current work was done at a private university, such that replication at public institutions will be important. Finally, the intervention courses were small, with relatively few students, such that generalizability is somewhat limited. Nevertheless, even with the challenge of lower statistical power, we were able to detect effects of change in appreciation of beauty, humility, and moral attentiveness in the Environmental L&C class relative to the control class.

8 Conclusion

We have sought in our study to understand whether education for sustainable development's (ESD) transformative aims can be fulfilled within the context of a traditional environmental studies course. Scholars like Matt Ferkany, David Orr, and Brian Treanor have argued that theoretical and pedagogical frameworks drawn from virtue ethics and character education can prove fruitful for a transformative, whole-person approach to ESD that combines intellectual, technical, and character goals. Our study provides empirical support for this, namely that a class that invites students to develop environmental practical wisdom through engagement with multiple traditions of ethical environmental leadership and strategies for cultivating character supports students' growth in wonder (i.e., appreciation of beauty), humility, and moral awareness. Implementing character education methods in a traditional environmental and sustainability studies classroom setting can cultivate persons who are not only knowledgeable about environmental topics but who are also defined by deep and enduring dispositions to foster environments that sustain many forms of life.

9 Declaration of Interest

The authors report there are no competing interests to declare.

10 Funding Details

This work was supported by the Lilly Endowment Inc. under Grant 2019 0951.

References

Barrable, A. (2019). Refocusing environmental education in the early years: A brief introduction to a pedagogy for connection. *Education Sciences, 9*(1). https://doi.org/10.3390/educsci9010061

Bouma-Prediger, S. (2020). *Earthkeeping and Character: Exploring a Christian Virtue Ethic*. Baker Academic.

Brady, E. (2018). Aesthetics, humility, and wonder: other-regarding ways of relating to environment. *La Pensée écologique [Ecological Thought], 2*(1). https://doi.org/10.3917/lpe.002.0044.

Cafaro, P. (2004). *Thoreau's living ethics: Walden and the pursuit of virtue*. University of Georgia Press.

Corral-Verdugo, V., Tapia-Fonllem, C., & Ortiz-Valdez, A. (2015). On the relationship between character strengths and sustainable behavior. *Environment and Behavior, 47*(8), 877–901. https://doi.org/10.1177/0013916514530718

Diessner, R., & Niemiec, R. M. (2022). Can beauty save the world? Appreciation of beauty predicts proenvironmental behavior and moral elevation better than 23 other character strengths. *Ecopsychology*. https://doi.org/10.1089/eco.2022.0047

Ferkany, M. (2021). Aristotelian virtue education and education for sustainable development. In Jason Kawall (Ed.), *The Virtues of Sustainability*. Oxford University Press. https://doi.org/10.1093/oso/9780190919818.001.0001

Ferkany, M., & Whyte, K. P. (2012). The importance of participatory virtues in the future of environmental education. *Journal of Agricultural and Environmental Ethics, 25*(3), 419–434. https://doi.org/10.1007/s10806-011-9312-8

Hsieh, H. –F., & Shannon, S. E. (2005). Three approaches to qualitative content analysis. *Qualitative Health Research, 15*(9), 1277–1288.

Hursthouse, R. (2007). Environmental virtue ethics. In R. L. Walker & P. J. Ivanhoe (Eds.), *Working virtue: Virtue ethics and contemporary moral problems* (pp. 155–171). Clarendon Press.

Ingman, B. C., & McConnell, C. (2021). Ecologically minded teachers and character education. *Journal of Character Education, 17*(2), 75–83.

Johnson, K. (2014). Creative connecting: Early childhood nature journaling sparks wonder and develops ecological literacy. *International Journal of Early Childhood Environmental Education, 2*(1), 126–139.

Jordan, K., & Kristjánsson, K. (2017). Sustainability, virtue ethics, and the virtue of harmony with nature. *Environmental Education Research, 23*(9), 1205–1229. https://doi.org/10.1080/13504622.2016.1157681

Kabir, R. S., & Wiium, N. (2021). Positive youth development and environmental concerns among youth and emerging adults in Ghana. In R. Dimitrova, & N. Wiium (Eds.), *Handbook of Positive Youth Development: Advancing Research, Policy, and Practice in Global Contexts*, (pp. 81–94). Springer Series on Child and Family Studies, Springer International Publishing. https://doi.org/10.1007/978-3-030-70262-5_6

Kawall, J. (Ed.). (2021). *The virtues of sustainability*. Oxford University Press.

Kristjánsson, K. (2019). Scientific practice, wonder, and awe. In C. Deane-Drummond, T. A. Stapleford, & D. Narvaez (Eds.), *Virtue and the Practice of Science: Multidisciplinary Perspectives*, (pp. 124–35). University of Notre Dame, Center for Theology, Science, and Human Flourishing.

Lamb, M., Brant, J., & Brooks, E. (2021). How is virtue cultivated? Seven strategies for postgraduate character development. *Journal of Character Education, 17*(1), 81–108.

Maiteny, P. (2005). Education for sustainability and development: Psycho- emotional blocks and catalysts. *The Development Education Journal, 11*(2), 1–4.
McKeon, M. (2012). Two-eyed seeing into environmental education: Revealing Its 'Natural' readiness to indigenize. *Canadian Journal of Environmental Education (CJEE), 17*, 131–147.
Merino, A., Valor, C., & Redondo, R. (2020). Connectedness is in my character: The relationship between nature relatedness and character strengths. *Environmental Education Research, 26*(12), 1707–1728. https://doi.org/10.1080/13504622.2020.1825630
Moore, K. D. (2005). The truth of the barnacles: Rachel Carson and the moral significance of wonder. *Environmental Ethics, 27*(3), 265–277.
Orr, D. W. (2004). *Earth in Mind: On Education, Environment, and the Human Prospect*. 2nd edn. Island Press
Podger, D. M., Mustakova-Possardt, E., & Reid, A. (2010). A Whole-person approach to educating for sustainability. *International Journal of Sustainability in Higher Education, 11*(4), 339–352. https://doi.org/10.1108/14676371011077568
Pulkki, J. (2022). Humility imparts the wonders of nature: A virtue-ethical elaboration of some of Michael Bonnett's thoughts. *Environmental Education Research*, 1–11. https://doi.org/10.1080/13504622.2022.2083082
Reynolds, S. J. (2008). Moral attentiveness: who pays attention to the moral aspects of life? *Journal of Applied Psychology, 93*, 1027–1041. https://doi.org/10.1037/0021-9010.93.5.1027
Ronen, T., & Kerret, D. (2020). Promoting sustainable wellbeing: Integrating positive psychology and environmental sustainability in education. *International Journal of Environmental Research and Public Health, 17*(19), 6968. https://doi.org/10.3390/ijerph17196968
Sandler, R. (2007). *Character and environment: A virtue-oriented approach to environmental ethics.* Columbia University Press.
Sterling, S. (2021). Concern, conception, and consequence: Re-thinking the paradigm of higher education in dangerous times. *Frontiers in Sustainability, 2.*
Taylor, B., Wright, J., & LeVasseur, T. (2020). Dark green humility: Religious, psychological, and affective attributes of pro environmental behaviors. *Journal of Environmental Studies and Sciences, 10*(1), 41–56. https://doi.org/10.1007/s13412-019-00578-5
Treanor, B. (2014). *Emplotting virtue: A narrative approach to environmental virtue ethics* (Vol. 12). SUNY Press.
Valor, C., Antonetti, P., & Merino, A. (2020). The relationship between moral competences and sustainable consumption among higher education students. *Journal of Cleaner Production, 248*(March), Article 119161. https://doi.org/10.1016/j.jclepro.2019.119161
Washington, H. (2018). Education for wonder. *Education Sciences, 8*(3), 125. https://doi.org/10.3390/educsci8030125
van Wensveen, L. (2000). *Dirty virtues: The emergence of ecological virtue ethics.* Humanity Books.
York, T., & Becker, C. (2012). Cultivating an ethic of environmental sustainability: Integrating insights from Aristotelian virtue ethics and pragmatist cognitive development theory. *Journal of College and Character, 13*(4). https://doi.org/10.1515/jcc-2012-1884

Ryan Juskus Ph.D. is Assistant Professor of the Practice in the Institute for Social Concerns at the University of Notre Dame. His research and teaching focus on the intersections of environmental studies, religion, social ethics, and politics, using humanistic and qualitative research methods to examine and develop just and transformative responses to environmental, climate, and energy challenges.

Heather M. Maranges holds a Ph.D. in Social and Personality Psychology and her work seeks to understand what facilitates cooperation, moral virtue, and equity. She was Co-Director of Research at the Program of Leadership and Character at Wake Forest University when this work was done and is now Assistant Professor in the Department of Psychology at Florida State University.

Kate Allman Ph.D. is Research Associate Professor in the Department of Education at Wake Forest University, Research Fellow in the Program for Leadership and Character, and the Executive Director of Winstan-Salem TEACH. She researches the social and moral development of adolescents and young adults and examines the connections between student learning, program evaluation, and organizational thriving in Institutions of Higher Education.

GPSR Compliance
The European Union's (EU) General Product Safety Regulation (GPSR) is a set of rules that requires consumer products to be safe and our obligations to ensure this.

If you have any concerns about our products, you can contact us on

ProductSafety@springernature.com

In case Publisher is established outside the EU, the EU authorized representative is:

Springer Nature Customer Service Center GmbH
Europaplatz 3
69115 Heidelberg, Germany

www.ingramcontent.com/pod-product-compliance
Ingram Content Group UK Ltd.
Pitfield, Milton Keynes, MK11 3LW, UK
UKHW021944240926
475648UK00012B/20
* 9 7 8 3 0 3 2 2 2 1 4 5 2 *